ENZYME CYTOLOGY

ENZYME CYTOLOGY

Edited by

D. B. ROODYN

Department of Biochemistry,

University College London,

England

1967

ACADEMIC PRESS · LONDON AND NEW YORK

A*

ACADEMIC PRESS INC. (LONDON) LTD.
Berkeley Square House
Berkeley Square
London, W.1.

U.S. Edition published by
ACADEMIC PRESS INC.
111 Fifth Avenue
New York, New York 10003

Library of Congress Catalog Card Number: 66-30440

PRINTED IN GREAT BRITAIN BY PAGE BROS. (NORWICH) LTD.

LIST OF CONTRIBUTORS

NORMAN G. ANDERSON, *Molecular Anatomy Section, Biology Division, Oak Ridge National Laboratory and Technical Division, Oak Ridge Gaseous Diffusion Plant, Oak Ridge, Tennessee, U.S.A.*

C. DE. DUVE, *The Rockefeller University, New York, U.S.A. and Université de Louvain, Laboratoire de Chimie Physiologique, Louvain, Belgium.*

DAVID ELSON, *The Weizmann Institute of Science, Rehovoth, Israel.*

G. P. GEORGIEV, *Institute of Molecular Biology, Academy of Sciences of the U.S.S.R, Moscow, U.S.S.R.*

JOHN G. GREEN, *Molecular Anatomy Section, Biology Division, Oak Ridge National Laboratory and Technical Division, Oak Ridge Gaseous Diffusion Plant, Oak Ridge, Tennessee, U.S.A.*

D. O. HALL, *Department of Botany, University of London, King's College, England.*

E. REID, *Department of Biochemistry, University of Surrey, London, England.*

D. B. ROODYN, *Department of Biochemistry, University College London, England.*

W. STRAUS, *Division of Metabolic Research, Institute for Medical Research, The Chicago Medical School, Chicago, Illinois, U.S.A.*

F. R. WHATLEY, *Department of Botany, University of London, King's College, England.*

EDITOR'S FOREWORD

The bringing together of enzymology and cytology is relatively recent and has occurred in a somewhat disjointed, and at times confusing manner. Although the techniques for staining cells for enzymes have been considerably improved in the last fifteen years, the majority of enzyme-catalysed reactions still cannot be satisfactorily followed by histochemical means. The greater part of our knowledge of the enzymology and metabolism of cell components is therefore derived from studies on fractions isolated from cell homogenates, almost invariably by centrifugal procedures. The technique of "differential centrifugation", as it is usually called, is readily carried out by relatively unskilled workers, who certainly need no special training in cytological technique. The flexibility of the method, and the ease with which the isolated sub-cellular fractions can be subjected to standard chemical and enzymological procedures, has resulted in a great outburst of new information about the properties of cell components. Improved methods of cell disruption and centrifugal fractionation have resulted in greater precision and reproducibility and the method has now been applied to a very wide range of biological materials.

Concurrently with these developments, the application of electron microscopy has resulted in a massive increase in our knowledge of cell structure. With improved instrument design, embedding, staining and sectioning techniques, the electron microscopist has led us into an intriguing world of fine detail that reaches down to the molecules themselves. Magnificent pictures are now obtained showing exquisite detail of the nuclear membrane, chloroplast lamellae, mitochondrial membranes, ribosomes, cytomembranes and so forth. The convergence of these two powerful techniques, electron microscopy and the centrifugal fractionation of homogenates, has brought about a revolution in our understanding of the cell.

However, the progress has not always been even. Biochemists and electron microscopists are often uneasy collaborators. The tempo of biochemical research is different from that of cytological investigation. The biochemist tends to convert the cell into the abstractions of instrument readings, tabulated data and graphs, whereas the cytologist is essentially concerned with visual phenomena, and may easily feel that Michaelis constants and metabolic maps are rather second hand, remote ways of describing the cell. Also, the electron microscopes themselves are frequently housed in anatomy, botany, zoology or cytology departments, and it is still rather unusual for biochemists to have very easy access to electron microscopy. Some research centres have deliberately worked for an integration of biochemical and cytological studies, with excellent results. Usually, however, the relationship is not so close, and detailed electron microscope characterization of isolated sub-cellular fractions is still far too infrequent.

There are other difficulties arising from the fact that the technique of centrifugal fractionation of homogenates is still relatively new. Rigorous attitudes, or even terminology, have not yet been worked out. There are few "standard" procedures, and in some fields (for example the isolation of nuclei) until very recently, almost every research worker seemed to use his own method. There are still serious problems of cross-contamination of fractions, of enzyme adsorption and elution during fractionation, of structural damage by homogenizers and isolation media as well as great technical difficulty in obtaining valid enzyme assays in the complex mixtures of enzymes present is isolated fractions. Many of the results obtained with radioactive tracers give essentially qualitative answers which are not amenable to accurate expression in "balance sheets". For reasons of convenience, there has been undue emphasis on certain tissues (such as rat-liver) and also on certain enzymes which are easy to assay but are not necessarily of great importance.

In spite of these difficulties, there is now little doubt that enzyme cytology is a new scientific discipline. As Professor de Duve says below; "It is a true hybrid, which after the inevitable growing pains and search for its own identity, has now come of age as a vigorous individual in its own right . . . . " One has only to survey the major findings reported in each chapter of this book to realize the magnitude of the advance that has been made in such a short period. The role of the nucleus in nucleic acid synthesis and protein synthesis has been revealed, and our knowledge of the molecular basis of gene action greatly enriched. In addition, the general enzymology of the nucleus has been explored in detail. There have been great advances in our understanding of the structure and function of the mitochondrion. Its role in ATP synthesis, lipid, amino acid and carbohydrate metabolism, ion transport and other processes has been elucidated in great detail in a flood of publications. The discovery of mitochondrial (and chloroplast) DNA has resulted in an awakening of interest in the problem of the mechanism of biogenesis of cell structures. Also, research on the structure and function of the chloroplast has been most fruitful, and in addition to the most detailed elucidation of the fine structure of this organelle, its role in CO_2 fixation, photophosphorylation and a wide range of other activities has been examined most fully. A whole class of lysosomal particles has been discovered and these particles have been shown to be involved in many different physiological and pathological processes. It is hard to believe that the discovery of the lysosome was as recent as 1955. The properties of cell membranes, and in particular the endoplasmic reticulum, have been studied in detail, and a great variety of membrane-associated enzymes has been revealed. The wealth of information that we now have of the structure and function of the ribosome is quite striking. When one considers the contemporary literature on ribosomes, polysomes, messenger RNA and the genetic code, again it is almost unbelievable that the term "ribosome" was only introduced into the literature in 1958! Finally, the soluble phase of the cell has been shown to have an important role in many metabolic processes, including glycolysis, fatty acid synthesis, nucleotide metabolism, and the pentose shunt pathway.

Thus, in spite of the many imperfections of technique and the obvious gaps in our knowledge, it cannot be denied that a series of findings of the utmost significance to biology have been made in this field. Most of these results have come in the last decade, and a good proportion in the last 2 or 3 years. This book is an attempt to bring these findings together, covering the whole field of enzyme cytology in one volume. After an introductory chapter on the principles of the subject, we have, so to speak, worked from the bottom of the centrifuge tube upwards, dealing with nuclei, mitochondria, peroxisomes, chloroplasts, lysosomes, membranes, ribosomes and cell sap. The general design of the chapters is more or less the same, in that an attempt has been made in each case to survey the whole field in a reasonably detailed fashion, but without excessive emphasis on the authors' own interests, or on last month's research reports. The text is brought together by a rather lengthy subject index, and a list of the enzymes discussed is added at the end of the book. Some suggestions for terminology in enzyme cytology have also been included as an appendix. Although the editor consulted the various contributors for their views on this, and indeed has taken some terminology directly from the text in some cases, these suggestions are best regarded as the private prejudices of the editor, put forward more to stimulate discussion on the important semantic problems in the field than to lay down firm rulings.

It is hoped that the reader will not be too impatient with the length of the book, and will appreciate the great difficulty in compressing what are now very active fields of research into chapters of a manageable size. The burden of new literature is heavy, and even as we write, our ideas must be modified to adapt to the ceaseless flow of new information. It is inevitable that much of what is written below will soon become outdated. However, if we are not to be overwhelmed by the outpourings of our electron microscopes, ultracentrifuges and automatic spectrophotometers, some attempt must be made, at one point of time, to stand back from the work and to assess the progress so far made. Perhaps such an effort is futile, and one should allow the subject to fragment into a series of specialities of an increasingly esoteric nature, presented in a language that is more and more incomprehensible to the general reader. However, it would be unfortunate if this process occurred simply because of the inability of the individual worker in the field to acquaint himself with the major findings in all aspects of the subject. However complex the cell appears to be, and however detailed and abstruse are the descriptions of its component parts, one must always hope that a unifying pattern exists, which will ultimately be comprehensible to a single individual. If this book is of some use in the quest for this pattern, the labours of its contributors will not have been in vain.

November, 1966 D. B. Roodyn

EDITOR'S ACKNOWLEDGEMENTS

The Editor would like to express his sincere thanks to the following for permission to reproduce material, the loan of plates and diagrams, and for helpful discussion and correspondence: Dr. A. D. Bangham, Professor B. Chance, Dr. J. B. Chappell, Dr. R. S. Criddle, Professor C. de Duve, Professor H. Fernandez-Moran, Professor S. Fleischer, Professor D. E. Green, Dr. R. Gustafsson, Dr. H. W. Heldt, Professor M. Klingenberg, Professor H. A. Krebs, Professor A. L. Lehninger, Dr. D. J. L. Luck, Professor F. Lynen, Dr. P. Mitchell, Dr. E. Mugnaini, Professor M. J. Nadukavukaren, Dr. M. M. K. Nass, Dr. L. Packer, Dr. L. D. Peachey, Dr. C. S. Rossi, Professor D. R. Sanadi, Dr. P. Siekevitz, Professor E. C. Slater, Dr. R. T. Ward, Dr. E. P. Whitehead and Dr. V. P. Whittaker. He is also grateful to Dr. R. Bellairs for electron micrographs of mitochondria.

The Editor is grateful to Dr. H. W. Heldt for the following recent comments. Since publication of the scheme given in Fig. 21 of Chapter 3, on "The Mitochondrion" [Fig 21. is adapted from Fig. 7 of Heldt (1966) and Fig. 15 of Klingenberg and Pfaff (1966)], it now appears that the inner compartment of rat-liver mitochondria contains a GTP - AMP - phosphate transferase, and Dr. Heldt and his colleagues no longer maintain that adenylate kinase is also present in the inner compartment [see Heldt, H. W. and Schwalbach, K. (1966). *Fed. Europ. Bioch. Soc. Abstr.* Warsaw, p. 134].

The following have kindly given permission to reproduce material in the various figures plates and diagrams: The American Chemical Society, The American Physiological Society, *Annual Review of Biochemistry*, W. A. Benjamin, Inc., The Biochemical Society, *Journal of General Microbiology*, The Elsevier Publishing Company, *Journal of Biological Chemistry*, The National Academy of Sciences and the Rockefeller University Press. In cases where the photographs were reproduced from printed halftone copy, the quality of the results should not be taken as representative of the originals, to which the reader is referred. Other acknowledgements are given in the text in the various chapters, and the source of all plates and figures is marked in the appropriate legends. The Editor trusts that due acknowledgements have been made in all cases and apologizes for any omissions that may have been made inadvertently.

The Editor is most grateful to his wife for arduous work in the preparation of the subject index and to the contributors to "Enzyme Cytology" who have kept their patience in what has been a rather complex and time-consuming affair. He is also most grateful to the staff of Academic Press who have coped with the production of quite complicated textual material.

D. B. R.

CONTENTS

Chapter 5

LYSOSOMES PHAGOSOMES AND RELATED PARTICLES

Chapter 8

ABBREVIATIONS USED IN THE TEXT

$\sim$	Energy-rich intermediate
Å	Angstrom unit
ACTH	adrenocorticotrophic hormone
ADP	adenosine diphosphate
δ-ALA	δ-amino laevulinic acid
AMP	adenosine monophosphate
ATP	adenosine triphosphate
ATPase	adenosine triphosphatase
CDP	cytidine diphosphate
CE	chloroplast extract
CMP	cytidine monophosphate
CMU	3-(4-chlorophenyl)-1,1-dimethyl urea
CoA	coenzyme A
CoA.SH	free coenzyme A
CTP	cytidine triphosphate
DEAE	diethylaminoethyl
DNA	deoxyribonucleic acid
DNase	deoxyribonuclease
DNP	deoxyribonucleoprotein
D-RNA	RNA capable of hybridizing with DNA
E_0'	standard oxidation reduction potential
EAA	extractable amino acids
EDTA	ethylene diamine tetra-acetate
ETP	electron transfer particle
FAA	free amino acids
FAD	flavin adenine dinucleotide
FDP	fructose diphosphate
FMN	flavin mononucleotide
GDP	guanosine diphosphate
GTP	guanosine triphosphate
I	intermediate in oxidative phosphorylation
IDP	inosine diphosphate
K_m	Michaelis constant
mRNA	messenger RNA
NAD	nicotinamide adenine dinucleotide
NAD^+	oxidized NAD
NADH	reduced NAD

NADP	nicotinamide adenine dinucleotide phosphate
$NADP^+$	oxidized NADP
NADPH	reduced NADP
P_i	inorganic phosphate
$P\text{-}P_i$	pyrophosphate
RNA	ribonucleic acid
RNA-P	phosphorus in RNA
RNase	ribonuclease
RNP	ribonucleoprotein
R-RNA	RNA with a base composition of ribosomal RNA
S	Svedberg unit
SP	soluble phase
SPAA	soluble phase amino acids
sRNA	soluble RNA
T_m	melting temperature
TMPD	tetramethylparaphenylenediamine
tRNA	transfer RNA
tris	2 amino-2 hydroxymethylpropane,1-3 diol
UDP	uridine diphosphate
UDPase	uridine diphosphatase
UDPG	uridine diphosphoglucose
UMP	uridine monophosphate
UTP	uridine triphosphate
UTPase	uridine triphosphatase
W∼	intermediate in oxidative phosphorylation
X,X∼I,X∼P	intermediates in oxidative phosphorylation.

Chapter 1

GENERAL PRINCIPLES

C. DE DUVE

The Rockefeller University, New York, U.S.A., and Université de Louvain, Laboratoire de Chimie Physiologíque, Louvain, Belgium

I. THE HISTORICAL SIGNIFICANCE OF ENZYME CYTOLOGY

Enzyme cytology was born some twenty-five years ago from the cross fertilization of two flourishing disciplines, dynamic biochemistry and descriptive cytology, which traditional differences in outlook and methodology had long kept entirely separated. It owes its inception largely to the vision and efforts of a few isolated investigators who laboured for many years to break down a barrier which was taken for granted by most of their contemporaries. The names of Martin Behrens, Robert Bensley, Albert Claude and George Gomori remain attached to this historical breakthrough.

At the time when these pioneering studies were being made, biochemistry was almost exclusively occupied with metabolic reactions occurring in soluble extracts, still riding the sweeping wave on which it had been launched at the turn of the century by the Büchner discovery that yeast juice carries out the complete fermentation of sugar. This preoccupation is betrayed in many publications from that period by the sentence: "The extract was clarified by centrifuging and the insoluble residue discarded." Few biochemists probably gave a second thought to this residue: cells and tissues obviously had to be kept together by something and the residue was "it", no more than the inert shell within which living processes take place under the influence of soluble enzymes.

It is true that Warburg had already demonstrated in 1913 that cellular respiration is associated with sedimentable particles, but this finding had attracted little attention and it took almost thirty years before the interest of biochemists became focused on the "residue". By 1948, the analysis of this system had already brought to light the existence of an integrated multi-enzyme complex, the "cyclophorase" of Green, Loomis and Auerbach, and it is conceivable that further studies along these lines might eventually have led to the correct biochemical concept of a mitochondrion. But it is difficult to guess how many false trails biochemists unaided by enzyme cytology would have explored, or might still be exploring, before appreciating the importance of isolating the particles in such a way as to preserve their elusive and apparently paradoxical metabolic inertness. Surely, the development at that time of methods specifically directed towards the isolation of intact intracellular components played a decisive role in opening what has now become one of the most important areas of research in biochemistry. The demonstration by Schneider and Potter (1949) and by Kennedy and Lehninger (1949) that the mitochondria are the main sites of cellular oxidations and of the associated phosphorylations, and the subsequent discovery by Kielly and Kielley (1951) and by Lardy and Wellman (1952) that mitochondrial respiration is controlled by the rate of regeneration of ADP from ATP are essential landmarks in our understanding of oxidative phosphorylation and of respiratory control. They can be traced back directly to the efforts of Bensley and Hoerr (1934), Claude (1946a, b) and Hogeboom, Schneider and Palade (1948).

Another question which may well be asked is how far our knowledge of protein synthesis would have progressed without cell fractionation. It is sometimes forgotten that the triumphant road leading to the genetic code starts with the early experiments of Borsook, Deasy, Haagen-Smit, Keighley and Lowy (1950a, b) and of Hultin (1950) which brought to light the important role of the microsomal fraction in the new formation of proteins. The later identification of the ribosomes as the main centres of protein synthesis was again the result of combined cytological and biochemical studies (Littlefield, Keller, Gross and Zamecnik, 1955; Palade and Siekevitz, 1956; Siekevitz and Palade, 1958).

Many other important developments of biochemistry owe a similar debt to cell fractionation techniques. In fact, the introduction of the four-fraction scheme has provided such spectacular returns that few biochemists have felt impelled to penetrate further into the intricacies of cell structure. Biochemistry has moved from the "extract-residue" era to that of "nuclei, mitochondria, microsomes and supernatant"; it has not, except in the specialized field of enzyme cytology, attempted to move much further. Far from reflecting unfavourably on the adventurousness of biochemists, this fact emphasizes the richness of the field which has been opened to biochemical research through the first contributions of enzyme cytology, as well as the significance of the latter discipline.

The impact of enzyme cytology on descriptive morphology is hardly less important. We need only think of how sterile our vision of even the most beautiful present-day electron micrographs would be, were it not illuminated, sometimes even with direct cytochemical illustration, by our knowledge of the enzymic attributes of intracellular structures. How revealing this knowledge has been to our understanding of mitochondria and of ribosomes has already been mentioned. Another example of this revelation is provided by the wealth of data which have been derived from studies on acid phosphatase (3.1.3.2), one of the first enzymes to be successfully detected in tissue sections. Thanks to a combination of all the resources of enzyme cytology, this enzyme has helped to clarify the nature of numerous subcellular entities and to trace down both in time and in space the intricate digestive processes in which they are involved (Novikoff, 1963a; de Duve and Wattiaux, 1966). The role of the endoplasmic reticulum and of the Golgi apparatus in protein secretion (Palade, Siekevitz and Caro, 1962), the origin of mitochondria (Luck, 1963a, b; 1965a, b), the function of microbodies (de Duve and Baudhuin, 1966) are other problems which morphological research alone would have been quite unable to solve.

An important conclusion suggested by this brief historical survey is that the function of enzyme cytology is not simply to serve as a connecting link between its parent disciplines. It is a true hybrid which, after the inevitable growing pains and search for its own identity, has now come of age as a vigorous individual in its own right, with specific properties that are much more than a simple combination of parental traits. It is a separate discipline with its own objectives, rules and methods, and it is capable of making unique contributions to our understanding of cell structure and function which neither biochemistry nor morphology could possibly achieve within their respective boundaries.

II. The Objectives and Methods of Enzyme Cytology

Enzyme cytology proposes to establish the exact intracellular localization of enzymes, justifying this objective by the rational argument that the working of a machine can be understood only from a knowledge of both the functional properties and the spatial organization of its component parts. In order to achieve this purpose, it has to draw heavily on the resources and acquisitions of dynamic biochemistry and of descriptive cytology; but it must refrain from any dogmatic adherence to all their pronouncements.

The reconstructions of the biochemist are not all established facts that stand no correction; neither are electron micrographs faithful representations of all that is to be known about the fine structure of cells. If any revision is needed in either area, enzyme cytology is best placed to do so, and its searches should be conducted with a sufficient lack of bias so as to allow this important role to be played. Nothing could be more stultifying, especially when exploring a field where so much remains to be discovered, than an approach that leaves nothing to the unexpected.

These aims and warnings can be translated into two simple rules. In the first place, one should aim as much as possible at localizing the site of single step enzymic reactions rather than of multi-enzyme systems. Truly, this may not always be feasible, nor even commendable, and one can think of many examples where work on complex systems has greatly helped our understanding. However, the danger in doing this is that the association of enzymes found by the biochemists to be operative in disrupted systems is implicitly taken to be biologically meaningful and even to have a morphological counterpart. Such is not always the case and the verification can come only from a careful biochemical dissection of the system into its individual enzymic units, followed by localization of the latter by means of the tools of enzyme cytology. This necessity was clearly realized already in the early days of centrifugal fractionation, when a number of investigators found the mitochondrial fraction to be the main site of oxidative processes. As pointed out by Schneider and Hogeboom (1951), what the experiments actually proved was the unique localization of the terminal oxidase in this fraction and not, for instance, that of all the dehydrogenases many of which are present outside the mitochondria (See Chapter 3). A similar mistake was made in visual cytochemistry when the tetrazolium method was introduced for the localization of dehydrogenases: the mitochondria were found to stain selectively with all substrates, not because they contain all the dehydrogenases, but simply because they happen to contain the diaphorase necessary for the reduction of tetrazolium (Novikoff, 1963b). These are not isolated instances and it is now known that many polyenzymic processes require the co-operation of two or more intracellular sites (de Duve, Wattiaux and Baudhuin, 1962).

The second rule is to consider the cell as being still largely uncharted territory. Our present picture of cellular structure is delineated by such molecular associations, most of them of lipoprotein nature, which when fixed and stained in certain ways provide suitable contrast in the electron microscope. Impressive as the image developed in this manner may seem, there is little doubt that it would look very gross against a detailed enzyme map of the cell in which, for instance, each individual enzyme were painted a different colour. No doubt, the basic delineation by cytomembranes would remain clearly recognizable, but it would appear in a variety of colours and shades. In addition, the uniform areas between the membranes might well reveal an unsuspected mosaic of multicoloured patterns. To establish such a map is the main purpose of enzyme cytology.

The most obvious and direct way of accomplishing this purpose is to let each individual enzyme signal its location by means of reliable staining techniques applicable to intact cells or tissue sections. Like the individual plates of multichrome prints, the images obtained in this manner should bring out one by one the elements of the pattern we wish to reconstruct. This approach has generally been followed by the investigators who have entered enzyme cytology from the side of morphology. Working against great odds, they have succeeded in developing staining procedures for a fairly wide variety of enzymes. Unfortunately, very few of these procedures answer the requirements of applicability at the level of the electron microscope, which is the only instrument providing the degree of resolution necessary for accurate enzyme cytology. The only detailed enzyme maps available today are those of some hydrolytic enzymes, especially phosphatases, and the resulting picture we have of the cell is still woefully sketchy and monochromatic. It is even fair to state that the images would be much less enlightening than they actually are, were it not for the additional information obtained by other methods and allowing a single enzyme to serve as a marker for a whole group of others. When such information is available, the staining procedures become extremely valuable tools, as illustrated by the usefulness of the acid phosphatase method in the study of lysosomes. They also have the advantage of being simple and easily applicable to a variety of biological materials and are thereby well suited for histological controls and for comparative work. But they have not yet been developed to a point where they can by themselves play any significant role in mapping out the enzymes of the cell. Up to the present time, most of this work has been done with the much more indirect procedure of cell fractionation which still remains the major tool of enzyme cytology.

The rest of this chapter will be devoted to a discussion of this technique. It constitutes a somewhat expanded version of a previous article on the subject (de Duve, 1964) and draws heavily on the frustrations and satisfactions experienced in our laboratories over the last fifteen years. The manner in which these experiences have progressively shaped our approach has been recounted in a recent review (de Duve, 1965).

Some of the views expressed may seem controversial, especially if they are misunderstood. I have taken some pains to avoid this, even at the risk of being repetitious, and wish to emphasize again that what I am discussing specifically is cell fractionation *as a method of enzyme cytology* and not all the applications of the technique, for instance in biochemistry where, as pointed out by Roodyn (1965), it has many important uses that do not require the same rigorousness as when it serves to localize enzymes in cells.

III. The Conduct of Fractionation Experiments

All fractionation experiments proceed in three successive steps: homogenization, fractionation and analysis. The sequence is obligatory in that order: we fractionate, not cells or tissues, but preparations obtained by disrupting them in an artificial medium; we analyse, not intracellular organelles, but fractions separated from homogenates.

A. FROM TISSUES AND CELLS TO HOMOGENATES

Homogenization is the necessary prerequisite of all mass fractionation experiments. It results in a considerable loss of morphological information, but such is the price we pay for being able to apply henceforth the innumerable techniques of biochemistry. Our problem is to conduct the operation in such a manner as to lose no more information than is needed, while making what is left retrievable with as little distortion as possible.

Unfortunately, homogenization is still the most empirical of all technical procedures used in tissue fractionation. Numerous mechanical devices have been constructed (see: Anderson, 1956; Allfrey, 1959), but the principles on which they operate are not clearly known and the manner in which they are handled varies greatly from one investigator to the other. For lack of a satisfactory theoretical basis and of adequate standardization methods, homogenization is still largely an art. This is unfortunate since it is the critical step which determines the quality of our starting material. Here is an area which undoubtedly would repay a more comprehensive investigation of what is, after all, largely an engineering problem.

The choice of an adequate medium creates an even more difficult problem since there is practically no other way of solving it except by empirical methods. Sucrose, first advocated by Hogeboom *et al.* (1948), is now widely used as the main ingredient, but there are innumerable individual recipes differing by such details as sucrose concentration, removal or neutralization of CO_2, addition of buffer, of EDTA, of traces of specific cations, of macromolecular substances, etc. In addition, it is now clear that sucrose solutions are not always superior to ionic media, as was originally believed (de Duve and Berthet, 1954). Salts tend to promote agglutination in liver homogenates, but the reverse is true with spleen preparations which actually show more agglutination in 0·25 M sucrose than in 0·2 M KCl (Bowers, 1964). Another controversial subject relates to the use of non-aqueous media, as introduced by Behrens (1932). The advantages and disadvantages of this method have been discussed in detail by Allfrey (1959).

In general it may be stated that each biological material poses a specific homogenization problem, to be solved largely by trial and error. But this requires that we define first the qualities which we should strive for in a homogenate as well as the control methods whereby they are best assessed. In discussions of this question, stress is often put on morphological criteria and the phase contrast microscope is mentioned as the main tool to be used for controlling homogenates. While recognizing the importance of microscopic examinations to ascertain the degree of disruption achieved as well as gross alterations of organelles or agglutination artifacts, I submit that the most useful instrument for estimating the quality of a homogenate is the centrifuge. The best homogenate is the one which lends itself most successfully to fractionation. Or to put it colloquially, the proof of the homogenate is in the fractionating.

This point is of more than trivial importance; it reflects a rather basic attitude which is that adopted by our group. In our opinion, cell fractionation is not a routine standardized technique, to be set up once and for all and then applied as such whenever desirable; it is a never-ending exploration of the organization of cells, every phase of which invites reappraisal as more information becomes available. When applied in a routine fashion, all it can do is to reveal the manner in which an enzyme is distributed amongst the isolated fractions; it rarely brings to light the true intracellular localization of the enzyme, the objective of enzyme cytology. To attain this objective, it is generally necessary to use a multiplicity of approaches, following the self-correcting technique of the chemist, whereby the information gained is continuously fed back into the experimental design until satisfactory resolution is achieved. This, at least, is how hepatic enzymes such as glucose-6-phosphatase (3.1.3.9) (Hers, Berthet, Berthet and de Duve, 1951), cytochrome oxidase (1.9.3.1) (Hogeboom, Schneider and Striebich, 1952) or acid phosphatase (Appelmans, Wattiaux and de Duve, 1955) were first identified as specific components of certain intracellular structures although their distribution between the fractions isolated by the routine procedure was quite heterogeneous. The resolution of catecholamine-containing granules from mitochondria in preparations from the adrenal medulla (Blaschko, Hagen and Hagen, 1957), that of nerve endings from other cytoplasmic components in brain homogenates (de Robertis, 1963; Whittaker, 1963), that of peroxisomes (microbodies) from mitochondria and lysosomes (Beaufay, Jacques, Baudhuin, Sellinger, Berthet and de Duve, 1964; Baudhuin, Beaufay and de Duve, 1965) and many other separations were brought to a successful conclusion by adhering to the same principle.

The control of homogenization procedures has become increasingly easier as our knowledge of cellular organization has grown. For instance, we now know enough about the general properties of mitochondria to feel entitled, even with an unknown material, to use cytochrome oxidase or succinate dehydrogenase (1.3.99.1) as marker enzymes for these particles, or respiratory control as a test of mitochondrial integrity. Again, the well-established distribution of lysosomes in a wide variety of animal cells allows the latency of acid hydrolases to serve as an indicator of the fate suffered by these fragile particles. The usefulness of this easy and convenient test as a means of controlling homogenization damage is illustrated by the work of Greenbaum, Slater and Wang (1960) on mammary gland and by those of Conchie, Hay and Levvy (1961) and of Bowers (1964) on spleen. The binding and latency of catalase (1.11.1.6) can serve a similar purpose in peroxisome-containing tissues such as liver and kidney (Baudhuin *et al.*, 1965).

These tests, together with suitable morphological examinations, may already go a long way in helping to devise a homogenization procedure appropriate to the material under study. However, it still remains to be seen whether the degree of disruption achieved and the physical environment offered are such as to allow satisfactory resolution. As already mentioned, this verification requires fractionation, preferably by several different techniques, and characterization of the fractions. Here again, much information can be gained by using marker enzymes for known organelles to appreciate the efficiency of fractionation, and thereby the quality of the homogenate.

Most fractionation experiments are performed on solid tissues and this introduces two additional complications which have not yet been considered. One is the cellular heterogeneity of the material, which multiplies the difficulties mentioned by the number of different cell types present. The other is due to the connective framework of the tissue, the disruption of which may necessitate mechanical stresses highly traumatic to subcellular constituents. A similar difficulty is encountered with single cells when they are encapsulated. In this respect, tissue fractionation may gain a great deal from the numerous attempts that are being made with increasing success to prepare suspensions of single cells from solid tissues by gentle mechanical or enzymic techniques. If the cells can be obtained in viable and non-leaky form, a result which seems very difficult to achieve, one can even consider the possibility of separating the cell populations before homogenization, thus reducing the problems created by cellular heterogeneity (see: Mateyko and Kopac, 1963). Even imperfect techniques of this type may sometimes help to settle an ambiguity created by the results obtained on the whole tissue.

It must be noted that the randomizing effect of disrupting a heterogeneous tissue does not necessarily lead to an irretrievable loss of the relevant information. For instance, homologous components of two different cell types may differ sufficiently in their physical properties to be separable by appropriate techniques; if so, it may be further possible to identify their respective host-cells by suitable morphological or biochemical criteria. For instance, recent work on spleen homogenates has disclosed the presence of several lysosome populations which appear to originate from different cell types (Bowers and de Duve, 1967).

B. FROM HOMOGENATES TO SUBCELLULAR FRACTIONS

1. Technical Procedures

The centrifuge is by far the most widely used instrument for the fractionation of homogenates. Thanks largely to its important applications in the analysis of macromolecules, centrifuging has received exhaustive theoretical treatment from both the engineering and the physico-chemical point of view.

Excellent instruments are available and efforts are constantly being made to construct better machines. The behaviour of particles in a centrifugal field and the various factors which affect their rate of sedimentation have been studied in great detail and excellent monographs and reviews have been published on the subject (Svedberg and Pedersen, 1940; Ogston, 1956; Schachman, 1959).

While it may not be necessary to master all the intricacies of analytical centrifuging to be able to perform good fractionation experiments, a thorough understanding of the physical processes involved and also of the construction characteristics of the instruments used must be considered essential. Several papers have dealt with these theoretical and practical aspects of the technique, as related more specifically to tissue fractionation (Pickels, 1943; de Duve and Berthet, 1953, 1954; Anderson, 1956; Allfrey, 1959; de Duve, Berthet and Beaufay, 1959; Beaufay and Berthet, 1963). The latter two papers and that of Anderson (1956) should be consulted particularly with respect to the use of density gradients. Workers searching for a suitable solute for the preparation of density gradients may be interested in the long list compiled by Mateyko and Kopac (1963).

A detailed discussion of techniques and procedures falls beyond the scope of this chapter, but one point of general interest deserves to be stressed. The centrifugal separation of subcellular particles takes advantage either of differences in density, or of differences in sedimentation coefficient, itself a function of the density, size and shape of the particles. All these parameters are dependent on the properties of the medium. This is of course true of the sedimentation coefficient which is an explicit function of the density and viscosity of the medium; but it also applies to the physical characteristics of the particles owing mainly to the influence exerted by the composition of the suspension medium on the exchanges of water and solutes between particles and medium. These exchanges are themselves regulated by such intrinsic properties of the particles as hydration, membrane permeability and internal osmotic pressure. Particles belonging to different populations may differ significantly in some of these properties and may therefore respond in a different manner to changes in the composition of the medium. Although they may not be separable in a given medium, they may become so in another and it is therefore extremely useful to try a variety of approaches. We have discussed this principle in several publications (de Duve *et al.*, 1959; Beaufay and Berthet, 1963; de Duve, 1964, 1965) and demonstrated its practical validity and usefulness (Beaufay *et al.*, 1964). An extension of this approach is to subject the animals or the isolated preparations to treatments which modify the physical properties of some particles. For instance, it is now known that the size and density of the lysosomes can be altered to a considerable extent by injecting the animals with substances which accumulate in these particles. Injection of egg-white (Straus, 1954), Triton WR-1339

(Wattiaux, Wibo and Baudhuin, 1963), sucrose (Wattiaux, Wattiaux-De Coninck, Rutgeerts and Tulkens, 1964) and dextran (Baudhuin *et al.*, 1965) have been found to produce such changes. The density and chemical composition of mitochondria can also be altered experimentally under certain conditions (Greenawalt, Rossi and Lehninger, 1964; Luck, 1965a, b).

Other techniques besides centrifuging have been applied to the fractionation of homogenates. Chromatography has been used successfully for the separation of viruses and also of cell particles (Riley, Hesselbach, Fiala, Woods and Burk, 1949). Albertsson (1960) has developed a number of two phase systems which have served mostly for the separation of macromolecules and viruses, but have also been applied to cell particles and even whole cells. The application of electrophoresis to fractionation of homogenates has been explored recently by Davenport (1964) and by Hannig (1964). On the whole, a wide choice of procedures and approaches is now available for the fractionation of homogenates. It is the task of the investigator to determine which is best suited to the solution of his particular problem. No rule can be set, except that more than a simple four-fraction separation is likely to be required.

2. The Preparative Approach

Fractionation methods were first developed with the aim of isolating specific intracellular constituents in pure form and then studying their chemical and functional properties. They are still most widely used for this purpose today, especially in the field of biochemistry where it is often advantageous to start with the preparation of "nuclei", "mitochondria" or "microsomes" as a preliminary to metabolic or enzymological studies. Even in the specific field of enzyme cytology, some workers, like Allfrey (1959), tend to consider the "analysis of single fractions isolated under conditions which permit an estimate of their nature and purity" as an approach allowing a "more definite answer to the question of the relationship between the function of an intracellular structure and its enzymatic composition" than the analytical approach as exemplified by quantitative enzyme distribution studies. Statements to the opposite effect have frequently been expressed by our group and the causes of this disagreement deserve some attention.

To examine this question, let us consider the electron micrograph of Fig. 1. It shows part of a rat liver cell and has been selected so as to illustrate most of the known constituents of this cell, including some of the rarer cytoplasmic particles. Assuming that we have no prior knowledge of the functional properties of any of the objects that we see, we may be tempted to seek such knowledge by the straightforward method of separating a chosen constituent from all the others and then determining its enzymic equipment. However, to attain this goal in practice, we must be able to monitor our purification attempts. Since we can only rely on morphological examination

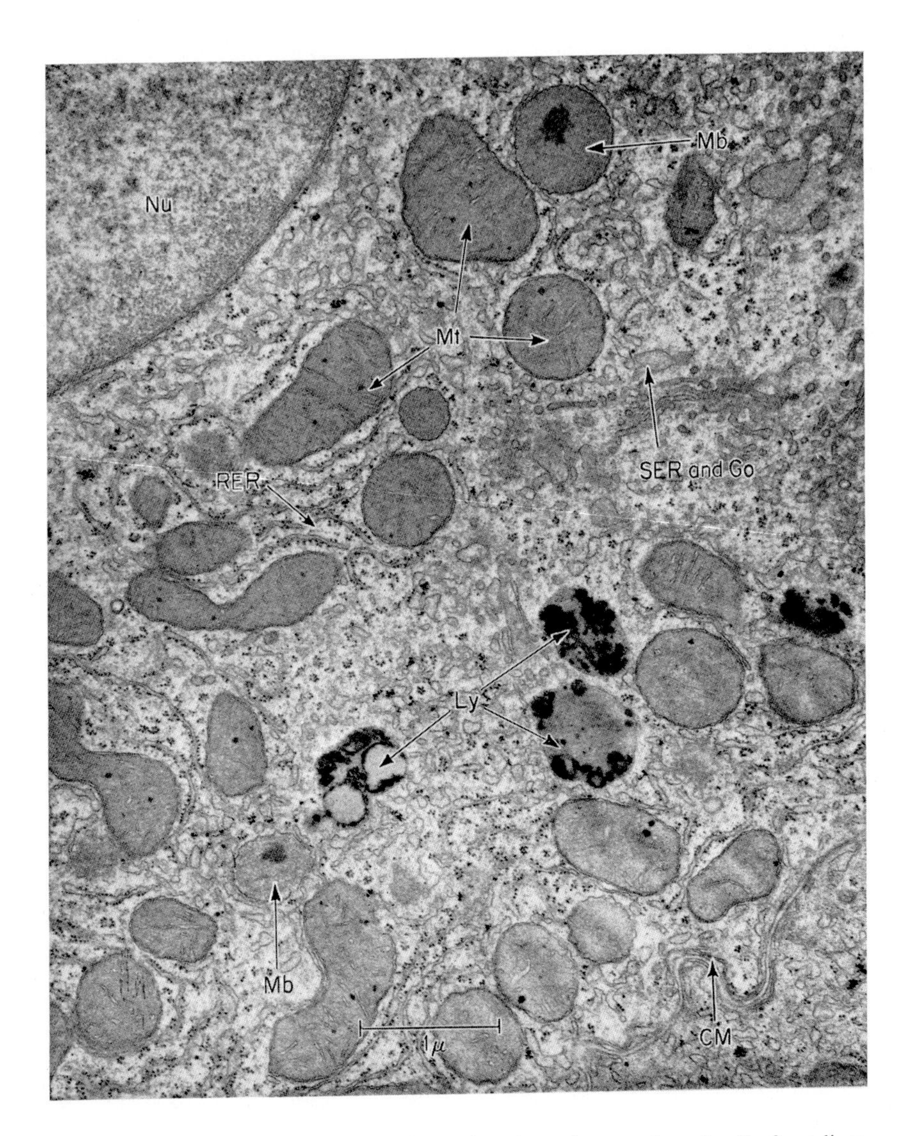

FIG. 1. Electron micrograph of ultrathin section through parenchymal cell of rat liver. Image shows part of the nucleus (Nu) in the upper left corner and of the cell membrane (CM) in the lower right corner, numerous mitochondria (Mt), several pericanalicular dense bodies or lysosomes (Ly), two microbodies or peroxisomes (Mb), areas of rough-surfaced endoplasmic reticulum (RER) and of smooth-surfaced endoplasmic reticulum (SER) and elements of the Golgi apparatus (Go). Small rosette figures are probably polysomes. Very rare rosettes made up of thicker particles are glycogen particles. Two particles with single membrane and finely granular matrix, amongst mitochondria in upper third and near cell membrane, are probably microbodies whose dense core lies outside the section. Courtesy of Dr. J. D. Jamieson, Rockefeller University, New York.

for this purpose, our search is virtually restricted to recognizable objects that are sufficiently large or numerous to be easily identified in complex mixtures. We may validly set out to purify nuclei or mitochondria, but should we wish to isolate the pericanalicular dense bodies or the microbodies, we would find it almost impossible to do so owing to the difficulty of assessing the frequency of our chosen material in the early steps of our purification attempt and consequently of devising a suitable preparative procedure. When it comes to subunits that cannot even be seen in electron micrographs—multi-enzyme systems like the fatty acid synthetase of Lynen (1961) could be of this type—there is of course no way of even defining the problem.

Even if we decide to aim our attempt at a relatively accessible target, such as for instance the purification of mitochondria, we will nevertheless encounter the limitations of morphological techniques when we become faced with the problem of controlling the purity of our preparations. Obviously, objects which for one reason or another do not lend themselves to purification on the sole basis of morphological examinations, will for the same reason escape accurate detection when they occur as contaminants.

Biochemical methods do not suffer from these limitations. Especially in the form of specific enzymic analyses, they can serve to detect and even estimate quantitatively any cell constituent, whether visible or invisible, rare or abundant, provided the enzyme measured is associated exclusively with this constituent and is known to be so. In our opinion, the latter information can come only from detailed studies of the distribution of enzymes amongst all the fractions that can be isolated from homogenates, in other words from analytical fractionations. For this reason, we consider purification an ultimate rather than an immediate aim in the field of enzyme cytology, and have in fact arrived at it in this manner in our own work.

It is true, as will be pointed out below, that morphological methods are open to considerable improvement and could be developed so as to provide relatively accurate quantitative information, at least for components that can be detected and identified under the electron microscope. However, even with such information available, it would still be very difficult to appreciate its significance with respect to the biochemical properties of the preparation without analytical data to draw on. If we find, for instance, that a preparation of mitochondria contains one microbody for every twenty mitochondria and also possesses some catalase activity, we will still be unable to correlate these two findings unless we can relate them to other observations bearing on the frequency of microbodies and on the catalase activity in the homogenate and in other subcellular fractions. This again subordinates the preparative to the analytical approach

The point at issue, therefore, is not whether the preparative approach can or cannot provide a more definitive answer to a problem than the analytical approach, but rather whether a preparative attempt can be validly undertaken "under conditions which permit an estimate of the nature and purity" of the resulting preparation without a considerable amount of prior analytical work to guide it. It has been our contention that it cannot. There will probably be little quarrel with this opinion as applied to the rarer cytoplasmic particles with which most of our investigations have been concerned. But it could easily be argued that we are strongly biased by this very experience and that if workers engaged in the study of the major cell constituents had shared our concern, biology would be much less advanced than it is today. Had they waited, for instance, until they had available biochemically pure mitochondria or nuclei, we would still be largely ignorant of oxidative phosphorylation or of nuclear protein synthesis. This is an obvious truth. But so is the fact that if nuclear or mitochondrial fractions had been taken "at their face value", we would still credit the nucleus with a remarkable array of metabolic activities, while our functional representation of mitochondria would include a strange association of oxidative phosphorylation, acid digestion and hydrogen peroxide metabolism.

If the tree is to be judged by its fruits, we can only conclude that both approaches have their merits as well as their limitations and that the richest returns may be expected if both are pursued simultaneously, drawing constantly on each other and on all the resources of morphology and cytochemistry for further improvement. Inadequately controlled preparative fractionation has been a highly successful shortcut to a great deal of important information; but it involves something of a calculated risk and is related in this respect to the usual methods of dynamic biochemistry. Analytical fractionation is slower and less immediately rewarding, but it surrounds itself with greater precautions before merging its aims with those of the preparative approach. It is more specifically a tool of enzyme cytology.

3. The Analytical Approach

Any quantitative fractionation, whether or not it is conducted on the basis of *a priori* morphological considerations, is an analytical experiment. Or at least it can be so if the investigator chooses not to commit himself as to the cytological composition of the fractions. It is perhaps easier to do so when the fractionation scheme is an arbitrary one, as in most forms of density gradient centrifuging, but the cuts do not have to be arbitrary to make the experiment an analytical one. Quantitative work and a certain mental attitude are the essential conditions.

In an analytical experiment, we divide a homogenate in a certain number of fractions and assay each fraction as well as the original homogenate for one or more biochemical constituents, mostly enzymes. For clearer visualization, the results of the measurements are represented graphically as distribution patterns, or preferably, if the experimental design permits it, as frequency distribution histograms or curves of the enzymes as a function of the parameter, sedimentation coefficient or density, which has served as a basis for the separation. Quantitative recovery is considered very important and the distributions are taken to be significant only if the sum of the activities of the fractions corresponds reasonably well to the activity of the original homogenate.

To proceed further, it is necessary to assume some kind of relationship between the enzymes measured and the subunits with which they are associated in the homogenate. The word subunit is used here on purpose in order not to confuse the issue by any premature introduction of cytological terms. The simplest hypothesis is to assume that all subunits of a given population have the same enzyme content, as related for instance to their mass or total protein. We have called this assumption the *postulate of biochemical homogeneity*, emphasizing its hypothetical nature, but pointing also to the theoretical and experimental arguments which tend to support its validity.

Taking this postulate as a working hypothesis, we may now tentatively consider the distributions of the enzymes as representing those of the subunits to which they are attached. If two enzymes show sufficiently different distributions, we conclude that they are likely to belong to different subunits. If they show the same distribution, we take this as an indication but by no means as proof that they may be associated together with the same subunit. In order to distinguish between coincidental superimposition and true physical association, the fractionation is repeated under a number of different conditions. If the association is found to persist under each new set of conditions, the conclusion that it is real and not spurious becomes progressively strengthened.

It will be clear from this schematic description of the analytical approach that it can in itself lead only to conditional interpretations whose acceptability depends on the validity of the underlying postulate. Their final verification comes when enough information has been gathered to warrant a preparative attempt and a correlation of the biochemical results with morphological observations. This aim has been achieved in a number of cases, thus providing an experimental justification for the approach and for the hypothesis on which it rests.

Obviously, the analytical approach can be successful only if the procedures employed do not cause the disruption of the subunits themselves. Otherwise, one would eventually be led to identify as many subunits as there are enzymes in the cell. However, additional information of a very valuable kind may often be obtained by the deliberate use of disruptive procedures which may bring to light other similarities or dissimilarities between enzymes, based this time on the relative fragility of the subunits to which they belong. For instance, we may by comparative latency studies obtain new sets of curves which express the manner in which individual bound enzymes are released by the treatment applied. Such curves complement the distribution curves and may be similarly interpreted, although with greater caution since there is no compelling reason why two enzymes associated with the same subunit should necessarily be released together when the subunit is subjected to graded disruption. However, the information obtained in this manner may be of great help in the interpretation of distribution data and may also serve to characterize further the subunits under study in terms of their physico-chemical properties.

As pointed out by many authors, analytical fractionation is by no means a fool-proof method and is exposed to many artifacts. However, most of the latter are due to the rearrangement of contituents as a consequence o homogenization and exposure to an artificial medium. We have avoided discussing them here by defining the objective of the fractionation as a characterization of the *subunits* present in the homogenate. How these subunits are related to cytological entities becomes a matter of interpretation; it will be examined later.

C. FROM SUBCELLULAR FRACTIONS TO ANALYTICAL RESULTS

1. Biochemical Measurements

Once the fractions are isolated, they fall almost entirely within the area of competence of the biochemist. Some cytological aspects have to be kept in mind, especially with regard to the influence structural integrity may have on enzyme activities. Many particle-bound enzymes are shielded by barriers which restrict their accessibility to the substrates used for their assay. This phenomenon must be investigated and methods ensuring the complete release of the enzymes for assay purposes must be worked out. The dissociation of structurally linked multi-enzyme systems may pose other problems, especially if we adhere to the rule that enzyme cytology should deal with single enzymes whenever it is at all possible. If those difficulties can be dealt with satisfactorily, we are left with straightforward enzymological problems. This does not mean that they are easily solved or free from hidden pitfalls.

The assay of enzymes in complex mixtures is rarely easy and thorough kinetic experiments are necessary in order to define conditions of temperature, pH, substrate concentration and medium composition under which the reaction velocity or its integral over a period of time provides a valid and reproducible measure of the number of enzyme molecules present. These kinetic experiments have to be repeated on every new biological material and also on the various isolated subfractions, for homologous enzymes often have different kinetic properties in different tissues and they may be accompanied in them or in separated subfractions by varying amounts of inhibitors or activators.

It is also of great importance to investigate the assay method from the point of view of the number of enzymes involved. It often happens, for instance, that a given substrate is attacked simultaneously by more than one enzyme. If the enzymes involved are associated with different subunits, complex distribution curves will be obtained, possibly leading to false interpretations unless the source of the complexity is correctly traced down to the enzymes themselves. As already mentioned, the reaction measured may involve two or more enzymes acting in succession, leading to other possible artifacts. These problems have been discussed in more detail by de Duve and Berthet (1954) and by Allfrey (1959).

2. Morphological Examinations

A number of investigators have successfully combined morphological examinations with enzymic measurements in analytical fractionations and have been able in this manner to correlate enzyme distributions with the observed frequency of certain structural components. As they have been applied until now, the morphological methods have been at best grossly semi-quantitative, but there is no reason why they could not, with a little effort, be improved very significantly.

Accurate morphological assessment of the composition of a fraction depends on three conditions: (1) Perfect random sampling, a very important requirement in view of the extreme minuteness of the sample actually examined; (2) Reliable criteria of identification of the various components present; (3) A measuring device allowing the investigator to put the findings in quantitative terms.

The condition of random sampling is not satisfied by current techniques, most of which rely on centrifuging for collecting the material and furnish stratified pellets. These can be examined at different levels, but this method is either inaccurate or very time-consuming. Baudhuin, Berthet and Evrard (1967) have recently developed a collection technique by Millipore filtration which seems to obviate this difficulty.

The condition of identification is likely to be the most difficult one to fulfill, especially since fractions isolated by centrifuging frequently contain damaged particles and variable amounts of unrecognizable debris. As to the means whereby the results can be quantitated, simple counting of profiles is not sufficient since the number of profiles furnished in thin sections by any given component depends on the volume and shape of the component. It is necessary, therefore, to make the measurements in such a manner as to be able to estimate the contribution of the component to the total volume or mass of the preparation. Loud, Barany and Pack (1965) have recently described a technique for measuring volume ratios on tissue sections which could also be applied to isolated fractions; these authors also discuss a number of technical problems associated with the use of the electron microscope for quantitative work. Dry mass distributions can be estimated by means of a method developed by Bahr and Zeitler (1962). This method does not allow identification of the individual particles, but it provides sufficient statistical information to reveal in some cases the presence of more than one particle population.

IV. The Interpretation of Fractionation Data

The three successive steps followed in the conduct of fractionation experiments have to be retraced backwards in the process whereby the analytical results are converted into information on the intracellular localization of enzymes. Each of the three steps requires the investigator to assume, so to speak, a different personality and it is very important that they should be taken in succession and in the right order.

A. From Analytical Results to Subcellular Fractions

1. Validity Control: the Balance-Sheet

Our first concern is to examine the results of the assays strictly from the point of view of the enzymologist who has been given a certain number of unknowns to analyse. The question raised here is whether the raw data are faithful expressions of enzyme activities. Attention has already been called to the necessity of performing adequate kinetic experiments and if this precaution has been taken the results should inspire confidence. But provision must be left for the unexpected or the accidental.

One of the most valuable controls is to add up the activities found on the various fractions and to compare this sum with the activity observed on the starting material. This presupposes that the fractionation has been conducted in a quantitative fashion and that no fraction has been discarded. Not all workers agree on the need of such a balance-sheet. Our own opinion coincides with that expressed by Schneider and Hogeboom (1951); we consider the balance-sheet essential.

In the first place, it provides an almost indispensable test of the validity of the results. When the recovery is distinctly deficient or excessive, something must have gone wrong somewhere and it is better to know it. The cause may be accidental and traceable, and then a mistake will have been avoided; or it may be untraceable, but it will at least be known that the results are not reliable. If the defect in the balance-sheet is systematic, it becomes very significant and may either help to correct an unsuspected error or reveal a complicating phenomenon which may turn out to be of great biological interest. The latency of lysosomal enzymes was originally made manifest by a 200% recovery (Berthet and de Duve, 1951).

The balance-sheet is also important at further stages of the interpretation, for instance in judging the validity of enzyme assays for purity assessment in preparative experiments. A purified fraction cannot be declared pure simply because it lacks the characteristic enzymes of contaminants; the enzymes must be shown to have been actually removed rather than simply inactivated. The form in which they have been removed is also important. For enzymes that are easily released in soluble form from their host-particles, evidence that they are still particle-bound in the discarded fractions is necessary before their absence in purified fractions can be taken as proof that a true separation of particles has actually been achieved.

Finally, the balance-sheet provides an essential element of information when it comes to deciding whether a given enzyme activity should be assigned to the subcellular component with which it is found associated or to a contaminant. The point at issue here is not whether one does or does not consider significant the association of let us say 10% of the total activity of an enzyme with 100% of the nuclei (Allfrey, 1959). This is a matter of opinion and the author is of course entitled to his own. But so is the reader, and he can only arrive at one if he is given the relevant facts. Surely it cannot be considered irrelevant to this question to know whether the activity displayed by isolated nuclei corresponds to 100, 10 or 0·1% of the activity found in the corresponding amount of whole tissue.

2. *Presentation of Results*

Graphical representation of the results often helps to bring out their significance more clearly than simple rows of figures. Especially in enzyme distribution studies, there is considerable advantage in following methods which may eventually allow the results to become represented in the form of histograms or frequency distribution curves which may serve to extract new information (see: IV. B. 2) and may even be amenable to statistical analysis. Examples of such methods are described in the appendix.

B. FROM SUBCELLULAR FRACTIONS TO HOMOGENATES

1. *Biochemical Characterization of Subunits*

Once a clear and reliable picture has been obtained of the manner in which enzymes and other biochemical components are distributed between the fractions, it now becomes necessary to interpret the results in terms of their association with one or more distinct populations of particulate entities. As already explained (III. B. 3), our main guiding principle here is the postulate of biochemical homogeneity. On the basis of this postulate, we identify as many subunits as there are clearly distinct distribution patterns, and keep in mind further the possibility that the procedure followed may not resolve all the subunits present and therefore that two enzymes showing the same distribution may nevertheless belong to different subunits. As evidence accumulates that two or more enzymes are truly inseparable, some sort of biochemical picture of the subunit as an enzyme complex begins to emerge. Although we have not yet reached the stage where the subunit can be regarded as a true cytological component, this picture may be very helpful in suggesting further experiments. In our own investigations, both the lysosome and the peroxisome concepts were first suggested by such enzyme associations and they then directed our further search towards acid hydrolases in the former case and towards oxidases in the latter.

Comparison of the distribution of enzymes with that of proteins and the computation of specific activities are also very helpful, indicating whether the hypothetical subunits represent a major or a minor component.

Finally, studies of enzyme latency may serve to characterize the subunits further by indicating the mode of binding of the enzymes. They may also, as has been mentioned, help to distinguish different populations of subunits.

2. Physical Characterization of Subunits

Having tentatively identified one or more subunits, we may try to derive additional information concerning their physical properties from the manner in which they become distributed under the influence of the fractionation procedure. In order to do this, it is necessary to know accurately the conditions under which the operation has been conducted, and also to ascertain that no artifacts such as convection or drop sedimentation have vitiated the experiment. Provided these conditions are met, the Svedberg equation can be used to estimate one or more physical properties of the subunits. If these calculations are based on enzyme distributions, it is obvious that their validity depends on the applicability of the postulate of biochemical homogeneity.

Since fractionations are based either on differences in sedimentation coefficient or on differences in density, the most immediate information which can be extracted from the observed distributions relates to the distribution of these two particle parameters. The manner in which such graphs can be constructed is explained in the appendix.

As shown elsewhere (Beaufay and Berthet, 1963; Beaufay *et al.*, 1964; de Duve, 1965), when sufficient data are available concerning the density and sedimentation coefficient of the subunits in different media, one may attempt to fit them to the theoretical equation worked out by de Duve *et al.* (1959) for a model which seems applicable in first approximation to several types of cytoplasmic particles. These calculations which are generally performed on the median values of the distribution diagrams make it possible to estimate numerous additional physical parameters of the subunits, including their size, their dry mass, their content in osmotically active solutes and the dimensions of their water compartments. Complementary data on the osmotic behaviour of the subunits are sometimes provided by the results of enzyme latency studies.

C. FROM HOMOGENATES TO CELLS AND TISSUES

This is the final and most critical step in the interpretation. Having, through the process outlined above, arrived at some kind of robot picture of certain particulate subunits present as resolvable components in the homogenate, we are now required to answer the question whether these subunits exist as such within the cells or whether they are to a greater or lesser extent the product of homogenization artifacts. Granting this question can be answered satisfactorily, we must finally identify the intracellular entities from which the subunits originate and locate them at the cytological and at the histological level.

The most direct way of achieving this result is through morphological examination of the fractions, especially if the analysis can be made quantitative (see: III. C. 2) or if a reasonable degree of purity has been attained. Sometimes, careful examination of electron micrographs of the intact tissue may already be quite revealing. One of the arguments supporting the identification of the subunits characterized as peroxisomes in hepatic homogenates with the microbodies seen in parenchymal cells was that no other intracellular organelle fitted the robot picture derived from the fractionation experiments (Baudhuin *et al.*, 1965). Cytochemical staining reactions may also be very helpful sometimes in correlating biochemistry with morphology, as illustrated by the results of Essner and Novikoff (1961) and Holt and Hicks (1961) which confirmed in a definitive manner the lysosomal nature of the hepatic pericanalicular dense bodies. However, the confirmation was to some extent mutual in this case, since the biochemical data also helped to establish the validity of the staining method. Finally, there are already a number of instances where morphological examination can be largely dispensed with and where enough evidence is available to validate the use of specific enzymes as markers for their host-particles.

All these procedures depend in their final accuracy on how successful we can be in eliminating or recognizing artifacts. There is no doubt that enzymes get reshuffled to a considerable extent and are also sometimes altered in their catalytic properties from the time cells and tissues are first removed from their natural environment up to the moment when subcellular fractions are separated and subjected to analysis. When complete translocations occur, they may escape recognition entirely until some new observation turns up to alert our suspicions. When the translocation is a partial one, the enzyme appears as a component of more than one population of subunits or is present partly in the supernatant fraction. The resulting experimental situation is identical with that which would obtain, without artifact, if the enzyme had a true multiple location. To distinguish between the two possibilities, the only logical attitude is always to suspect an artifact when a complex distribution is encountered, since it is the only attitude which results in positive action. Guided by this working hypothesis which we have called the *postulate of single location*, we can then carry out a variety of experiments designed to reveal an artifact if there is one. This may lead us to experiment with different grinding devices and suspension media, to try new fractionation procedures, to investigate further the mode of binding of the enzymes, to search for biochemical differences between homologous activities associated with different subunits, and so on. If all these approaches fail, the ambiguity will remain, but at least we will have eliminated some possibilities and thereby strengthened the hypothesis that the observed distribution actually reflects multiple location. The rule is to accept this interpretation only to the extent that the alternative explanation invoking an artifact can be excluded on the basis of the experimental facts.

V. Conclusion

The pathway described in this chapter is long, arduous and circuitous. It is not likely to attract the pragmatist who sees many shortcuts leading to what appears to be the same destination. Neither is it our contention that it is the only road to success. Our attempts at reconstructing the cell all proceed by a series of approximations and the approach presented is no exception. Our final goal is the understanding of cellular function, and enzyme cytology itself is only one of the multiple ways leading to it. To be fruitful, its findings must be integrated with all the information that can be obtained through the other biological disciplines. They become truly validated only to the extent that the behaviour of living cells is found to be both compatible with and illuminated by our enzyme maps.

However, there is no doubt that enzyme cytology has a particularly important role to fulfill in this joint venture and it is essential therefore that those who practise this discipline remain critically aware of the nature of the risks which they accept when adopting a given approach. We believe that the analytical approach, as it has been outlined here and followed in our own investigations, involves fewer risks than more direct approaches and that its laborious character is justified on this account. Whether one chooses to follow it or not, the principles set out may help to define the limitations which are accepted and to facilitate communication between investigators engaged in connected areas of research. As has been pointed out elsewhere (de Duve, 1964), the confusion of operational with cytological terms has created something of a nomenclature problem in enzyme cytology and has led in some cases to errors of interpretation. This is easily avoided provided cytological terms such as nuclei or mitochondria are never used in reference to subcellular fractions unless adequate evidence is available that the entities referred to are truly the organelles so designated in descriptive cytology.

VI. Appendix: Graphical Presentation of Results

GENERAL PRINCIPLE

Let Q be the absolute amount of an enzyme or other biochemical component found in a subcellular fraction (eventually corrected for known sampling losses, see: Beaufay *et al.*, 1964).

Provided the sum ΣQ of the amounts found on all the fractions is not too different from the amount found on the starting material (satisfactory recovery);

Then $Q/\Sigma Q$ may be taken as representing the fractional amount of the component present in the fraction (percentages ($Q \cdot 100/\Sigma Q$) may be used as an alternative).

Graphs are constructed as shown schematically in Fig. 2, by aligning in a suitable order dictated by the experimental conditions along an abscissa x rectangles of width Δx and of height ($Q/\Sigma Q \cdot \Delta x$) [or ($Q \cdot 100/\Sigma Q \cdot \Delta x$)].

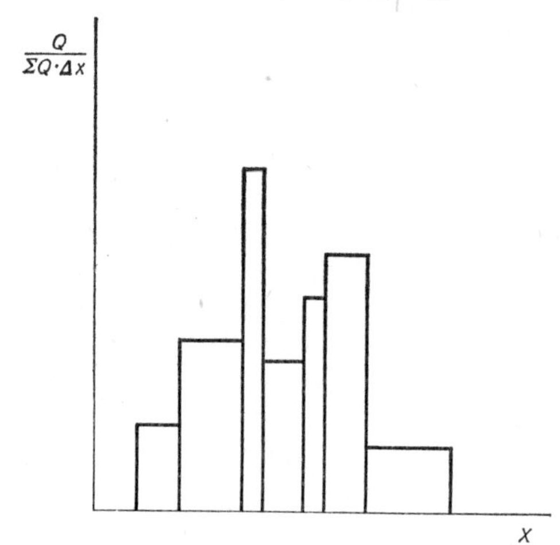

FIG. 2. Figure illustrates principle of graphical representation of distributions. For signficance of symbols, see text.

The surface area of each rectangle is then equal to the fractional (or percentage) amount of the component. The surface area of the whole diagram is always equal to unity (or to 100) if all the fractions are represented. With sufficient fractions, the diagram can be smoothened out to a continuous curve with the same total unitary integral and with fractional integrals equalling as closely as possible the surface areas of the individual rectangles. It is possible to program a computer to do this and also to differentiate the curve when it has a cumulative significance (differential density gradient centrifuging: boundary method). If x is a particle parameter, the diagrams correspond to frequency distribution histograms or curves. If of normal shape, the latter may be expressed in statistical terms.

APPLICATIONS WITHOUT PHYSICAL CHARACTERIZATION

1. Any fractionation

Δx is arbitrary and equal for all fractions. This type of representation is not very informative, but may be necessary, for instance when type 2 cannot be used because of too high relative specific activity values or for lack of information on protein or nitrogen distribution.

2. Fractionation by differential sedimentation

Δx is the fractional (or percentage) amount of protein or nitrogen present in the fraction: $P/\Sigma P$ [or $(P.100)/\Sigma P$]. Then the ordinate $(Q.\Sigma P)/(\Sigma Q.P)$ represents the *relative specific activity* of the enzyme and measures the degree of purification achieved over the homogenate. The fractions are aligned along the abscissa in the order, from left to right, in which they are isolated, i.e. in the order of decreasing sedimentation coefficient (For examples, see: de Duve, Pressman, Gianetto, Wattiaux, and Appelmans, 1955).

3. Density gradient centrifuging

Δx is the fractional (or percentage) volume occupied by the fraction in the gradient, expressed in terms of the total volume of the gradient: $V/\Sigma V$ [or $(V.100)/\Sigma V$]. With cylindrical tubes, fractional heights may be used similarly, with the total height taken as the distance between the meniscus and the bottom of the tube, minus one-third of the radius to correct for the sphericity of the bottom (Beaufay, Bendall, Baudhuin, Wattiaux and de Duve, 1959; Beaufay *et al.*, 1964). The ordinate $(Q.\Sigma V)/\Sigma Q.V)$ represents the relative concentration C/C_i of the component, C_i being the initial concentration when the material is homogenously distributed in the gradient at the beginning of the experiment, or the concentration it would have should it become so distributed, when the material is initially layered on top of the gradient. The fractions are aligned along the abscissa in the order in which they occur in the tube (For examples, see: Beaufay *et al.*, 1964).

APPLICATIONS WITH PHYSICAL CHARACTERIZATION

1. Density distribution

If the final shape of the gradient is known in experiments of isopycnic centrifuging, the intervals of density $\Delta\rho$ and their limits are easily computed graphically. The diagram is plotted on a density scale with $\Delta x = \Delta\rho$. It represents a density distribution histogram or the corresponding frequency distribution curve, provided the conditions of the experiment are such that density equilibration has been achieved (For examples, see: Beaufay *et al.*, 1964).

2. Distribution of sedimentation coefficient

Δx equals Δs , the interval of sedimentation coefficients (in 0·25 M sucrose at 0°) spanned by the fraction. To compute Δs_s and its absolute limits, it is necessary to know accurately the conditions of centrifuging, especially the geometry of the rotor and tubes and the time integral of the squared angular velocity $W = \int_0^t \omega^2 \, dt$ (de Duve *et al.*, 1959). In ordinary differential sedimentation experiments, one can calculate from the conditions of isolation of each fraction the quantity s_{min} (de Duve and Berthet, 1953; Appelmans *et al.*, 1955), the sedimentation coefficient of the lightest particle to be sedimented completely. If the fractions have been washed twice or more, Δs_s may be taken in first approximation as being limited by the s_{min} of the fraction and by that of the fraction preceding it. In practice, such diagrams are not very useful owing to the small number of fractions and to the width of the intervals Δs_s. Density gradient experiments provide much more accurate information, but the computation of Δs_s is more difficult. It has been treated in detail by de Duve *et al.* (1959) (For examples, see: Beaufay *et al.*, 1959).

3. Size distribution

Assuming a uniform density and a spherical shape for the particles, one can convert a sedimentation coefficient distribution into a size distribution by means of the Svedberg equation. Alternatively, one can also convert a density distribution into a size distribution by assuming a uniform sedimentation coefficient. Since both parameters are statistically distributed, we have not constructed such diagrams and have performed further calculations using the median values (or exceptionally the modes) of the distribution histograms (Beaufay and Berthet, 1963; de Duve, 1965). An example of a size distribution computed from sedimentation values can be found in the paper by Kuff, Hogeboom and Dalton (1956).

REFERENCES

Albertsson, P.-A. (1960). "Partition of Cell Particles and Macromolecules" John Wiley, New York.

Allfrey, V. (1959). *In* "The Cell" (J. Brachet and A. E. Mirsky, eds.) Vol. I, p. 193, Academic Press, New York and London.

Anderson, N. G. (1956). *In* "Physical Techniques in Biological Research" (G. Oster and A. W. Pollister, eds.) Vol. III, p. 300, Academic Press, New York and London.

Appelmans, F., Wattiaux, R. and de Duve, C. (1955). *Biochem. J.* **59**, 438.

Bahr, G. F. and Zeitler, E. (1962). *J. Cell Biol.* **15**, 489.

Baudhuin, P., Beaufay, H. and de Duve, C. (1965). *J. Cell Biol.* **26**, 219.

Baudhuin, P., Berthet, J. and Evrard, P. (1967). *J. Cell Biol.* (In press).

Beaufay, H., Bendall, D. S., Baudhuin, P., Wattiaux, R. and de Duve, C. (1959). *Biochem. J.* **73**, 628.

Beaufay, H. and Berthet, J. (1963). *Biochem. Soc. Symp.* **23**, 66.

Beaufay, H., Jacques, P., Baudhuin, P., Sellinger, O. Z., Berthet, J. and de Duve, C. (1964). *Biochem. J.* **92**, 184.

Behrens, M. (1932). *Z. Physiol. Chem.* **209**, 59.

Bensley, R. R. and Hoerr, N. L. (1934). *Anat. Rec.* **60**, 449.

Berthet, J. and de Duve, C. (1951). *Biochem. J.* **50**, 174.

Blaschko, H., Hagen, J. M. and Hagen, P. (1957). *J. Physiol.* (*London*) **139**, 316.

Borsook, H., Deasy, C. L., Haagen-Smit, A. J., Keighley, G. and Lowy, P. H. (1950a). *J. biol. Chem.* **184**, 529.

Borsook, H., Deasy, C. L., Haagen-Smit, A. J., Keighley, G. and Lowy, P. H. (1950b). *J. biol. Chem.* **187**, 839.

Bowers, W. E. (1964). *J. Cell Biol.* **23**, 13A.

Bowers, W. E. and de Duve, C. (1967). *J. Cell Biol.* (In press).

Claude, A. (1946a). *J. exp. Med.* **84**, 51.

Claude, A. (1946b). *J. exp. Med.* **84**, 61.

Conchie, J., Hay, A. J. and Levvy, G. A. (1961). *Biochem. J.* **79**, 324.

Davenport, J. B. (1964). *Biochim. biophys. Acta* **88**, 177.

de Duve, C. (1964). *J. theoret. Biol.* **6**, 33.

de Duve, C. (1965). *In* "The Harvey Lectures" Series **59**, p. 49, Academic Press, New York and London.

de Duve, C. and Baudhuin, P. (1966). *Physiol. Rev.* **46**, 323.

de Duve, C. and Berthet, J. (1953). *Nature, Lond.* **172**, 1142.

de Duve, C. and Berthet, J. (1954). *Int. Rev. Cytol.* **3**, 225.

de Duve, C., Berthet, J. and Beaufay, H. (1959). *Prog. Biophys. biophys. Chem.* **9**, 325.

de Duve, C., Pressman, B. C., Gianetto, R., Wattiaux, R. and Appelmans, F. (1955). *Biochem. J.* **60**, 604.

Chapter 2

THE NUCLEUS

G. P. GEORGIEV

Institute of Molecular Biology, Academy of Sciences of the USSR, Moscow, USSR

I Introduction

The metabolic processes which are known to occur in the cell involve four main types of reaction: firstly, reactions concerned with energy transfer;

secondly, reactions of intermediary metabolism which result in the construction of building blocks for the synthesis of macromolecules and coenzymes including the synthesis of biopolymers that do not require a template; thirdly, catabolic reactions; and finally reactions of polymer synthesis that require templates. The most important of these is the synthesis of nucleic acids and proteins.

Reactions of the fourth type may in turn be divided according to Spiegelman and Hayashi (1963) into three groups, namely those related with the autoreproduction or duplication of genetic information (synthesis of DNA), with the transcription of genetic information (RNA synthesis) and with translation of this information (synthesis of proteins). This conditional classification is useful in the discussion of the functions of subcellular structures.

The reactions mentioned above are localized in different cellular structures. Thus mitochondria and chloroplasts are mainly concerned with the reactions of energy metabolism. Many enzymes of intermediary metabolism are present and function in the cell sap. The reactions of protein synthesis occur in ribosomes and associated structures and finally the duplication and transcription of genetic information take place in chromosomal structures of the nucleus. Therefore the synthesis of nucleic acids is the basic and specific function of the nucleus. This division of functions among cellular structures is, however, somewhat imprecise and not absolute. The synthesis of nucleic acids is not the only function of the nucleus, since reactions of protein synthesis, phosphorylation, synthesis of coenzymes and other reactions of intermediary metabolism also occur in the nucleus. But it is the synthesis of DNA and RNA that distinguishes the activities of the nucleus from those of other cell structures.

The complexity of the function of nucleus is reflected in its complex organization. The nucleus from all types of cell is known to have chromosomes, nucleoli, nuclear sap and a nuclear membrane. Different nuclear structures in turn have different and specific functions as will be evident from the ensuing discussion.

II. Methods of Isolation and Fractionation of Nuclei

It is well known that no fractionation procedure can completely exclude the possibility of artifacts, e.g. the acquisition or loss of components or the inactivation of enzymes. The nucleus is not an exception in this regard. The principal sources of artifacts are the following: (1) contamination of nuclear preparations with whole cells; (2) contamination with cytoplasmic enzymes: (3) loss of some of the nuclear proteins and nucleic acids; (4) degradation and inactivation of nuclear components. In practice, there is no method available at present which is free from all these drawbacks and different procedures have to be used for isolation of nuclei to avoid different artifacts.

Several excellent reviews on the methods of isolation of nuclei have been published in the past few years (Dounce, 1955; Allfrey, 1959; Roodyn, 1959, 1963; and Siebert and Humphrey, 1965). Therefore, we will only consider the principles which one should adopt in looking for a method to use for a specific purpose and only briefly characterize the procedures most often employed.

There are two principal groups of procedures used for isolation of nuclei: those which involve the use of aqueous solutions, mainly containing sucrose, and those making use of organic solvents. We will consider the use of organic solvents first.

A. PROCEDURES INVOLVING THE USE OF ORGANIC SOLVENTS

A non-aqueous method for the isolation of cell nuclei was first put forward by Behrens (1932, 1938). Further modification of the technique was made by Dounce, Tishkoff, Barnett and Freer (1950), Allfrey, Stern, Mirsky and Saetren (1952) and recently by Siebert (see Siebert, 1961 and Siebert and Humphrey, 1965). There are no important differences between the various modifications of non-aqueous methods of nuclear isolation. All of them include (1) lyophilization of tissue; (2) homogenization of lyophilized material by treatment in ball mill (as suspension medium, petroleum ether or some other light organic solvent is usually used) and (3) differential centrifuging. Mixtures of two organic solvents with high and low density are used to perform the centrifuging (for instance, CCl_4-benzene or brombenzene-cyclohexane). By varying the ratio of solvents one can obtain mixtures of different density. Subsequent use of these mixtures causes the nuclei to sediment or conversely to float, thus gradually separating them from cytoplasmic contaminants and whole cells which differ from nuclei in density. In Fig. 1(a) a preparation of clean nuclei obtained by non-aqueous methods is presented.

In practice only non-aqueous methods are completely guaranteed from the loss of nuclear proteins and nucleoprotein components. In any aqueous method some of the nuclear protein and RNA may be washed out of nuclei. Therefore the results obtained with non-aqueous methods should be taken as the most reliable guide to the final localization of a particular enzyme in cellular sub-structures. The most obvious example supporting the use of non-aqueous methods in the investigation of enzyme localization is the study of the distribution of DNA polymerase (2.7.7.7.) in the cell. While after isolation of nuclei by aqueous methods, this enzyme was found exclusively in the cytoplasm, the application of non-aqueous procedure gave quite opposite results: most of the DNA-polymerase activity was found in the nuclei (Keir, Smellie

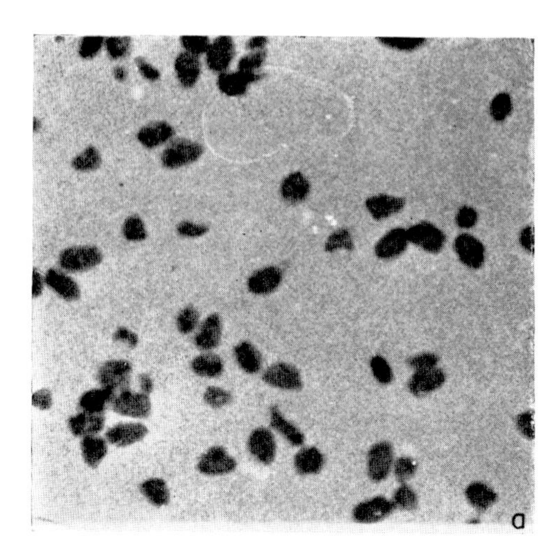

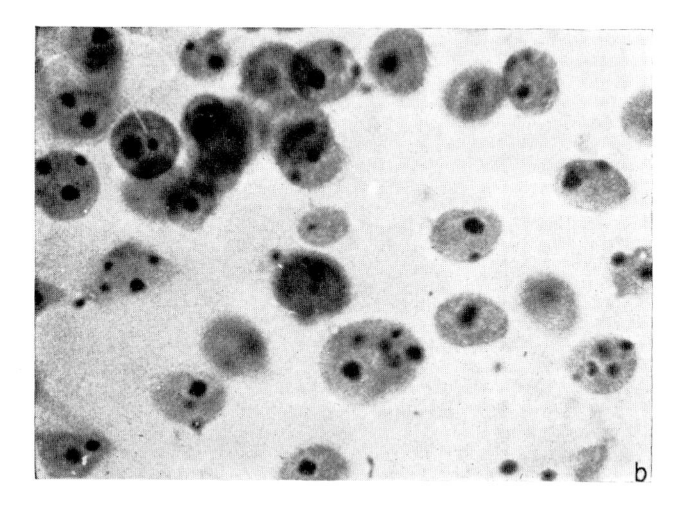

FIG. 1. Smears of isolated nuclei in light microscope. (a) Nuclei of avian erythrocytes obtained by non-aqueous method and stained with haematoxylin-eosin. × 1750 (Allfrey, 1959). (b) Rat liver nuclei isolated in concentrated sucrose—glycerophosphate medium and stained with methylene blue. Numerous nucleoli are clearly visible. × 1500 (Georgiev *et al.* 1960).

and Siebert, 1962). Furthermore the use of non-aqueous methods prevents chemical binding of soluble proteins by nuclear nucleoproteins. Finally these methods are quite useful in the study of intracellular distribution of such small molecular weight substances as intermediary glycolytic metabolites, free nucleotides and amino acids.

The principal drawbacks of non-aqueous procedure are as follows: (1) it is not always possible to isolate nuclei without contamination; (2) the ultrastructure of nucleus may often be damaged thus interfering with subsequent fractionation of sub-nuclear components; (3) the denaturation of enzymes in organic solvents is not completely ruled out. (Although the activity of an enzyme studied may still be high after isolation, one should always determine the effect of the isolation method on the enzymic activity), (4) non-aqueous methods cause the solubilization of nuclear lipids, (5) finally, in comparison to aqueous methods, the non-aqueous technique is always less reproducible and more troublesome.

Nevertheless for the most part only the use of non-aqueous methods allows one to draw final conclusions as to the presence or absence of an enzyme in nuclei.

B. AQUEOUS METHODS OF NUCLEAR ISOLATION

Aqueous methods of nuclear isolation are more commonly used than non-aqueous, and more have been described. They do not suffer from the above mentioned drawbacks. Since they are not so tedious and time-consuming they are more often used in everyday practice. An example of nuclei isolated in an aqueous medium is given in Fig. 1b.

Two main stages are distinguished in any procedure which involves the use of aqueous media: homogenization and differential centrifuging. As a rule, the disruption of tissues and cells (homogenization) is carried out in the same medium that is used later in the procedure. To disrupt cells, earlier workers used either a high-speed blender (Waring Blendor) or a bacterial mill (see Dounce, 1955). Since considerable breakage of nuclei and mitochondria was always observed in these procedures, these machines are now only rarely used. However, in some exceptional cases, nuclei prepared in the Waring Blendor are of good quality (Allfrey, Mirsky and Osawa, 1957). The glass Potter–Elvehjem homogenizer (Potter and Elvehjem, 1936), often fitted with a teflon pestle, and the somewhat less popular Dounce homogenizer (Dounce, Witter, Monty, Plate and Cottone, 1956) are widely used now. Together with these simple tools, more complicated apparatus is also in use; for example, the Bonner apparatus for the disruption of plant tissues between two rotating rolls (Rho and Chipchase, 1962), or devices where tissues are forced through narrow apertures (Emanuel and Chaikoff, 1957).

These mechanical methods gave quite satisfactory results when applied to different tissues but in almost no case was it possible to achieve complete disruption. They were found to give unsatisfactory results in the disruption of free cells (ascites tumours, cell cultures). The search for some other method has shown that hypotonic shock may be more successful: when placed in hypotonic medium or water, these cells swell and burst (Samarina, 1961, Fisher and Harris, 1962; and Martin, Malec, Coote and Work, 1961). Depending on the cell resistance, these procedures may involve only hypotonic shock in water, additional more or less intensive homogenization of cells or, finally, the use of detergents such as Tween-80 or saponin.

The homogenization is usually followed by differential centrifuging at relatively low speed (500–1000 *g*) to separate nuclei from cytoplasmic material. Many media originally used for the isolation of nuclei contained citric acid (Dounce, 1943). This however led to aggregation and condensation of nuclear material as well as to a considerable extraction of nuclear proteins (Dounce, 1955). As a more mild procedure for general tissue fractionation, sucrose media of low ionic strength were introduced by Hogeboom, Schneider and Palade (1948) and Schneider (1948) and later used for nuclear isolation with some modifications by most investigators. The addition of Ca^{2+} ions to the medium was shown to have a favourable effect during the isolation procedure (Schneider and Petermann, 1950). Other workers also suggested layering the suspension of nuclei over more concentrated sucrose solutions before centrifuging (Wilbur and Anderson, 1961; Hogeboom, Schneider and Striebich, 1952; Roodyn, 1956). The use of low ionic strength solutions of sucrose that have relatively high osmotic pressure significantly reduces the loss of nuclear proteins. There is no difference between the content of protein and RNA in thymus nuclei isolated in sucrose media or by non-aqueous methods (Stern and Mirsky, 1953). On the other hand, intact ultrastructural organization of nuclei, as observed by electron microscopy, is also preserved in sucrose isolation media (Georgiev and Chentsov, 1960, 1963; Maggio, Siekevitz and Palade, 1963a). (See Figs. 2–4.)

Further progress in the development of sucrose methods was made by Chauveau, Moulé and Rouiller (1956), who used concentrated sucrose solutions (2·2 M) for nuclear isolation. The density of these solutions is lower than that of nuclei but exceeds the density of whole cells and cytoplasmic structures. Hence during centrifuging of homogenates in these solutions nuclei move down the centrifuge tube, whereas all the cytoplasmic contaminants go up. Using the original Chauveau technique or its modifications (Georgiev, Yermolaeva and Zbarsky, 1960; Gvozdev, 1960) it became possible to isolate intact nuclei without contamination (Fig. 2b). A further development of the Chauveau procedure was the use of a layering technique in which a homogenate prepared in more dilute sucrose (0·88 M, for example) is then layered on 2·2 M sucrose. In this method, nuclei reach the bottom of the tube during centrifuging, and whole cells and cytoplasmic contaminations are left in the upper layer or interphase, (Maggio *et al.*, 1963a) (Fig. 3). The use of concentrated sucrose solutions for the isolation of nuclei is so far the only method by which to obtain a nuclear preparation with no whole cells present.

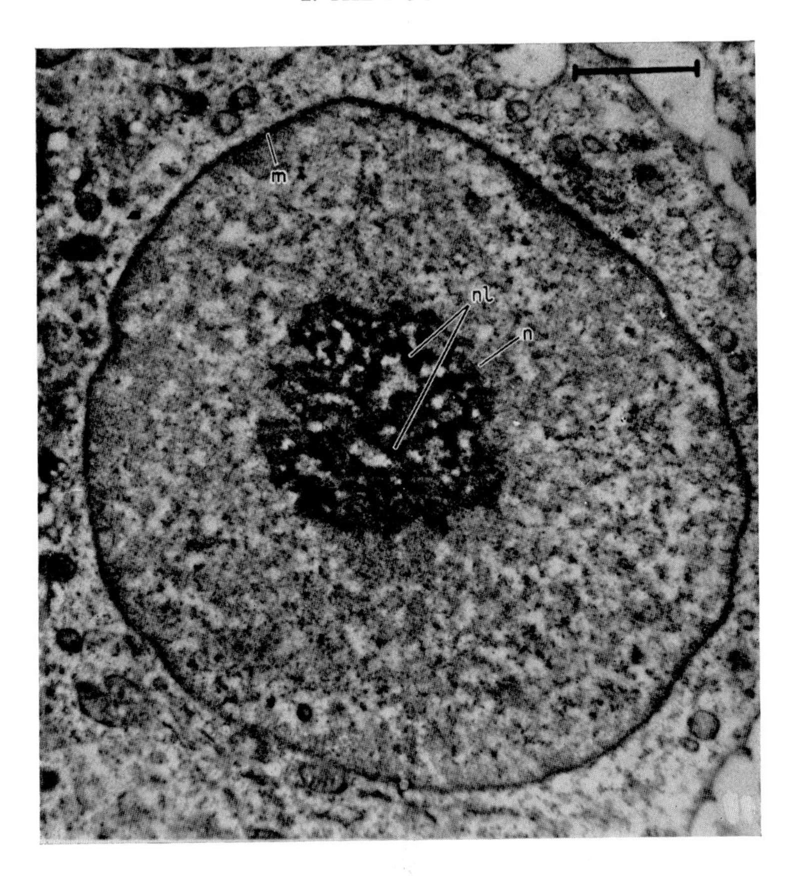

FIG. 2. Electron micrograph of intact nucleus of Walker carcinoma. One can see a nucleolus (n), and double membrane (m). The nucleolus consists of nucleolonemata (nl) that are made of numerous small granules. Nucleoplasma contains elements of chromosomes and nuclear sap, not separated one from another. Microfibrils and granules 100–200Å in diameter are visible. Chromosomal elements are present in the nucleolar zone. Nuclei fixed according to Palade. × 16,500 (Muramatsu *et al.* 1963).

In addition to sucrose media glycerol media have been described for isolation of nuclei (Schneider, 1955). These somewhat reduce the loss of nuclear proteins. The high viscosity of these solutions, however, makes it difficult to isolate pure nuclei.

The diversity of aqueous methods for isolation of nuclei, and the ease and simplicity which make it possible to obtain relatively pure and intact nuclei suitable for further fractionation into sub-nuclear components strongly favours the use of aqueous procedures for studies of nuclear metabolism. However, to exclude completely the danger of the loss of water-soluble components in these procedures, one should perform control experiments with nuclei isolated with organic solvents.

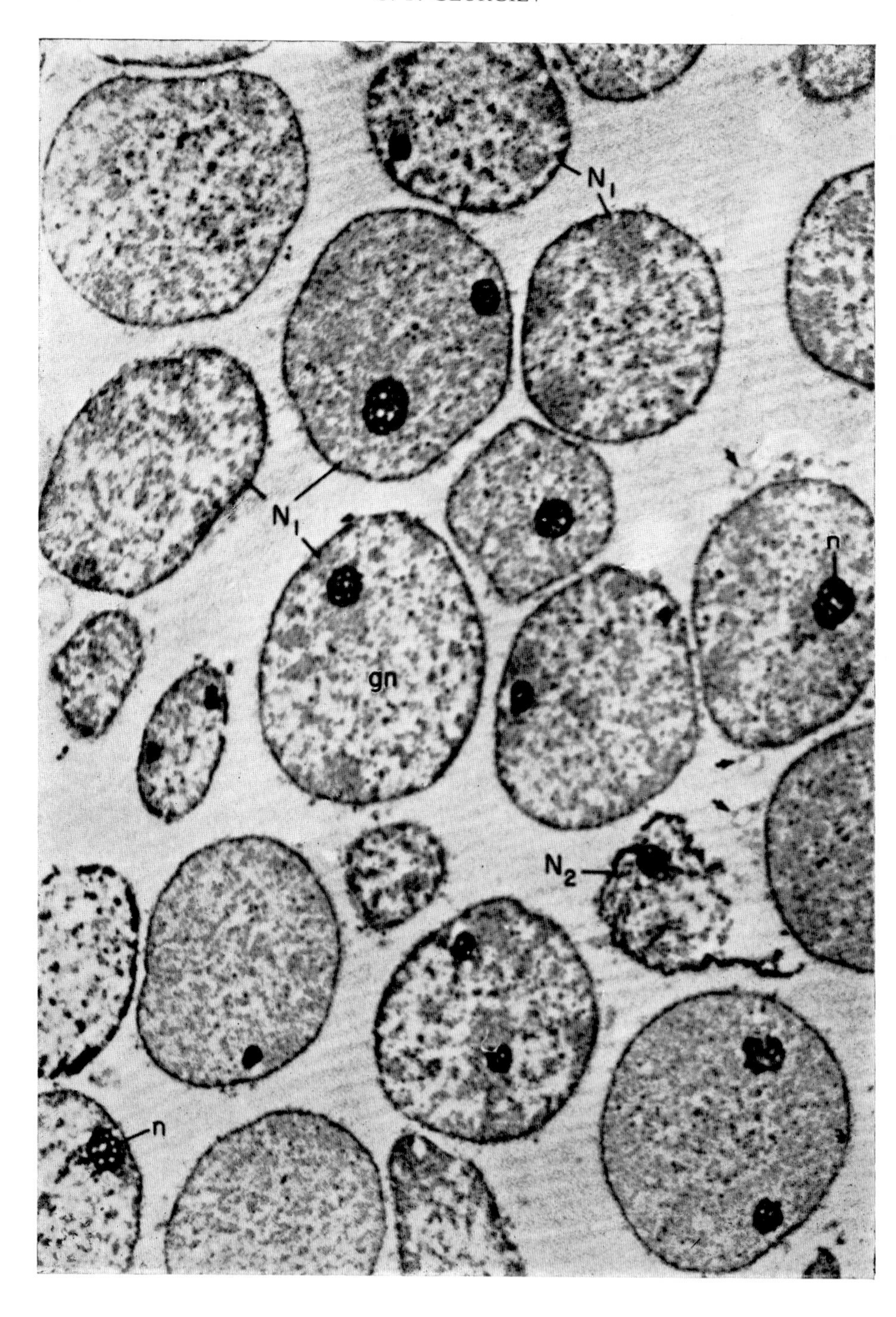
N1
N1
gn
n
N2
n

III. Structure and Methods of Sub-fractionation of the Nucleus

In addition to experiments with intact nuclei the problem of further fractionation of nuclei and the isolation of subnuclear structures often arises. This problem was first studied many years ago. In 1943 Claude and Potter isolated nuclear material which was shown by microscope studies to be chromatin threads. Later these threads were identified as interphase chromosomes (Mirsky and Ris, 1951), but this conclusion is not completely correct and the above mentioned threads are probably broken pieces of chromosome. In 1952 Vincent described a technique for isolation of nucleoli from oocyte nuclei of the starfish. In 1960 methods for the isolation of nuclear ribosomes were developed in a number of laboratories (Frenster, Allfrey and Mirsky, 1960; Samarina and Georgiev, 1960). Recently quite a few papers have been published on systematic studies of nuclear fractionation with the purpose of isolating of various nuclear structures. The composition of the fractions is usually controlled chemically and by electron microscopy.

Two main groups of fractionation procedures are being developed at present: chemical fractionation and mechanical fractionation.

A. Structure and Ultrastructure of Nuclei

To evaluate these methods let us first briefly recall what is known about the structure and ultrastructure of nuclei. By light microscopy it is possible to distinguish in the interphase nucleus nucleoli, chromatin (chromosomes) and nuclear sap. However, if unfixed preparations are used, it is not possible to differentiate chromatin and nuclear sap in interphase nuclei. Uncoiled chromosomes fill the whole nucleus and one can only see the nucleolus and the nucleoplasm (Ris and Mirsky, 1949) (see Fig. 1). The same situation is observed in the study of nuclear structure by electron microscopy. After fixation of the tissue with OsO_4 the nucleolus zone may be clearly seen in electron micrographs. The rest of the nucleus is filled with fibrillar or granular material whose organization is not clear (Fig. 2). Aldehyde fixation (formaldehyde, acrolein, glutaraldehyde) or fixation in acid media lead to the separation of chromosome material and nuclear sap (Swift, 1963; Maggio *et al.*, 1963a) (Fig. 4). This picture however does not apparently reflect the distribution of nuclear material in the intact cell.

The nuclear envelope is easily seen in electron micrographs of nuclei. It

Fig. 3. Electron micrograph of liver nuclei obtained by centrifuging homogenate in 0·88 M sucrose/1·5 mM $CaCl_2$ in discontinuous density gradient (the homogenate was layered over 2·2 M sucrose-0·5 mM $CaCl_2$). Intact (N_1) and partly damaged (N_2) nuclei and rare cytoplasmic impurities are recorded. (n) — -Nucleoli. Pellet fixed in 3·3 M formaldehyde in 0·1 M phosphate buffer, pH 7·6 and stained in $Pb(OH)_2$ × 4900 (Maggio *et al.*, 1963a).

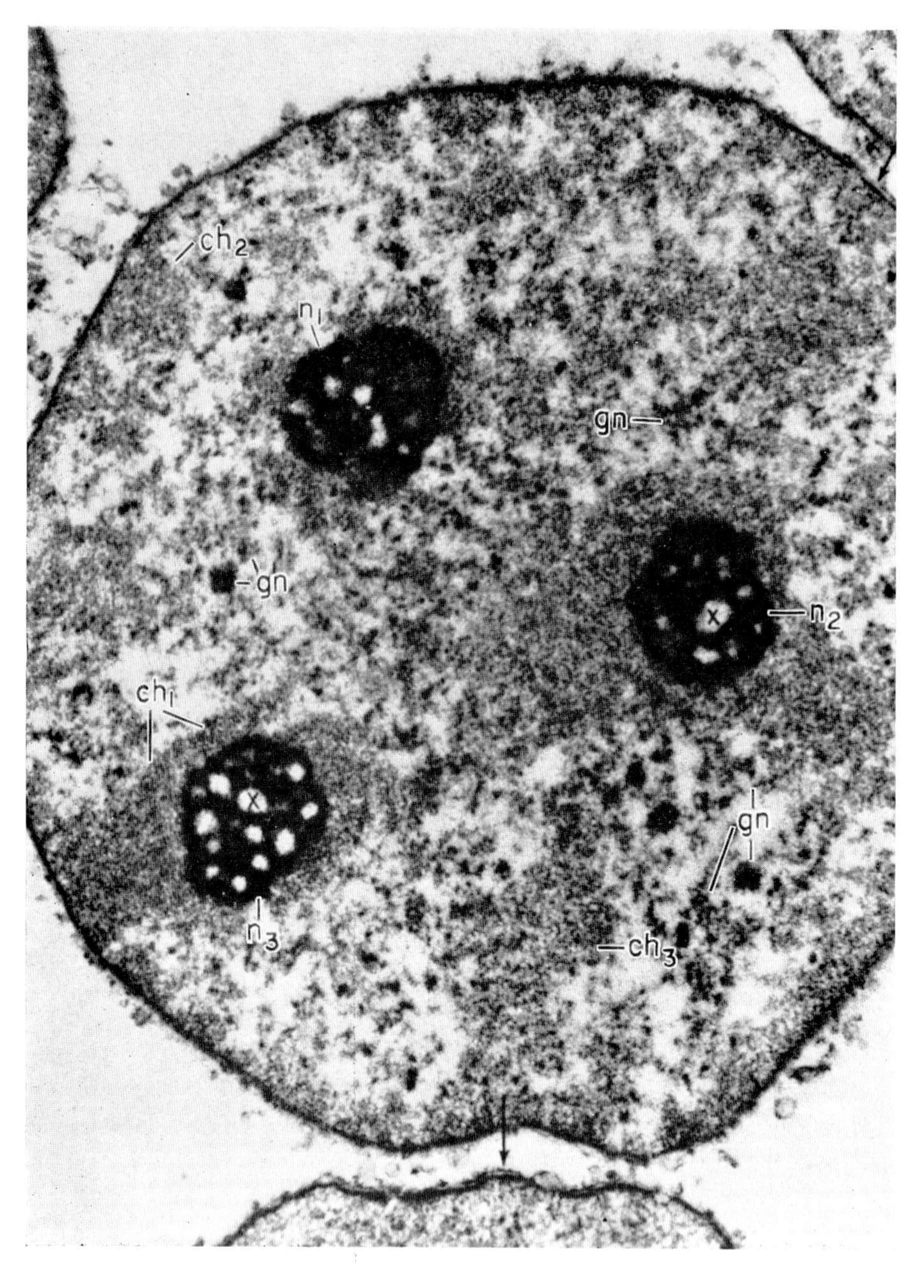

FIG. 4. The same as in Fig. 3 at higher magnification. One can see (1) nucleoli (n) with well defined nucleolonemata; (2) chromatin masses (ch) localized around the nucleolus and intranucleolar, at the periphery of nucleus and scattered throughout the rest of nucleus; (3) interchromatin regions (nuclear sap) occupied by aggregated small granular material (gn); and (4) nuclear membranes (shown in arrows) × 17,000 (Maggio *et al.*, 1963a).

consists of two membranes. The external membrane is essentially an extension of the endoplasmic reticulum (Hartmann, 1953; Watson, 1955). So-called pores are seen in zones where the two membranes of nuclear envelope are joined together (Callan and Tomlin, 1950; Watson, 1959; (Fig. 5)).

Although electron microscopy of nuclei did not help much in the elucidation of the three-dimensional structure of chromosomes and nucleoli it led to the characterization of the main structural elements which form these structures. The basic structural elements of the nucleus are more or less coiled strands about 100 Å in diameter, called elementary fibrils (Ris, 1957, 1961; Yasuzumi, 1955; Kaufmann and McDonald, 1956; Bopp-Hassenkamp, 1959; Georgiev and Chentzov, 1960, 1963; Swift, 1963). These strands have been observed by practically all the authors who have studied nuclear structure *in situ* as well as in isolated nuclei (Figs. 5–7). The fibrils are found

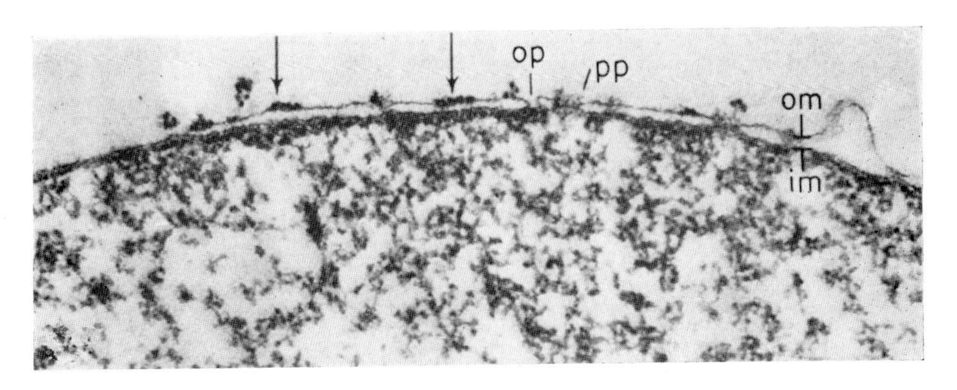

FIG. 5. Electron micrograph of normal section through nuclear envelope. Om—outer, im—inner nuclear membrane. PP, op—pores of nuclear membrane ("plugged" or "open"). Inside the nucleus DNP fibrils. × 47,000 (Maggio *et al.* 1963a).

distributed in the nucleoplasm after osmium fixation or concentrated in chromosomal zones after aldehyde fixation. They may also penetrate the nucleolar zone (Figs. 4 and 6). The diameter, degree of coiling and aggregation of elementary fibrils depend on their source and on the method of preparation. The ionic strength of solutions used during the fixation of nuclei plays a most important role. The fibrils are localized in the zones rich in DNA, they are sensitive to DNase, dissolve in strong salt solutions and aggregate in the presence of 0·14 M NaCl and bivalent cations. These facts demonstrate their deoxyribonucleoprotein nature (Ris, 1957, 1961; Georgiev and Chentsov, 1960, 1963). Therefore they will be referred to below as "DNA-protein fibrils" or "DNP fibrils."

The molecular structure of "DNP fibrils" is still quite obscure. Zubay and Doty (1959) have shown that the diameter of a DNA protein (DNP) molecule

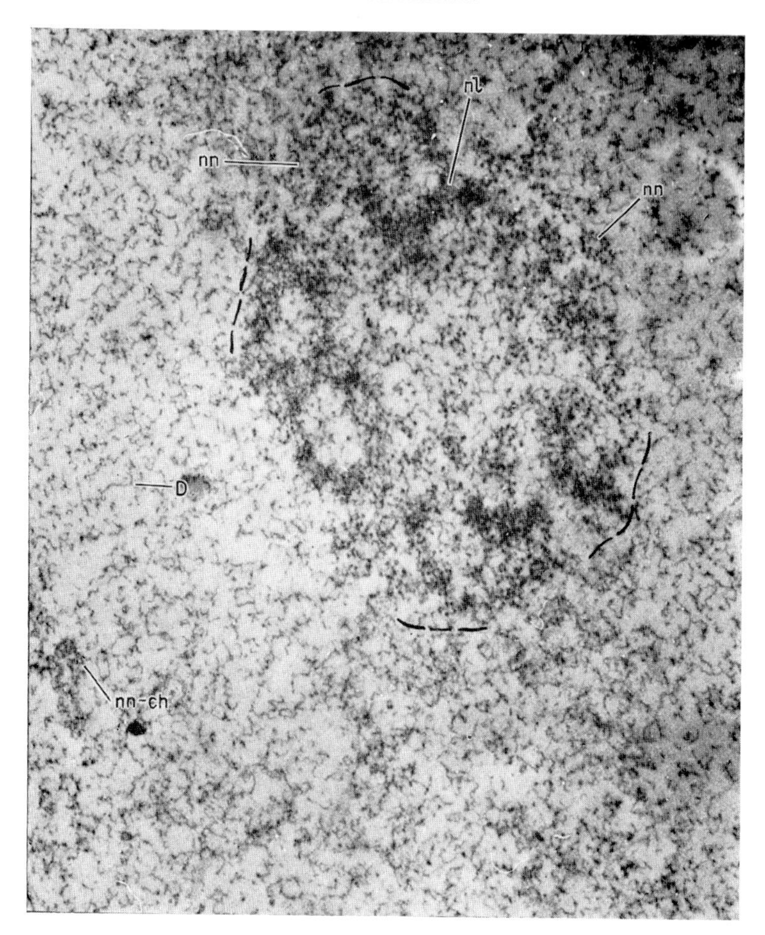

FIG. 6. Electron micrograph of rat liver nucleus, from which nuclear sap is removed by salt extraction. One can see a nucleolar zone, that consists predominantly of nucleonemata (nn). Bundles of nucleonemata composed of nucleolonemata (nl). DNP fibrills (D) penetrate into the nucleolar zone. They fill the whole nucleus. In some part of it nucleonemata of chromosomes (nn-ch) are shown. Nuclei isolated in 2·2 M sucrose-1% glycerophosphate, extracted by 0·14 M NaCl, then placed in sucrose-glycerophosphate medium again and fixed by OsO_4. Slices are stained in uranyl acetate. × 26,500 (Georgiev and Chentszov, 1960, 1963).

FIG. 7. Electron micrographs of deoxynucleoprotein fibres. (a) Nuclei of red blood cells of *Trituras pyrohogaster* disintegrated by ultrasonic vibrations, smeared on planchets and shadowed. f—100Å fibrils, N, F—fibrils of the higher order of complexity. 10,000 × (Yasuzumi, 1955). (b), (c)—Calf thymus nucleohistone fibrils from chromosomes prepared in 0·07 M NaCl-0·001 M EDTA pH 6·5 and dispersed in water. 100Å-fibrils are visible (b) × 85,000). (d) The same as in (b), (c), but DNP isolated was performed at pH 8·0. One can see the separation of 100Å fibril into two 40Å fibrils. × 130,000 (Ris, 1960). (e) DNP fibril of rat thymocyte, stained with uranylacetate. The separation of DNP fibril is visible (Struchkov, 1964).

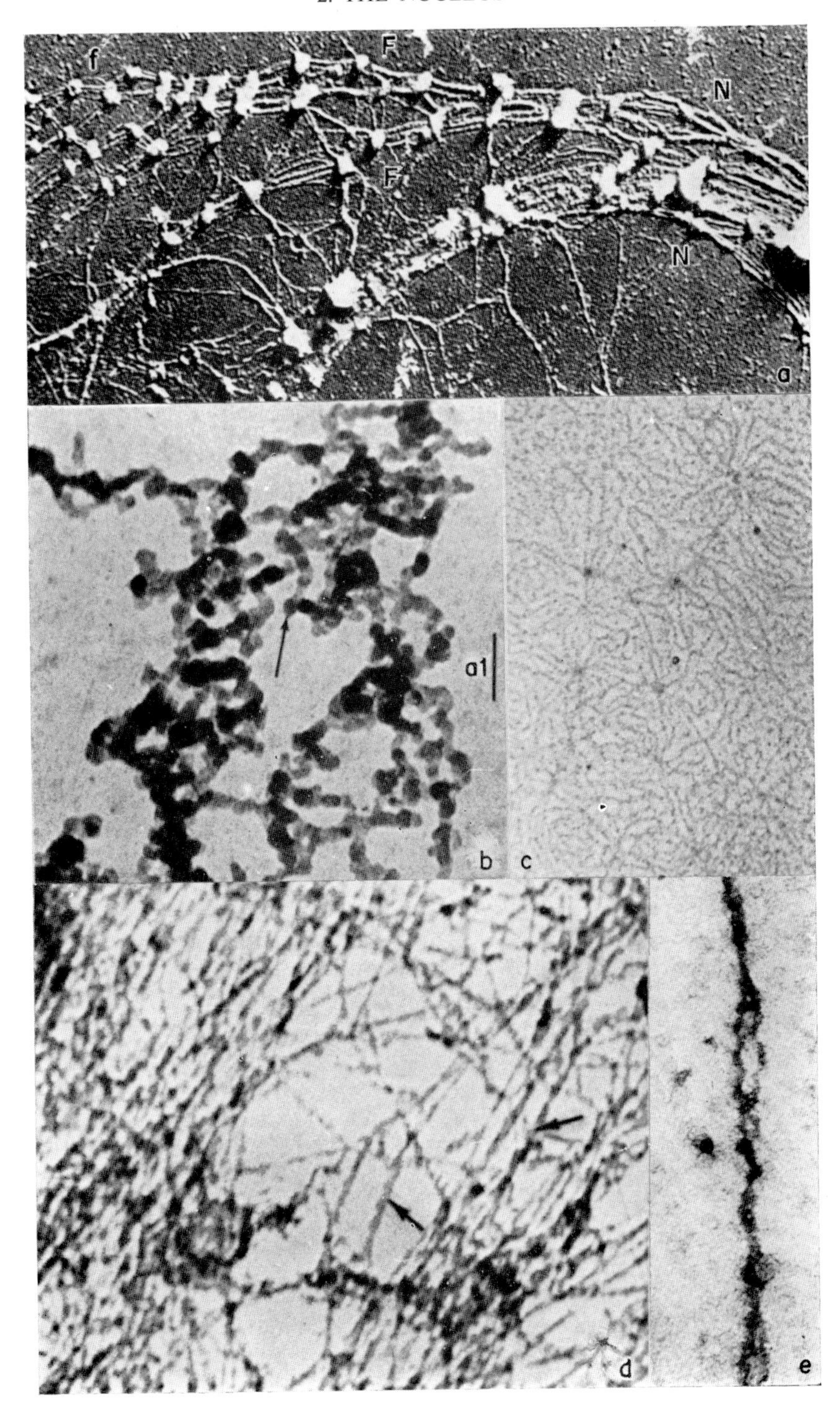
f
F
N
F
N
a
a1
b
c
d
e

containing DNA and histone in the ratio 1:1 is significantly less (30 Å) than the diameter of DNP fibrils (Fig. 8). However nuclear DNP contains more protein than purified nucleo-histone and includes some proteins of non-histone nature (Mirsky and Pollister, 1946; Zbarsky and Debov, 1951).

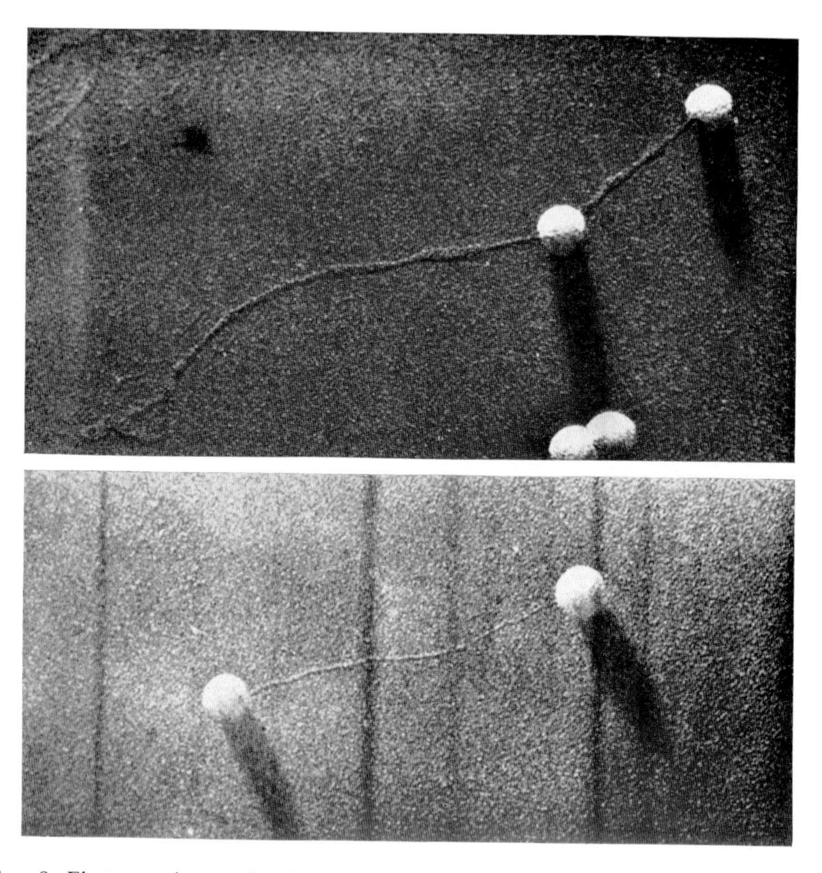

FIG. 8. Electron micrograph of nucleohistone molecules (obtained by the method of Zubay and Doty (1959) and of free DNA. Diameter of fibrils are DNP 30 Å, (top) and DNA 20 Å, (bottom). Zubay and Doty (1959).

According to Ris (1961) 100 Å strands separate into 30–40 Å strands under the conditions favouring the removal of some protein from nucleo-protein (e.g. at high pH, or after removal of histone by extraction with HCl). In electron micrographs 30–40 Å strands resemble nucleo-histone molecules (Fig. 7). It is possible that each DNP strand consists of two nucleo-histone molecules held together by additional histone or non-histone protein.

In addition to DNP fibrils some other structures may be distinguished in the nucleus. First of all, there are ribosome-like granules 150–200 Å in diameter. Some of these are not associated with fibrillar structures and lie freely in the nucleoplasm (Georgiev and Chentzov, 1963). They are easily observed in interchromosomal regions (Swift, 1960) (Fig. 4). These free granules might correspond to the ribosomes which have been isolated from nuclei (see below).

Similar granules are also found in nucleoli and chromosomes. In electron micrographs, using OsO_4 stain, the nucleolus looks like a tangle of thick densely stained threads 500–1000 Å in diameter, called "nucleolonemata" (Zivago, 1948; Estable and Sotelo, 1951) (Fig. 2, 4). Nucleolonemata in turn consist of more thin fibrils (approximately 100 Å in diameter) and ribosome-like granules. Fixation of isolated nuclei with OsO_4 at low ionic strength causes the nucleolonemata to loosen, thus disclosing their inner structure (Fig. 6). The basic structural elements of the nucleolonemata are 100 Å strands with attached ribosome-like granules (Georgiev and Chentzov 1960, 1963). The existence of granules as well as fibrils (which could resemble granules in thin section) is supported by the fact that after removal of osmium, the strands are no longer visible and the nucleolus looks like an accumulation of granules (Swift, 1960). The term "nucleonemata" was suggested for this structural complex of 100 Å strands associated with ribosome-like granules, and we will use this term in our review. Although nucleonemata are mainly located in the nucleolus, chromosomes also contain them. After specific extraction procedures chromosomal nucleonemata are aggregated and become visible in the light microscope (Georgiev and Chentzov, 1960, 1963) (Fig. 9).

Thus one can distinguish the following structural elements by electron microscopy: (1) DNP fibrils; (2) nucleonemata; (3) free ribosome-like granules; (4) double-layer membranes with pores. In addition (5) large granules (300–350 Å in diameter) have been described in chromosomes (Swift, 1963). Finally, as mentioned earlier, (6) non-structural or small grained material—(pars amorpha)—is found in the nucleolus associated with nucleolonemata. At present, the nature of the last two components is unknown.

Data on the distribution of these components among basic nuclear structures are summarized in Table I. It is evident that the same elements, of apparently similar chemical composition, are found in various nuclear structures. It therefore follows that there are two principal approaches to the problem of the fractionation of the nucleus:

(1) The chemical approach for the separation of nuclear ultrastructures which differ in chemical composition.

(2) The mechanical approach—for the separation of fundamental nuclear macrostructures which in themselves are highly complex in organization.

TABLE I. *Chemical fractionation of nuclear ultrastructural elements*

Name of fraction	Extraction medium	Basic ultra-structural components released from nucleus	Chemical constituents	Nuclear localization (major localization and any additional site)
Globulin fraction	0·05–0·2 M neutral salts pH 6·0–7·0	Free particles 150–200 Å in diameter	Ribosomes (single or aggregates) sRNA soluble proteins	In nuclear sap (sRNA partly, from nucleolus)
	pH 7·8–8·0		ribonucleo-proteins containing mRNA	Chromosomes
Deoxynucleo-protein fraction	2–2·5 M NaCl	DNP fibrils	DNP consists of DNA, histone and non-histone protein (2:3:1)	In chromosomes (including material penetrating the nucleolar zone)
Residual fraction	—	Nucleonemata: fibrils (100 Å diam.) and ribosome-like granules attached to them	"Acid" protein Ribosomal RNA. Probably in ribosomes or ribosomal precursors	In nucleolus (Partly in chromosomes "residual chromosomes")
		Nuclear membrane	"Residual" protein (may be purified by extraction of nucleonemata by dilute alkali)	

→

FIG. 9. Electron micrograph of liver nucleus from which nuclear sap and deoxynucleoprotein have been removed. One can see nucleoli (n) consisting of nucleonemata (nn) and chromosomal nucleonemata (nn-ch). Aggregated chromosomal nucleonemata are visible in the light nicroscope as "residual chromosomes". The double membrane (m) with pores (p) is visible. Nuclei were treated successively by DNase, 0·14 M NaCl and 2 M NaCl and fixed by OsO_4 in 2 M NaCl. (Georgiev and Chentsov, 1963).

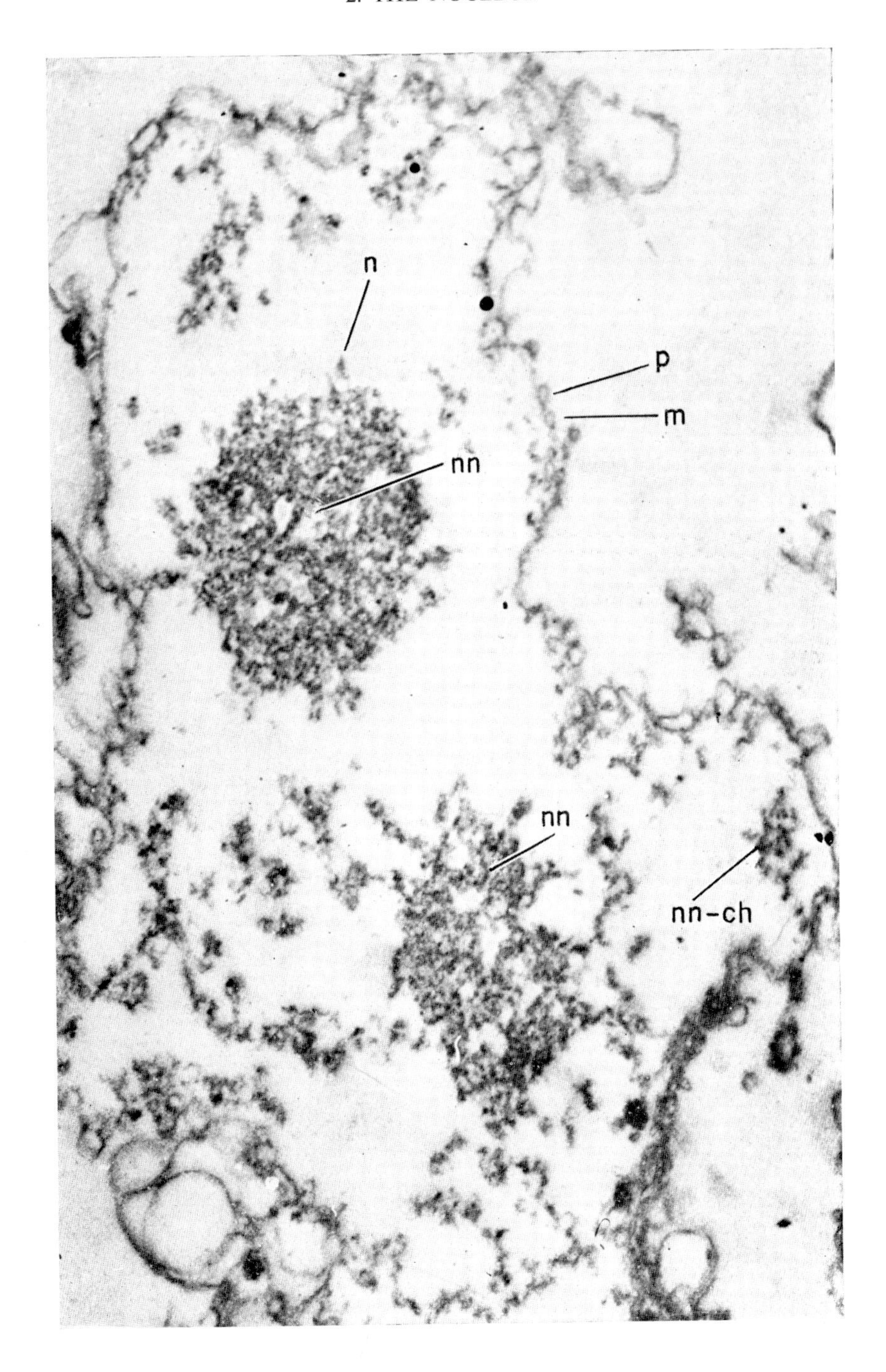
n
p
m
nn
nn
nn-ch

B. CHEMICAL FRACTIONATION

The chemical method of fractionation includes extraction of isolated nuclei with dilute salt solutions (0·05 M *tris* or phosphate buffer; 0·15 M NaCl and other more complex mixtures) and strong salt solutions (1·0–2·5 M NaCl). Various modifications of this method were used for a long time for the separation of proteins and nucleic acids of the nucleus (Mirsky and Pollister, 1946; Zbarsky and Debov, 1951; Allfrey *et al.*, 1957; Georgiev *et al.*, 1960). The initial extraction removes free particles from the nucleus, and thread-like elements (DNP fibrils and nucleonemata) are left (Georgiev and Chentsov, 1960, 1963). A certain amount of RNA and soluble nuclear proteins are present in the extracts obtained. Ribonucleoprotein particles were isolated from nuclear extracts by ultracentrifuging (Frenster, Allfrey and Mirsky, 1960; Samarina and Georgiev, 1960). These were later identified as ribosomes by their structural and metabolic properties (Pogo *et al.*, 1962; Wang, 1963) (Fig. 10). sRNA was also found in this fraction. The fraction also contains a

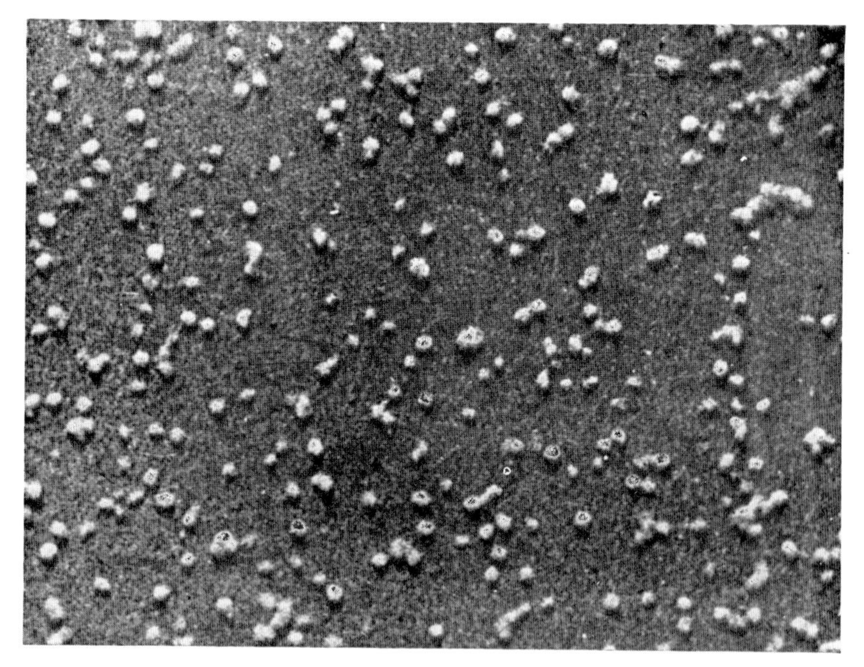

FIG. 10. Purified ribosomes from nuclear sap of calf thymus nuclei. × 76,800 (Wang, 1961).

considerable amount of protein which corresponds in composition to soluble cytoplasmic protein (see Section VI). Hence the extract of nuclei obtained by dilute salt solution originally called the globulin fraction (Dounce, 1955) resembles cytoplasmic sap in its composition. In other words this globulin fraction corresponds to "the nuclear sap fraction" (see Appendix: "Some Suggestions on Terminology"). This statement, of course, is somewhat arbitrary since it is not yet known from which part of the nucleus the components of the nuclear sap fraction originate. For example, ribosomes present in the nuclear sap fraction are not only derived from free ribosomes of the nuclear sap but may come from the nucleolus which also contains some ribosomes not firmly associated with nucleonema strands. This possibility also holds true for other components which may be associated *in situ* with some structures and are thus unequally distributed throughout the non-structural material. For example, some components although not structurally bound, may still only be found in the region of the nucleolus. Finally some components of the fibrillar elements may be cleaved off and brought into solution during the extraction procedure. Thus one may obtain in this fraction ribonucleo-proteins containing messenger RNA (mRNA) originating from chromosomes by suitable modification of the extraction procedure (Samarina *et al.*, 1965). All these factors complicate the comparison of results where extracting mixtures of different composition were used. Nevertheless the fact that the composition of the nuclear "globulin" fraction and that of cytoplasmic sap are similar proves that the globulin fraction is not an artifact but that it is derived mainly from the soluble phase of nuclei (i.e. nuclear sap).

After extraction of nuclear sap, DNP fibrils and nucleonemata are left in the nucleus. By extraction of this residual material with 1·0–2·5 M NaCl, DNP fibrils from chromosomes and the nucleolus region may be removed. In the extract DNA, histones and non-histone protein are present in the ratio 2:3:1. Some RNA is also present but the greater part of chromosomal RNA is recovered in the dilute salt extract. Unfortunately, strong salt solutions cause the DNP fibrils to dissociate and the DNP fraction thus obtained is not actually a suspension of DNP fibrils but rather a suspension of the chemical components of which these fibrils consist. Simple procedures have been developed for further fractionation of DNP fraction into histones, non-histone protein and DNA (Mirsky and Pollister, 1946; Zbarsky and Debov, 1951; Allfrey *et al.*, 1957; Georgiev *et al.*, 1960).

The material left after removal of chromosomal material with strong salt solutions represent structures consisting of nucleonemata (Georgiev and Chentsov, 1960): nucleoli and "residual chromosomes" (Mirsky and Ris 1951) and was termed the "nucleolus fraction" by, Allfrey *et al.* (1957), although this definition is not quite correct. Nucleonemal granules contain ribosomal RNA and seem to be ribosomes or their precursors (Lerman *et al.*, 1964). By extraction with 0·01 N KOH nucleonemal material can be separated from membranes but this is accompanied by partial degradation.

Thus by salt fractionation it is possible to separate and study the composition of the basic structural components of nuclei: nuclear sap, DNP fibrils and nucleonemata. However, this extraction procedure is not free from artifacts since some extraction of material from other nuclear components may occur and there is a certain disruption of the structural organization of the nuclei.

C. MECHANICAL METHODS OF FRACTIONATION

This group of methods are carried out as follows: nuclei are generally suspended in a medium of low ionic strength and relatively high pH (around 7·5). Most of the authors usually add Ca^{2+} ions since otherwise some destruction of nucleoli may occur after disruption of the nuclei. Various techniques of mechanical disruption are then used: sonication, treatment in a Waring Blendor or forcing through a small hole. The suspension of broken nuclei so formed is fractionated by differential centrifuging in dilute or concentrated sucrose solutions. Several fractions may be obtained after centrifuging: (1) intact nuclei; (2) the nucleolus fraction; (3) one or two fractions containing broken chromosomes (the chromatin fraction); (4) free ribosomes; (5) and the final supernatant fraction (Vincent, 1952; Johnston, Setterfield and Stern, 1959; Birnstiel, Chipchase and Bonner, 1961; Birnstiel, Rho and Chipchase, 1962; Muramatsu, Smetana and Busch, 1963; Maggio, Siekevitz and Palade, 1963b). These methods are of special importance for the isolation of nucleoli since their ultrastructure and particularly their nucleolonemal structure are well preserved under these conditions (Fig. 11).

One should remember, however, that material so obtained contains not only nucleolonemata but all components of the "nucleolus zone" and the chromatin adjacent to nucleolus (Fig. 11). In other words nucleolar ribonucleoproteins are isolated together with adjacent chromatin or DNP fibrils. This is apparently a result of the close juxtaposition RNP and DNP-containing nuclear ultrastructures. Hence isolated nucleoli usually contain approximately equal amounts of DNA and RNA, as well as histone that is completely derived from the DNP fibrils. It is not yet known whether chromatin associated with nucleoli is different from the rest of the cellular chromatin.

RNA isolated from mechanically prepared nucleoli contains typical ribosomal RNA components (Chipchase and Birnstiel, 1963), in good agreement with the data on the nature of the RNA of nucleonemata. Furthermore, after treatment of nucleoli with deoxycholate, Birnstiel, Chipchase and Hyde (1963) were able to isolate ribonucleoprotein particles with properties similar to ribosomes. The sedimentation coefficient of these particles was found to be approximately 80 S and they could be reversibly dissociated into 40 S and 60 S subunits by changing the Mg^{2+} ion concentration in the medium. There were more 60 S than 40 S subunits. Rates of protein synthesis as measured by amino acid incorporation in a cell-free system with nucleolar ribosomes are lower than those observed with cytoplasmic ribosomes.

Nucleonemal ribonucleoprotein particles of the nucleolus are thus actually ribosomes and/or ribosomal precursors (Birnstiel *et al.*, 1963a; Lerman, Mantieva and Georgiev, 1964).

In addition to ribosomal RNA nucleoli contain a considerable amount of soluble RNA (sRNA) (10 to 30% of the total nucleolar RNA) as shown by its physical-chemical properties, its metabolic behaviour and its ability to accept amino acids (Vincent and Baltus, 1960; Birnstiel, Fleissner and Borek, 1963b). Since sRNA has not been detected in nucleonemata it is probably localized in the amorphous part of the nucleolus.

The information on the chromatin fractions which are obtained after removal of the nucleoli is quite scanty. During homogenization in media of low ionic strength some breakage of the original structure of DNP-fibrils

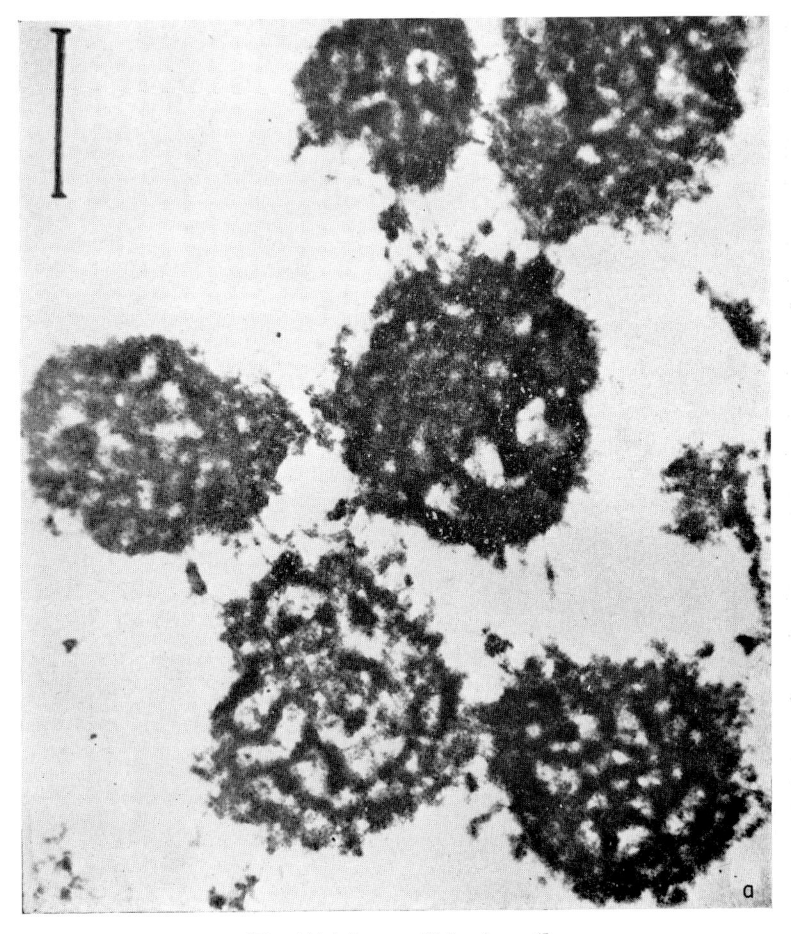

Fig. 11(a) [see p. 49 for legend]

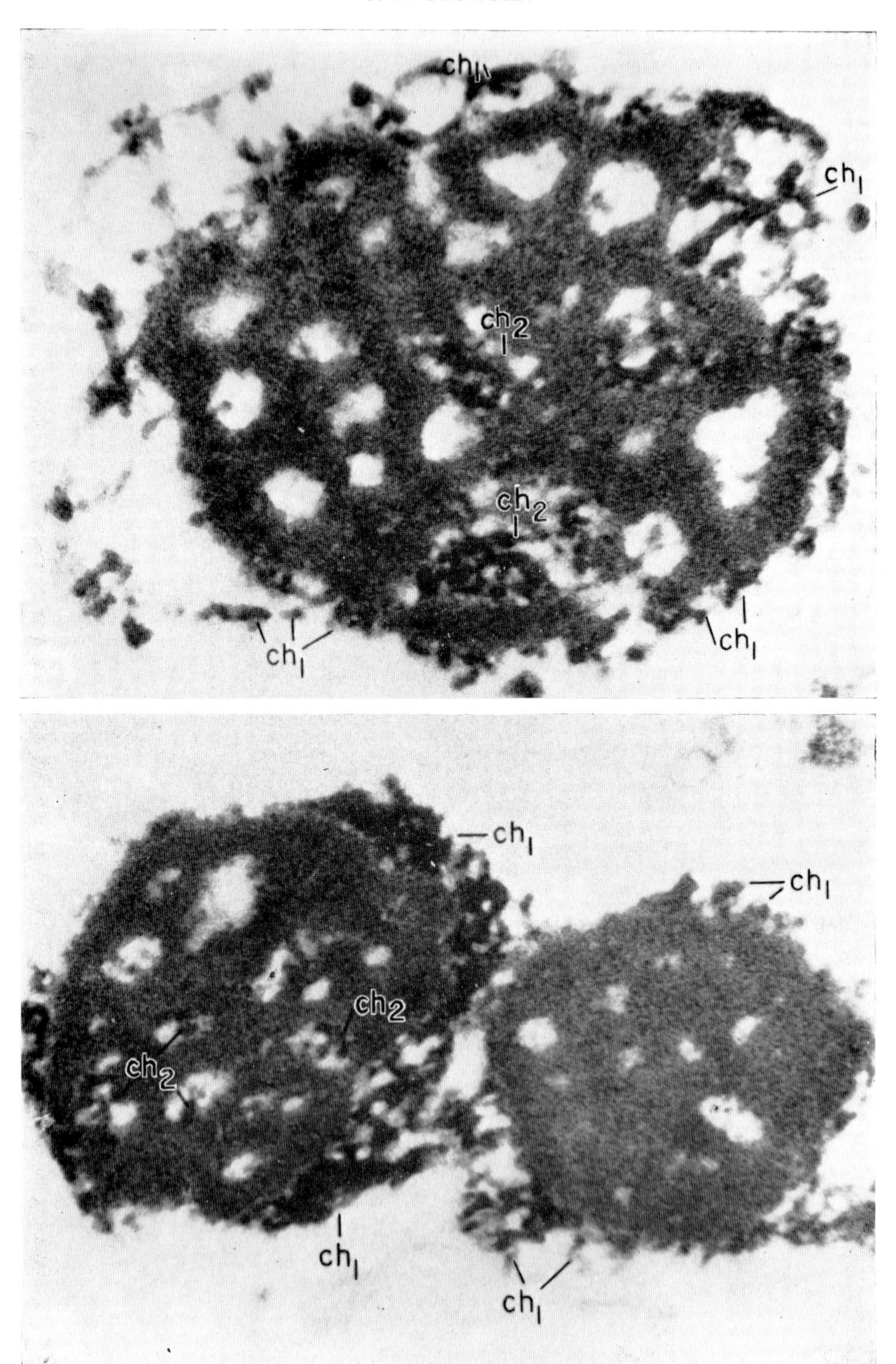
ch1
ch1
ch2
ch2
ch1
ch1
ch1
ch1
ch2
ch2
ch1
ch1

with subsequent aggregation probably occurs, since thick threads are present in the precipitate (Maggio *et al.*, 1963b). Almost all the DNA and a significant part of the RNA of the nucleus are present in the chromatin fraction. Even less is known about the composition of the other fractions obtained by mechanical fractionation of nuclei.

It is evident that mechanical methods of fractionation of nuclei are not sufficiently developed. Nevertheless they have already made possible the study of the function of the nucleolus and particularly of nucleolar enzymes. The principal drawback of mechanical methods is that it is not possible to isolate the different subnuclear structures free from contamination one with another. Also it is not possible to rule out the dangers of extraction and adsorption of various substances on isolated structures during fractionation as occurs during isolation of nuclei in aqueous media. Further improvement of methods of fractionation of nuclear structures and control of their purity is therefore needed. Some combination of mechanical and chemical methods of fractionation of nuclei may probably be the most fruitful future development.

IV. Auto-Reproduction (Duplication) (DNA Biosynthesis)

The primary process which provides for the transmission and continuity of hereditary information in the cell is the duplication of the cellular genetic apparatus or, in biochemical terms, the synthesis of DNA. Apart from oocytes and certain plant cells the greater part of the DNA of the cell is located in the chromosomes of the nucleus. It is to be expected, therefore, that enzymes participating in DNA synthesis would also be localized in chromosomal structures. Numerous experiments *in vivo* with labelled DNA precursors, and particularly ^{3}H-thymidine, have demonstrated that these precursors are specifically incorporated into the nuclear chromosomes. It is now known that incorporation of thymidine into DNA also takes place in mitochondria and chloroplasts. This incorporation, however, does not reflect the synthesis of a chromosomal DNA precursor, but rather the synthesis of the DNA of these structures. The base composition of cytoplasmic DNA differs from chromosomal DNA and it probably carries information for the synthesis of at least some of the proteins of the mitochondrion or chloroplast, as the case may be. In cells where the DNA content of the cytoplasm is extremely low incorporation of ^{3}H-thymidine is almost ex-

Fig. 11. Electron micrographs of liver nucleoli isolated (a) by Muramatsu *et al.* (1963) and (b) by Maggio *et al.* (1963b) (× 60,000). The nucleolonemal structure of nucleoli is well defined. In preparation (b) stained by uranyl acetate and $Pb(OH)_2$ nucleolus-associated chromatin located around nucleoli (ch_1) or in nucleolar lacunae (ch_2) is visible.

clusively into the nucleus. (See Chapter 3 for a fuller account of mitochondrial DNA.)

Since chromosomal incorporation of ^{3}H-thymidine is an established fact, let us consider the nature of enzymes which participate in DNA synthesis.

A. NUCLEAR DNA POLYMERASE

In 1956 Kornberg demonstrated the presence of an enzyme in bacterial extracts which catalysed the synthesis of DNA from deoxyribonucleoside triphosphates. This enzyme was called DNA polymerase or DNA deoxynucleoside triphosphate nucleotidyltransferase (Kornberg, Lehmann, Bessman and Simms, 1956). The enzyme catalyses the following reaction:

$$n\,(\text{deoxyribonucleoside triphosphate}) \underset{}{\overset{\text{DNA, Mg}^{2+}}{\rightleftharpoons}} (\text{deoxyribonucleoside monophosphate})_n + n\ \text{pyrophosphate}$$

It was found that for the synthesis of DNA to proceed all four deoxynucleoside triphosphates and DNA had to be present. DNA has a template function in the reaction, that is the sequence of bases in the synthesized DNA is determined by that of the added template DNA (Kornberg, 1960). In other words, complementary synthesis is observed. The synthesis of a complementary strand and "repair" of single-stranded DNA into double-stranded form take place. Double-stranded DNA is less effective as a primer in the reaction. Although net synthesis of DNA in the presence of double-stranded DNA does occur, anomalous DNA and in particular branched molecules are formed (Richardson, Schildkraut and Kornberg, 1963).

Not long after the discovery by Kornberg and co-workers of bacterial DNA polymerase a similar enzyme was isolated from animal tissues (Mantsavinos and Canellakis 1959a,b; Bollum, 1959, 1960a,b; Smellie, Keir and Davidson, 1959; Smellie *et al.*, 1960). It was demonstrated that tissue extracts of different origin stimulate the incorporation of ^{14}C or α-^{32}P labelled deoxynucleoside triphosphates into an acid insoluble product with the properties identical to native DNA. As with the bacterial extracts, the reaction required all four nucleoside triphosphates, Mg^{2+} ions and DNA (as a primer). The composition of the product was also determined by the nucelotide composition of the primer DNA.

In experiments with partially purified animal enzymes it was found that there was an obligatory requirement for single-stranded DNA as primer (for example heat denatured DNA, or DNA from the ϕX 174 virus). Native double-stranded DNA does not possess primer activity in the reaction (Bollum, 1959, 1960b). Synthesis is again complementary and results in the "repair" of single-stranded DNA into a double-stranded molecule.

When this is achieved the synthesis ceases. Thus in contrast to *E. coli* DNA polymerase, purified polymerase from higher organisms possesses absolute specificity with respect to single-stranded primer. Apparently some factor is lost during purification of DNA polymerase which is responsible for the uncoiling of DNA strands.

At first it appeared paradoxical that althought all DNA is to be synthesized in the nucleus all the DNA-polymerase activity was present in the supernatant of cytoplasmic extracts after centrifuging at 105,000 *g* (Mantsavinos and Cannelakis, 1959a,b; Bollum, 1959, 1960a, b; Smellie *et al.*, 1959). Nuclei did not reveal any polymerase activity. Furthermore Smellie and Eason (1961) studied calf thymus nuclei isolated by non-aqueous method and found no polymerase activity in these nuclei either. Using the same material, Krakow, Coutsogeorgopoulos and Canellakis (1962) demonstrated the presence of an enzyme which catalyses the attachment of deoxynucleotides to the end of DNA strands ("end DNA nucleotidyl transferase"), but again DNA polymerase was absent from the nuclei.

However, it was later shown that in different extracts of nuclei in addition to DNA polymerase there are deoxyribonucleases whose activity depends on the method of extraction. These deoxyribonucleases mask the presence of the DNA polymerase (Keir and Aird, 1962). Taking these facts into account, new experiments were performed with nuclei isolated in non-aqueous media from regenerating rat liver (Keir, Smellie and Siebert, 1962) hepatoma, (Behki and Schneider, 1963) and calf thymus (Smith and Keir, 1963).

It was found that in addition to the enzyme catalysing end attachment, nuclei do contain a true DNA polymerase and nuclear extracts which had no DNase activity or from which DNase had been removed, were found to have a higher activity than cytoplasmic extracts. The activity of DNA polymerase in crude extracts may vary depending upon the method of extraction but after removal of DNase all nuclear extracts had approximately the same activity. DNA polymerase is readily soluble in water and practically any solution (with pH $\geqslant$ 4·5) removes this enzyme completely from the nuclei. In this respect the original failure to localize the DNA polymerase in the nucleus is a good example of the necessity of control experiments with nuclei isolated in non-aqueous media. At the same time Mazia and Hinegardner (1963) have demonstrated that nuclei of sea urchin oocytes isolated in the presence of mM $MgCl_2$ contained practically all the cellular DNA polymerase. Bivalent cations were also shown to prevent the washing of this enzyme out of calf thymus nuclei (Main and Cole, 1964). It is of great interest that a significant part of the DNA polymerase in the bacterial cell is closely associated with DNA primer (Billen, 1962, 1963).

Thus nuclei contain the enzyme DNA polymerase which catalyses the complementary repair of single-stranded DNA into a double-stranded molecule. Although no rigorous proof is yet available, it is very likely that

DNA polymerase participates in DNA replication *in vivo*. It is worth mentioning that the activity of this enzyme correlates well with the rate of DNA synthesis in different tissues. In order to discuss this question more thoroughly, let us consider some macromolecular aspects of the reproduction of the cellular genome.

B. MECHANISM OF DNA SYNTHESIS IN CHROMOSOMES

In 1953 Watson and Crick put forward the hypothesis of a semi-conservative mechanism of DNA replication. According to this idea the DNA molecule formed contains one old and one newly synthesized strand. This concept was proved in the well-known experiments of Meselson (1960). He showed the presence of one preformed and one newly formed strand in *E. coli* DNA by combining the heavy isotope technique with density gradient centrifuging in caesium chloride of isolated native and denatured DNA's. The experiment of Meselson was repeated with DNA from other organisms with the same result (Sueoka, 1960; Simon, 1961).

Recently this mechanism of DNA replication has acquired new support from the experiments carried out by Cairns (1963). By using the methods of molecular autoradiography and electron microscopy it was possible to demonstrate that the bacterial genome is a continuous double-stranded structure 400 Å long. Synthesis of DNA is accompanied by the formation of a "fork", i.e. the division of one chain into 2 strands with both daughter strands containing newly formed material (see Fig. 12).

The proposed semiconservative mechanism for DNA replication *in vivo* is in good agreement with the data existing on the mechanism of the DNA-polymerase reaction. The latter is known to involve the formation of double-stranded from single-stranded molecules, i.e. each new double-stranded molecule contains one old and one newly formed strand.

If DNA polymerase does participate in DNA reproduction the question arises as to what factors are responsible for the unwinding of double-stranded DNA that must occur before the nucleotidyl-transferase reaction. First of all, does unwinding of a considerable length of the DNA molecule occur before the replication starts? Some indirect data, for example the action spectra of u.v.-light at different stages of phage formation in bacteria (Setlow and Setlow, 1960) may be interpreted as some proof for the existence of single-stranded DNA. On the other hand, all direct attempts to isolate single-stranded DNA from different organisms have failed (Georgiev and Mantieva, 1962; Rolfe, 1963). Experiments by Cairns (1963) also show that a region of intensive incorporation or DNA synthesis exists following the point of unwinding of DNA molecule into two strands. There is no lag between separation of the two strands and the subsequent formation of a double-stranded structure. Probably these two processes are coupled, and not

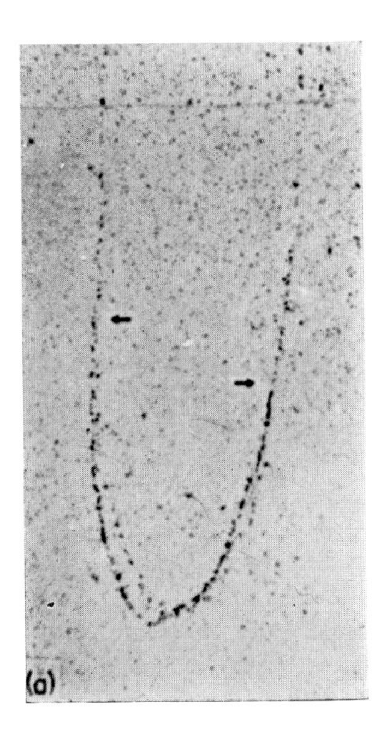

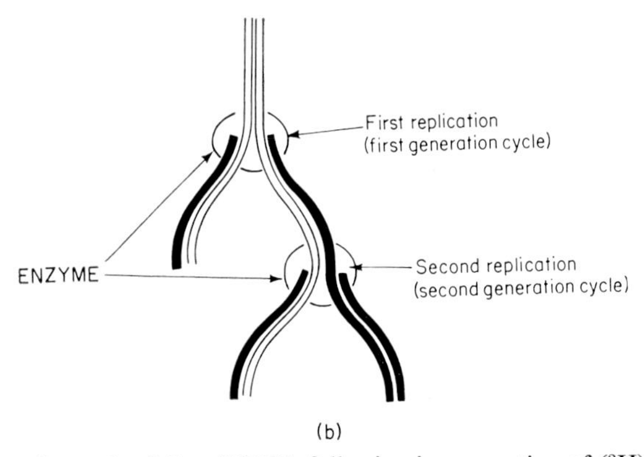

FIG. 12a. Autoradiograph of *E. coli* DNA following incorporation of (^{3}H) thymidine for a period of two generations. The arrow shows the point of replication. One of newly formed strand is twice as active as the other preformed strand. Scheme explaining the sequence of labelling is given in Fig. 12(b). × 2200 (Cairns, 1963).

separated in time and space. However the concomitance of the unwinding and doubling processes does not resolve the nature of factors controlling the unwinding. It has been suggested that nuclear deoxyribonucleases may be these factors. The nature of DNases present in the nucleus is at present quite uncertain. Experiments with nuclei isolated in non-aqueous media demonstrate that there is no DNase I (3.1.4.5) in nuclei (Stern and Mirsky,

1952). However, according to Keir and co-workers (Keir and Aird, 1962; Keir and Smith, 1963) nuclei isolated in non-aqueous media exhibit other DNase activities, and the DNases of nuclei from different tissues are different. Some DNases are known to enhance the activity of primer DNA (Bollum, 1963). As a rule, however, their action is not specific. This activation largely results from the appearance of single-stranded DNA and oligonucleotides in the course of DNA degradation by DNase. Single-stranded DNA, as well as oligonucleotides, are both good primers. It is therefore probable that nuclear DNases are not functionally connected with DNA unwinding, although such a possibility cannot be completely ruled out at the moment.

It is more likely that the factors controlling unwinding *in vivo* are closely related to DNA polymerase *per se*. It is possible that the interaction of DNA polymerase with DNA or the proteins in the DNP-complex results in local unwinding of the double strand and rapid subsequent synthesis of complementary strands. An attractive hypothesis for the possible participation in this process of iron atoms with variable valency was recently suggested by Ivanov (1965). According to this Fe^{2+} ions are bound to adenine of DNA and may cause the transition of adenine from the amino- into the imino-form and as a result a weakening of the double helix occurs.

What are the cytological ultrastructural features of chromosome reproduction in the cell? DNA synthesis is known to occur at the so-called S phase of the cellular cycle. This is a phase of interphase which is preceded and followed by inactive states: the presynthetic G_1 phase and postsynthetic G_2 phase. During S phase different chromosomes and even different chromosomal regions duplicate in certain specific order so as to double their DNA content. Since incorporation of ^3H-thymidine is observed in mitotic chromosomes and DNA synthesis occurs during interphase, experiments on the order of DNA synthesis in chromosomes do not allow us to correlate DNA synthesis with definite changes in chromosome structure. In the S-phase of the interphase nucleus ^3H-thymidine is incorporated in all regions of the nucleus with some concentration of label in the region adjacent to the nucleolus. A more convenient experimental situation is observed in the macronucleus of *Euploteus* where the nucleus is elongated and a narrow zone of DNA synthesis moves from one end of the nucleus to the other. It is important to note that in the region where DNA synthesis occurs, RNA synthesis is completely suppressed. In other words, there is competition between DNA polymerase and RNA polymerase (2.7.7.6) for the template (Prescott and Kimball, 1961). In contrast to the rest of the nucleus, DNP fibrils in the zone of synthesis appear in a de-spiralized and extended state (Fig. 13) (Furé-Fremiet, Rouiller and Ganchery, 1957). Hence it is necessary for DNA autoreproduction that there is transition of DNP from the condensed to the extended state. Similar results were obtained by Littau, Allfrey, Frenster and Mirsky (1964), who used ^{14}C-thymidine to label thymus nuclei;

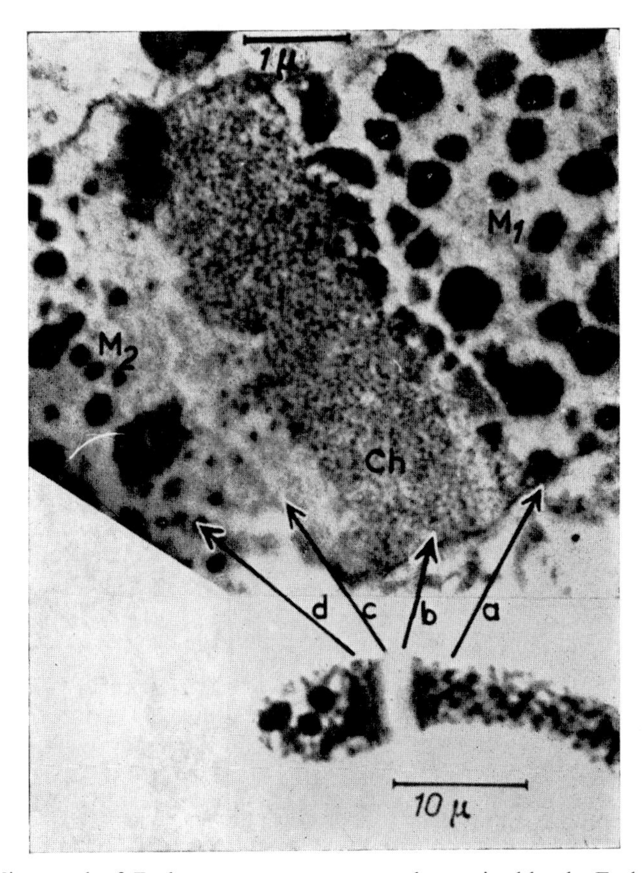

FIG. 13. Micrograph of *Euplotes eurystomus* macronucleus stained by the Feulgen method and an electron micrograph of the same nucleus. The following zones are seen: (a) zone of condensed non-duplicated chromatin; (b) zone of chromatin reorganization; (c) non-stained zone and (d) second zone of condensed chromatin (replicated). Arrows indicate the corresponding zones in the electron micrograph. M_1—aggregates of condensed deoxy-nucleoprotein; Ch—transformation of DNP material into the extended state; DNA duplication takes place in zone (c) (as shown by (^{3}H), thymidine incorporation (Prescott and Kimball, 1961); M_2—aggregates of condensed DNP in post-duplication zone (Furé-Fremiet *et al.* 1957).

after labelling, the nuclei were disrupted and separated into material which contained condensed and extended chromatin DNP fibrils. Almost all the label was found in the extended fraction. It is of interest to note that DNA polymerase is inhibited by excess histone (Bazil and Philpot, 1963 and Billen and Hnilica, 1964). These are only the first experiments in the elucidation of factors responsible for the proper interaction of DNA polymerase with the DNA template.

In conclusion it may be seen that the auto-reproduction of genetic material (or DNA synthesis) is one of the basic metabolic functions of the nucleus.

DNA synthesis in the nucleus is carried out by the DNA polymerase present in high concentrations in the nucleus. Synthesis is likely to be semiconservative. However a number of questions are still not answered. First of all it is not known whether only one enzyme participates in the synthesis of DNA or whether the process of repair into the double-stranded form by DNA polymerase is preceded by the action of another enzyme that controls the unwinding of the double helix. What is the function of deoxyribonucleases present in the cell nucleus? What factors determine the entry of DNP into the duplicative phase? What is the role of the interaction of the soluble enzyme with the DNA primer in this process? This list of unsolved problems, although far from complete, demonstrates that the principal stages of the process of auto-reproduction are not understood at present, except for the last stage carried out by DNA polymerase.

V. Transcription (RNA Biosynthesis)

In addition to the system which provides for the duplication of genetic material there exists another system in the cell nucleus responsible for supplying genetic information in a form suitable for controlling the synthesis of proteins. This role of information transfer from the genetic apparatus to the protein-synthesizing centres is played by messenger RNA (mRNA). The enzymic system participating in mRNA synthesis is the system that provides the supply of information from the genome. As suggested by Spiegelman and Hayashi (1963) this process is called "transcription".

However, mRNA alone is not enough for the expression of genetic information. A system is required which provides for the transition from the nucleotide sequence of mRNA to the amino acid sequence of the polypeptide chain that is synthesized. This is the stage of "translation". Two other types of RNA take part in the process of translation: ribosomal RNA (organized into the RNP particle or ribosomes) and soluble RNA (sRNA) which functions in this process as an "adaptor" or "translator". The synthesis of these RNA molecules, or in other words the creation of the translation system for protein synthesis, is also the function of the cell nucleus. All this demonstrates that RNA synthesis is one of the basic and specific functions of the cell nucleus which determines its role as regulator of cellular function. We will therefore discuss this question in detail. (Note that the mechanism of protein synthesis is also discussed in Chapter 7.)

A. Intracellular Localization of RNA Synthesis

The nuclear origin of RNA was first suggested by Caspersson (see Caspersson, 1950). This hypothesis became widely accepted when evidence accumulated that the hereditary information for protein structure is encoded in

DNA molecules while protein synthesis *per se* does not require DNA but is controlled by RNA. It was natural to suggest that DNA determines the structure of proteins not directly but rather through some intermediary RNA. In other words RNA molecules are synthesized on DNA in the nucleus and then RNA moves to protein-synthesizing structures primarily in the cytoplasm. The question became somewhat more complex when the heterogeneity of cellular RNA was established. In particular ribosomal, soluble and informational RNA's were shown to exist in the cell. However, it was further demonstrated that all three types of RNA are synthesized in the nucleus, in the same reaction catalysed by RNA polymerase.

Let us recall briefly the principal facts which demonstrate the nuclear origin of RNA, before we start the discussion on RNA-polymerase itself. First of all there are kinetic data on the incorporation of labelled RNA precursors into the nucleus and cytoplasm. Bergstrand *et al.* (1948) and Marshak and Calvet (1949) were the first to show that after short exposure of cells to $^{32}P_i$ the specific activity of nuclear RNA was more than ten or even hundred times higher than that of cytoplasmic RNA. This important fact was later confirmed by many investigators using various cells and different precursors (Jenner and Szafarz, 1950; Smellie, McIndoe and Davidson, 1953; Smellie, McIndoe, Logan, Davidson and Dawson, 1955; Fresco and Marshak, 1953).

Similar results were obtained in biochemical investigations of RNA fractions, isolated from nuclear and cytoplasmic preparations, as well as in autoradiographic experiments with tritiated precursors (Fitzgerald and Vinijchaikul, 1959; Goldstein and Micou, 1959; Pelling, 1959; Woods, 1959; Zalokar, 1959, 1960). In the majority of animal and plant cells after short incubation (1 hr or less) with labelled RNA precursors more than 90% of label incorporated into RNA is found in the nucleus and in some cases no label at all is found in cytoplasmic RNA.

It was further shown that isolated nuclei retain the ability to incorporate purines, pyrimidine nucleosides and even $^{32}P_i$ into RNA (Allfrey and Mirsky, 1957; Rho and Bonner, 1961) under conditions where phosphorylation of nuclear nucleotides takes place. These data unequivocally demonstrate the existence of RNA synthesis in the nucleus although they do not solve the problem of the origin of cytoplasmic RNA.

More recent studies on the incorporation of labelled precursors into nuclear RNA have dealt with the question of what type of RNA is synthesized in the nucleus. It has been found that in nuclei there is rapid incorporation of label into all three types of RNA (informational, ribosomal and soluble) so that synthesis of all three types of RNA does occur in the nucleus (Georgiev and Samarina, 1961; Georgiev and Mantieva, 1962).

What kind of experiments demonstrate the origin of cytoplasmic RNA? Numerous biochemical and autoradiographic studies on the kinetics of precursor incorporation into RNA fractions have been carried out using conditions of cold chase, i.e. after pulse labelling (for 10 min to 2 hr) the cells were transferred to non-radioactive medium and incubated for a further period of time. The principal drawback of this method is that the labelled precursor stays in the nucleus and continues to be incorporated into RNA a long time after transfer of the nuclei into the non-radioactive medium. This necessarily complicates the analysis of the transport of label. Kinetic data on specific and total radioactivity obtained in this kind of experiment are consistent with the assumption that cytoplasmic RNA is synthesized in the nucleus. This holds true not only for total RNA, but also for the main types of RNA: mRNA, ribosomal RNA, and transfer RNA (tRNA) (Amano and Leblond, 1960; Perry, 1960; Georgiev, 1961). However, the interpretation of kinetic data is based on a number of assumptions (for example that there is a common pool of precursors for nucleus and cytoplasm) and hence such results cannot be considered as conclusive. Thus on the basis of kinetic experiments some authors came to the conclusion that the synthesis of cytoplasmic RNA and the degradation of nuclear RNA were independent (Harris, 1961).

Another approach to the study of the origin of cytoplasmic RNA is to investigate cytoplasmic synthesis after inactivation or removal of the nucleus. In experiments with amoeba and tissue culture of human amnion cells it was demonstrated (Prescott, 1959; Goldstein, Micou and Crocker, 1960) that the removal of the nucleus leads to immediate cessation of RNA synthesis in the anuclear cellular fragments. In fragments containing nuclei, normal synthesis of RNA was found to occur. The cessation of RNA synthesis in anuclear fragments of cells is a specific reaction, since the incorporation of amino acids into the proteins of anuclear fragments continues normally for a long time. In experiments by Perry and Errera (Perry, 1960; Perry, Hell and Errera, 1961) it was found that the inactivation of the nucleolus by a microbeam of u.v. light results in 70% decrease in the incorporation of labelled precursors into the RNA of the cytoplasm.

On the other hand enucleation of *Acetabularia* does not result in cessation of RNA synthesis in the anucleated fragments (Brachet, 1957). It was later demonstrated that this synthesis is completely dependent on the presence of chloroplasts (Naora, Naora and Brachet, 1960), structures of plant cells which are quite autonomous organelles and which contain their own DNA.

A third possible approach to the problem is to prepare artificial systems of labelled nuclei and normal cytoplasm. Goldstein and Plaut (1955) transferred nuclei containing labelled RNA into non-labelled amoeba and followed the flow of label from the nucleus to cytoplasm. Similar phenomena were observed by Scholtissek and Potter (1960) and Schneider (1959) who mixed isolated rat liver-nuclei and non-labelled cytoplasm. They were able to show the flow of label into free RNA and into ribonucleoprotein particles in the cytoplasm. These experiments, however, are not very conclusive since the procedure of homogenization and transfer may disturb the normal functions of the nucleus.

The fact that for all types of RNA molecule there are complementary DNA molecules in the cellular genome strongly favours the nuclear origin of RNA. This was demonstrated in experiments where specific double-stranded hybrids between mRNA, ribosomal RNA, sRNA and denatured homologous DNA were obtained (Goodman and Rich, 1962; Yankofsky and Spiegelman, 1962; Hoyer, McCarthy and Bolton, 1963). Since almost all cellular DNA is localized in the nucleus these experiments speak in favour of the nuclear origin of cytoplasmic RNA.

Finally a method has been suggested recently that uses actinomycins which specifically inhibit DNA-primed RNA synthesis (Goldberg, Rabinovitz and Reich, 1962; Harbers and Muller, 1962; Reich, Franklin, Shatkin and Tatum, 1962). It was shown that these inhibitors completely stop the incorporation of label into all three types of RNA in the nucleus as well as in the cytoplasm. Hence the synthesis of cytoplasmic RNA depends on DNA and occurs on a DNA template. In later experiments cells, after pulse labelling, were treated with actinomycin D in doses which completely blocked RNA synthesis ("actinomycin chase-experiments"). The purpose of these experiments was to follow the fate of newly synthesized RNA under conditions where there was no further synthesis of cellular RNA. It was found however that the inhibition of RNA synthesis lead to disturbances in its transport and removal from the site of formation. Nevertheless, in some cases, it was possible to follow the transfer of labelled RNA from nucleus to cytoplasm under conditions of actinomycin D block, although the transfer was not complete (Georgiev, Samarina, Lerman and Smirnov, 1963; Girard, Penman and Darnell, 1964).

While each group of experiments is not conclusive *per se* taken together they unequivocally prove the nuclear origin of at least a considerable part of all three types of cytoplasmic RNA.

Two points, however, require special mention. Firstly, not all the RNA synthesized in the nucleus is transferred to the cytoplasm. As "cold-chase" and "actinomycin-chase" experiments demonstrate, a certain amount of the newly synthesized RNA is degraded in the nucleus (Harris, 1963a, b; Lieberman, Abrams and Ove, 1963). The amount broken down may vary considerably in different cells. This phenomenon will be discussed below. Secondly, some RNA synthesis occurs in those cytoplasmic structures which contain DNA. Thus RNA is synthesized in chloroplasts which have been shown to have DNA. A similar situation probably exists in mitochondria (see Chapter 3). One may think that RNA synthesized in chloroplasts is necessary for the synthesis of chloroplast proteins. However, the main part of cellular RNA which participates in the synthesis of cellular proteins is of nuclear origin. (It is only after viral infection that most RNA synthesis is localized in the cytoplasm, but under these conditions viral and not cellular RNA is synthesized.)

In bacteria, which do not usually possess well defined nuclei, more active incorporation of labelled RNA precursors is also observed into the nuclear zone. After short exposures to radioactive precursors the greater part of the label was found in the nuclear region or in bacterial "nucleoids" in autoradiographic experiments (Caro and Forro, 1961), as well as after fractionation of bacterial cells into nuclear and cytoplasmic material (Godson and Butler, 1962). Thus in different organisms the nucleus has been shown to be the organelle whose specific function is the synthesis of cellular RNA. In other words, the nucleus determines the amount of protein synthesis (via the ribosome-soluble RNA system) and its specificity (via the messenger RNA).

B. RNA POLYMERASE AND ITS INTRACELLULAR LOCALIZATION

Thus different RNAs are synthesized in the nucleus and then are transported to the cytoplasm. What enzymes participate in these processes? In 1959 Weiss and Gladstone obtained an enzyme preparation from rat liver nuclei containing practically all the nuclear DNP. The enzyme catalyses the transfer of labelled uridine 5′ monophosphate from uridine 5′ triphosphate into internucleotide positions in RNA strands. The incorporation occurred only if all four nucleoside-triphosphates were present. Incorporation was inhibited by DNase (Weiss and Gladstone, 1959; Weiss, 1960). Soon after these experiments enzyme preparations with similar properties were isolated from different sources: thymus nuclei (Biswas and Abrams, 1962), HeLa cells (Goldberg, 1961), pea seedlings (Huang, Maheshwari and Bonner, 1960) and various bacteria (Furth, Hurwitz and Anders, 1962; Chamberlin and Berg, 1962). All these preparations had similar properties regardless of the source of isolation. The enzyme from *E. coli* was studied more fully than the others.

The rational systematic name for the enzyme is DNA-dependent ribonucleic acid-nucleoside-5′-triphosphate-nucleotidyl-transferase (2.7.7.6) since it catalyses the following reaction

$$n\,(\text{nucleoside-5}'\text{-triphosphate}) \rightleftharpoons (\text{nucleoside-5}'\text{-monophosphate})_n + n\ \text{Pyrophosphate}$$

As a trivial name the term RNA nucleotidyltransferase is recommended. However use of this term may lead to confusion of two different enzymes, namely the DNA-dependent cellular enzyme and the RNA-dependent enzymes induced by RNA viruses. Therefore we will use the well-known term "RNA polymerase" for the cellular enzyme in distinction from the

"RNA synthetase" induced by viruses. Recently the term "transcriptase" has been suggested for RNA polymerase, which more precisely reflects its functional significance.

There is an absolute requirement for the following components for the action of RNA polymerase: (1) all four nucleoside triphosphates (if DNA and not a homodeoxypolynucleotide is used as template), (2) DNA and (3) Mg^{2+} and/or Mn^{2+} ions. The pH optimum of the enzyme is in the slightly alkaline region (pH 7·5–8·0). The presence of DNA is absolutely essential for RNA polymerase. If the purified enzyme preparation does not contain DNA the reaction occurs only when DNA is added to the reaction mixture (Furth *et al.*, 1962; Furth and Loh, 1963, 1964; Ballard and Williams-Ashman, 1964).

The product of the reaction is a polyribonucleotide with 3′,5′-phosphodiester bonds having quite a high sedimentation constant (10–20 S) (Hurwitz and August, 1963; Bremer and Konrad, 1964).

The nucleotide sequence of the product is wholly determined by template DNA. In fact, RNA synthesized in the reaction has a base composition analogous to that of the DNA added. It can form double-stranded hybrids with this DNA and it has the same nearest-neighbour frequencies as the DNA primer. Thus the RNA synthesized is complementary to the DNA present in the reaction mixture (Geiduschek, Nakamoto and Weiss, 1961; Hurwitz and August, 1963).

RNA polymerase is localized predominantly in the nucleus. In the study of RNA polymerase from animal and plant sources the enzyme was purified from nuclear preparations isolated by various aqueous methods. Treatment of isolated nuclei with dilute salts (0·1–0·4 M) does not lead to the extraction of the enzyme from nuclei (Weiss, 1960; Goldberg, 1961; Biswas and Abrams, 1962). After disruption and subsequent fractionation of nuclei most of the enzyme is present in the fraction containing chromosomal fragments (Huang and Bonner, 1962). Thus the enzyme is probably a part of chromosomal structures. Solubilization of DNP in 1 M NaCl makes RNA polymerase come into solution; if water is added to the solution the enzyme is precipitated with DNP (Goldberg, 1961). In the experiments by Huang and Bonner (1962) chromosomal DNP of pea seedlings was centrifuged in 4 M caesium chloride. In concentrated CsCl, DNP dissociated and most of the proteins including all the histones came to the surface. Only a small part of the undissociated protein sedimented with DNA (the protein: DNA ratio of the sediment was 1:24). This non-histone type of protein attached to DNA was found to contain most of the RNA polymerase activity. Hence nuclear RNA polymerase is a part of DNP fibrils and appears to be firmly bound to DNA. In the experiments of Huang and Bonner (1962) the possibility of contamination of nuclei with cytoplasmic enzyme was minimized since the CsCl used in the centrifuging effectively removes absorbed proteins from nucleic acids or nucleoprotein structures. Thus the nuclear localization of RNA polymerase seems to be well established at present. Recently Reid, El-Aaser, Turner and Siebert (1964) have demonstrated the presence of RNA polymerase in nuclei isolated in non-aqueous media.

In addition to RNA polymerase bound to DNP (the so-called "aggregated enzyme") some activity is found in cellular extracts (Furth and Loh, 1963, 1964). Recently RNA polymerase was partially purified from extracts of seminal vesicles; these enzyme preparations do not contain DNA and addition of DNA is necessary to activate the enzyme (Ballard and Williams-Ashman, 1964). "Soluble RNA polymerase" is probably bound to the nucleus (or is in the nuclear sap) but no conclusive evidence for this is yet available. It is likely that DNA-bound enzyme is "functional" RNA polymerase and the soluble enzyme represents "reserve" RNA polymerase of the nucleus.

That most of the RNA-polymerase activity is associated with DNA has also been demonstrated for bacteria. Treatment by streptomycin or protamine was needed to dissociate the RNA polymerase from DNA (Hurwitz and August, 1962). Therefore *in vivo* a part of the RNA polymerase is always bound to the template which takes part in transcription.

Let us now consider some aspects of the mechanism of the RNA-polymerase reaction. It is very important for the elucidation of chromosome function to establish the mechanism of the participation of DNA in the RNA-polymerase reaction. Does it follow the conservative or semiconservative mechanism and what happens to the DNA molecule during RNA synthesis?

It is now established that DNA strands do not separate along any considerable length of the molecule in the course of RNA synthesis. DNA participates in RNA synthesis as a double-stranded molecule. This statement is based on the following facts.

In systems studied *in vitro*, RNA synthesis by bacterial RNA polymerase is primed by native as well as by denatured DNA (Chamberlin and Berg, 1962). When native DNA is used in this system it is not altered in the course of reaction; no intermediary complexes consisting of one DNA strand and one RNA strand are formed (Hurwitz and August, 1963). The reaction follows the conservative scheme of replication. The DNA template remains unchanged after completion of the cycle. When denatured DNA is used a DNA–RNA hybrid is first formed which then serves as template or primer in RNA synthesis. Continuous replacement of RNA in the hybrid molecules occurs, i.e. the semiconservative mechanism is operating, and at the end of the cycle, the double-stranded template contains one old and one new strand (Chamberlin and Berg, 1963). However, *in vivo* it was not possible to find double-stranded hybrids containing one RNA and one DNA strand and in which the RNA was characterized by a high rate of turnover (Mantieva,

1963). The only report of natural hybrids of this kind was from a bacteriophage-bacterial system (Spiegelman, Hall and Storck, 1961), but this was not later confirmed by Konrad and Stent, (1964).

In experiments with soluble RNA polymerase from animal sources it was shown that *only* native double-stranded DNA is used as primer (Furth and Loh, 1964). Single-stranded DNA is not able to serve as primer.

Finally actinomycin D was found to inhibit RNA synthesis only when double-stranded DNA was used as primer in experiments *in vitro*. Synthesis of RNA on single-stranded DNA is considerably less sensitive to this antibiotic (Strelzoff, 1963). At the same time RNA synthesis *in vivo* is very sensitive to actinomycin and the effect is comparable to the sensitivity of RNA synthesis on double-stranded DNA observed *in vitro*.

Thus double-stranded DNA performs the function of primer in the RNA-polymerase reaction *in vivo* and this process is not accompanied by unwinding of DNA along any significant part of the DNA molecule. Since, however, strand separation is obligatory for complementary synthesis, one may postulate that some restricted separation (e.g. of several nucleotide pairs) does take place and a separation wave moves along the DNA molecule participating in the reaction. Hypothetical mechanisms of this kind for RNA synthesis have been suggested by a number of authors, for instance by Butler (1963), Fig. 14.

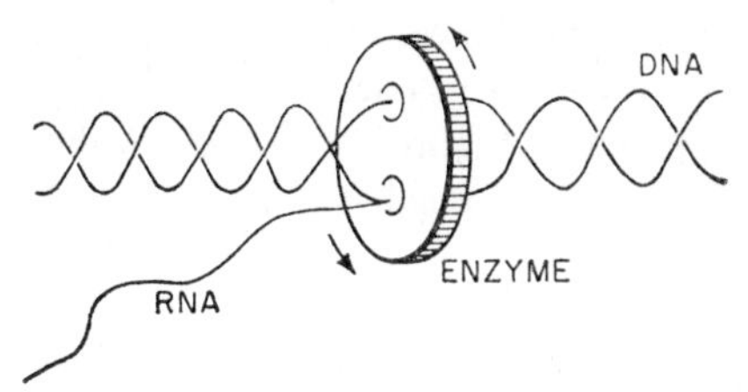

FIG. 14. A schematic mechanism for the RNA polymerase reaction (Butler, 1963).

According to this scheme, during the process of RNA synthesis in chromosomes *in vivo* a complex is formed containing double-stranded DNA and single-stranded RNA, linked by some protein (probably by RNA polymerase). The existence of such complexes *in vivo* and *in vitro* has been shown by several investigators. Schulman and Bonner (1962) have isolated from *Neurospora* material containing native DNA and RNA which could not be separated by various methods. In experiments by Bonner, Huang and Maheshwari (1961) the incorporation of label into nuclei from pea seedlings was studied *in vitro*. They found that newly synthesized RNA is bound to the chromosomes and may be detached from them by treatment with detergent, DNase or heating to 60°. Since phenol does not extract newly formed mRNA at temperatures below 55–60° (Georgiev and Mantieva, 1962), it is likely that the thermal phenol fractionation of cellular RNA is also based on the existence of this type of complex. In the course of studying RNA synthesis with a cell-free system, Bremer and Konrad (1964) observed the formation of a complex containing DNA, RNA polymerase and newly formed RNA. Each DNA molecule was associated with several molecules of enzyme, each of which in turn formed one RNA molecule.

One may speculate that just as the ribosome moves along the mRNA molecule "reading off" the information for the synthesis of protein, the RNA-polymerase molecule moves along the length of the DNA, sequentially copying it. This process results in the gradual increase of RNA chain length. If several molecules of RNA polymerase participate in the transcription of the DNA molecule, a structure should result containing a DNA strand (or DNP) with molecules of RNA polymerase and synthesized RNA attached to it. This structure should resemble polysomes in its morphology. The molecular weight of RNA polymerase is approximately 6×10^5. This means that RNA-polymerase molecules can be readily visualized by electron microscopy. Although such structures have not been found in the nucleus, so far nobody has searched for them. In electron micrographs of DNP fibrils dense zones along the length of the molecules are often observed (see Fig. 7). The above arguments show the importance of detailed study of the organization of chromosomal DNP fibrils in general in order to elucidate the structural localization of RNA polymerase. Recently, Fuchs, Zillig, Hofschneider and Preuss (1964) have examined such a complex by electron-microscopy. They used a cell-free system with T2 phage DNA as a primer, and *E. coli* RNA polymerase. According to these authors, RNA polymerase consists of a cylindrical array of 6 sub-units. DNA probably passes through the axial hole of this cylinder.

The problem of whether one or two strands of double-stranded DNA serve to generate genetic messages is of great importance. A number of observations demonstrate that *in vivo* only one DNA strand is transcribed into RNA. First of all, all mRNA isolated from phage-infected bacteria, or newly formed RNA isolated from nuclei of animal cells is single-stranded. Attempts to obtain RNA-RNA hybrids from naturally occurring RNA have not been successful either (Bautz, 1963; Samarina, Lerman, Tumanjan, Ananieva and Georgiev, 1965). This demonstrates that no complementary strands are present in RNA as would be expected if both DNA strands had participated in RNA synthesis. Furthermore, mRNA synthesized in bacteria after phage infection is complementary only to one strand of phage DNA, and no hybrids can be formed between mRNA and the second strand (Spiegelman and Hayashi, 1963; Tocchini-Valentini *et al.* 1963; Marmur and Greenspan, 1963). In hybridization experiments with agar-trapped DNA it was found that only half of the bacterial DNA could form hybrids with homologous mRNA. (McCarthy and Bolton, 1964). Thus RNA synthesis *in vivo* has an assymmetrical character, i.e. only one DNA strand is transcribed.

In experiments *in vitro* one can observe both types of synthesis: asymmetrical and symmetrical. Geiduschek, Moohr and Weiss (1962) observed the transcription of both strands of DNA and the accumulation of double-stranded RNA. Hayashi, Hayashi and Spiegelman (1964) were able to demonstrate an asymmetrical RNA synthesis by using the "replicative form" of DNA of phage ϕX 174 to prime the reaction. It is interesting that ϕX 174 replicative DNA is a circular, double-stranded molecule. However, with sonicated DNA (containing opened strands) both DNA strands were copied. Hence in this case the cyclic state of the DNA molecule is necessary for the asymmetrical character of RNA synthesis. This fact is probably not of universal significance, however. Geiduschek, Tocchini-Valentini and Sarnat (1964) were able to observe synthesis of RNA on one strand of DNA using opened strands of T2 phage DNA and a crude enzyme preparation from *Bacillus megaterium*. On the other hand, Green (1964) and Luria (1965) obtained asymmetrical synthesis with a purified enzyme from *E. coli*. Only DNA denaturation altered the character if the RNA synthesis in these experiments. The studies *in vitro* of factors that determine asymmetric RNA synthesis hinge on the problem of the initial sites of transcription. Some experiments on complex formation between DNA and RNA polymerase may indicate a role of free 3′-OH ends in this process. RNA polymerase competes with DNA polymerase for the template. It also inhibits the action of DNA of 5′-exonucleases but does not inhibit the action of endonucleases (Berg, Kornberg, Fancher and Dieckmann, 1965). It is probable that only one strand of DNA has available initial sites of transcription and this would explain why only one strand has template activity. Modifications of DNA or of the enzymes that takes place in the above experiments may render the corresponding sites of another strand available to the enzyme. The studies on the problem of the initial sites and precise character of transcription seem to offer great possibilities for understanding the morphological organization of the functional unit of the genome.

We have devoted so much attention to the macromolecular aspects of the RNA-polymerase reaction and to the role of DNA in this reaction because this enzyme plays a key role in RNA synthesis and consequently in the most important and specific function of the nucleus, namely the generation of genetic messages. This important role of RNA polymerase will become especially evident when we consider the types of RNA which are made in the RNA-polymerase reaction.

It is relatively certain that RNA polymerase catalyses mRNA synthesis since the RNA synthesized in this system exhibits many characteristic properties of mRNA. Firstly it has a DNA-like nucleotide composition and shows complementarity towards DNA (see above). Also, this RNA stimulates amino acid incorporation into proteins in ribosomes (Chamberlin and Berg, 1962) and synthesis of specific protein in ribosomes (Bonner, Huang and

Gilden, 1963). Participation of RNA polymerase in the synthesis of ribosomal RNA and sRNA is less certain. However some results indicate that these RNA types are also synthesized on DNA by the RNA-polymerase reaction. These are for example, the experiments that show the inhibition of all RNA synthesis by actinomycin (which inhibits the functioning of the DNA template in the RNA-polymerase reaction) and experiments demonstrating the complementarity of all RNA types to DNA (Reich *et al.*, 1962; Yankofskyand Spiegelman, 1962, and Goodman and Rich, 1962).

Thus normally all types of cellular RNA are probably synthesized by the RNA-polymerase reaction although it is not yet known whether there is only one enzyme for the synthesis of all RNA or whether several different isoenzymes of RNA polymerase exist in the cell. We will now consider the role of different nuclear structures in the synthesis of the various types of RNA.

C. CHROMOSOMES AND mRNA SYNTHESIS

Let us first consider experiments which demonstrate the chromosomal synthesis of mRNA. It was already noted that the base composition of chromosomal RNA differs from that of ribosomal RNA and is similar to that of DNA (Edström, 1960; Edström Grampp and Schörr, 1961; Edström and Gall, 1963). DNA-like RNA isolated from nuclei by phenol fractionation is also of chromosomal origin (Georgiev *et al.*, 1963).

Further data on the intracellular localization of mRNA synthesis are obtained from the study of the nature of newly synthesized RNA. It has been shown that all newly formed RNA (i.e. RNA labelled after short incubation of cells with suitable radioactive precursors of RNA) is localized in the nucleolo-chromosomal complex (Sibatani, Yamana, Kimura and Okagaki, 1959; Georgiev and Mantieva, 1960). Newly formed or pulse-labelled RNA is a mixture of molecules with DNA-like base composition (D-RNA) and RNA with base composition of ribosomal RNA (R-RNA) (Georgiev and Mantieva, 1962; Lerman, Mantieva and Georgiev, 1963).

By its properties (base composition, ability to form hybrids with homologous DNA, and its stimulatory effect on the amino acid incorporation activity of isolated ribosomes) D-RNA has been identified as newly formed messenger RNA (Samarina *et al.*, 1965b). Consequently the localization of newly formed D-RNA would correspond to the site of mRNA synthesis. In autoradiography experiments it was demonstrated that the synthesis of nucleolar RNA is much more sensitive to actinomycin D than RNA synthesis in the chromosomes (Perry, 1962; Sirlin, Jacob and Kato 1962). On the other hand, low concentrations of actinomycin D first cause inhibition of R-RNA synthesis (Lerman *et al.*, 1963). It is possible to find a concentration of actinomycin such that all the label would be found only in the chromosomes; under these conditions all newly formed RNA would be D-RNA (Georgiev *et al.*, 1963). These facts clearly demonstrate that mRNA is synthesized in chromosomes. However, the possibility cannot be excluded that other types of RNA are also synthesized in the chromosomes (see below).

It is well known that in higher organisms only a certain part of the genome is active in mRNA synthesis. Most of the genome is suppressed and these cistrons do not participate in RNA synthesis. This allows differentiation of cells in an organism so that there may be somatic differences between various cellular types with the same genome. This is well illustrated cytologically by the following facts. In the giant chromosomes of the salivary glands of *Drosophila* so-called "puffs" appear at different stages of morphogenesis in some chromosomal disks. Although puffs differ in their chemical characteristics and metabolic properties, in many cases the essential feature of puffs is that they are the site of intensive RNA synthesis (Fig. 15) (Rudkin and

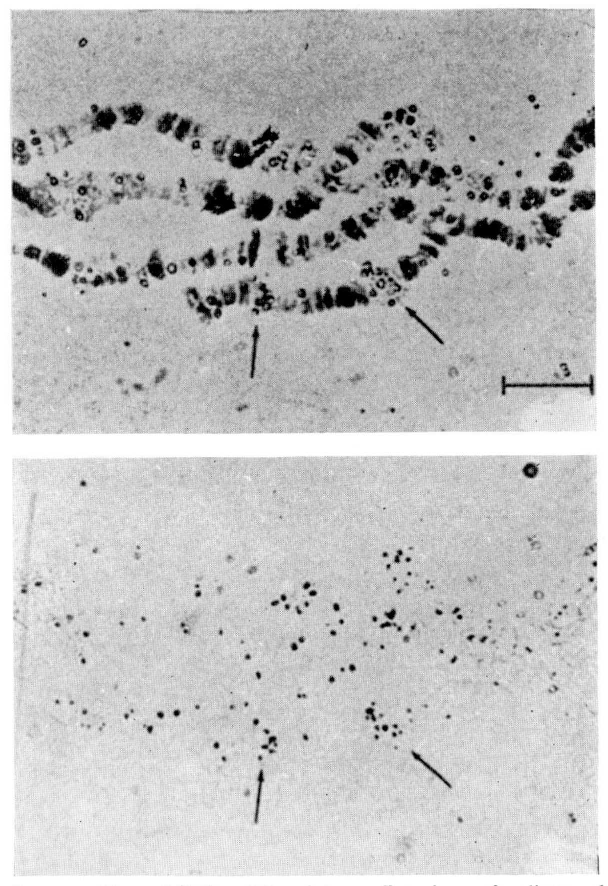

FIG. 15. The incorporation of (^{3}H) cytidine into puff regions of salivary gland chromosomes of *Drosophila melanogaster*. The position of two puffs in sections 58 and 60 of chromosome 2R are shown by arrows. × 1250 (Rudkin and Woods, 1959). One figure is stained and the other shows the autoradiogram.

Woods, 1959). It is of special significance that at certain stages of cellular differentiation puff formation always takes place in the same chromosomal region. On the other hand differences in puff localization are observed in cells at different stages of development (Breuer and Pavan, 1955). Consequently at different stages of morphogenesis RNA is synthesized in different parts of the genome and on different molecules of DNA.

Another illustration of this idea is given by experiments on the hybridization of DNA from animal organisms with mRNA isolated from different organs of the same species and on the competition between these mRNAs for DNA cystrons. By this approach it is possible to find out whether same or different genes are active in different tissues. It was shown that a significant proportion of active genes are indeed different in different tissues (McCarthy and Hoyer, 1964).

The question therefore arises as to what factors are responsible for the regulation of DNA activity during RNA synthesis. Let us again consider the cytological aspect of the problem. It seems to be well established that the difference between mitotic and interphase chromosomes lies in the fact that the former represent completely spiralized and condensed structures whereas the latter have at least partially lost this configuration. RNA is not synthesized during mitosis (Prescott and Bender, 1962). It is possible therefore that the spiralized and condensed state of the deoxynucleoprotein fibrils in the mitotic chromosome prevents the participation of DNA in RNA synthesis.

A very attractive system to study chromosome function are the so-called "lampbrush chromosomes" which are meiotic chromosomes of amphibian oocytes. During growth and development of the oocytes, side loops are branched off the axis of the chromosomes which consists of chromomeres. In these loops the DNP fibrils are in a stretched state and have despiralized whereas in the chromomeres the fibrils are condensed. The loops are formed in a sequential order and different chromomeres form loops of varying length. Thus in different chromomeres different DNP fibrils undergo the transition from the condensed into the extended configuration. As soon as growth and development of oocytes is over these loops disappear. Gall and Callan (1962) have found that RNA synthesis takes place in these loops. (Fig. 16). Thus the study of lampbrush chromosomes demonstrates that some regions of chromosome may become stretched while the main part of chromosome stays in relatively condensed state. It is stretched and non-spiralized zones in loops that participate in RNA synthesis, i.e. are "active cistrons".

In experiments by Frenster, Allfrey and Mirsky (1963) thymus nuclei were incubated with labelled RNA precursors, then disrupted by sonication and eventually preparations of "condensed" and "extended" chromatin were isolated by differential centrifuging. It was found that the specific and total radioactivity of the "stretched" chromatin was considerably higher than that

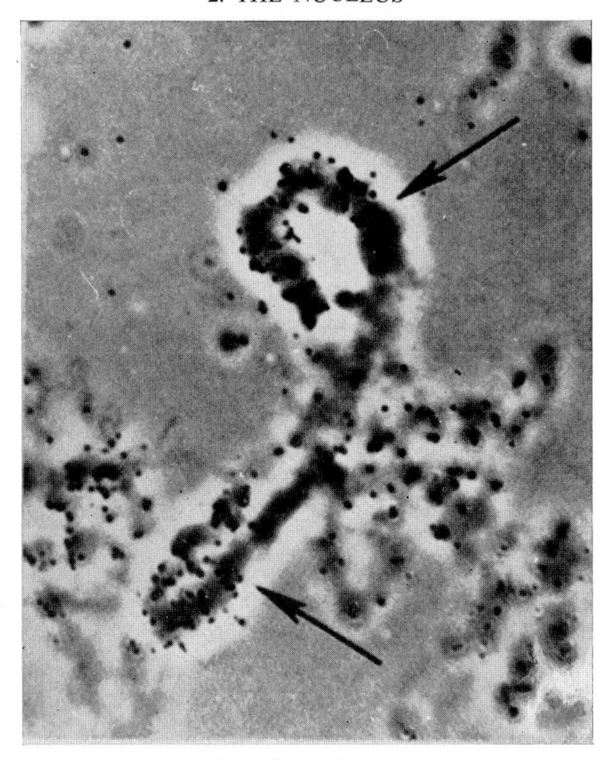

FIG. 16. The incorporation of uridine-H^3 into loops of lampbrush XII chromosome of *Triturus cristatus*. The giant granular loops are shown by arrows. The incorporation is concentrated in loops and is negligible in axis. × 2000 (Gall and Callan, 1962).

of the condensed chromatin. A similar distribution of label between condensed and extended chromatin was observed in autoradiography studies (Littau *et al*., 1964).

Thus the transition from the state of repression to the state of active function and RNA synthesis is accompanied by cytological changes i.e. by transition of DNP fibrils from the condensed state with a high degree of spiralization into the extended, despiralized form.

What is the molecular basis of this phenomenon? What chemical changes lead on to the one hand to despiralization and on the other to "switching on" of the cistron? In 1947, Stedman and Stedman suggested that inhibition of gene activity or gene "switching off" may be explained by the action of histones, principal constituents of chromosomal structure. The first experimental support to this idea came from experiments by Huang and Bonner (1962) who demonstrated that histones inhibited RNA-polymerase activity in chromosomes from pea-seedlings. The removal of histones from the chromosomes led to 25-fold increase in their RNA-polymerase activity. The addition of histones had the opposite effect.

Similar results, i.e. 200% increase in the RNA synthesis was observed by Allfrey, Littau and Mirsky (1963) with isolated calf thymus nuclei after digestion of histones with trypsin. RNA synthesis in isolated nuclei was inhibited by histones. The site of histone action is the DNA template. No immediate effect of histones on enzyme activity itself was observed (Bonner and Huang, 1963).

In order to demonstrate that the histone mechanism of repression is used in the selection of active genes *in vivo*, the following experiments were carried out (Bonner *et al.*, 1963). Chromatin from pea-seedlings was added to *E. coli* RNA polymerase and *E. coli* ribosomes. The synthesis of mRNA occurred which in turn led to protein synthesis. Chromatin from parts of the plant which are known to synthesize globulin *in vivo* was shown to induce the synthesis of globulin *in vitro* whereas chromatin isolated from non-synthesizing parts did not stimulate globulin formation. If DNA isolated from both parts of the plant was added to the system, little synthesis of globulin was observed in either case. Thus proteins of the nucleoprotein complex may be considered as specific inhibitors of the template in the RNA-polymerase system or in biological terms as the repressor of the genetic activity of DNA.

However, the problem is actually much more complex than it seems at first sight. Histones are known to be a very heterogeneous group of proteins, and there are a large number of individual histones, usually divided into three main groups: arginine-rich, moderately lysine-rich and lysine-rich. (See Mirsky and Osawa, 1961). Even the question of which of these types of histone inhibit RNA synthesis is not yet answered. According to Bonner and Huang (1963) it is the lysine-rich histones that inhibit, but Allfrey *et al.* (1963) ascribe this function to the arginine-rich histones. Recently, Huang and Bonner (1965) have observed that histone molecules are bound to a new type of RNA, characterized by a very low molecular weight (equivalent to only 40 nucleotides) and a high content of dihydrouracil. The authors suggested that the complex between this RNA and histone is a true repressor. The presence of RNA allows for the recognition of base sequences in DNA.

If histones do regulate gene activity it is necessary to assume the existence of two types of DNP in chromosomes: one more saturated and one less saturated with histone. Nucleohistone was found to melt at a temperature 10–12° higher than DNA. Chromatin isolated from different sources gives a melting profile consisting of two phases: the first small increase in optical density at the melting temperature (*Tm*) of DNA and the larger second increase at the *Tm* of nucleohistone. The probable explanation of these results is that DNP contains a small proportion of free DNA, or DNA with less than the normal amount of histone (Bonner and Huang, 1963; Bonner *et al.* (1963).

Soluble DNP isolated from nuclei according to Zubay and Doty (1959) behaves on melting like pure nucleohistone, i.e. "free" DNA seems to remain in the nuclear residue. The latter indeed contains less histone (Bonner and Huang, 1963).

It is necessary to point out however that the fractionation of DNP and histones needs further study. In particular, individual histones, and not the total histone fraction, should be studied. In addition it is important to remember that not only histones but also certain basic compounds of low molecular weight considerably affect the RNA-polymerase system. Thus some polyamines stimulate RNA synthesis on double-stranded DNA (Krakow, 1963).

Finally the most complex problem is the mechanism of the "switching on" and "off" of genes during cellular differentiation and even during the cell cycle. Experimental data concerning this problem are very scanty. A possible approach to the question has been made in experiments by Allfrey, Faulkner and Mirsky (1964), who showed that the arginine-rich histone fraction retains the ability to combine with DNA and to change its melting temperature after chemical acetylation or methylation, but is no longer able to inhibit the RNA-polymerase reaction. Acetylation and methylation reactions were also discovered in the nucleus. Whether these reactions are related to the process of "switching on" of the gene is not known at the moment.

Frenster (1965) suggested that non-histone proteins of the chromosomes or some part of the chromosomal DNA may act as de-repressors by making DNA cistrons available to the action of RNA polymerase. Finally one must consider the results that indicate that the nucleus of a cell that is not actively synthesizing nucleic acids begins to be so after it has been transplanted into the cytoplasm of active cells. For example, erythrocyte nuclei begin to synthesize not only RNA but also DNA after association with tumour cells (Harris 1965). These results indicated the role of cytoplasmic factors in determination of gene activity.

In conclusion, let us consider the problem of the removal of mRNA from the template-enzyme complex and its transfer into the cytoplasm. Samarina *et al.* (1965a., 1966) isolated from cell nuclei ribonucleoprotein particles containing mRNA. The particles sedimented as a very homogeneous peak with sedimentation constant of 30 S. It was also possible to observe the formation of an mRNA-protein complex *in vitro* by addition of free mRNA to isolated nuclear 30 S particles. Some indirect experiments indicate that these particles take part in the transfer of mRNA from the chromosomal template to the sites of protein synthesis. Similar particles with sedimentation constant of 40 S have been observed in the cytoplasm of animal cells (Joklik and Becher, 1965). It was suggested that these were small ribosomal sub-units involved in the transport of mRNA (Samarina *et al.*, 1966; Henshaw *et al.*, 1965; Joklik and Becker, 1965).

However, the possibility cannot now be excluded that the 30–40 S particles involved in the transport of mRNA are special protein entities and not ribosomal sub-units. Recent results show this to be more probable.

D. SYNTHESIS OF RIBOSOMAL RNA AND sRNA AND THE FUNCTION OF THE NUCLEOLUS

In addition to mRNA the synthesis of R-RNA, i.e. RNA with ribosomal RNA base composition also occurs in the nucleolo-chromosomal apparatus. The molecular weight of newly formed R-RNA is higher than that of ribosomal RNA and therefore it sediments faster (35–45 S) than ribosomal RNA (18 S and 28 S) (Hiatt, 1962, Scherrer and Darnell, 1962; Georgiev *et al.*, 1963).

A number of investigators have shown that the heavy R-RNA (35–45 S RNA) is a metabolic precursor of ribosomal RNA. Thus if a short pulse of radioactive precursor is given, the major part of the newly formed RNA is heavy R-RNA. If this is followed by addition of actinomycin to inhibit further RNA synthesis, the heavy non-ribosomal R-RNA completely disappears after several hours and all the label is then found in typical ribosomal RNA (Perry, 1962; Sherrer *et al.* 1963). By thermal phenol fractionation it can be demonstrated that there are two components in heavy R-RNA with sedimentation coefficient 45 S and 35 S. The 35 S component is probably the precursor of 28 S RNA and the other (45 S) is the precursor of 18 S ribosomal RNA (Georgiev and Lerman, 1964; Lerman, Vladimirzeva, Terskich and Georgiev, 1965). Thus there are two stages in the synthesis of ribosomal RNA: synthesis of heavy R-RNA precursors, and their subsequent cleavage into ribosomal RNA. Both processes take place in the nucleolo-chromosomal apparatus.

In what part of this structure do these synthetic processes occur? A number of experiments demonstrate that synthesis of R-RNA and its transformation into ribosomal RNA takes place in the nucleolus. As already mentioned (Section III) the nucleolus contains ribosome-like particles and the greater part of nucleolar RNA is ribosomal RNA. Inactivation of the nucleolus by a microbeam of u.v. light strongly inhibits incorporation of label into cytoplasmic RNA (Perry, 1960; Perry *et al.*, 1961). These data suggest that the nucleolus is concerned in the synthesis of ribosomal RNA. Recently, the following more direct evidence has been obtained:

(1) Low concentrations of actinomycin only inhibit the synthesis of R-RNA and this is accompanied by selective inhibition of incorporation into nucleolar and not chromosomal RNA (Perry, 1962; Lerman *et al.*, 1963).

(2) Newly formed RNA of nucleoli (or more precisely of the nucleonemal material) was prepared by subsequent salt extraction of isolated nuclei and its base composition was studied. It was found that newly formed RNA of nucleonemata is actually R-RNA. Since after short pulses it has the highest specific activity of the various nuclear RNA fractions, nucleonemata are the original site of localization of newly formed R-RNA (Lerman *et al.*, 1964; Steele, Okamura and Busch, 1965).

(3) Our recent autoradiography experiments carried out with nuclei that had been treated with phenol-0·14 M NaCl at different temperatures indicate that solubilization of heavy R-RNAs (35 S and 45 S) was followed by loss of label from nucleoli. Thus the localization of the bulk of newly formed precursors of ribosomal RNA is the nucleolus.

(4) Mutants of *Xenopus laevis* which do not have nucleoli cannot synthesize ribosomal RNA, although mRNA synthesis occurs normally (Brown and Gurdon, 1964).

(5) The distribution has been studied of DNA complementary to ribosomal RNA in chromatin fractions obtained after mechanical disruption and fractionation of cell nuclei. In HeLa cells the content of complementary DNA in the nucleolar fraction was 5–6 times higher than in the rest of the chromatin (McConkey and Hopkins, 1964). Most of the DNA with base composition similar to that of ribosoma. RNA was also found in the nucleolar fraction of the aquinus fungus *Blastogladuella emersonii* (Comb and Katz, 1964). Ritossa and Spiegelman (1965) observed in mutants of *Drosophila* that lack of one, two or three nucleoli results in a corresponding decrease in the amount of DNA complementary to ribosomal RNA. Hence in these cells, R-RNA is apparently synthesized in nucleoli (on DNP fibrils of the nucleolar region) or in chromatin adjacent to the nucleolar surface. On the other hand, in pea-seedlings, cistrons complementary to R-RNA are equally distributed amongst the various nuclear fractions (Chipchase and Birnstiel, 1963). One should emphasize, however, that nucleonemata are concentrated not only in nucleoli but are found in considerable amount in chromosomes. Probably synthesis of ribosomal RNA takes place not only in the nucleolus but in *nucleonemata* in general (see above) and this may explain these contradictions.

(6) Liau, Hnilica and Hurlbert (1965) have observed the synthesis of R-RNA in nucleoli isolated from hepatoma cells.

(7) It was already pointed out that nucleonemata contain newly formed R-RNA. Actinomycin chase experiments show that the transformation of the R-RNA precursor into true ribosomal RNA occurs in those structures in which synthesis of R-RNA took place; newly formed ribosomal RNAs are present in the same fractions in which heavy precursors of ribosomal RNA were found prior to addition of actinomycin (Lerman *et al.*, 1965).

The following conclusions may be drawn from the above. The accumulation of newly formed R-RNA occurs in nucleoli and probably in chromatin nucleonemata. The conversion of precursor R-RNA into true ribosomal RNA also takes place in the nucleolus. The synthesis of R-RNA occurs on the DNA template of chromosomal DNP fibrils of the nucleolar region or of the adjacent zone and possibly on DNP fibrils bound to extra-nucleolar nucloenemata, although the latter point is not yet rigorously proved. Thus the nucleolus is the organelle participating in the synthesis of ribosomal RNA.

The next question is the localization of sRNA synthesis in the nucleus. A number of facts suggests that the nucleolus has a certain role in this process. Nuclear sRNA exhibits high specific activity after pulse-labelling (Georgiev and Samarina, 1961). A considerable part of this nuclear sRNA is found in the nucleolus. (Birnstiel *et al.*, 1963 a, b). [^{3}H] Pseudo-uridine, a specific precursor of sRNA, is incorporated initially into the nucleolus (Sirlin, Kato and Jones 1961). A specific form of low molecular weight RNA which differs from sRNA by the absence of methylated bases and is considered to be the precursor of sRNA, is found in the nucleolus. (Comb and Katz, 1964). These data suggest that, in addition to the synthesis of ribosomal RNA, sRNA is also synthesized in the nucleolus.

In addition, the nucleolus has been shown to participate in the methylation of sRNA and probably ribosomal RNA. It is well known that in addition to the four principal bases some rare or minor bases (mostly methylated derivatives of purines and pyrimidines) are encountered in nucleic acids. sRNA is especially rich in such minor components and their content in ribosomal RNA is considerably less. The process of base methylation occurs at the level of the macromolecules and is catalysed by the enzyme adenosylmethionine-RNA-methyltransferase (Fleissner and Borek, 1962). This enzyme was found in preparations of nucleoli from pea-seedlings by Birnstiel *et al.* (1963) and the nucleolar enzymic activity was much higher than that of the chromosomes. The incorporation of methyl groups occurred mainly into low molecular weight RNA, evidently sRNA. The formation of the methylated bases 1-methylguanine and 6-methyl-adenine was demonstrated.

Sirlin, Jacob and Tandler (1963) have studied incorporation *in vivo* of ^{14}C-methionine into the salivary gland cells of *Chironomus* in which protein synthesis has been inhibited by puromycin. The label incorporated disappeared after treatment with RNase. This fact demonstrated that this incorporation was due to methylation of RNA. It was found that after short incubation times the label is found mainly in the nucleolus.

Thus from experiments *in vivo* and *in vitro* one can conclude that at least the final stages of ribosomal RNA formation and methylation of sRNA take place in the nucleolus.

E. OTHER ENZYMES INVOLVED IN SYNTHESIS AND BREAKDOWN OF POLYNUCLEOTIDES

In addition to RNA polymerase which plays a key role in the synthesis of different RNAs a number of enzyme catalysing the incorporation of nucleoside phosphates into polynucleotides have also been found in the nucleus.

These enzymes are found in nuclear extracts and stay in the supernatant after ultracentrifuging. They catalyse the incorporation of the nucleotidyl residue of nucleoside triphosphates into homopolynucleotides (i.e. polynucleotides that contain only one type of base). The most thoroughly studied is the enzyme polyadenylate-adenosinotriphosphate-nucleotidyltransferase (polyadenylate-synthetase: 2.7.7.19.) found by Edmunds and Abrams (1960, 1962) in the supernatant fraction of nuclear extracts. The presence of primer (RNA or polyadenylate) is obligatory for the enzyme activity. Similar enzymes which can catalyse the synthesis of homopolymers and also the addition of nucleotidyl residues on to the end position of polynucleotide chains have been described by other authors (Burdon, 1963; Smellie, 1963). The functional role of these enzymes is not yet understood.

It is of interest that RNA degradative enzymes have been discovered in the nucleus, since RNA breakage apparently takes place *in vivo*. Harris and co-workers (Harris, 1961; Harris, 1963a, b, Harris, Fisher, Rodgers, Spencer and Watts, 1963) and other authors (Lieberman *et al.*, 1963) have demonstrated that a fraction of RNA synthesized in the nucleus also undergoes breakage there. In different cells, and in similar cells under different conditions, variable amounts of the newly formed RNA were degraded.

An attempt to elucidate the nature of the enzyme responsible for the degradation of this RNA in HeLa cells was made by Harris *et al.* (1963). The enzyme was found in chromosomes and it appears to hydrolyse only newly formed RNA. 5′-nucleotides were found as the only reaction products. Orthophosphate seems to activate the reaction whereas ADP exerts the opposite effect. If RNA labelled with ^{14}C adenine is used in the reaction, labelled ADP is formed. RNA degradation is accompanied by incorporation of ^{32}P into ADP. These results suggest that degradation of newly formed RNA in cells is the result of polynucleotide-phosphorylase activity (polynucleotide-nucleosidediphosphate-nucleotidyltransferase) (2.7.7.8). Data on the presence of polynucleotide-phosphorylase in rat liver cell nuclei were obtained earlier by Hilmoe and Heppel (1957). Recently ribonuclease activity (2.7.7.16) has been observed in nuclei isolated in non-aqueous media (Reid *et al.*, 1964).

The functional significance of RNA degradation in the nucleus is not yet clear. The most probable explanation is that the degraded RNAs do not differ fundamentally from those RNAs which are not degraded, but are transformed into various stable RNAs (Harris, 1964; Lerman *et al.*, 1965). It has been suggested that the bottleneck in the formation of stable cellular polynucleotides is not the synthesis of RNA, but their subsequent incorporation into ribonucleoproteins and transport to their functional sites. Therefore these latter processes regulate the transfer of RNA into the cytoplasm and any excess of RNA synthesized in the nucleus is degraded. It cannot be excluded that this "excess" nuclear RNA plays some specific role.

At present the problem of degradation of nuclear RNA and the characterization of the enzymes involved is not yet solved, but it is undoubtedly of great interest. The possible role of nuclear polynucleotide-phosphorylase in the process of degradation might throw some light on the role of this enzyme in other systems.

VI. Translation (Protein Biosynthesis)

The information sent by the genome is encoded in the nucleotide sequence of mRNA and for its expression it must be converted into an amino acid sequence by the process of protein synthesis in the ribosomes. To denote this process the term "translation" was therefore suggested.

The process of protein synthesis takes place in the ribonucleoprotein particles, the ribosomes, and is described in Chapter 7. Here we will consider the experiments which show that protein synthesis occurs not only in the cytoplasm but also in the nucleus and will discuss some details of this nuclear protein synthesis.

The first indications of protein synthesis in nuclei were obtained by Caspersson and by Kedrowsky (see Caspersson, 1950; Kedrowsky, 1959). Later, experiments with labelled amino acids showed that the rate of incorporation *in vivo* into nuclear proteins is similar to that into cytoplasmic proteins (Daly, Allfrey and Mirsky, 1952; Smellie *et al.*, 1953; Zbarsky and Perevoshchikova, 1956). In non-proliferating cells little incorporation of label into histones was observed while the rate of incorporation into histones of proliferating cells was much higher. Independent synthesis of proteins in nuclei was finally demonstrated by the study of protein synthesis in isolated nuclei. Incorporation of labelled amino acids into protein was found after incubation of nuclei isolated in sucrose from thymus, lymphosarcoma and some other tissues. In control experiments this incorporation was demonstrated to represent net protein synthesis. In particular, the distribution of activity between various nuclear fractions was the same in experiments *in vivo* and with isolated nuclei. Peptides containing labelled amino acids were isolated from hydrolysates of nuclear proteins (Allfrey *et al.*, 1957).

Further it was demonstrated that the protein synthesis observed does occur in nuclei and is not due to contamination of nuclear preparations with whole cells or cytoplasmic constituents. In contrast with protein synthesis in cell suspensions, amino acid incorporation into isolated nuclei is sensitive to DNase and requires Na^+ ions. Na^+ is necessary for amino acid transport into nuclei. Degradation of DNA by DNase addition interferes with ATP synthesis in nuclei necessary for protein synthesis. This may explain the somewhat unusual requirements for amino acid incorporation in isolated nuclei. On the other hand these requirements demonstrate that the synthesis does occur in nuclei themselves (Allfrey, *et al.*, 1957). That the label was really incorporated into nuclei free of contamination was also shown by autoradiography (Logan, Ficq and Errera, 1959). Labelled amino acids are incorporated into protein in the nuclear sap as well as in the nucleolo-chromosomal complex.

A. PROTEIN SYNTHESIS IN NUCLEAR SAP RIBOSOMES

As already mentioned we mean by the term "nuclear sap" nuclear proteins and nucleoproteins not bound firmly to thread-like structures of nucleolo-chromosomal-complex and therefore extracted from nuclei by dilute salt solutions.

Protein synthesis is known to involve two principal stages. The first is activation of an amino acid and its attachment to sRNA.

$$\text{ATP} + \text{Amino acid} \leftrightharpoons \text{E-AMP} \sim \text{Amino acid} + \text{PP}_i \quad (1)$$

$$\text{E} - \text{AMP} \sim \text{Amino acid} + \text{sRNA} \leftrightharpoons \text{sRNA} - \text{Amino acid} + \text{E} + \text{AMP} \quad (2)$$

Enzymes (E) catalysing this reaction are called aminoacyl sRNA-synthetases (6.1.1.1–11) (Hoagland, 1960).

Hopkins (1959) has found that calf thymus nuclei and pig liver nuclei could catalyse amino acid-dependent ATP-pyrophosphate exchange, reaction (1). It is to be noted that amino acid activation was also observed with nuclei isolated from pig kidney by a non-aqueous method; this seems to prove the nuclear origin of the enzyme. In the same report it was demonstrated that transfer of amino acid to sRNA (2) also takes place in nuclei since the incorporation of labelled amino acids into RNA was shown to occur. Somewhat later amino acid activation was demonstrated for plant cell nuclei (Webster, 1960) and rat liver nuclei (Gvozdev, 1960).

All the aminoacyl-sRNA synthetase activity of nuclei may be extracted with dilute salt solutions. After ultracentrifuging of these extracts the activity stays in supernatant and may be precipitated on lowering the pH to pH5. In other words, the enzyme is similar in its behaviour to the corresponding enzyme from the cytoplasm (Hopkins, 1959).

We have already noted that nuclear sap contains low molecular weight sRNA (see Section III). It is this RNA to which amino acids are transferred by the aminoacyl-sRNA-synthetase of cell nuclei. No complex is formed between amino acids and the nucleolo-chromosomal RNA.

Cytoplasmic and nuclear aminoacyl-sRNA-synthetases, as well as sRNAs, do not differ from each other. In particular cell-free systems consisting of nuclear (or cytoplasmic) enzymes and sRNA, nuclear enzymes and cytoplasmic sRNA and cytoplasmic enzymes and nuclear sRNA all have the same activity, so that no specificity has been found (Khessin, Gvozdev and Astaurova, 1961).

To establish the exact localization of the enzyme in nuclei autoradiography experiments were performed on the distribution of the enzyme amongst nuclear structures of oocytes. The incorporated amino acid removed by hydroxylamine or ribonuclease treatment was used as a measure of the amount of sRNA-amino acid complex. Homogeneous distribution of label in the nucleus was observed without any relation to structural elements (Platova, 1962). This fact is consistent with data on the localization of the enzyme in the soluble nuclear phase or nuclear sap.

The next stage of protein synthesis is the transfer of the aminoacyl-sRNA to the ribosome and the subsequent attachment of amino acid to the —COO-terminal end of the growing polypeptide chain. The incorporation into proteins seems to require two different enzymes, GTP, ATP and Mg^{2+} ions. It was mentioned above that ribonucleoprotein particles found in nuclear sap (Frenster *et al.*, 1960; Samarina and Georgiev, 1960) were later identified as ribosomes (Frenster *et al.*, 1961; Pogo *et al.*, 1962; Wang, 1961). The particles isolated from nuclear sap by ultracentrifuging originally contain 15–20% of RNA but by treatment with deoxycholate particles identical to cytoplasmic ribosomes were obtained (Wang, 1961; Allfrey, 1963).

Ribosomes of the nuclear sap actively incorporate label *in vivo* (Samarina and Georgiev, 1960) and also during incubation *in vitro* of isolated nuclei (Frenster *et al.*, 1961). The nuclear origin of ribosomes found in nuclear sap is well established. Although extracted from nuclei, the ribosomes retain the ability to incorporate amino acids and the requirements of the system are similar to those for cytoplasmic ribosomes. In addition to sRNA, amino acids and aminoacyl-sRNA-synthetase (or i.e. in addition to aminoacyl sRNA) this system requires a cytoplasmic or nuclear supernatant enzyme, ATP, GTP and a ATP-generating system. (Frenster *et al.*, 1961).

Thus nuclear sap contains a complete system for protein synthesis, including ribosomes, sRNA and aminoacyl-sRNA synthetase, which is active *in vivo* and *in vitro*. Probably *in vivo* this system participates in the synthesis of soluble proteins of the nuclear sap. Thus, for example, if slices from rat liver labelled *in vivo* with radioactive amino acid are incubated in a medium containing non-radioactive amino acid, one observes a fall in radioactivity of nuclear sap ribosomes and an equivalent increase in the radioactivity of the soluble proteins of the nuclear sap (Zbarsky and Samarina, 1962).

On the other hand some non-nuclear proteins and particularly some cell specific functional proteins are probably synthesized in the nuclear sap. Thus in erythropoetic cells, haemoglobin synthesis takes place in nuclei, the haemoglobin being transferred later to the cytoplasm (Hamel and Bessman, 1964). In a number of other cells with a high nuclear/cytoplasmic ratio nuclei also appear to participate in the synthesis of cytoplasmic proteins (Brodsky, 1961).

The amount of protein synthesis in nuclear sap may vary quite significantly (Hopkins, 1959; Brodsky, 1961). This synthesis is apparently not a basic and specific function of the nucleus and is largely analogous to protein synthesis in cytoplasmic ribosomes.

B. PROTEIN SYNTHESIS IN THE NUCLEOLO-CHROMOSOMAL COMPLEX

The existence of protein-synthesizing enzymes in chromosomes and the nucleolus is not yet established. Both *in vivo* (Daly *et al.*, 1952; Smellie *et al.*, 1953; Zbarsky and Perevoshchikova, 1956; Samarina, 1961) and in experiments with isolated nuclei incubated with labelled amino acids (Allfrey, *et al.*, 1957) the label is incorporated into proteins of the nucleolo-chromosomal-complex. The highest rate of incorporation is observed in the non-histone protein of DNP and in "acidic protein" ("residual protein"). In non-dividing differentiated cells incorporation into histones is lower than in dividing cells, though some labelling of histones does occur.

After chemical fractionation of cell nuclei, the ability of the nucleolo-chromosomal-complex to synthesize proteins is lost. However some synthetic ability is still left after mechanical fractionation of nuclei (Birnstiel *et al.*, 1961; Birnstiel and Hyde, 1963).

The first stage of protein synthesis, i.e. the formation of aminoacyl-sRNA needed for protein synthesis in the nucleolo-chromosomal-complex probably takes place in the nuclear sap. Since the soluble phase is distributed throughout the nucleus, aminoacyl-sRNA molecules are available for all nuclear structures. The second stage of protein synthesis, the formation of peptide bonds, is probably localized in fibrillar structures of the nucleolochromosomal-complex. Since this activity of the nucleolo-chromosomal-complex has not yet been studied in a cell-free system, it is not possible to discuss the enzymic aspects of the problem, and we shall only briefly comment on some important aspects of the question.

1. Role of Ribosomes

Since the nucleolo-chromosomal complex has a complicated structural and chemical organization and contains proteins associated with DNA and at least some of the proteins are synthesized synchronously with it (Bloch and Godman, 1955; Chalkley and Maurer, 1965) the question arises: what is the mechanism of protein synthesis in this complex and particularly in the DNP fraction? Is it related to the mechanism involving ribosomes or does it follow some non-ribosomal scheme?

It is well known that the antibiotic, puromycin, inhibits protein synthesis in ribosomes and it may be used to rule out artificial incorporation. It was found that puromycin is equally effective towards all proteins of cell nuclei including histones (Schweiger, Master and Alivisatos, 1964; Gvozdev, 1963). Although lysine-rich histone synthesis is somewhat more resistant to puromycin its synthesis is also inhibited. These observations strongly suggest a ribosomal mechanism of synthesis for all nuclear proteins. In other experiments the transfer of labelled amino acid from aminoacyl-sRNA to nuclear protein in the presence of ATP has been studied. With isolated liver nuclei, the incorporation occurred in all proteins except the histones. However, in

Erlich ascites tumour cells damaged by hypotonic shock there was transfer of amino acid from aminocyl-sRNA to histones (V. A. Gvozdev, personal communication).

One can conclude therefore that proteins of nucleolo-chromosomal-complex are probably synthesized on ribosomes and this process involves enzymes similar to cytoplasmic enzymes.

2. Role of mRNA

Protein synthesis in ribosomes is known to require the participation of mRNA. A possible approach to this question is to study protein synthesis after inhibition of RNA synthesis. This is difficult to carry out in animal cells because of the stability of mRNA. However, two hours after actinomycin inhibition of RNA synthesis, a 50% decrease in incorporation of amino acids into nuclear DNP (including histones and non-histone protein) occurred (Honig and Rabinovitz, 1964; Zbarsky and Gauze, 1964). It is possible that the life-time of mRNA for proteins of the DNP-complex is comparatively short.

The fact that cell nuclei contain large amounts of informational (DNA-like) RNA with high metabolic heterogeneity is of interest (Samarina *et al.*, 1965b). One may suppose therefore that nuclear mRNA contains mRNA for chromosomal proteins in addition to newly formed mRNA due to be transferred into the cytoplasm. Since the synthesis of chromosomal proteins and particularly histones is related to the regulation of gene expression (see Section IV) further investigation of this problem is of considerable importance.

3. Nucleolar Protein Synthesis

Nucleolar proteins are actively labelled by radioactive amino acids both *in vivo* and *in vitro* during the incubation of isolated nuclei with label. This demonstrates the existence of intensive protein synthetic activity in the nucleolus (Allfrey, *et al.*, 1957; Samarina, 1961; Birnstiel *et al.*, 1961; Birnstiel and Hyde, 1963). Which proteins are formed in the nucleolus? It was already mentioned that it has been found that ribosomal RNA is synthesized in the nucleolus. Also, ribosomes attached to nucleonemal fibrils are one of the major nucleolar components. It is natural to suggest that not only synthesis of ribosomal RNA, but also the synthesis of ribosomal protein and the formation of ribosomes occur in the nucleolus. Some indirect experiments support this view. Thus ribosomes isolated from the nucleolus are heterogeneous and incorporate amino acids less actively than cytoplasmic ribosomes. This may be a consequence of their "incompleteness" (Birnstiel *et al.*, 1963a). A significant part of the nucleolar ribosomal RNA was found in particles smaller than ribosomes which could not be collected by centrifuging at 105,000g (Zbarsky and Samarina, 1962). It was also demonstrated by Tamaoki and Mueller (1965) that in chase experiments after addition of actinomycin some label in RNA and in proteins of the nucleolochromosomal complex was transferred to RNA and proteins in free ribosomes. These authors also demonstrated the presence in nuclei of ribonucleoproteins containing heavy R-RNA. This indicates that R-RNA combines with protein before it is cleaved into the shorter chains of ribosomal RNA. It is evident of course that more direct experiments are needed to elucidate the function of the nucleolus in the synthesis of ribosomal proteins and the formation of ribosomes.

In conclusion it should be emphasized that the study of protein synthesis in the nucleolo-chromosomal-complex is extremely important since it is here that the synthesis of proteins that probably regulate gene activity and the synthesis of ribosomal proteins occur, i.e. the basic and specific function of the nucleus is related to the synthesis of proteins in the nucleolochromosomal-complex.

Unfortunately no conclusive information on the mechanism and localization of these processes is at present available.

VII. Enzymes of Energy Metabolism

It has been shown in the preceding sections that many synthetic reactions in the cell nucleus, such as DNA, RNA and protein formation require a constant supply of energy in the form of high energy bond nucleoside triphosphates. Energy is consumed by many other processes as well, for example the complex nuclear reorganizations that occur during the cell cycle. Therefore the question arises as to where these high energy bond compounds (primarily ATP) are formed and by which mechanism. Do they travel to the nucleus from the cytoplasm, for example from mitochondria, or does their synthesis take place in the nucleus itself? The first question cannot be answered definitely at the present, although one should take into account that the membrane of the isolated nucleus is impermeable to ATP. The second question may be answered in the affirmative, since isolated nuclei when incubated at 37° without any additions of nucleotides to the medium can phosphorylate nucleoside monophosphates to nucleoside triphosphates (Osawa *et al.*, 1957). A number of investigations have been undertaken in order to identify the system supplying nuclei with energy.

A Glycolysis

It has been shown in a number of papers that cell nuclei from various tissues contain a number of glycolytic enzymes (Dounce and Beyer, 1948; Dounce, 1953; Stern and Mirsky, 1952, 1953; Stern, Allfrey, Mirsky and Saetren, 1952; Roodyn, 1956a, b). It was shown that nuclei could catalyse the entire sequence of glycolytic reactions, that is the formation of lactic acid from glucose. (Le Page and Schneider, 1948). Lang and Siebert (1951) however, observed lactic acid formation only with fructose-1,6-diphosphate as a substrate but not with metabolites preceding it in the glycolysis sequence. In these experiments nuclei were isolated in an aqueous medium and therefore loss of a number of enzymes could not be excluded. Decisive experiments with nuclei isolated in non-aqueous media were performed by Siebert and his collaborators (Siebert, 1960, 1961, a, b; Siebert, Bassler, Hannover, Adloff and Beyer, 1961) and later by McEwen, Allfrey and Mirsky (1963b). It was found that nuclei possess a complete system of glycolysis and that they can catalyse the formation of lactic acid from glucose. The presence of a number of glycolytic enzymes in nuclei was demonstrated: these were hexokinase (2.7.1.1), phosphofructokinase (2.7.1.11), aldolase (4.1.2.7), triosephosphate isomerase (5.3.1.1), triosephosphate dehydrogenase (1.2.1.13), glycerophosphate dehydrogenase (1.1.1.8), phosphoglycerate kinase (2.7.2.3), lactate dehydrogenase (1.1.1.27), and pyruvate kinase (2.7.1.40).

A number of important features in the distribution of these enzymes have been observed. The specific activities of glycolytic enzymes in nuclei and in cytoplasm are approximately equal. Consequently the relative activities of different enzymes in nuclei and in cytoplasm are very close. The high specific activity of enzymes in nuclei excludes the possibility of contamination with cytoplasmic enzymes, especially if one takes into account that the nuclei were isolated in non-aqueous media. These enzymes are found in the soluble fraction both in the nuclei and in the cytoplasm. In the cytoplasm glycolytic enzymes are found in the final supernatant after ultracentrifuging (see chapter 8). A portion of the nuclear glycolytic enzymes can be extracted by solutions of very low ionic strength, and another portion can be extracted with saline. After ultracentrifuging of nuclear extracts these enzymes are found in the supernatant (Roodyn, 1956b, Siebert, 1960).

Thus all nuclear glycolytic enzymes are localized in the nuclear sap. The specific activities of nuclear and cytoplasmic extracts based on their nitrogen content are equal, as mentioned above. The presence of glycolytic enzymes in nuclei is characteristic of all organisms studied, both animal and plant (Siebert, 1961a; McEwen *et al.*, 1963b; Stern and Mirsky, 1952). Nuclear glycolytic enzymes are inhibited if isolated nuclei are treated with deoxyribonuclease. There is no inhibition if whole cells or cytoplasmic extracts are treated with DNase. This is an additional proof of the true intranuclear localization of these glycolytic enzymes. The phenonenon is probably due to the liberation of histones by DNase treatment, since subsequent addition of DNA or other poly-anions releases the inhibition (McEwen *et al.*, 1963b).

The presence of virtually all the glycolytic enzymes does not prove, however, that glycolysis really proceeds in nuclei. In order to prove this Siebert, *et al.* (1961) and McEwen *et al.* (1963b) studied the content of low molecular metabolites of glycolysis in nuclei isolated in non-aqueous media. It was found that all the metabolites studied could be detected in nuclei in approximately the same concentrations as in cytoplasm. This was demonstrated for such metabolites as glucose-6-phosphate, fructose-1,6-diphosphate, dihydroxyacetone phosphate, 3-phosphoglyceric acid, glycerol-1-phosphate, phosphoenolpyruvic acid and lactic acid, as well as for AMP, ADP and ATP.

Thus one may conclude that glycolysis really proceeds in cell nuclei *in vivo*. Further proof for the existence of glycolysis in nuclei, as well as of its importance in nuclear energy metabolism, comes from experiments with inhibitors (McEwen, Allfrey and Mirsky, 1963c). Iodoacetic acid, an inhibitor of triosephosphate dehydrogenase, and NaF, an inhibitor of phosphopyruvate hydratase (or enolase: 4.2.1.11), decrease nuclear respiration, formation of $^{14}CO_2$ from [6-^{14}C]-glucose, as well as the phosphorylation of nucleotides. In anaerobic conditions when oxidative reactions in nuclei are suppressed, addition of glucose to isolated nuclei stimulates glycolysis and leads to the partial restoration of phosphorylation. An indirect indication of the funtional importance of glycolysis *in vivo* is the increase in the content of glycolytic enzymes during the activation of nuclear macromolecular syntheses, preceding cell division (Siebert, 1961a).

Another enzyme system for carbohydrate breakdown, the hexose monophosphate shunt, is also present in cell nuclei. Such enzymes as glucose-6-phosphate dehydrogenase (1.1.1.49) and 6-phosphogluconate dehydrogenase (1.1.1.43) are found in nuclei isolated in non-aqueous media, (Siebert, 1961; McEwen *et al.*, 1963b). These enzymes play a part in the formation of a pool of pentose compounds necessary for nucleic acid synthesis and for this reason this metabolic pathway may be important.

B. CITRIC ACID CYCLE

The presence of a number of enzymes of the citric acid cycle has been described in nuclei isolated both in aqueous and non-aqueous media. These enzymes are isocitrate dehydrogenase (1.1.1.41), malate dehydrogenase (1.1.1.37) and succinate dehydrogenase (1.3.99.1) (Siebert, 1961b; McEwen *et al.*, 1963b). The specific activities of these enzymes and ratios of their activities are identical in nuclei and in cytoplasm. Similarly, corresponding cytoplasmic enzymes found in the supernatant, i.e. such enzymes as isocitrate dehydrogenase or malate dehydrogenase, are found in the nuclear sap. Since they have been demonstrated in nuclei isolated in organic solvents their origin is certainly nuclear. Less certain is the situation with succinate dehydrogenase, a typical mitochondrial enzyme. In organic solvents this dehydrogenase is inactivated, while in work with nuclei isolated in aqueous media the danger of enzyme adsorption from the cytoplasm always exists.

McEwen *et al.* (1963) presented some evidence for the existence of a truly nuclear succinate dehydrogenase. For example, unlike the enzyme in whole cells or mitochondria, the nuclear enzyme is inhibited by deoxyribonuclease.

The results presented, however, are not decisive and new experiments are necessary. The concept of the existence of a complete and functional tricarboxylic acid cycle in nuclei is supported by the following facts: the presence of corresponding metabolites such as α-oxoglutarate, succinate and malate in nuclei isolated in non-aqueous media (McEwen *et al.*, 1963b); the sensitivity of nuclear phosphorylation to fluoroacetic acid, an inhibitor of aconitate hydratase (4.2.1.3), and to iodoacetic acid, an inhibitor of succinate dehydrogenase; the ability of isolated nuclei to catalyse the conversion of 2-^{14}C-acetate and 3-^{14}C-pyruvate to $^{14}CO_2$. The latter is possible only if all the stages of the citric acid cycle are present. This process is suppressed by the inhibitors mentioned above.

Finally one should mention the high content of NAD in nuclei, which is a co-factor of many reactions in the above mentioned metabolic sequences (Stern and Mirsky, 1952). In addition to NAD, a peptide-bound flavin, a coenzyme of succinate dehydrogenase, was found in nuclei isolated in non-aqueous media (McEwen *et al.*, 1963b). One therefore can draw the general conclusion that cell nuclei contain enzymes of glycolysis, the citric acid cycle and the hexose monophosphate shunt; these enzymes enable them to accomplish the first stages of carbohydrate oxidation, up to the formation of succinate, $NADH_2$ and $NADPH_2$ and the formation of high energy bond compounds, such as ATP.

Most of the enzymes mentioned, except succinate dehydrogenase, are found in the nuclear sap, that is in the soluble phase of the cell nucleus which can be extracted from nuclei with dilute salt solutions or even with solutions of low ionic strength. In the cytoplasm these enzymes are found in the final supernatant fraction, that is in soluble phase also (see Chapter 8). Some of them are also found in the matrix of the mitochondrion, but not attached to lipoprotein membranes—the main structural and functional component of mitochondria (see Chapter 3). Thus the nuclear sap is the functional analogue of other soluble cellular components, primarily of the hyaloplasm. It has been suggested that nuclear sap is a continuation into the nuclei of the soluble phase of the cell (Siebert, 1961a). This question however is rather complex and will be discussed in the following section.

C. OXIDATIVE PHOSPHORYLATION

In this section the question will be considered as to whether enzymes of terminal oxidation exist in nuclei. These enzymes effect phosphorylation during transfer of electrons from $NADH_2$ and succinate to oxygen via a number of stages.

The bulk of older results, obtained with purified nuclei, indicated that enzymes of this group are absent from the nuclei. The presence of small amounts of succinate oxidase, NAD-cytochrome *c* oxidoreductase (1.6.99.3.), cytochrome *c* and cytochrome oxidase (1.9.3.1) was explained by mitochondrial contamination. Additional purification of nuclei from mitochondria and whole cells sharply decreased the activities of these enzymes (Stern and Timonen, 1954; Schneider and Hogeboom, 1950; Hogeboom *et al.*, 1952; Schneider, 1946; Siebert and Smellie, 1957; Roodyn, 1956a, 1959). Residual activity can be explained by the adsorption of protein onto the nuclei, since Beinert (1951) has shown that nuclei in sucrose medium strongly adsorb added cytochrome *c*. Nuclei isolated in non-aqueous media are unsuitable for studies of oxidative enzymes since some of them are inactivated under such conditions. Nevertheless the absence of a number of them from nuclei isolated in non-aqueous media has been demonstrated (for example of NAD-cytochrome *c* oxidoreductase, Stern and Timonen, 1954).

There is no doubt, therefore, that the chain of electron transfer enzymes described for mitochondria is absent from nuclei. An exception are the nuclei of erythrocytes, which contain a number of oxidative enzymes, for example the cytochrome oxidase system (Rubinstein and Denstedt, 1953). It was suggested in the above papers that anaerobic metabolism is a characteristic property of nuclei in general. However, this conception was revised after the work of Allfrey *et al.* (1957), who showed that the phosphorylation of nucleotides in nuclei depends on respiration. Inhibitors of electron transport and uncoupling agents (cyanide, antimycin, dinitrophenol and anaerobic conditions of incubation) inhibited phosphorylation of nucleotides in isolated calf thymus nuclei. Somewhat later it was shown that isolated calf thymus nuclei consume oxygen; however, for purified nuclear preparations a qo_2 of 0·5 was obtained (McEwen *et al.*, 1963a; Gaizhoki, 1963). Further experiments are necessary however, since much of the observed qo_2 was due to cytoplasmic contamination.

In Mirsky's laboratory some experiments have shown that oxidative phosphorylation in nuclear preparations is really nuclear and is not due to contamination. Although the nuclear and the mitochondrial processes are similar in their sensitivity to certain inhibitors, considerable differences were observed in other properties. For example, carbon monoxide, Ca^{2+} ions, methylene blue, Janus green and dicumarol inhibit mitochondrial but do not affect nuclear oxidative phosphorylation. Also, nuclear oxidative phosphorylation, like many other intranuclear processes, is sensitive to DNase. Destruction of DNA by DNase strongly inhibits nuclear phosphorylation, and subsequent addition of DNA or poly-anions restores it. The inhibition is probably due to histones liberating during DNA digestion. DNase does not affect phosphorylation either in whole cells or in isolated mitochondria. Unlike the mitochondrial process, the addition of substrates, such as succinate, does not stimulate nuclear oxidation. This is due to the impermeability of the nuclear membrane to dicarboxylic acids (Allfrey and Mirsky, 1957;

McEwen *et al.*, 1963, a, b, c; Rees and Rowland, 1961; Rees, Ross and Rowland, 1962). Thus oxidative phosphorylation observed in nuclear preparations is very probably a truly nuclear process, different from mitochondrial oxidative phosphorylation.

However there is no information at the present time about the enzymes taking part in this process. The nuclear structures participating in this process are also obscure. It is possible that the nuclear membrane is somehow involved. The question of the universality of the oxidative phosphorylation system described for calf thymus nuclei is also uncertain. Rees and Rowland (1961) and Rees *et al.* (1962) described a similar enzymic system in liver nuclei and identified a number of individual enzymes, e.g. NAD-cytochrome *c* oxidoreductase and cytochrome oxidase. In these experiments however there were no controls for purity of the nuclei. On the other hand Creasy and Stocken (1959) have confirmed the presence of oxidative phosphorylation in thymus nuclei and also in nuclei of spleen, lymph glands, bone marrow and intestine mucosa, but failed to detect it in nuclei of liver, kidney and pancreas. This however may be due to greater damage to these nuclei during their isolation. It is of interest that nuclear oxidative phosphorylation is extremely sensitive to X-ray irradiation.

Thus, the cell nucleus possesses all the machinery necessary for the breakdown of carbohydrates via glycolysis and the citric acid cycle to CO_2, NADH and succinate. An enzymic system of electron transport from NADH and succinate to molecular oxygen probably also exists in nuclei. However there is no information on its components. Recently, Conover and Siebert (1965) have observed cytochrome b_5 and cytochrome *c* in nuclei isolated in aqueous and non-aqueous media, but the authors suggested that the presence of these cytochromes may be an artifact, due to cytoplasmic contamination.

VIII. Other Enzymes (Enzymes of Intermediary Metabolism)

Apart from enzymes necessary for the reproduction and expression of genetic information and enzymes of energy supply, a large group of enzymes of intermediary metabolism are also found in nuclei. Among them are enzymes participating in the formation of building blocks such as activated amino acids and nucleotides for polymer synthesis; in the biosynthesis of coenzymes; in the synthesis of cell polymers by non-template mechanisms, and some other enzymes. Studies of the distribution of these enzymes between subcellular structures have only rarely clearly established their nuclear localization. Many of them are found in nuclei isolated from one kind of tissue but not from another so that their nuclear localization is not a general rule. Probably these enzymes are not concerned with specific nuclear functions. Therefore results with this group of enzymes will be discussed briefly and only enzymes of nucleotide metabolism will be described in more detail.

A. NUCLEOTIDE METABOLISM

An important group of enzymes of intermediary metabolism found in nuclei is that involved in nucleotide metabolism. In the preceding section devoted to oxidative phosphorylation we have already mentioned the presence in nuclei of enzymes catalysing the phosphorylation of nucleoside monophosphates to di- and triphosphates. This group of enzymes has not been studied in detail, but it is clear that the formation of the triphosphates of different nucleosides requires not only a system of energy generation but also the presence of suitable kinases. The presence of adenylate kinase (2.7.4.3.) in nuclei isolated in non-aqueous media was demonstrated by Miller and Goldfeder (1961). Reid *et al.* (1964) have reported active phosphorylation of UMP to UDP and UTP by rat-liver nuclei isolated in non-aqueous media.

Cell nuclei from various sources contain high concentrations of another enzyme of nucleotide metabolism—nucleoside phosphorylase (2.4.2.1.) catalysing the reaction: X-ribose + $P_i \leftrightharpoons$ X + ribosephosphate, where X is a purine or pyrimidine base. The concentration of nucleoside phosphorylase in nuclei is much higher than in the cytoplasm. Results obtained with nuclei isolated in aqueous and non-aqueous media were similar (Stern *et al.*, 1952; Stern and Mirsky, 1953).

Experiments on the localization of NAD synthesizing enzymes are of considerable interest. It has been shown that the synthesis of NAD from nicotinamide proceeds via the following stages (Preiss and Handler, 1957; Kornberg, 1950):

$$\text{Nicotinamide} + 5'\text{-Phosphoribosylpyrophosphate} \rightleftharpoons \text{Nicotinamide mononucleotide} + \text{Pyrophosphate.}$$

$$\text{Nicotinamide mononucleotide} + \text{ATP} \leftrightharpoons \text{NAD} + \text{Pyrophosphate.}$$

The first of these enzymes (nicotinamide phosphoribosyl-transferase (2.4.2.12) is found in mitochondria but not in nuclei (Morton, 1961). The second enzyme, NAD pyrophosphorylase (2.7.7.1) or NAD synthetase is found exclusively in cell nuclei. This was first shown by Hogeboom and Schneider (1952) with rat liver nuclei isolated in a sucrose medium. Later NAD pyrophosphorylase was found in nuclei from various sources, isolated by different methods, including the non-aqueous technique (Branster and Morton, 1956, 1957; Morton, 1958, 1961). This enzyme was not found in the cytoplasm. The above mentioned experiments proved unequivocally the exclusive nuclear localization of NAD pyrophosphorylase. Morton purified this enzyme from the hog liver nuclei and obtained a crystalline preparation with very high specific activity (Morton, 1961). It should be noted that nuclei contain high concentrations of NAD (Stern and Mirsky, 1952). The precise intranuclear localization of this enzyme is at present still unknown. Treatment of nuclei in sucrose media with sonic vibrations or extraction with dilute salt solutions (Morton, 1961) solubilizes NAD pyrophosphorylase; this probably means that it is not bound to nucleolo-chromosomal structures and is localized in the nuclear sap, and that the nuclear membrane is impermeable to this enzyme. On the other hand Baltus (1956) detected NAD pyrophosphorylase in nucleoli from the oocytes of starfish. The problem deserves further investigation but in any case it is certain that NAD pyrophosphorylase, the terminal enzyme of NAD synthesis, is concentrated exclusively in the cell nucleus.

It is known that NAD is a coenzyme of a large number of dehydrogenases, catalysing the early stages of oxidation of many substrates. Therefore the cell nucleus may regulate the level of many cellular oxidative processes via NAD synthesis.

Among other enzymes of coenzyme synthesis one should note the finding in nuclei of glucose-1-phosphate uridylyl transferase (2.7.7.9.). This enzyme catalyses the synthesis of UDP-glucose:

$$\text{Glucose-1-phosphate} + \text{UTP} \leftrightharpoons \text{UDPG} + \text{Pyrophosphate}$$

(Smith, Münch-Petersen and Mills, 1953; Mills, Ondarza and Smith, 1954). It is known that UDPG is a substrate necessary for many synthetic reactions, including glycogen synthesis.

The nucleus also contains up to 80% of the total cellular adenylate cyclase, an enzyme catalysing the conversion of ATP to adenosine-2′-3′ cyclic phosphate, which is a regulator of phosphorylase activity (Sutherland, Rall and Menon, 1962).

A large number of investigations in enzyme cytochemistry are devoted to phosphatases. Older experiments demonstrated a high level of phosphatases in the nucleus, but this was subsequently found to be due either to enzyme adsorption during isolation of nuclei in aqueous media or to redistribution of the reaction product during the histochemical reaction. Experiments with nuclei isolated in non-aqueous media (Stern *et al.*, 1952) have shown that acid phosphatase (3.1.3.2), alkaline phosphatase (3.1.3.1.) and ATPase are absent from the nuclei. Somewhat later, however, Fisher, Siebert and Adloff (1959) and Siebert (1961a) demonstrated the presence of ATPases (3.6.1.–) in nuclei. In these experiments nuclei were also isolated in non-aqueous media. These workers found two ATPases A and B, differing somewhat in certain properties such as solubility, pH optimum and affinity to the substrate. ATPases, especially ATPase B, are tightly bound to the DNP chromosomal complex; during salt reprecipitation (dissolving in 1 M NaCl with subsequent dilution) the enzyme remains in complex with DNA. Apart from its chromosomal localization, nuclear ATPase differs from mitochondrial by its low substrate-specificity. The reaction is equally rapid with the triphosphates of adenosine, guanosine, cytidine and uridine as well as with deoxynucleoside triphosphates. Moreover the affinity of enzyme for the deoxynucleoside triphosphates is higher than that for the ribonucleoside triphosphates. Therefore nuclear ATPases are non-specific nucleoside triphosphatases. However the existence of several enzymes with different specificity cannot be excluded. The role of these enzymes in the nucleus is obscure. Their wide action spectrum and chromosomal localization suggest that they may regulate nucleic acid synthesis in chromosomes (Siebert, 1961). It is possible, however, that the phosphatase activity is a side effect of some other unknown reaction.

It is clear from the previous comments that cell nuclei are rich in enzymes catalysing the release of pyrophosphate from a range of substrates (e.g. various nucleotidyl transferases, nucleoside triphosphatases and amino-acyl sRNA synthetase). It has been found that the levels of inorganic pyrophosphatase (3.6.1.1) and other polyphosphatases (3.6.1.–) in the nucleus are higher than in the cytoplasm (Grossman and Lang, 1962). The removal of pyrophosphate by the pyrophosphatases makes the reactions mentioned above practically irreversible.

One should note that the majority of the enzymes listed were studied with the nuclei isolated from a limited number of sources so that the universal nuclear occurrence of these enzymes is not well proven. Studies with nuclei isolated from various sources have only been made with NAD-ATP nucleotidyl transferase (NAD pyrophosphorylase) and in all cases this enzyme was found to be strictly nuclear.

In summary, one should stress that nuclei contain many enzymes of nucleotide metabolism, and among them enzymes catalysing the synthesis of a number of key compounds of cellular metabolism. These findings enhance our concept of the regulatory function of the nucleus in cellular metabolism.

B. "SPECIAL" ENZYMES AND WIDELY DISTRIBUTED ENZYMES

The major experiments devoted to enzymes described in previous sections were usually done with nuclei isolated from a variety of sources. Data on the distribution of these enzymes are in general in accordance with our concept of the metabolic function of the cell nucleus. The nuclear localization of these enzymes has been proved by different methods corroborating each other. However there are no such decisive results for many other enzymes. The experimental material is often scant, controls with non-aqueous nuclei have been made in only a few cases and the data on the distribution of these enzymes are difficult to correlate with the functions of nuclei common to different types of cells.

For this reason we shall not discuss these experiments in detail. Only two general aspects will be considered. The first is whether the nuclear enzymic machinery depends on the tissue from which the nuclei were isolated. In other words, are nuclei differentiated? The second question is: which enzymes characteristic of cytoplasmic structures are present in nuclei and which are not. For other works on the enzymology of the nucleus one may consult the summary table in the review by Roodyn (1959).

The first of the questions raised has been studied in detail by Stern *et al.* (1952) and by Stern and Mirsky (1953), who worked exclusively with nuclei isolated in non-queous media so that adsorption or loss of enzymes was excluded. Nuclei were isolated from different sources and both "special enzymes", that is, enzymes present only in certain types of organs and absent from others, and "widely distributed enzymes" present in most of the organs studied were investigated.

The distribution of "special" enzymes indicated (1) that the cell nucleus, like the cytoplasm, is differentiated; (2) the differentiation of the nucleus and the cytoplasm are not necessarily similar and hence (3) the enzymic pattern of the cell nucleus may not reflect the enzymic pattern of cytoplasm. Arginase (3.5.3.1), for example, is present in the nuclei of calf, horse and chicken liver as well as in calf kidney, but it is absent from chicken kidney nuclei, while it is found in the chicken kidney cytoplasm. Catalase (1.11.1.6) is present in the nuclei of horse liver but not of horse kidney. Haemoglobin is found in the nuclei of the avian erythrocyte and mammalian erythroblasts (Carvalho and Wilkins, 1954), while another haemprotein, myoglobin is absent from heart nuclei. The nuclei of pancreas do not contain lipase (3.1.1.3), amylase (3.2.1.1) and deoxyribonuclease I. In other words "special enzymes" may be present in nuclei or may not. It is difficult to draw any general conclusion, since even in nuclei of a particular organ from different species a given enzyme may be present or not (see for example arginase).

Another group of enzymes, studied in the above mentioned work from Mirsky's laboratory, are those enzymes that are widely distributed in different tissues. Again there are no general rules as to the distribution of this group of enzymes. Some enzymes were absent in the nuclei of all the tissues studied; these were alkaline phosphatase, adenylate 3′- and 5′-phosphatases, β-glucuronidase (3.2.1.31). Other enzymes were absent from the majority of types of nuclei studied but were found in a few. For example esterase was found in calf heart and liver nuclei but it was absent from kidney and thymus nuclei. DNase II (3.1.4.6) was found in liver nuclei, but not in nuclei of a number of other organs. Finally several enzymes, such as adenosine deaminase (3.5.4.4) and nucleoside phosphorylase are found in high concentrations in the nuclei from many organs (liver, heart, pancreas) but they are absent in nuclei from a few other organs (e.g. nuclei of intestinal mucosa).

Thus cell nuclei contain a number of enzymes of intermediary metabolism, as well as catabolic enzymes. Nuclei from different tissues and from different species have different enzyme patterns. Sometimes, but not always, these differences reflect the differences between tissues themselves. Generally the picture is rather confused, however, and no fruitful conclusions can be made at present. These enzymes are probably unrelated to the specific nuclear functions common to all different tissues.

C. OCCURRENCE OF ENZYMES OF DIFFERENT CYTOPLASMIC STRUCTURES IN THE NUCLEUS

It is clear from the other chapters in this book that different cytoplasmic structures differ sharply in their enzymic pattern. Let us now consider how frequently enzymes characteristic of different cytoplasmic structures are found in the cell nucleus. The most evident fact is that enzymes typical of the structural elements (or membranes) of mitochondria are absent from nuclei (see Chapter 3). Thus, nuclei are devoid of the enzymes of the mitochondrial chain of electron transfer. The only exception to this is succinate dehydrogenase found in nuclei from a variety of tissues (McEwen *et al.*, 1963). The enzymes of lysosomes are also absent from nuclei; these are such catabolic enzymes as alkaline nucleases, hyaluronidase, phosphatases and some others. Cell nuclei are also devoid of enzymes of secretory granules: amylase, lipase, trypsinogen, chymotrypsin and others (Stern *et al.*, 1952). Nuclei do not contain many enzymes, typical of the endoplasmic reticulum, for example alkaline phosphatase, which is abundant in Golgi bodies. However a typical membrane enzymes, glucose-6-phosphatase (3.1.3.9) is found in the cell nucleus but in a lower concentration than in the cytoplasm (Moule and Chauveau, 1959; Roodyn, 1959). This may be explained by the fact that the nuclear membrane, at least in its outer layer, is a continuation of the endoplasmic reticulum and hence contains reticulum enzymes. It is also possible that the synthesis of membranes of the reticulum begins in the nucleus.

It may be interesting in this connection that the ribosomal pellet isolated from the nuclear sap contains a large amount of protein. The metabolic activity of this protein, as shown from amino acid incorporation experiments is different from that of the supernatant protein of the nuclear sap (Allfrey, 1963). Like proteins of the microsomal membranes this protein can be solubilized and separated from the ribosomes by deoxycholate treatment (Allfrey, 1963; Wang, 1963). The questions of the site of the synthesis of reticulum membranes and the nature of the nuclear membrane are very important and therefore the study of the presence and localization in the nucleus of various proteins and enzymes typical of cytoplasmic structures is of considerable interest.

Finally enzymes present in the soluble phase of the cytoplasm are widely distributed in nuclei. It has been already mentioned that enzymes of glycolysis, the hexosemonophosphate shunt and the citric acid cycle are present in the nucleus. These enzymes are usually found in the soluble phase of the cytoplasm (many enzymes of the citric acid cycle are found in mitochondria, but in the matrix and not in the membranes). Most of the enzymes discussed in the preceding section (B.) are also in the soluble phase.

Thus, the cell nucleus contains many enzymes usually found in the soluble phase of the cytoplasm. In the nucleus they are also found in the soluble phase and either come into solution during extraction of nuclei with solvents of low ionic strength, or may be extracted with dilute salt solutions having ionic strengths of 0·1–0·2. This is why the soluble phase of the cell nucleus may be regarded as analogous to that of the cytoplasm. Siebert (1961a) regards the nuclear sap as a continuation of the cytoplasmic soluble phase into the nucleus, so that there is free exchange between nuclear sap and soluble phase of the cytoplasm. This suggestion was based on the following facts. Firstly the enzymic pattern of the nuclear sap and the hyaloplasm, as shown by the specific activities of the enzymes and the ratios of activities of different enzymes, and the content of metabolites are roughly similar. Secondly it was suggested that the nuclear membrane is permeable to proteins. The first point has already been discussed and only some exceptions will be mentioned. For example, Sibert *et al.* (1961) have shown that at certain stages of liver regeneration the activity of nuclear glycolytic enzymes increases while that of the cytoplasmic enzymes does not. This indicates that immediate exchange between nuclear sap and cytoplasm may be difficult. Also it has been shown that nuclear and cytoplasmic lactate dehydrogenases are not identical enzymes but are in fact isoenzymes (Vesell and Bearn, 1962). This raises some doubt about the identity of nuclear sap and hyaloplasm.

The second point about the permeability of nuclear membranes to proteins is rather difficult to prove. The membranes of isolated nuclei are certainly permeable to proteins, since some soluble proteins are lost even in solutions of low ionic strength. The association of many soluble proteins with nuclei in media of low ionic strength depends on electrostatic interaction between these proteins and deoxynucleoprotein but not on the impermeability of nuclear membrane. Roodyn (1957) has shown that if nuclei are disrupted by ultrasonic vibrations in a solution of low ionic strength aldolase remains in the nuclear sediment, but if undamaged nuclei are extracted with 0·14 M NaCl they lose this enzyme. If nuclei extracted with 0·14 M NaCl are placed in a sucrose extract of cytoplasm they reversibly bind cytoplasmic protein. The amount of bound protein is roughly equal to the protein content of the nuclear globulin fraction (Burton, 1960). The permeability of the nuclear membrane to protein in isolated nuclei can hardly be an artifact, due to its damage, since this membrane is at the same time impermeable to many low molecular substances (see below). On the other hand, if oocytes are placed in the solution of fluorescent γ-globulin or albumin these proteins readily accumulate in the cytoplasm, but not in the nucleus. This means that the nuclear membrane *in vivo* is not permeable to all proteins (Harding and Feldherr, 1959; Feldherr and Feldherr, 1960). The reasons for the difference between the permeability *in vitro* and *in vivo* have not been fully investigated.

To summarize, nuclear sap may be regarded as analogous to the cytoplasmic supernatant. Free exchange between these two soluble phases of the cell may occur under certain conditions but not under others. The interrelation between protein components or enzymes of the nuclear sap and the hyaloplasm probably depends on many factors. Among these are whether a particular enzyme is synthesized in nuclear or cytoplasmic ribosomes, whether it can migrate rapidly through the nuclear membrane, its electrostatic properties, and the size of the soluble phase in the particular type of nucleus. These factors may differ considerably in nuclei of different tissues and this is why the enzymic pattern of nuclear sap is extremely variable. However in spite of all these limitations, it may be concluded that nuclear sap is probably functionally and structurally similar to the hyaloplasm.

D. THE NUCLEAR MEMBRANE AND PERMEASES

In the preceding section the passage of proteins through the nuclear membrane was discussed. At the present time, the mechanism of protein and nucleoprotein transport from the nucleus to the cytoplasm is unknown; it is not even certain whether it is an energy dependent process. We have already mentioned that the passage of proteins through the membrane of isolated nuclei is a non-enzymic process. On the other hand, the penetration of some low molecular substances through the nuclear membrane probably involves some enzymic reactions. For example, as mentioned above, the membrane of isolated nuclei is impermeable to such substances as dicarboxylic acids and nucleotides. Amino acids do penetrate into isolated nuclei, but this process is an endergonic one and depends on the presence of sodium ions in the medium. Only L-amino acids can penetrate, and competition between different amino acids for incorporation into the intranuclear pool has been described. Similar phenomena have been described for a number of other low molecular weight compounds, such as nucleosides and purine bases (Allfrey *et al.*, 1957; Allfrey *et al.*, 1961). Thus one can consider the possibility that the transport of low molecular weight precursors of proteins and nucleic acids into the nucleus depends on permeases, attached to the nuclear membrane. Future investigation into the existence of different kind of permeases and study of their properties and specificity will help us to understand the precise role of the nuclear membrane in cell metabolism.

IX. Conclusion: The Function of the Cell Nucleus and Nuclear Structures

Consideration of the nuclear enzymes, the nuclear structures and their related metabolic processes makes it possible to draw certain conclusions about the role of the nucleus in cellular metabolism.

The most important and specific structural feature of the nucleus is the nucleolo-chromosomal complex, the functional genetic apparatus of the cell. Accordingly the major and specific aspect of nuclear metabolism is the conservation, reproduction and expression of genetic information. The key nuclear enzymes are therefore DNA polymerase, for DNA duplication and RNA polymerase, for DNA transcription. Both processes take place in the nucleolo-chromosomal-complex. The synthesis of certain cellular proteins, for example protein components of the ribosomes and deoxynucleoprotein fibrils, and other proteins bound to nucleic acids, also takes place there.

It has been mentioned above that chromosomes and nucleoli probably have different functions. The formation of the non-specific machinery of protein synthesis, i.e. the synthesis of ribosomal RNA, assembling of ribosomes, and probably the synthesis of sRNA takes place in nucleoli and probably also in nucleonemata of the chromosomes. The synthesis of informational RNA programming this machinery proceeds in the chromosomes. Thus the nucleolo-chromosomal complex is a complete system, producing the protein synthesizing machinery of the cell. The cell nucleus may regulate both the rate of protein synthesis and its specificity.

Another fundamental nuclear component, nuclear sap, has non-specific functions and this is why its enzymic pattern may differ in different cells. Nuclear sap is a medium in which reactions take place such as the synthesis of nuclear and some cellular proteins, high energy bond formation and many steps of intermediary metabolism. Some of these, for example, the final stages of NAD synthesis, may proceed exclusively in nuclei and this increases the extent of nuclear control over cell metabolism.

The third nuclear component—the nuclear membrane—plays a part in nucleo-cytoplasmic exchange of both low molecular weight metabolites and high molecular weight proteins and nucleoproteins.

One may conclude that the cell nucleus is a regulatory organelle in cell metabolism. The main channel for this regulation is through the control of genetic information, the transcription of which finally determines the total enzymic profile of a given cell.

REFERENCES

Allfrey, V. G. (1959). *In* "The Cell" (J. Brachet and A. E. Mirsky, ed.), Vol. 1, p. 193, Academic Press, New York and London.
Allfrey, V. G. (1963). *Expl Cell Res.* Suppl. 9, 183.
Allfrey, V. G., Faulkner, R., and Mirsky, A. E. (1964). *Proc. natn. Acad. Sci. U.S.A.* **51**, 786.
Allfrey, V. G., Littau, V. C. and Mirsky, A. E. (1963). *Proc. natn. Acad. Sci. U.S.A.* **49**, 414.
Allfrey, V. G., Meudt, R., Hopkins, J. W. and Mirsky, A. E. (1961). *Proc. natn. Acad. Sci. U.S.A.* **47**, 907.
Allfrey, V. G. and Mirsky, A. E. (1957). *Proc. natn. Acad. Sci. U.S.A.* **43**, 821.
Allfrey, V. G., Mirsky, A. E. and Osawa, S. (1957). *J. gen. Physiol.* **40**, 451.
Allfrey, V. G., Stern, H., Mirsky, A. E. and Saetren, H. (1952). *J. gen. Physiol.* **35**, 529.
Amano, M. and Leblond, C. P. (1960). *Expl Cell Res.* **20**, 250.
Ballard, P. and Williams-Ashman, W. G. (1964), *Nature, Lond.* **203**, 150.
Baltus, E. (1956). *Arch. int. Physiol.* **64**, 124.
Bautz, E. (1963). *Proc. natn Acad. Sci. U.S.A.* **49**, 68.
Bazil, G. W. and Philpot, J. S. L. (1963). *Biochim. biophys. Acta* **76**, 223.
Behki, R. M. and Schneider, W. C. (1963). *Biochim. biophys. Acta* **68**, 34.
Behrens, M. (1932). *Z. physiol. Chem.* **209**, 59.
Behrens, M. (1938). *Z. physiol. Chem.* **253**, 185.
Beinert, H. (1951). *J. biol. Chem.* **190**, 287.
Berg, P., Kornberg, R. D., Fancher, H. and Dieckmann, M. (1965). *Biochem. biophys. Res. Commun.* **18**, 932.
Bergstrand, A., Eliasson, N. A., Hammarsten, E., Norberg, B., Reichard, P. and von Ubich, H. (1948). *Cold Spr. Harb. Symp. Quant. Biol.* **13**, 22.
Billen, D. (1962). *Biochem. biophys. Res. Commun.* **7**, 179.
Billen D. (1963). *Biochim. biophys. Acta* **68**, 342.
Billen, D. and Hnilica, L. (1964), *In* Nucleohistones" (J. Bonner and P. Tso, eds). p. 289. Holden-Day, San Francisco.
Birnstiel, M. L., Chipchase, M. and Bonner, J. (1961). *Biochem. biophys.Res Commun.* **6**, 161.
Birnstiel, M. L., Chipchase, M. I. H. and Hyde, B. B. (1963a). *Biochim. biophys. Acta* **76**, 454.
Birnstiel, M. L., Fleissner, E. and Borek, E. (1963b.) *Science N.Y.* **142**, 1577.
Birnstiel, M. L. and Hyde, B. B. (1963). *J. Cell Biol.* **18**, 41.
Birnstiel, M. L., Rho, J. H. and Chipchase, M. I. H. (1962). *Biochim. biophys. Acta* **55**, 734.
Biswas, B. B. and Abrams, R. (1962). *Biochim. biophys. Acta* **55**, 827.
Bloch, D. P. and Godman, G. C. (1955). *J. biophys. biochem. Cytol.* **1**, 53.
Bollum, F. J. (1959). *J. biol. Chem.* **234**, 2733.
Bollum, F. J. (1960a). *J. biol. Chem.* **235**, 2399.
Bollum, F. J. (1960b). *In* "The Cell Nucleus", p. 60, Butterworth and Co. London.
Bollum, F. J. (1963). *In* "Progress in Nucleic Acid Research" (J. N. Davidson and W. E. Cohn, eds., Vol. 1, p. 1. Academic Press, New York and London.
Bonner, J. and Huang, R. C. (1963), *J. mol. Biol.* **6**, 169.
Bonner, J., Huang, R. C. and Gilden, R. V. (1963), *Proc. natn. Acad. Sci. U.S.A.* **50**, 893.
Bonner, J., Huang, R. C. and Maheshwari, N. (1961). *Proc. natn. Acad. Sci. U.S.A.*, **47**, 1548.
Bopp-Hassenkamp, G. (1959). *Z. Naturf.* **14b**, 188.

Brachet, J. (1957). "Biochemical Cytology", Academic Press, New York and London.
Branster, M. J. and Morton, R. K. (1956). *Biochem. J.* **63**, 640.
Branster, M. J. and Morton, R. K. (1957). *Nature, Lond.* **181**, 540.
Bremer, H. and Konrad, M. W. (1964). *Proc. natn. Acad. Sci. U.S.A.* **51**, 801.
Breuer, M. E. and Pavan, C. (1955/56). *Chromosoma* **7**, 371.
Brodsky, V. B. (1961). *Cytologiya* **3**, 312.
Brown, D. D. and Gurdon, J. B. (1964). *Proc. natn. Acad. Sci. U.S.A.* **51**, 139.
Burdon, R. H. (1963). *Biochem. biophys. Res. Commun.* **13**, 37.
Burton, A. D. (1960). *In* "The Cell Nucleus", p. 142. Butterworth and Co. London.
Butler, J. A. V. (1963). *Nature, Lond.*, **199**, 68.
Cairns, J. (1963). *J. mol. Biol.* **6**, 208.
Callan, H. G. and Tomlin, S. G. (1950). *Proc. R. Soc.* B **137**, 367.
Caro, L. G. and Forro, F. (1961). *J. biophys. biochem. Cytol.* **9**, 555.
Carvalho, S. and Wilkins, M. F. H. (1954). *Proc. 4th Int. Congr. Intern. Soc. Haemotol.* p. 119.
Caspersson, T. (1950). "Cell Growth and Cell Function", Academic Press, New York and London.
Chamberlin, M. and Berg, P. (1962). *Proc. natn. Acad. Sci. U.S.A.* **48**, 81.
Chamberlin, M. and Berg, P. (1963). *Cold Spr. Harb. Symp. Quant. Biol.* **28**, 67.
Chalkley, G. R. and Maurer, H. R. (1965). *Proc. natn. Acad. Sci. U.S.A.* **54**, 498.
Chauveau, J., Moule, Y. and Rouiller, Ch. (1956). *Expl Cell Res.* **11**, 317.
Chipchase, M. I. H. and Birnstiel, M. L. (1963), *Proc. natn. Acad. Sci. U.S.A.* **50**, 1101.
Comb, D. G., Brown, R. and Katz, S. (1964). *J. mol. Biol.* **8**, 781.
Comb, D. G. and Katz, S. (1964). *J. mol. Biol.* **8**, 790.
Conover, T. E. and Siebert, G. (1965). *Biochim. biophys. Acta* **99**, 1.
Creasy, W. A. and Stocken, L. A. (1959). *Biochem. J.* **72**, 519.
Daly, M. M., Allfrey, V. G. and Mirsky, A. E. (1952). *J. gen. Physiol.* **36**, 173.
Dounce, A. L. (1943). *J. biol. Chem.* **147**, 685; **151**, 221.
Dounce, A. (1953) *Int. rev. Cytol.*, **3**, 199.
Dounce, A. L. (1955). *In* "The Nucleic Acids", E. Chargaff and J. N. Davidson, (eds.) Vol. 2, p. 93. *Academic Press, New York and London.*
Dounce, A. L., and Beyer, G. T. (1948) *J. biol. Chem.* **173**, 159.
Dounce, A. L., Tishkoff, G. H., Barnett, S. R. and Freer, R. M. (1950). *J. gen. Physiol.* **33**, 629.
Dounce, A. L., Witter, R. F., Monty, K. J., Plate, S. and Cottone, M. A. (1956). *J. biophys. biochem. Cytol.* **2**, 139.
Edmonds, M. and Abrams, R. (1960). *J. biol. Chem.* **235**, 1142.
Edmonds, M. and Abrams, R. (1962). *J. biol. Chem.* **237**, 2636.
Edström, J.-E. (1960). *J. biophys. biochem. Cytol.* **8**, 47.
Edström, J.-E. and Gall, J. G. (1963). *J. Cell Biol.* **19**, 279.
Edström, J.-E., Grampp, W. and Schor, N. (1961). *J. biophys. biochem. Cytol.* **11**, 549.
Emanuel, C. F. and Chaikoff, I. L. (1957). *Biochim. biophys. Acta* **24**, 254.
Estable, C. and Sotelo, J. R. (1951). *Inst. Invest. clin. Biol.* **1**, 105.
Feldherr, C. M. and Feldherr, A. B. (1960). *Nature, Lond.* **185**, 250.
Fisher, H. W. and Harris, H. (1962). *Proc. Roy. Soc.*. B **156**, 521.
Fisher, F., Siebert, G. and Adloff, E. (1959), *Biochim. Z.* **332**, 131.
Fitzgerald, P. J. and Vinijchaikul, K. (1959), *Lab. Invest.* **8**, 319.
Fleissner, E. and Borek, E. (1962). *Proc. natn. Acad. Sci. U.S.A.* **48**, 1199.

Frenster, J. H. (1965). *Nature, Lond.*, **206**, 680.
Frenster, J. H., Allfrey, V. G. and Mirsky, A. E. (1960). *Proc. natn. Acad. Sci. U.S.A.* **46**, 432.
Frenster, J. H., Allfrey, V. G. and Mirsky, A. E. (1961). *Proc. natn. Acad. Sci. U.S.A.* **47**, 130.
Frenster, J. H., Allfrey, V. G. and Mirsky, A. E. (1963). *Proc. natn. Acad. Sci. U.S.A.* **50**, 1026.
Fresco, R. and Marshak, A. (1953). *J. biol. Chem.* **205**, 585.
Fuchs, E., Zillig, W., Hofschneider, P. H. and Preuss, A. (1964). *J. mol. Biol.* **10**, 551.
Furé-Fremiet, E., Rouiller, C. and Ganchery, M. (1957). *Expl Cell Res* **12**, 135.
Furth, J. J., Hurwitz, J. and Anders, M. (1962). *J. biol. Chem.* **237**, 2611.
Furth, J. J. and Loh, P. (1963). *Biochem. biophys. Res. Commun.*, **13**, 100.
Furth, J. J. and Loh, P. (1964). *Science*, **145**, 161.
Gaizhoki, V. S. (1963). *Vopr. Med. Khimii* **9**, 537.
Gall, J. C. and Callan, H. G. (1962). *Proc. natn. Acad. Sci. U.S.A.* **48**, 562.
Geiduschek, E. P., Moohr, J. W. and Weiss, S. B. (1962). *Proc. natn. Acad. Sci. U.S.A.* **48**, 1078.
Geiduschek, E. P., Nakamoto, T. and Weiss, S. B. (1961). *Proc. natn. Acad. Sci. U.S.A.* **47**, 1405.
Geiduschek, E. P., Tocchini-Valentini, G. P. and Sarnat, M. T. (1964). *Proc. natn. Acad. Sci. U.S.A.* **52**, 486.
Georgiev, G. P. (1961). *Biokhimiya* **26**, 1095.
Georgiev, G. P. and Chentzov, J. S. (1960). *Dokl. Akad. Nauk. SSSR* **132**, 199.
Georgiev, G. P. and Chentzov, J. S. (1963). *Biophysika* **8**, 50.
Georgiev, G. P. and Lerman, M. I. (1964). *Biochim. biophys. Acta* **91**, 678.
Georgiev, G. P. and Mantieva, V. L. (1960). *Biokhimiya* **25**, 143.
Georgiev, G. P. and Mantieva, V. L. (1962). *Biokhimiya* **27**, 949.
Georgiev, G. P. and Samarina, O. P. (1961). *Biokhimiya* **26**, 454.
Georgiev, G. P., Samarina, O. P., Lerman, M. I. and Smirnov, M. N. (1963), *Nature, Lond.* **200**, 1291.
Georgiev, G. P., Yermolaeva, L. P. and Zbarsky, I. B. (1960). *Biokhimiya* **25**, 318.
Girard, M., Penman, S. and Darnell, J. E. (1964). *Proc. natn. Acad. Sci. U.S.A.* **51**, 205.
Godson, G. N. and Butler. J. A. V. (1962). *Nature, Lond.* **193**, 655.
Goldberg, I. H. (1961). *Biochim. biophys. Acta* **51**, 201.
Goldberg, I. H., Rabinovitz, M. and Reich, E. (1962). *Proc. natn. Acad. Sci. U.S.A.* **48**, 2094.
Goldstein, L. and Micou, J. (1959). *J. biophys. biochem. Cytol.* **6**, 1, 301.
Goldstein, L., Micou, J. and Crocker, T. T. (1960), *Biochim. biophys. Acta* **45**, 82.
Goldstein, L. and Plaut, W. (1955). *Proc. natn. Acad. Sci. U.S.A.* **41**, 874.
Goodman, H. M. and Rich, A. (1962). *Proc. natn. Acad. Sci. U.S.A.* **48**, 2101.
Green, M. (1964). *Proc. natn. Acad. Sci. U.S.A.* **52**, 1288.
Grossman, D. and Lang, K. (1962). *Biochim. Z.* **336**, 351.
Gvozdev, V. A. (1960). *Biokhimiya* **25**, 920.
Hamel, C. L. and Bessman, S.P. (1964). *J. biol. Chem.* **239**, 2228.
Harbers, E. and Müller, W. (1962). *Biochem. biophys. Res. Commun.* **7**, 107.
Harding, C. V. and Feldherr, C. (1959). *J. gen. Physiol.* **42**, 1155.
Harris, H. (1961). *Proc. Vth Int. Congr. Biochem.*, Moscow Symp. 2.
Harris, H. (1963a). *Nature, Lond.* **198**, 184.
Harris, H. (1963b), *In* "Progress in Nucleic Acid Research" (J. N. Davidson and W. E. Cohn, eds.), Vol. 2, p. 20, Academic Press, New York and London.
Harris, H. (1964). *Nature, Lond.* **201**, 863.

Harris, H. (1965). *Nature, Lond.* **206**, 583.
Harris, H., Fisher, H. W., Rodgers, A., Spencer, T, and Watts J. W. (1963). *Proc. R. Soc.* B. **157**, 177.
Hartmann, J. F. (1953). *J. comp. Neurol.* **99**, 201.
Hayashi, M., Hayashi, M. N. and Spiegelman, S. (1964). *Proc. natn. Acad. U.S.A.* **51**, 351.
Henshaw, E. C., Revel, M. and Hiatt, H. H. (1965). *J. mol. Biol.*, **14**, 241.
Hiatt, H. (1962). *J. mol. Biol.*, **5**, 217.
Hilmoe, R. J. and Heppel, L. A. (1957). *J. Am. chem. Soc.* **79**, 4810.
Hoagland, M. B. (1960). *In* "The Nucleic Acids" Vol. 3, p. 349, Academic Press. New York and London.
Hogeboom, G. H. and Schneider, W. C. (1952). *J. biol. Chem.* **197**, 611.
Hogeboom, G. H., Schneider, W. C. and Palade, G. E. (1948). *J. biol. Chem.* **172**, 619.
Hogeboom, G. H., Schneider, W. C. and Striebich, M. J. (1952). *J. biol. Chem.* **196**, 111.
Honig, G. R. and Rabinovitz, M. (1964). *Fed. Proc.* **23**, 268.
Hopkins, J. W. (1959). *Proc. natn. Acad. Sci. U.S.A.* **45**, 1461.
Hoyer, B. H., McCarthy, B. J. and Bolton, E. T. (1963). *Science, N.Y.* **140**, 1408.
Huang, R. C. and Bonner, J. (1962). *Proc. natn. Acad. Sci. U.S.A.* **48**, 1216.
Huang, R. C. and Bonner, J. (1965). *Proc. natn. Acad. Sci. U.S.A.* **54**, 960.
Huang, R. C., Maheshwari, N. and Bonner, J. (1960). *Biochem. biophys. Res. Commun.* **3**, 689.
Hurwitz, J. and August, J. T. (1963). *In* "Progress in Nucleic Acid Research" (J. N. Davidson and W. E. Cohn, eds.), Vol. 1, p. 59, Academic Press, New York and London.
Ivanov, V. (1965). *Biophysika* **10**, **11**.
Johnston, F. B., Setterfield, G. and Stern, H. (1959). *J. biophys. biochem. Cytol.* **6**, 53.
Jenner, R. and Szafarz, D. (1950). *Archs Biochem. Biophys.* **26**, 54.
Joklik, W. K. and Becker, Y. (1965). *J. mol. Biol.* **13**, 511.
Kaufmann, B. P. and McDonald, M. K. (1956). *Cold Spr. Harb. Symp. Quant. Biol.* **21**, 233.
Kedrowsky, B. V. (1959). "Cytology of Protein Synthesis in Animal Cell", Nauka, Moscow.
Keir, H. M. and Aird, G. L. (1962). *Biochem. J.* **84**, 44 p.
Keir, H. M., Smellie, R. M. S. and Siebert, G. (1962). *Nature, Lond.* **196**, 752.
Keir, H. M. and Smith, S. M. J. (1963). *Biochim. biophys. Acta* **68**, 589.
Khessin, R. B., Gvosdev, V. A. and Astaurova, O. B. (1961). *Biokhimiya* **26**, 807.
Konrad, M. W. and Stent, G. C. (1964). *Proc. natn. Acad. Sci. U.S.A.* **51**, 647.
Kornberg, A. (1950). *J. biol. Chem.* **182**, 779.
Kornberg, A., Lehman, I. R., Bessman, M. J. and Simms, E. S. (1956). *Biochim. biophys. Acta* **21**, 197.
Kornberg, A. (1960). *Science*, **131**, 1503.
Krakow, J. S. (1963). *Biochem. biophys. Acta* **72**, 566.
Krakow, J., Coutsogeorgopoulos, C. and Cannellakis, E. S. (1962). *Biochim. biophys. Acta* **55**, 639.
Lang, K. and Siebert, G. (1951). *Biochem. Z.* **322**, 196.
LePage, G. A. and Schneider, W. C. (1948). *J. biol. Chem.* **176**, 1021.
Lerman, M. I., Mantieva, V. L. and Georgiev, G. P. (1963). *Doklady Akad. Nauk. SSSR* **152**, 744.
Lerman, M. I., Mantieva, V. L. and Georgiev, G. P. (1964). *Biokhimiya* **29**, 518.
Lerman, M. I., Vladimirzeva, E. A., Terskich, V. V. and Georgiev, G. P. (1965). *Biokhimiya* **30**, 375.

Liau, M. C., Hnilica, L. S. and Hurlbert, R. B. (1965). *Proc. natn. Acad. Sci. U.S.A.* **53**, 626.
Lieberman, I., Abrams, R. and Ove, P. (1963). *J. biol. Chem.* **238**, 2141.
Littau, V. C., Allfrey, V. G., Frenster, J. H. and Mirsky, A. E. (1964), *Proc. natn. Acad. Sci. U.S.A.*, **52**, 93.
Logan, R., Ficq, M. and Errera, M. (1959). *Biochim. biophys. Acta* **31**, 402.
Luria, S. E. (1965). *Biochem. biophys. Res. Commun.* **18**, 735.
Maggio, R., Siekevitz, P. and Palade, G. E. (1963a). *J. Cell Biol.* **18**, 267.
Maggio, R., Siekevitz, P. and Palade, G. E. (1963b). *J. Cell Biol.* **18**, 293.
Main, R. K. and Cole, L. J. (1964). *Nature, Lond.* **203**, 646.
Mantieva, V. L. (1963). *Vopr. Med. Chimii* **9**, 282.
Mantsavinos, R. and Canellakis, E. S. (1959a). *Cancer Res.* **19**, 1239.
Mantsavinos, R. and Canellakis, E. C. (1959b). *J. biol. Chem.* **234**, 628.
Marmur, J. and Greenspan, C. M. (1963). *Science, N.Y.* **142**, 387.
Marshak, A. and Calvet, F. (1949). *J. cell. comp. Physiol.* **34**, 451.
Martin, E. M., Malec, J., Coote, J. L. and Work, T. S. (1961). *Biochem. J.* **80**, 606.
Mazia, D. and Hinegardner, R. T. (1963). *Proc. natn. Acad. Sci. U.S.A.* **50**, 148.
McCarthy, B. J. and Bolton, E. T. (1964). *J. mol. Biol.* **8**, 184.
McCarthy, B. J. and Hoyer, B. H. (1964). *Proc. natn. Acad. Sci. U.S.A.*, **52**, 915.
McConkey, E. H. and Hopkins, J. W. (1964). *Proc. natn. Acad. Sci. U.S.A.*, **51**, 1197.
McEwen, B. S., Allfrey, V. G. and Mirsky, A. E. (1963a), *J. biol. Chem.* **238**, 758.
McEwen, B. S., Allfrey, V. G. and Mirsky, A. E. (1963b). *J. biol. Chem.* **238**, 2564.
McEwen, B. S., Allfrey, V. G. and Mirsky, A. E. (1963c). *J. biol. Chem.* **238**, 2571.
Meselson, M. (1960). *In* "Cell Nucleus", p. 240. Butterworth & Co., London.
Miller, L. A. and Goldfeder, A. (1961). *Expl Cell Res.* **23**, 311.
Mills, G. T., Ondarza, R. and Smith, E. E. B. (1954). *Biochim. biophys. Acta* **14**, 159.
Mirsky, A. E. and Osawa, S. (1961). *In* "The Cell" (J. Brachet and A. E. Mirsky, eds.), Vol. 2, p. 677. Academic. Press, N.Y.
Mirsky, A. E. and Pollister, A. W. (1946). *J. gen. Physiol.* **30**, 117.
Mirsky, A. E. and Ris, H. (1951). *J. gen. Physiol.* **34**, 475.
Morton, R. K. (1958). *Nature, Lond.* **181**, 540.
Morton, R. K. (1961). *Aust. J. Sci.* **24**, 260.
Moulé, Y. and Chauveau, J. (1959). *Expl Cell Res.* **7**, 156.
Muramatsu, M., Smetana, K. and Busch, H. (1963). *Cancer. Res.* **23**, 510.
Naora, H., Naora, H. and Brachet, J. (1960). *J. gen. Physiol.* **43**, 1083.
Osawa, S., Allfrey, V. G. and Mirsky, A. E. (1957). *J. gen. Physiol.* **40**, 491.
Pelling, G. (1959). *Nature, Lond.* **184**, 655.
Perry, R. P. (1960). *Expl Cell Res.* **20**, 216.
Perry, R. P. (1962). *Proc. natn. Acad. Sci. U.S.A.* **48**, 2179.
Perry, R. P., Hell, A. and Errera, M. (1961). *Biochim. biophys. Acta* **49**, 47.
Platova, T. P. (1962). *Cytologiya* **4**, 238.
Pogo, A. O., Pogo, B. G. T., Littau, V. C., Allfrey, V. G., Mirsky, A. E. and Hamilton, M. G. (1962). *Biochim. biophys. Acta* **55**, 849.
Potter, V. R. and Elvehjem, C. A. (1936). *J. biol. Chem.* **114**, 495.
Preiss, J. and Handler, P. (1957). *J. biol. Chem.* **225**, 759.
Prescott, D. M. (1959). *J. biophys. biochem. Cytol.* **6**, 203.
Prescott, D. M. and Bender, M. A. (1962). *Expl Cell Res.* **26**, 260.
Prescott, D. M., Kimball, R. F. (1961). *Proc. natn. Acad. Sci. U.S.A.* **47**, 686.
Rees, K. R., Ross, H. F. and Rowland, G. F. (1962). *Biochem. J.* **83**, 523.

Rees. K. R. and Rowland, G. F. (1961). *Biochem. J.* **78**, 89.
Reich, E., Franklin, R. M., Shatkin, A. J. and Tatum, E. L. (1962). *Proc. natn. Acad. Sci. U.S.A.* **48**, 1238.
Reid, E., El-Aaser, A. B. A., Turner, M. K. and Siebert, G. (1964). *Z. Physiol. Chem.* **339**, 135.
Rho, J. H. and Bonner, J. (1961). *Proc. natn. Acad. Sci. U.S.A.* **47**, 1611.
Rho, J. H. and Chipchase, M. I. (1962). *J. Cell Biol.* **14**, 183.
Richardson, C. C., Schildkraut, C. L. and Kornberg, A. (1963). *Cold Spr. Harb. Symp. Quant. Biol.* **28**, 9.
Ris, H. (1957). *In* "Chemical Basis of Heredity", p. 23, Baltimore.
Ris, H. (1961). *Can. J. Genet. Cytol.* **3**, 95.
Ris, H. and Mirsky, A. E. (1949). *J. Gen. Physiol.* **32**, 489.
Rolfe, R. (1963). *Proc. natn. Acad. Sci. U.S.A.* **49**, 386.
Ritossa, F. M. and Spiegelman, S. (1965), *Proc. natn. Acad. Sci. U.S.A.* **53**, 737.
Roodyn, D. B. (1956a). *Biochem. J.* **64**. 361.
Roodyn, D. B. (1956b) *Biochem. J.* **64**, 368.
Roodyn, D. B. (1957). *Biochim. biophys. Acta* **25**, 129.
Roodyn, D. B. (1959). *Int. Rev. Cytol.* **8**, 279.
Roodyn, D. B. (1963). *In* "Methods of Separation of Subcellular Structural Components", Biochem Soc. Symp. 23, p. 20. Cambridge University Press.
Rubinstein, D. and Denstedt, O. F. (1953). *J. biol. Chem.* **204**, 623.
Rudkin, G. T. and Woods, P. S. (1959). *Proc. natn. Acad. Sci. U.S.A.* **45**, 997.
Samarina, O. P. (1961). Biokhimiya **26**, 61.
Samarina, O. P. and Georgiev, G. P. (1960) *Dokl. Akad. Nauk SSSR* **133**, 694.
Samarina, O. P., Asrijan, I. S. and Georgiev, G. P. (1965a). *Dokl. Akad. Nauk SSSR*, **163**, 1510.
Samarina, O. P., Krichevskaya, A. A. and Georgiev, G. P. (1966) *Nature, Lond.* (In press).
Samarina, O. P., Lerman, M. I., Tumanjan, V. D., Ananieva, L. N. and Georgiev, G. P. (1965b), *Biokhimiya*, **30**, 880.
Scherrer, K. and Darnell, J. E. (1962). *Biochem. biophys. Res. Commun.* **7**, 486.
Scherrer, K., Latham, H. and Darnell, J. E. (1963). *Proc. natn. Acad. Sci. U.S.A.* **49**, 240.
Schneider, J. H. (1959). *J. biol. Chem.* **234**, 2728.
Schneider, R. M. and Petermann, M. L. (1950). *Cancer Res.* **10**, 751.
Schneider, W. C. (1946). *Cancer Res.* **6**, 685.
Schneider, W. C. (1948). *J. biol. Chem.* **176**, 259.
Schneider, W. C. (1955). *Expl Cell Res.* **8**, 24.
Schneider, W. C., Hogeboom, G. H. (1950). *J. natn. Cancer. Inst.* **10**, 969.
Scholtissek, C. and Potter, V. R. (1960). *Z. Naturf.* **15**b, 453.
Schulman, H. M. and Bonner, D. B. (1962). *Proc. natn. Acad. Sci. U.S.A.* **48**, 53.
Schweiger, H. G., Master, R. W. P. and Alivisatos, S. G. A. (1964). *Fed. Proc.* **23**, 382.
Setlow, J. K. and Setlow, R. B. (1960). *Proc. natn. Acad. Sci. U.S.A.* **46**, 791.
Sibatani, A., Yamana, K., Kimura, K. and Okagaki, H. (1959). *Biochim. biophys. Acta* **33**, 590.
Siebert, G. (1960). *In* "Cell Nucleus" p. 176. Butterworth & Co., London.
Siebert, G. (1961a). *Vth Int. Congr. Biochem.* Symp. II.
Siebert, G. (1961b). *Biochem. Z.*, **334**, 369.
Siebert, G. and Humphrey, G. B. (1965). *Adv. Enzymol.* **27**. 239.
Siebert, G., Bassler, K. H., Hannover, R. Adloff, E. and Beyer, R. (1961). *Biochim. Z.* **334**, 388.

Siebert, G. and Smellie, R. (1957). *Int. Rev. Cytol.* **6**, 383.
Simon, E. H. (1961). *J. mol. Biol.* **2**, 101.
Sirlin, J. L., Jacob, J. and Kato, K. I. (1962). *Expl Cell Res.* **27**, 355.
Sirlin, J. L., Jacob, J. and Tandler, C. J. (1963). *Biochem. J.* **89**, 447.
Sirlin, J. L., Kato, K. I. and Jones, K. W. (1961). *Biochim. biophys. Acta* **48**, 421.
Smellie, R. M. S. and Eason, R. (1961). *Biochem. J.* **80**, 39P.
Smellie, R. M. S. (1963). *In* "Progress in Nucleic Acid Research" (J. N. Davidson and W. E. Cohn, eds.) Vol. 1, p. 27. Academic Press, New York and London.
Smellie, R. M. S., Gray, E. D., Keir, H. M., Richards, J., Bell, D. and Daivdson, J. N. (1960). *Biochim. biophys. Acta* **37**, 243.
Smellie, R. M. S., Keir, H. M. and Davidson, J. N. (1959) *Biochim. biophys. Acta* **35**, 389.
Smellie, R. M. S., McIndoe, W. N. and Davidson, J. N. (1953). *Biochim. biophys. Acta* **11**, 559.
Smellie, R. M. S., McIndoe, W. M., Logan, R., Davidson, J. N. and Dawson, I. M. (1955), *Biochem. J.* **54**, 280.
Smith, S. H. J. and Keir, H. M. (1963). *Biochim. biophys. Acta* **68**, 578.
Smith, E. E. B., Munch-Petersen, A. and Mills, G. T. (1953). *Nature, Lond.* **172**, 1038.
Spiegelman, S., Hall, B. D. and Storck, R. (1961). *Proc. natn. Acad. Sci. U.S.A.* **47**, 1135.
Spiegelman, S. and Hayashi, M. (1963). *Cold Spr. Harb. Symp. Quant. Biol.* **28**, 161.
Stedman, E. and Stedman, E. (1947). *Cold Spr. Harb. Symp. Quant. Biol.* **12**, 224.
Steele, W. J., Okamura, N. and Busch, H. (1965). *J. biol. Chem.* **240**, 1742.
Stern, H., Allfrey, V., Mirsky, A. E. and Saetren, H. (1952). *J. gen. Physiol.* **35**, 559.
Stern, H. and Mirsky, A. E. (1952). *J. gen. Physiol.* **36**, 181.
Stern, H. and Mirsky, A. E. (1953). *J. gen. Physiol.* **37**, 177.
Stern, H. and Timonen, S. (1954). *J. gen. Physiol.* **38**, 41.
Struchkov. V. A. (1964) *In* "Physikalische Chemie Biogen. Macromolecüle" p. 401. Akad. Verlag, Berlin.
Strelzoff, E. (1963). *Fed. Proc.* **22**, 462.
Sueoka, N. (1960). *Proc. natn. Acad. Sci. U.S.A.* **46**, 83.
Sutherland, E. W., Rall, T. W. and Menon, T. (1962). *J. biol. Chem.*, **237**, 1220.
Swift, H. (1960). *Brookhaven Symp. Biol.* **12**, 134.
Swift, H. (1963). *Expl Cell Res.* Suppl. **9**, 54.
Tamaoki, T. and Mueller, G. C. (1965). *Biochim. biophys. Acta* **108**, 73, 81.
Taylor, J. H., Woods, P. S. and Hughes, W. L. (1957). *Proc. natl. Acad. Sci. U.S.A.* **73**, 122.
Tocchini-Valentini, G. P., Stodolsky, M., Aurisicchio, A., Sarnat, M., Graziosi, F., Weiss, S. B. and Geidushek, E. P. (1963). *Proc. natn. Acad. Sci. U.S.A.* **50**, 935.
Vesell, E. S. and Bearn, A. G. (1962). *Proc. Soc. exp. biol. Med.* **111**, 100.
Vincent, W. S. (1952). *Proc. natn. Acad. Sci. U.S.A.* **38**, 139.
Vincent, W. S. and Baltus, E. (1960). *In* "The Cell Nucleus", p. 18. Butterworth & Co. London.
Wang, T.-Y. (1961). *Biochim. biophys. Acta.* **51**, 180.
Wang, T. Y. (1963). *Expl Cell Res. Suppl.* **9**, 213.
Watson, M. L. (1955). *J. biophys. biochem. Cytol.* **1**, 257.
Watson, M. L. (1959). *J. biophys. biochem. Cytol.* **6**, 147.
Watson, J. D. and Crick, F. H. C. (1953). *Nature, Lond.* **171**, 964.
Webster, G. C. (1960). *Biochem. biophys. Res. Commun.* **2**, 56.
Weiss, S. (1960). *Proc. natn. Acad. Sci. U.S.A.* **46**, 1020,

Weiss, S. and Gladstone, L. (1959). *J. Am. chem. Soc.* **81**, 4118.
Wilbur, K. M. and Anderson, N. G. (1961). *Expl Cell Res.* **2**, 47.
Woods, P. S. (1959). *Brookhaven Symp. Biol.* **12**, 153.
Yankofsky, S. A. and Spiegelman, S. (1962). *Proc. natn. Acad. Sci. U.S.A.* **48**, 1069, 1466.
Yasuzumi, G. (1955). *Biochim. biophys. Acta.* **16**, 322.
Zalokar, M. (1959). *Nature, Lond.* **183**, 1330.
Zalokar, M. (1960). *Expl Cell Res.* **19**, 184.
Zbarsky, I. B. and Debov, S. S. (1951). *Biokimiya* **16**, 390.
Zbarsky, I. B. and Gauze, G. G. (1964). *Vopr. Med. Khimii* **10**, 446.
Zbarsky, I. B. and Perevoshchikova, K. A. (1956). *Biokhimiya* **107**, 285.
Zbarsky, I. B. and Samarina, O. P. (1962). *Biokhimiya* **27**, 557.
Zivago, P. I. (1948). *Dokl. Akad. Nauk SSSR* **59**, 953; **60**, 445.
Zubay, G. and Doty, P. (1959). *J. mol. Biol.* **1**, 1.

Chapter 3

THE MITOCHONDRION

D. B. ROODYN

Department of Biochemistry,
University College, London, England

I. INTRODUCTION

The intention of this article is first to conduct a broad survey of the major metabolic activities of the mitochondrion, and then to consider the relationship between mitochondrial structure and function. It is on this question that the great sciences of biochemistry and cytology have come closest together, and the problem of the elucidation of the organization of mitochondrial enzyme systems has become identical to that of the elucidation of the structure of mitochondrial membranes.

An aspect that is not considered in detail here is the behaviour of the mitochondrion in the living cell. Unfortunately, the mitochondrion has been considered as a static structure for too long. In spite of many clear indications that mitochondria are dynamic, changing structures, capable in certain cases of growth and differentiation, the major interest of the great schools carrying out research into mitochondria has until very recently been almost exclusively into the elucidation of the detailed mechanisms of mitochondrial oxidations, energy transfer reactions, ion transport and related phenomena. For example it is most surprising that even now we have only fragmentary evidence as to the half-life of mitochondria in the cell and no idea of the cellular site of synthesis of such important enzymes as cytochrome oxidase (1.9.3.1.) or cytochrome *c*. The recent discovery of DNA in mitochondria, however, has awakened interest in the challenging problems of mitochondrial growth and replication and there is little doubt that there will be greater emphasis on these aspects of the problem in the near future. The question of the mechanism of biogenesis of the mitochondrion is considered in detail elsewhere (Roodyn and Wilkie (1966a).

A short addendum on peroxisomes (microbodies) has been included at the end of the chapter. The results with these particles are still very new and no doubt our views of their function may well change in the near future; however, the fascinating implications of their role in extra-mitochondrial oxidations fully merit inclusion of a section on them in an article on mitochondria.

Because of the sheer number of publications on the subject, it is now no longer possible to give a comprehensive and detailed account of the enzymology of the mitochondrion in the space of one chapter of a book. As a result, in this article many important aspects have been dealt with in a rather cursory manner and have often been illustrated merely by reference to one or two typical experiments. The detailed aspects of such sections are best followed in the many excellent books, monographs and review articles cited in the text. It is hoped that the reader will therefore regard this article as a rather subjective essay, or excursion, into a vast rapidly expanding field of contemporary biology, and will excuse the many factual omissions that are inevitable in such an approach.

II. Survey of the Morphology and Metabolic Activity of Mitochondria

This section is intended to give a general picture of the major aspects of mitochondrial metabolism. To provide a basis for the account, however, a very brief summary of the essentials of mitochondrial structure is given at the beginning. The mitochondrion is described in a most erudite and comprehensive fashion in the excellent book by Lehninger (1965). Recent developments in research into the mitochondrion are in the account on the Symposium on "The Regulation of Metabolic Processes in Mitochondria" held recently in Bari, Italy (see Tager, Papa, Quagliariello, and Slater, 1966) and recent accounts of oxidation and phosphorylation are given by Ernster and Lee (1964) and Slater (1966). A more general account of the mitochondrion, with particular emphasis on cytological aspects, is given Novikoff (1961) in a rather extensive article. In order to obtain a comprehensive list of mitochondrial enzymes, with specification of the tissues studied, the reader is recommended to consult the general survey of the distribution of enzymes in subcellular fractions of animal tissues by de Duve, Wattiaux and Baudhuin (1962). A quantitative survey of the sub-cellular distribution of enzymes in mammalian liver is given in Roodyn (1965a) and may be of use on obtaining precise information for the distribution of mitochondrial enzymes in this tissue. (Enzymes are listed according to their classification numbers of the Commission on Enzyme Nomenclature in both these articles, which usefuly complement each other.) There are many other reviews and monographs on mitochondria, respiratory enzymes and oxidative phosphorylation (see for example Ernster and Lindberg, 1958; Green and Fleischer, 1960; Chance, 1963; Sanadi, 1965) There is certainly no shortage of reading matter in this subject.

A. ESSENTIALS OF MITOCHONDRIAL MORPHOLOGY

The essential structural features of the mitochondrion are the double membrane, and the characteristic cristae. The reader is referred to Palade (1953) for what may now be regarded as a "classical" analysis of the main features of mitochondrial structure. As first stressed by Palade (1953) the number of cristae and their conformation varies considerably from one type of mitochondrion to another. However, the elements that appear to be common to all mitochondria are an outer membrane, an inner membrane, cristae, mitochondrial "sap" or "matrix", dense granules and DNA. The mitochondrial "sap" or "matrix" is generally regarded as being the space between the cristae and the inner membrane, and is sometimes called the "inner compartment". The space between the inner and outer membranes is best called the "outer compartment". The cristae are double-membrane structures, and the space between these membranes is in connection with the outer compartment. During mitochondrial swelling, this space and the outer compartment are greatly expanded. The nomenclature just used is based on that of Whittaker (1966) and a diagram showing these relationships is given in Fig. 1.

The variations in the morphology of cristae is discussed in detail in Lehninger (1965). They may have the appearance of simple stacked membranes, as

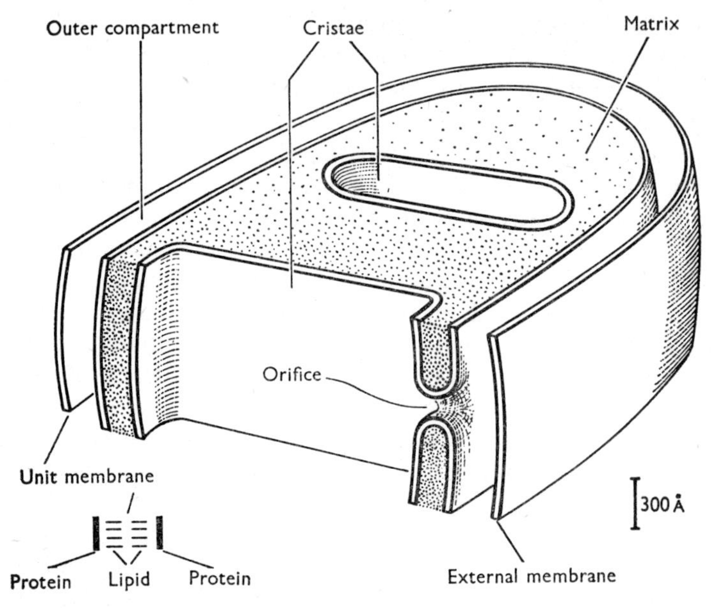

FIG. 1. Diagram illustrating the main structural features of the mitochondrion (from Whittaker, 1966).

in liver and kidney mitochondria, or may be tube-like, branched or have a variety of other shapes. A particularly clear electron micrograph showing stacked cristae in isolated heart muscle mitochondria is given in Fig. 2. The inner layer of the mitochondrial membrane is generally assumed to be in continuity with the membranes of the cristae, which may be regarded as invaginations of this inner membrane. However, these invaginations may only occur at certain regions of the inner membrane. The connection between the inner and the cristal membrane is particularly well shown in the section of a chicken oocyte mitochondrion given in Fig. 3. The sectioning technique, of course, will only give a true picture of the shape of the mitochondrion if it is lying in one plane, and is sectioned through its long axis. Whittaker (1966) has made elegant use of the technique of negative staining by applying it to fixed, intact mitochondria. He was able to demonstrate very clearly the elongated form of the liver mitochondrion (Fig. 4) It is interesting that this form is retained in mitochondria isolated in 0·44 M sucrose, but not in 0·25 M sucrose.

DNA may now be considered to be a universal constituent of mitochondria as a result of the extensive studies of Nass, Nass and Afzelius (1965) who showed the occurrence of DNA fibrils in mitochondria from a very diverse range of biological materials (The matter is discussed more fully in Section III. B. below). The dense granules are most probably insoluble aggregates

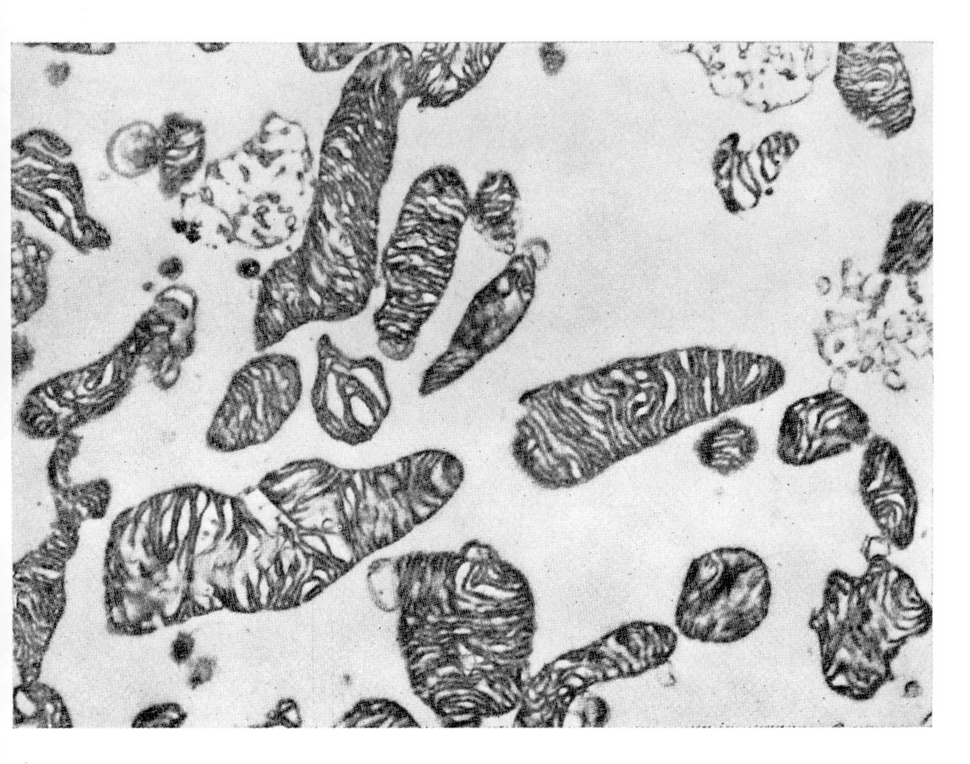

FIG. 2. Stacks of cristae in rabbit-heart muscle mitochondria isolated in 0·88 M sucrose-0·01 M EDTA. ×25,000 (from Deshpande, Hickman and Von Korff, 1961).

consisting mainly of calcium and magnesium phosphates, and this is discussed fully in Section III. C. below.

B. PROBLEMS IN SURVEYING METABOLIC ACTIVITY OF MITOCHONDRIA

There is great difficulty in giving an accurate, but reasonably concise account of the metabolic activity of the mitochondrion for several reasons. A rigorous account would have to consider the many differences in biochemical constitution of mitochondria from different sources, since there is little doubt that the range of morphological variation mentioned above is reflected by a similar range of metabolic variation. Although a great deal of work has been done with rat-liver mitochondria, it is certainly unsafe to assume that all these findings are of general significance. Indeed, even liver mitochondria isolated from different mammals are by no means identical. For example there are differences between rat-liver and rabbit-liver mitochondria. Certainly there are major differences between the enzymic constitution of, for

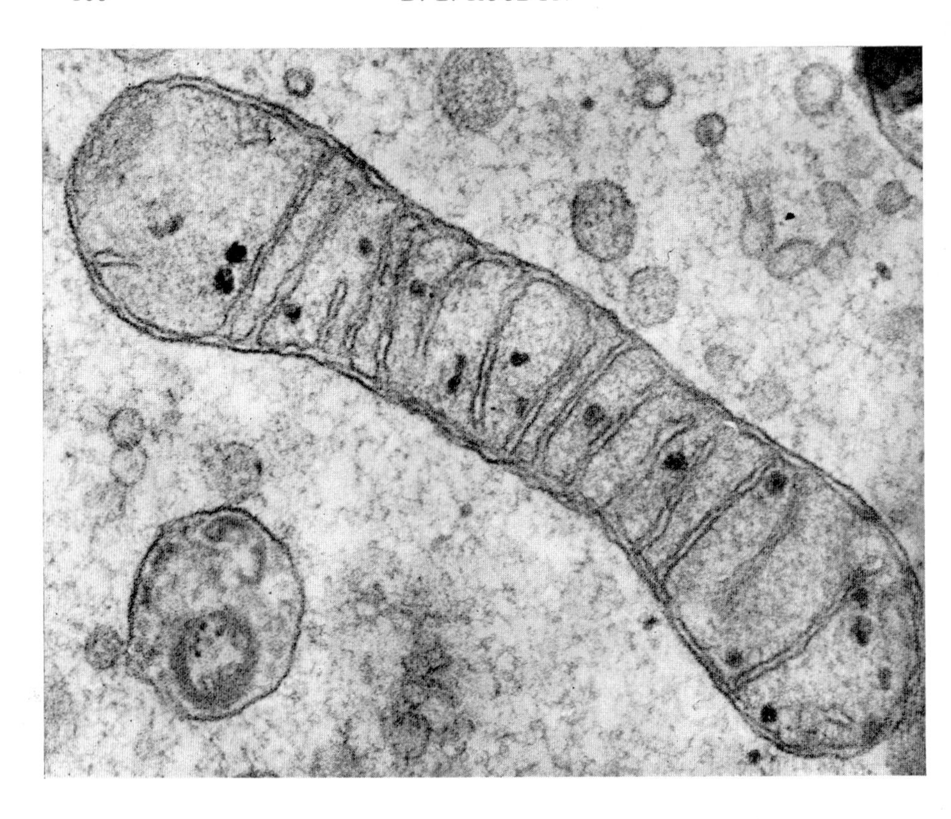

FIG. 3. Mitochondria taken from an oocyte in the ovary of a domestic fowl (*Gallus bankiva*). Fixed in OsO_4, stained in lead citrate. Note invaginations of inner membrane to form cristae, and also frequent dense granules. ×72,000. (This electron micrograph was kindly supplied by Dr. R. Bellairs.)

example, yeast, insect flight muscle and rat-liver mitochondria. The fact that the general features of the respiratory and phosphorylative systems appear to be similar in mitochondria of diverse origin has tended to obscure the many differences that exist.

A second difficulty is essentially due to technical reasons. Because of the ease of disruption of liver, and the abundance of mitochondria in the liver cell, it is relatively easy to obtain liver mitochondria in reasonable yield, without excessive damage or non-mitochondrial contamination. (The problem of contamination of liver mitochondria with other sub-cellular components should not be under-estimated, however. Neither should the possible presence

FIG. 4. Elongated rat-liver mitochondrion isolated in 0·44 M sucrose and negatively stained in sodium phosphotungstate after fixation. The electron micrograph is a composite one of a number of high magnification micrographs of different portions of the mitochondrion (from Whittaker, 1966).

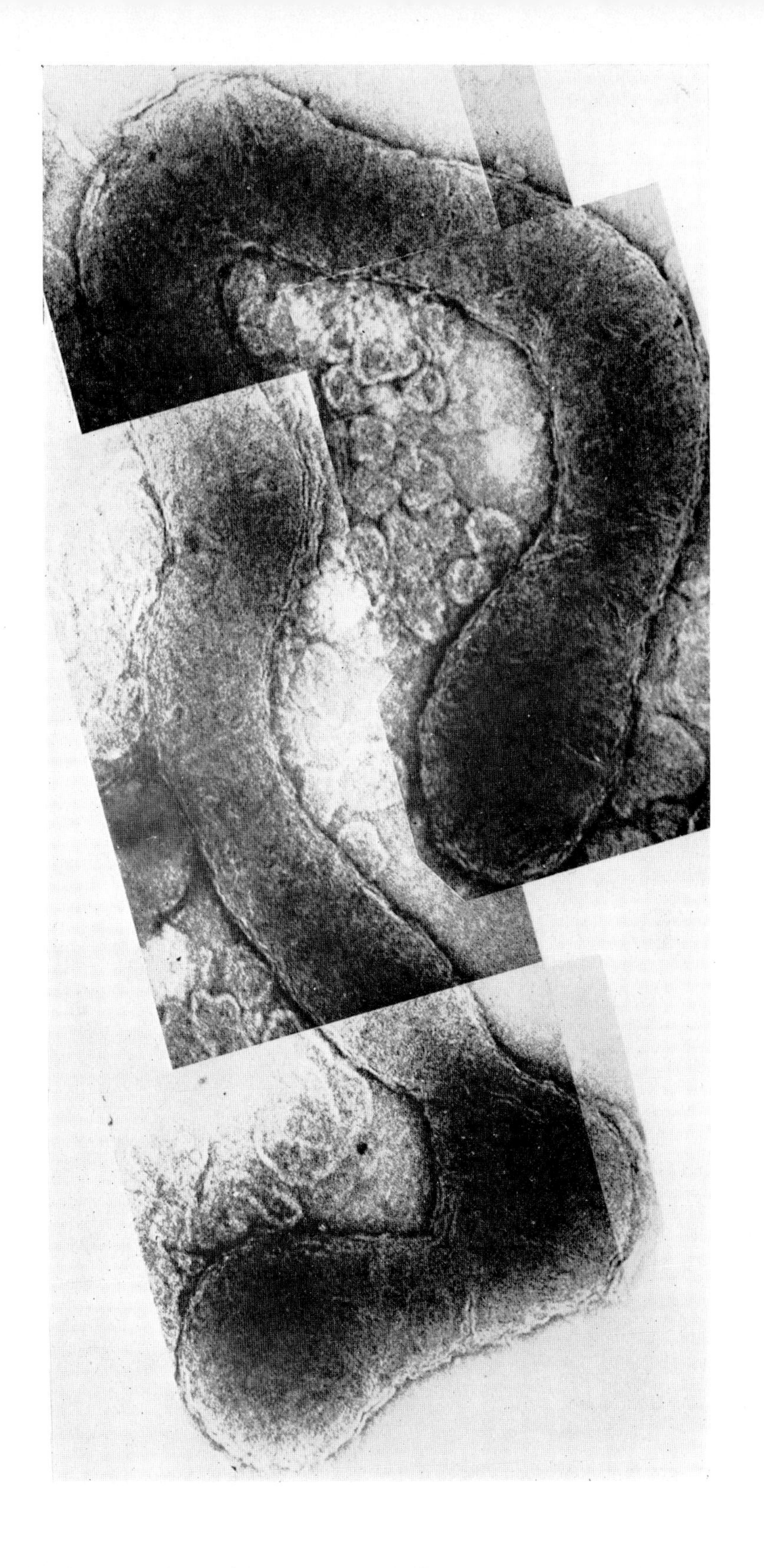

of bacteria be ignored: Roodyn, Reis and Work 1961). With other tissues, such as brain or ascites tumours, it is far more difficult to obtain reasonable preparations of mitochondria. Some early preparations of so-called "brain mitochondria" were in fact grossly contaminated with other structures, and special procedures are needed to obtain even a moderately pure preparation. The techniques used for the disruption of tumour cells, such as osmotic shock, may very easily result in osmotic shock to the mitochondria as well. Also, liver is a relatively homogeneous tissue as regards cell-type, since it consists mainly, but by no means entirely, of columnar cells. Preparations of mitochondria, for example, from chick embryos, or plant tissues are clearly derived from mixed populations of cells and interpretation of metabolic studies in such cases is difficult.

The third difficulty is more fundamental, and has been discussed extensively elsewhere (Roodyn, 1965a). It is that there is considerable variation in the aims and attitudes of the various investigators of mitochondrial metabolism. The early pioneer work of Claude, Schneider, Hogeboom and their associates was aimed at establishing the mitochondrial localization of certain enzymes. However, much of the subsequent work on mitochondria has lost contact with the aim of establishing intra-cellular localizations of the enzymes studied, but has concentrated on isolated mitochondria as if they were a tissue *per se*. This work is generally carried out without any concomitant examination of other sub-cellular fractions, or even of the total homogenate from which the mitochondria were derived. As a result, we have a truly massive literature describing the properties of isolated mitochondria, but a far smaller literature concerned with establishing whether the enzymes studied are in fact truly localized in the mitochondrion. An excellent example of the dangers of such an approach is given by studies on catalase (1.11.1.6.). There are a large number of papers published between 1950 and 1960 concerned with the catalase activity of mitochondria. However, it is now clear that this enzyme is in fact present in peroxisomes (microbodies) and not in mitochondria (see Section IV.)

In order to avoid such difficulties in the future, the author has suggested the adoption of five criteria that should be met before it can be said with certainty that an enzyme is truly associated with a sub-cellular component isolated from a tissue homogenate (Roodyn, 1965a). These are as follows:

1. The assay should be such that a single enzyme is measured. If a multi-enzyme system is studied, the components of the system should be known.
2. The enzymic activity should be compared in some way with the activity of the homogenate.
3. During the fractionation of the homogenate, no serious activation or inactivation of the enzyme should have occurred. Thus only those fractionations in which a reasonable recovery of enzyme is obtained are valid.

4. It should be shown that the enzymic activity of the isolated cell component is not due to contamination with some other component.

5. A clear association between the sub-cellular component in question and the enzyme should be demonstrated in a variety of preparations of varying purity.

It may be noted that these criteria are insufficient to establish that the observed activity of the isolated cell component corresponds to the activity of that component in the living cell, but only refer to the structure present in the homogenate. (See Chapter 1 for a general discussion of these issues.)

It is a disturbing fact that the greater part of the enzyme studies on isolated mitochondria do not satisfy one or other of these criteria. In some cases, none of the criteria are met at all. As a result, whilst we can say in many cases that such and such an enzyme is probably in the mitochondria, it is far more difficult to establish this without any reasonable doubt. Some may argue that there is no certainty in biology. Nevertheless there is sufficient information on the sub-cellular distribution of cytochrome oxidase to satisfy all the above criteria and so to establish without doubt that this is a truly mitochondrial enzyme. If similar information were available for other enzymes, our knowledge of the enzymology of the mitochondrion would be on a firmer footing than at present.

As a result of these considerations, it is certain that a rigourous account of the properties of the mitochondrion would require a careful examination, for each tissue used, of the evidence establishing which enzymes are truly mitochondrial. Space does not allow such an account, which would probably run to many hundreds of pages. The alternative is therefore to present a general picture of the major metabolic activities that appear to be common to mitochondria.

C. TRICARBOXYLIC ACID CYCLE

The details of the tricarboxylic acid cycle are given in Fig. 5 and the reader is referred to Krebs and Lowenstein (1960) for an account of the essential features of the cycle. Demonstration of the presence of a functional citric acid cycle in isolated mitochondria was one of the early achievements in the field. The basis of these findings may be found in Schneider and Hogeboom (1956). If was found that isolated mitochondria could completely oxidize pyruvate and various citric acid cycle intermediates to carbon dioxide and water when incubated in relatively simple media (for example in the presence of buffer, Mg^{2+} ions, sucrose and inorganic phosphate). Thus the entire complex of enzymes and co-factors necessary for the cycle is present in the mitochondrion.

The striking nature of these findings has tended to distract attention from the important quantitative question as to whether the enzymes of the citric

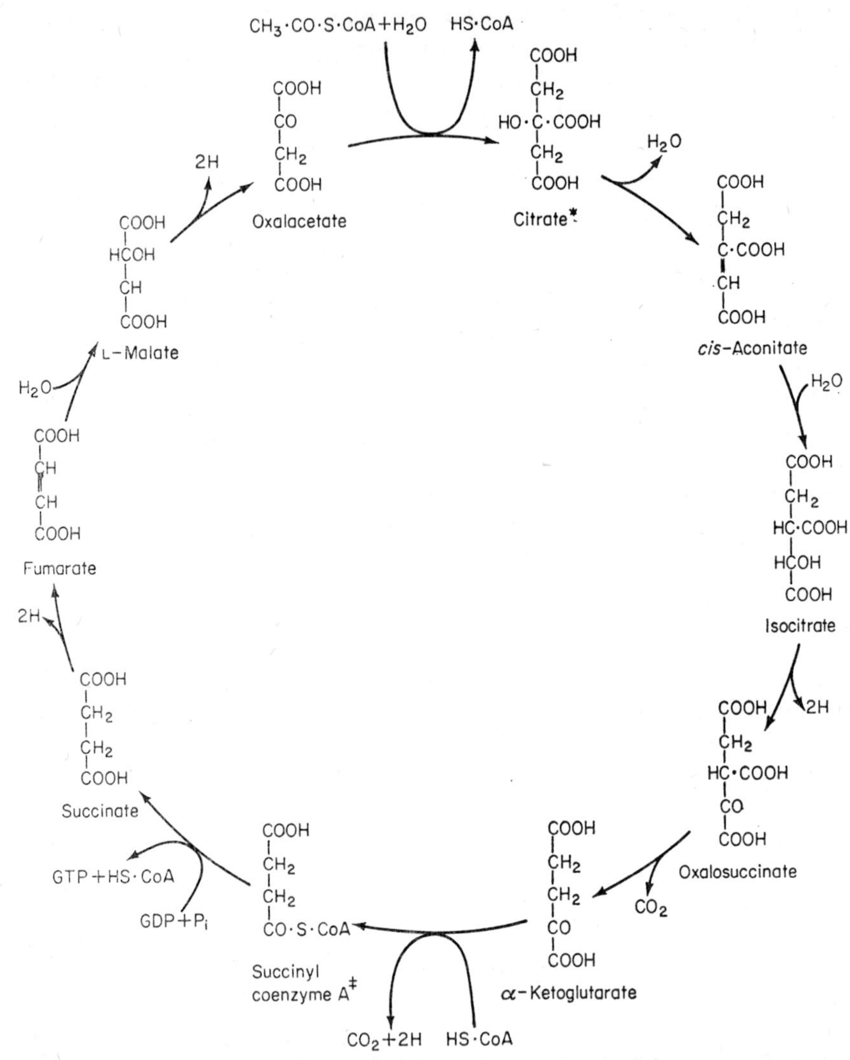

FIG. 5. The tricarboxylic acid cycle:* Citrate can yield isocitrate without *cis*-aconitate as a free intermediate;† isocitrate can yield α-ketoglutarate without oxalosuccinate as a free intermediate;‡ Succinyl-CoA can also yield succinate by the action of succinyl-CoA deacylase (3.1.2.3), and succinyl CoA transferases (from Krebs and Lowenstein, 1960: see the original text for full references to the above comments).

acid cycle are exclusively localized in the mitochondrion. It is interesting that the "dogma" of the unique mitochondrial localization of these enzymes was questioned as early as 1956 by Schneider and Hogeboom in a review at that time (Schneider and Hogeboom, 1956). Although there is little evidence at the moment for the existence of a functional tricarboxylic acid cycle in a cell component other than the mitochondrion, many of the enzymes of the cycle show a dual localization, appearing both in the soluble fraction and the mitochondrial fraction. The relative partition of enzyme observed between these two fractions may depend to a certain extent on the degree of damage inflicted on the mitochondria during isolation, if the enzyme is present in the mitochondrial sap. However, it is certain that not all the activity in the soluble fraction can be ascribed to leakage from the mitochondrion. The surest example of this is that of malate dehydrogenase (1.1.1.37.) where the "soluble" and "mitochondrial" enzymes have been shown to differ in many respects and are undoubtedly different proteins. For example, soluble and mitochondrial malate dehydrogenases from beef-heart have very different amino acid compositions (Siegel and Englard, 1962). The dual location of liver malate dehydrogenase is discussed in Roodyn, Suttie and Work (1962). Examples of other enzymes of the cycle that show such a dual location are aconitase (4.2.1.3.), iso-citrate dehydrogenase (1.1.1.41), fumarase (4.2.1.2.) and possibly citrate synthase (4.1.3.7.) although it has not been established in every case that the mitochondrial and soluble enzymes are different proteins.

The existence of "mitochondrial" and "soluble" enzymes that catalyse the same reaction but are different proteins is by no means confined to the citric acid cycle. (See Section II. I. and Chapter 8). The widespread nature of this duality has recently been stressed by Greville (1966) in a discussion of the factors affecting the utilization of substrates in the cell sap by mitochondria. Dare one speculate that the reason for this duality lies in the evolutionary origin of mitochondria, i.e. that they were originally symbiotic parasites within a host cell, and that the soluble enzymes are those of the original host? Arguments in favour of the "bacterial" theory of mitochondrial origin have become more credible recently and the matter is discussed critically in Roodyn and Wilkie (1966a).

The reason for our confidence in assigning the functional tricarboxylic acid cycle to the mitochondrion is that certain key enzymes of the cycle, namely the α-oxoglutarate dehydrogenase system, and succinate dehydronase (1.3.99.1.) appear to be uniquely associated with mitochondria. The enzymes of α-oxoglutarate oxidation (and the closely related enzymes of pyruvate oxidation) are arranged in organized enzyme assemblies of considerable particle weight. The complex processes involved in these oxidations require coenzyme A, lipoic acid, thiamine pyrophosphate and NAD amongst the co-factors. The detailed enzymology and organization of these systems is described fully by Sanadi (1963) and Massey (1963), and is also discussed in Sections II. F. and III. E. below. (See Figs. 13 and 34, below). Succinate dehydrogenase is the marker enzyme par excellence for mitochondria, and has been demonstrated in mitochondria from a very wide range of tissues. Since the venerated Keilin-Hartree muscle preparation is capable of oxidizing succinate, it is probably safe to say that succinate dehydrogenase was the first enzyme of the citric acid cycle to be assigned to a respiratory particle.

An interesting exception to the unique mitochondrial location of succinate dehydrogenase is the finding of Rossi, Hauber and Singer (1964) that there are at least three enzymes in yeast that catalyse the conversion of succinate to fumarate. One is the "classical" mitochondrial succinate dehydrogenase and is absent from yeast grown under anaerobic conditions and from respirtory deficient ("petite") mutants of yeast (see also Roodyn and Wilkie, 1966b). The other two are non-mitochondrial enzymes and are still present in anaerobic and respiratory-deficient yeast. They probably function more as fumarate-reductases than succinate dehydrogenases, and are possibly concerned in supplying succinate for synthetic reactions in the absence of a functional citric acid cycle. These findings illustrate very clearly the dangers pointed out above of arguing exclusively from the results with mammalian liver

In conclusion, it may be said that mitochondria contain all the enzymes of the tricarboxylic acid cycle, but that some of these are not tightly bound to the mitochondrial structure and also exhibit a dual distribution between the mitochondrial and soluble fractions.

D. OXIDATION OF FATTY ACIDS

Another "classical" property of the mitochondria is the ability to oxidize fatty acids, again when incubated in relatively simple reaction media. The details of the β-oxidation "spiral" of fatty acid oxidation are given in Fig. 6, and the reader is referred to Lynen (1955) for an account of the enzymes involved in this process. Since the terminal product of the spiral, acetyl-Co A, feeds directly into the tricarboxylic acid cycle by means of the enzyme

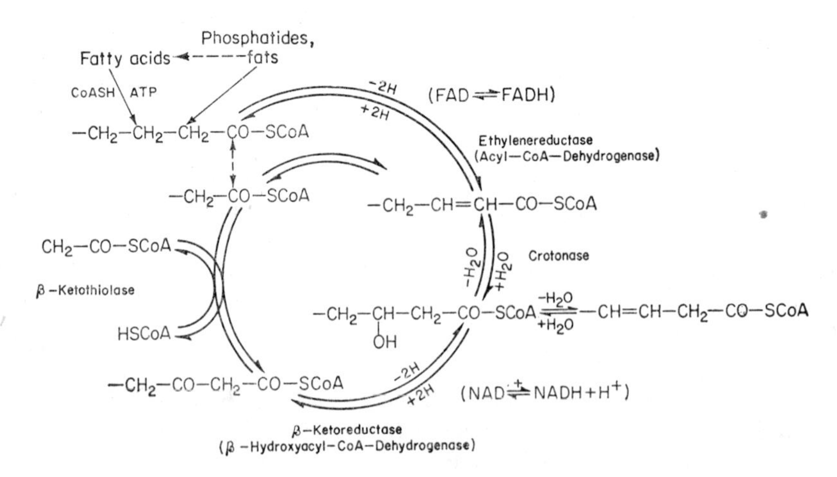

FIG. 6. The β-oxidation "spiral" for fatty acid oxidation (from Lynen, 1955).

citrate synthase, mitochondria can completely degrade the added fatty acid to carbon dioxide and water. They also provide a magnificent experimental tool for examination of the mechanism of interaction of carbohydrate and fat metabolism. The work of Lehninger and his colleagues, in particular, revealed in the 1950's the complex metabolic interactions that can still occur in isolated mitochondria.

Recently, there has been considerable interest in the mechanism of transfer of fatty acids and two-carbon fragments across the mitochondrial membrane. Interest has focused on the growth factor carnitine which can form esters with fatty acids. For example, Bremer (1963) suggested that palmityl-CoA formed from the result of fatty acid activation by the reaction:

$$\text{Palmitate} + \text{CoA} + \text{ATP} \xrightarrow{Mg^{2+}} \text{Palmityl-CoA} + \text{AMP} + \text{P-P}_i$$

can then react with carnitine to form palmityl-carnitine:

$$\text{Palmityl-CoA} + \text{carnitine} \rightarrow \text{palmityl-carnitine} + \text{CoA}$$

A palmityl-CoA: carnitine palmityl transferase (2.3.1.–.) was isolated from calf-liver mitochondria by Norum (1964) Acyl carnitine esters and palmityl carnitine transferase are apparently concerned in the transfer of fatty acyl derivatives across mitochondrial membranes. A scheme showing the possible intra-mitochondrial location of some of these reactions was suggested by Fritz and Marquis (1965) and is given in Fig. 7. This is the first example in this article in which we have had to consider the problem of the mechanism

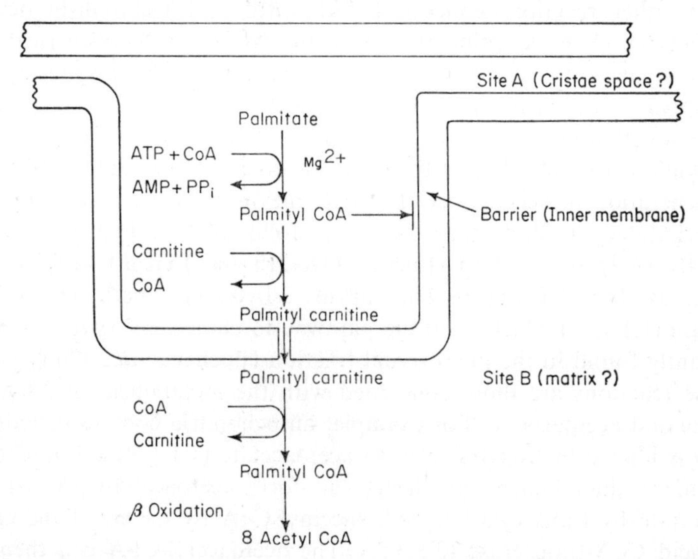

FIG. 7. Possible intra-mitochondrial location of reactions of carnitine metabolism. The "barrier" is permeable to palmitylcarnitine, but not to palmityl-CoA (adapted from Fritz and Marquis, 1965).

of transport of metabolites across the mitochondrial membrane. In connection with such problems, Chappell and Crofts (1965b) have examined the effect of atractylate, which interferes with the transport of adenine nucleotides into the mitochondrion (see Section II. I. below). The scheme suggested by these authors for the role of carnitine in transport across the membranes is similar to that of Fritz and Marquis, but in addition gives the site of inhibition of atractylate (Fig. 8).

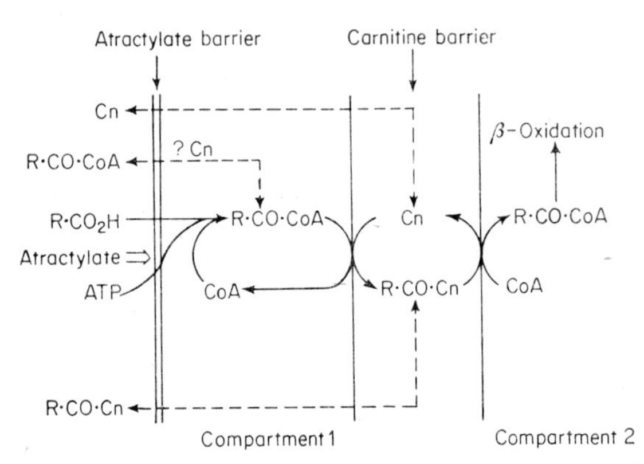

FIG. 8. Site of action of atractylate in fatty acid oxidation. Cn: carnitine, ⇒ site of inhibition by atractylate (from Chappell and Crofts, 1965b).

Many other reactions concerned with fatty acid metabolism occur in mitochondria. One interesting example is the cyclic process concerned in the conversion of acetoacetyl-CoA to free acetoacetate. The acetoacetyl-CoA is not apparently hydrolysed directly to the free acid under normal conditions but reacts with a molecule of acetyl-CoA to form the compound 3-hydroxy-3-methylglutaryl-CoA. This is then broken down to form acetoacetate, with the regeneration of acetyl-CoA. The two enzymes involved are 3-hydroxy-3-methyl-glutaryl-CoA synthase (4.1.3.5.), followed by 3-hydroxy-3-methyl-glutaryl-CoA lyase (4.1.3.4.) (Bucher, Overath and Lynen 1960). The cycle operates as shown in Fig. 9. The enzyme 3-hydroxy-3-methylglutaryl-CoA reductase (1.1.1.34) which is on the pathway to cholesterol synthesis, is predominantly found in the microsomal fraction (Bucher *et al.*, 1960). Related to these reactions are those concerned with the metabolism of 3-hydroxy-butyrate and acetoacetate. For example, mitochondria contain dehydrogenases to oxidize 3-hydroxybutyrate to acetoacetate (1.1.1.30.). Mitochondria from mammalian kidney and heart can form acetoacetyl-CoA from this acetoacetate by transacylation with succinyl-CoA by means of the enzyme 3-ketoacid CoA-transferase (2.8.3.5.). The acetoacetyl-CoA can then react with free CoA in the terminal "thioclastic" reaction of the β-oxidation spiral and thus be oxidized to carbon dioxide and water. Liver mitochondria, however, cannot form acetoacetyl-CoA from free acetoacetate, and hence oxidize 3-hydroxybutyrate to acetoacetate and no further. This fact was of considerable use in determining the yield of ATP from the oxidation of NADH in liver mitochondria, since 3-hydroxybutyrate could act as an

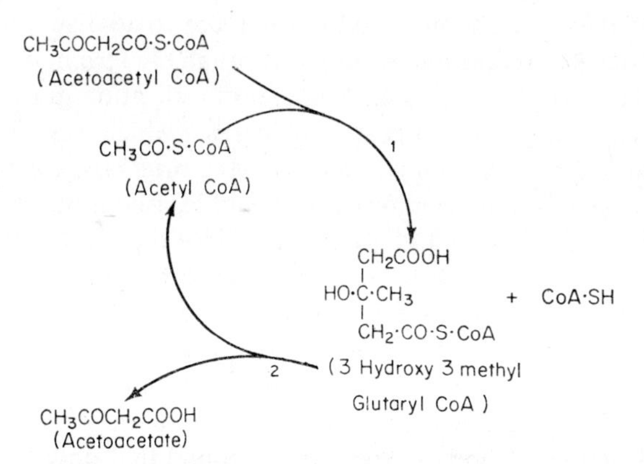

FIG. 9. The hydroxy-methylglutaryl cycle for the conversion of acetoacetyl-CoA to acetoacetate in mitochondria (adapted from Bucher *et al.*, 1960).

excellent source of NADH, uncomplicated by further oxidations. (The difference between liver and heart mitochondria is an important example of differences between mitochondria of different origin).

E. RESPIRATORY CHAIN

The respiratory activity of the mitochondrion is generally regarded as its central property. However, again we must be wary against assuming that all cellular oxidations occur in the mitochondrion. There are several more or less defined extra-mitochondrial oxidative systems. For example, it appears from recent evidence (given in Section IV.) that the peroxisome is an important example of such a system. The reader is referred to the recent review of de Duve and Baudhuin (1966) for an account of the role of extra-mitochondrial oxidation pathways.

Fortunately, however, the terminal enzyme of the respiratory chain in mitochondria, namely cytochrome oxidase, is established without doubt as being exclusively localized in the mitochondria. The pioneer experiments of Schneider and Hogeboom rigorously established that the enzyme was associaated with mammalian liver mitochondria. (See for example Schneider, 1946, and Schneider and Hogeboom, 1950). In addition it was shown by quantitative fractionation experiments that the cytochrome oxidase activity in the nuclear fraction was entirely due to mitochondrial contamination (Hogeboom, Schneider and Striebich, 1952). Studies of this enzyme in a wide range of tissues have consistently shown that it is exclusively concentrated in the mitochondrial fraction. In addition, the application of more sophisticated fractionation methods, in particular by the Louvain School, has firmly established cytochrome oxidase as a primary enzyme marker for mitochondria. (Beaufay, Bendall, Baudhuin, Wattaiux and de Duve, 1959).

Because of this fact, it must follow that any oxidative process that is measured by oxygen uptake and proceeds through the terminal enzyme of the respiratory chain will have the same apparent localization in the cell as cytochrome oxidase. Thus many so-called "oxidases", which were in fact systems of enzymes linked to the terminal oxidase, were originally found to have an exclusively mitochondrial location. For example, isocitrate oxidase activity was detected by Hogeboom and Schneider (1950) only in the mitochondrial fraction. When they measured the isocitrate dehydrogenase directly, however, they found activity both in the mitochondrial and soluble fractions. The fact that the sub-cellular distribution of the component enzymes of a multi-enzyme system need not be identical to that of the terminal enzyme of the system is often not fully appreciated. In a recent survey of enzyme distributions in sub-cellular fractions (Roodyn 1965a) it was found that only a small proportion of the studies on multi-enzyme systems satisfied the first criterion mentioned in Section II. B. above. This is that it is only possible to obtain quantitative distribution patterns if single enzymes, or multi-enzyme systems of known constitution, are studied.

It is therefore necessary to exercise a certain amount of caution when interpreting the massive literature on mitochondrial oxidations. If, for example, glutamate oxidation is only observed in mitochondria, it cannot be inferred that the primary dehydrogenase, glutamate dehydrogenase (1.4.1.3,4) is necessarily a mitochondrial enzyme. A small amount of the total cellular enzyme may well have been sufficient to catalyse the first stage of the oxidation to carbon dioxide and water, which would be the actual reaction measured by the "oxidase" assay. Unfortunately, as stated above, a critical assessment of the literature on the sub-cellular distribution of oxidases would be far too long for this article. It must therefore be sufficient to stress that the oxidation of a large number of substances has been demonstrated with isolated mitochondria. These include many fatty acids of different chain length, ketoacids, dicarboxylic and tricarboxylic acids, several amino acids, glycerol-1-phosphate, various steroids, choline and various other N-methyl derivatives, amines and other compounds. These oxidations usually proceed by interaction of a primary dehydrogenase with the respiratory chain. However, some oxidases may react directly with oxygen. (For example, this may apply to mono-amine oxidase (1.4.3.4.) which has been demonstrated in mitochondria by Baudhuin *et al.*, 1964).

The respiratory chain is formulated by Lehninger (1965) as shown in Fig. 10. The various substrates oxidized by it either react primarily with

NAD or NADP linked dehydrogenases or with various flavoproteins (FP) that then interact with the chain of cytochromes and ultimately with oxygen. It is difficult at the present time to present a formulation of the respiratory chain that is completely free of controversy, but that given in Fig. 10 is similar

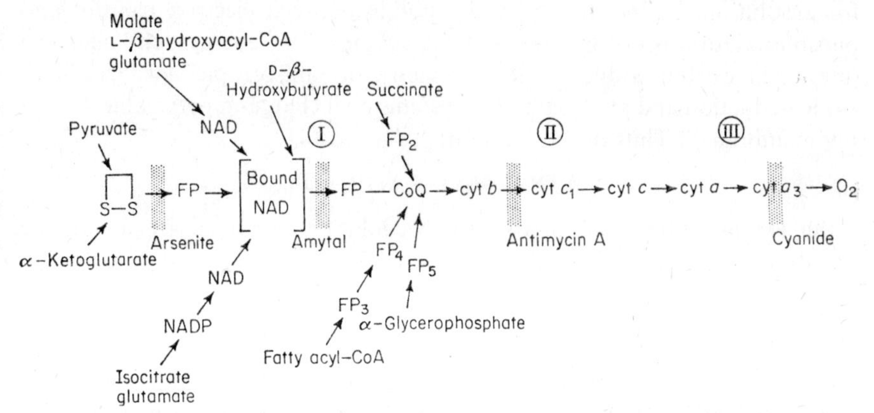

FIG. 10. Respiratory chain, sites of ATP synthesis and sites of action of important inhibitors as formulated in Lehninger (1965) (FP: flavoprotein).

to the recent formulation given by Sanadi (1965). Perhaps the most serious areas of contention are concerned with the roles of non-haem iron and of ubiquinone. The purification and characterization of many of the components of the respiratory chain has proved to be extremely difficult, because of their very tight binding to the mitochondrial membrane, As a result, there is still some uncertainty as to the relationship between the various enzyme preparations obtained by different workers. For example, Machinist and Singer (1965a, b) have suggested that the enzymes NADH-cytochrome *c* reductase (De Bernard, 1957) and NADH-ubiquinone reductase (Pharo and Sanadi, 1964) are really degraded fragments of the enzyme NADH-dehydrogenase (1.6.99.3.). In spite of these difficulties, the scheme presented in Fig. 10 probably gives a sufficiently accurate summary of the constitution of the respiratory chain as it is at present understood.

F. OXIDATIVE PHOSPHORYLATION

It has been established for many years that the oxidation of various substrates by isolated mitochondria is accompanied by the disappearance of inorganic phosphate from the medium, with the formation of ATP. The detailed mechanism of this phenomenon has engaged the interest of many active groups for the past 10–15 years. The reaction can be formulated in a deceptively simple fashion by the formula:

$$AH_2 + \tfrac{1}{2}O_2 + ADP + P_i \rightarrow A + H_2O + ATP$$

(where AH_2 is the substrate oxidized by the respiratory chain). The precise intermediates involved in this reaction have not been isolated, or even conclusively characterized. Perhaps the search has been misguided and one should look to theories similar to the chemi-osmotic theory of Mitchell (1961) for a solution. In this theory, the coupling between electron transfer and phosphorylation is not by means of the synthesis of "energy-rich" chemical intermediates, but is due to the asymmetric or "anisotropic" localization of oxido-reduction and phosphokinase systems on a charge-impermeable "coupling membrane". Thus in the reaction:

i $$ATP + H_2O \rightarrow ADP + P_i$$

f the H^+ and OH^- ions of the water could be separated by a suitable membrane, the reaction could be reversed, viz:

$$ADP + P_i \rightarrow ATP + H^+ + [OH^-$$

where the symbol [represents separation by a membrane. A model for such an anisotropic ATPase (3.6.1.3.) is given in Fig. 11. which shows how a OH^- ion participating in the reverse process (i.e. ATP synthesis) may only be

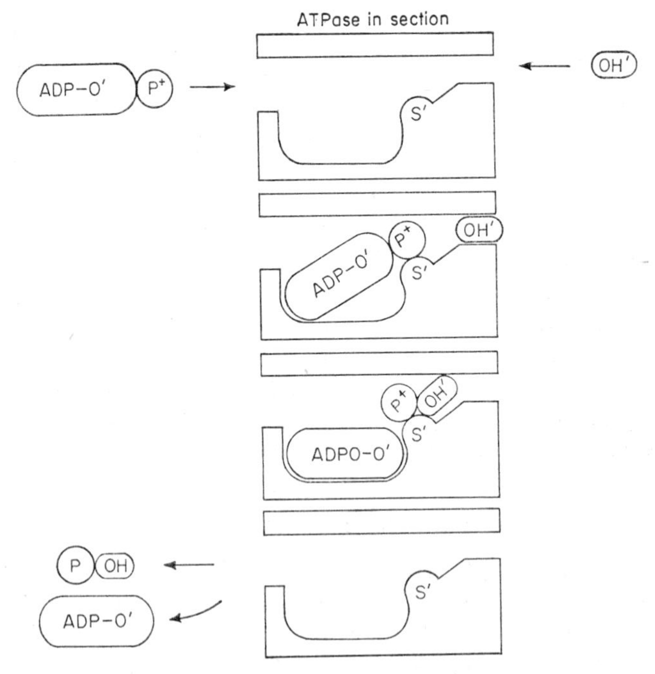

FIG. 11. A proposed model for an "anisotropic" ATPase. The channel to the active centre on the left allows passage of ATP (ADP-O′-P^+), phosphate (POH) and H^+ but not OH′. The right hand channel allows the passage of OH′ only. (The phosphorylium group P^+ is thought to form an intermediate with a negative group S′ in the enzyme) (from Mitchell, 1962).

generated on one side of the "enzyme". If such a system is present in a membrane and is coupled to an oxido-reduction system that is similarly anisotropic, except that it generates OH^- ions on the opposite side of the membrane, the oxidations may be used to drive the ATP synthesis, as shown in Fig. 12 (adapted from Mitchell, 1966a). In addition to the charge-impermeable

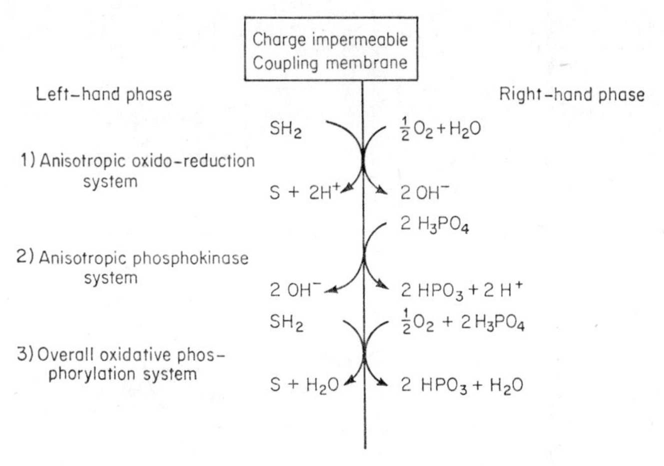

FIG. 12. Essentials of chemi-osmotic theory of oxidative phosphorylation. The over all P/O ratio in the reaction is 2·0 (adapted from Mitchell, 1966a).

coupling membrane, and the two anisotropic enzyme systems, Mitchell proposes the existence of an exchange-diffusion system. This is to maintain the internal ionic composition and pH of the mitochondrion against the production of H^+ ions on one side of the membrane, and OH^- ions on the other. The elegance of Mitchell's theory is that studies on oxidative phosphorylation, ion transport, water transport and membrane structure are all brought together as investigations of different facets of the same system, namely a functional membrane. (A full account of this theory is given in Mitchell, 1961, 1966a, b).

Although discussion of Mitchell's theory may well cause some heart-searching amongst those seeking to isolate defined chemical intermediates of oxidative phosphorylation, they may well be encouraged by the great dividends yielded by the "classical" approach to complex systems i.e. subfractionation, purification of enzymes and isolation of intermediates as applied to closely related problems. The detailed elucidation of the mechanism of ATP synthesis in the glycolysis system is usually taken as the most outstanding example of such success, However, it should not be forgotten that the mechanism of complex oxidation and energy transfer reactions occuring in the pyruvate oxidase and α-oxoglutarate oxidase complexes has been revealed in considerable detail (see Sanadi, 1963). In fact, the so-called "substrate-level phosphorylation" associated with the conversion of α-oxoglutarate to succinate proceeds by a series of reactions that are now relatively well defined, with known co-factors and enzymes being involved. The formation of acyl-CoA derivatives in these reactions is shown in Fig. 13. These

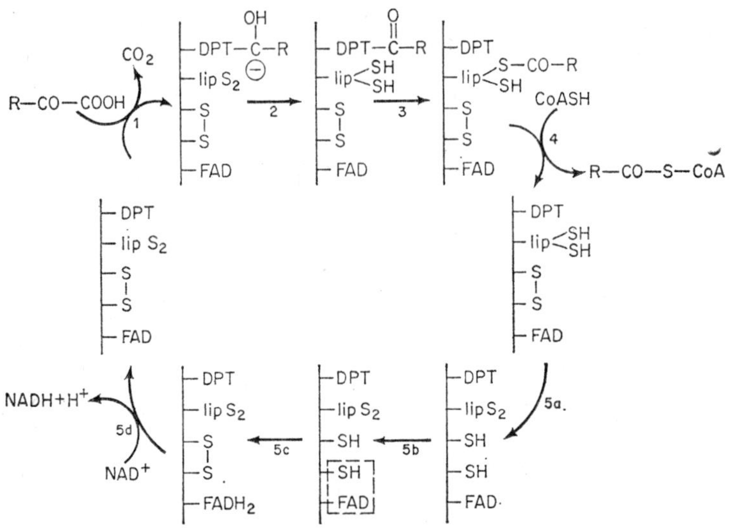

FIG. 13. Possible mechanism of oxidative decarboxylation of α-keto acids. DPT: diphosphothiamine.1)DPT-dependent decarboxylation,2),3) reductase-transacylase step, 4)transacylase,5 a-d) lipoamide dehydrogenase (from Sanadi, 1963).

derivatives are then transferred to ATP. Thus in the case of succinyl-CoA, energy transfer reactions occur in the sequence:

$$\text{Succinyl-CoA} \rightarrow \text{Phosphohistidine} \rightarrow \text{GTP} \rightarrow \text{ATP}$$

the latter reaction proceeding via nucleoside diphosphate kinase (2.7.4.6.). These reactions are also discussed in Section III. E. below (see Fig. 21).

The role of phosphohistidine, and phospho-proteins, is the subject of active investigation at the moment. Heldt (1966) showed that phosphohistidine was labelled with ^{32}P before GTP, ATP and ADP in substrate-level phosphorylation occurring during the oxidation of α-oxoglutarate. Sperti, Pinna, Lorini, Moret and Siliprandi (1964) had previously found evidence to suggest a role of phosphohistidine and phosphoproteins in substrate-level phosphorylation, although they could not exclude the possibility that some of the phosphoprotein was formed by phosphorylation of protein with ATP derived from respiratory chain phosphorylation. The role of phosphohistidine in respiratory chain phosphorylation has been the subject of some controversy, because of the relatively slow rate of turnover of this compound. However, Boyer (1965) has argued that this does not necessarily preclude the possibility that at least some of the phosphohistidine is involved in the major path of ATP synthesis. This is because of difficulties of interpretation of kinetic studies when complex pools exist. These conclusions were disputed by Slater (1965) however, Whatever the final outcome of these difficulties, there is now little doubt that phosphohistidine in particular, and phosphoproteins in general, are of considerable metabolic importance in mitochondria.

The example of phosphohistidine illustrates the difficulties that arise in the search for defined chemical intermediates of oxidative phosphorylation. However, if we accept the view that such intermediates do in fact exist, one may present a minimum hypothesis for the mechanism of oxidative phosphorylation. This is based on a large number of observations, best given in the many excellent reviews on the subject (see for example Chance and Williams, 1956; Racker, 1961; Lehninger, 1965; Slater, 1966). This highly specialized topic is still in a very dynamic state. Nevertheless, it seems that the essential basis of most "classical" theories of oxidative phosphorylation is as follows. During electron transfer along the respiratory chain, a suitable energy carrier (or carriers) reacts with the chain. There have been many symbols suggested for this carrier, and in this article we will use the most frequently quoted symbol I (see Chance and Williams, 1956). It is very likely that the formation of the complex between I and the respiratory carrier results in inhibition of electron transport. Whether I reacts with the carrier in its reduced or oxidized state is not so certain, although Chance and Williams favour the former possibility. There are probably three sites of interaction of I with the chain, and these are between NAD and flavoprotein, between cytochrome *b* and c_1 and between cytochrome *c* and oxygen. Using the symbol $\sim$ for any energy-rich intermediate, and assuming that there are different species of I at each site, and that these react with the reduced carriers, the first stage of oxidative phosphorylation is the formation of NADH $\sim I_1$, ferrocytochrome $b \sim I_2$ and ferrocytochrome $c \sim I_3$. It is believed that a second type of energy carrier reacts with these complexes. Again, the nomenclature is very confusing, but the symbol X is widely used (see Chance and Williams, 1956). It is not known if there are one, or three, species of X. The second reaction is then the formation of $X \sim I_1$, $X \sim I_2$ and $X \sim I_3$. These $X \sim I$ compounds (or non-phosphorylated energy rich intermediates) then react with inorganic phosphate to form $X \sim P$, a phosphorylated carrier. Finally, $X \sim P$ reacts with ADP to form ATP and liberate X. The essentials of such a scheme is given in Fig. 14 (from Packer 1961). The system is formulated slightly differently by Lehninger (1965) in that inorganic phosphate is thought to react directly with the complex between the respiratory carrier and I. (see Fig. 20 below).

It must be stressed that none of the various $\sim$ compounds have yet been

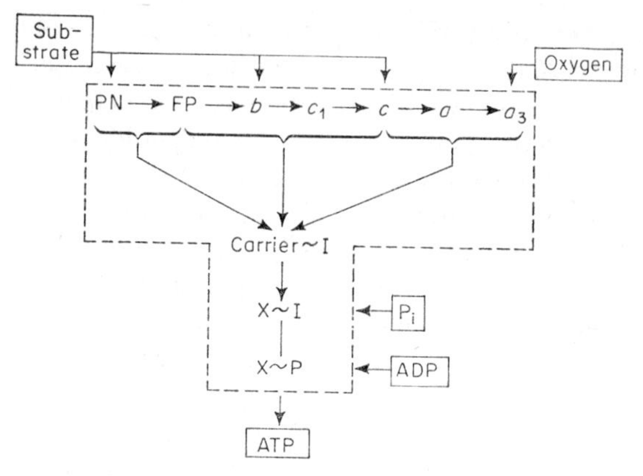

FIG. 14. A scheme for oxidative phosphorylation. PN: pyridine nucleotides (NAD, NADP); FP: flavoprotein; *b*, c_1, *c*, *a*, a_3: cytochromes; I, X: intermediates in oxidative phosphorylation; ~ indicates high-energy state (from Packer, 1961).

isolated, or even identified, and there is probably little purpose in this article to delve into the fine arguments for and against various theories. The point that is generally agreed is that non-phosphorylated energy-rich intermediates are generated during oxidative phosphorylation. Although these intermediates may be used to form ATP, they may also be used for a wide variety of energy-requiring reactions. These reactions have been discovered largely by the use of the antibiotic oligomycin which prevents the formation of ATP, but not of the non-phosphorylated energy rich intermediates. A detailed study of the mode of action of oligomycin, and particularly the inhibition of ATPase induced by a wide variety of agents, has recently been made Lardy, Connelly and Johnson (1964). The possibility was suggested that oligomycin combines with X ~ P, and so prevents phosphate transfer to any acceptor. These authors also proposed the existence of a special category of high-energy intermediate (called W ~) which is specifically concerned in ion transport, swelling and contraction phenomena, and is distinct from intermediates of oxidative phosphorylation, such as X ~ P and X ~ I. The variety of energy transfer reactions that occur in mitochondria are amply described in the report of a "Symposium on Energy-linked Functions of the Mitochondrion" (Chance, 1963).

Interest in energy transfer reactions within mitochondria was greatly stimulated by the important discovery of the reversal of oxidative phosphorylation and electron transport (see for example Klingenberg and Schollmeyer, 1960; Chance and Hollunger, 1961). It is possible, for example, to bring about the thermodynamically unlikely reaction:

$$\text{Succinate} + \text{NAD}^+ \rightarrow \text{Fumarate} + \text{NADH} + \text{H}^+$$

either by supplying exogenous ATP, or by using ATP or even non-phosphorylated energy-rich intermediates generated by the mitochondria themselves. If the energy is generated by the mitochondria, the energy-linked reduction of NAD^+ is lost after treatment that damage ~ production. By using suitable inhibitors and oxidizable substrates, it is possible to demonstrate that energy derived from electron transport along one part of the chain may be used to reduce another part of the chain. The effect of inhibitors on the respiratory chain and energy-transfer process was examined in detail by Chance and Hollunger (1963) and a summary of their findings is shown in Fig. 15.

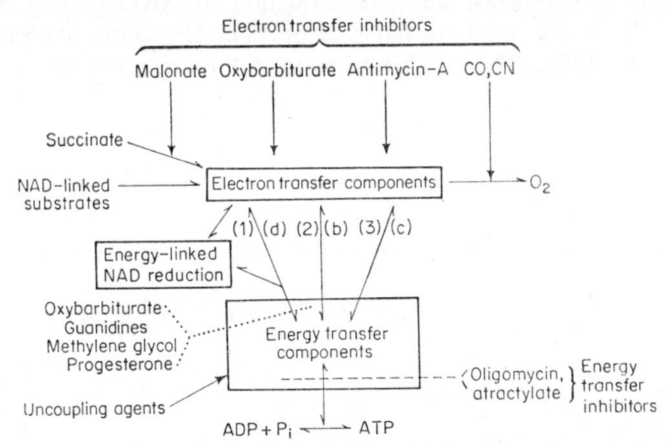

FIG. 15. Site of action of inhibitors of electron transport and energy transfer reactions. (1)(d), (2)(b) and (3)(c) represent the three sites of energy production, i.e. sites I, II and III (Atractylate is now thought to inhibit the transport of adenine nucleotides into mitochondria: see text) DPN = NAD (from Chance and Hollunger, 1963).

An elegant example of experiments on energy-transfer reactions in mitochondria is given by the work of Tager and Slater (1963) on the synthesis of glutamate by isolated mitochondria. By following the formation of glutamate from α-oxoglutarate and NADH plus NH_3, they were able to examine the energy-linked formation of NADH, e.g. by reduction of NAD^+ with succinate. The energy for NAD^+ reduction could be provided by ATP, by aerobic oxidation of succinate, or by the aerobic oxidation of tetramethylparaphenylene diamine (TMPD) which is oxidized by the terminal part of the respiratory chain. Tager and Slater examined many related energy-linked reactions, and perhaps it would be useful to illustrate this type of work with a specific example given by Slater and Tager (1963) as follows:

The respiratory chain is represented in an abbreviated form by the diagram:

$$NAD^+ \longleftrightarrow A \longleftrightarrow B \longrightarrow O_2$$

$$\begin{array}{cc} \updownarrow & \updownarrow \\ \text{Succinate} & \text{TMPD} \end{array}$$

where $\sim$ are energy-rich intermediates, A is a respiratory carrier in the flavo-protein-ubiquinone region, B is a carrier in the cytochrome *c* region and TMPD is the substrate tetramethy-paraphenylene diamine. In the presence of antimycin, succinate oxidation is prevented, However, succinate dehydrogenase can still reduce the carrier A by the reaction:

$$\text{Succinate} + \text{A} \leftrightharpoons \text{Fumarate} + \text{AH}_2$$

The reduced form of A is then used to reduce NAD^+. However, this is an energy requiring process, since the oxidation of NADH by A normally results in the formation of one molecule of ATP. The source of energy is the oxidation of TMPD. This is achieved by the following reaction:

$$\text{TMPD} + \text{B} \leftrightharpoons \text{oxidized TMPD} + \text{BH}_2$$

BH_2 is then oxidized by the terminal oxidase of the respiratory chain with the formation of an energy-rich compound with the carrier I_3, which acts at the third site of phosphorylation in the chain:

$$\text{BH}_2 + \tfrac{1}{2}\text{O}_2 + \text{I}_3 \rightarrow \text{B} \sim \text{I}_3 + \text{H}_2\text{O}$$

This compound may then be used to provide energy for the reduction of NAD^+ by AH_2 by means of the following reactions

$$\text{B} \sim \text{I}_3 + \text{NAD}^+ + \text{I}_1 \leftrightharpoons \text{B} + \text{I}_3 + \text{NAD} \sim \text{I}_1$$

(where I_1 is the carrier acting at the first site of phosphorylation in the chain), followed by:

$$\text{NAD} \sim \text{I}_1 + \text{AH}_2 \leftrightharpoons \text{A} + \text{NADH} + \text{I}_1 + \text{H}^+$$

The overall reaction is therefore:

$$\text{Succinate} + \text{TMPD} + \tfrac{1}{2}\text{O}_2 + \text{NAD}^+ \rightarrow \text{Fumarate} + \text{oxidized TMPD} + \text{NADH} + \text{H}^+$$

so that the reduction of one component of the chain has been driven by energy produced by oxidation in another.

These, and other reactions studied by Tager and Slater are excellent examples of the subtle interactions of energy-yielding and energy-requiring processes that may be demonstrated in the mitochondrion. This is clearly shown in a figure from Slater and Tager (1963) which shows competition for energy-rich compounds by 2,4-dinitrophenol, ATP synthesizing reactions and NAD^+ reducing systems (Fig. 16).

Perhaps a word of caution is needed in the interpretation of experiments with oligomycin, which is frequently used to study energy-transfer reactions. Although oligomycin effectively prevents the formation of ATP from added ADP and inorganic phosphate, it should not be assumed that if a reaction

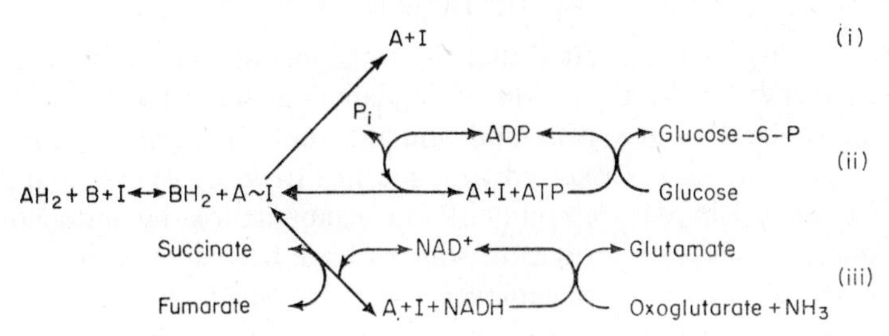

FIG. 16. Competing reactions for energy-rich compound (~) formed during oxidative phosphorylation. (i) decomposition by 2,4-dinitrophenol, (ii) synthesis of ATP (as followed by formation of glucose-6-phosphate in the presence of glucose and hexokinase), (iii) the energy driven reduction of NAD^+ by succinate (as followed by the formation of glutamate from α-oxoglutarate plus NH_3). A and B are adjacent carriers of the respiratory chain (from Slater and Tager, 1963).

observed in the intact mitochondrion is not inhibited by this agent, it must derive its energy from non-phosphorylated intermediates. A good example of this is found in some recent experiments of Kroon (1966). He had observed previously that amino-acid incorporation into protein in isolated mitochondria was not seriously inhibited by oligomycin. This had also been observed by Bronk (1963). However, if arsenite was added to inhibit substrate-level phosphorylation, it was possible to demonstrate inhibition by oligomycin. The failure to demonstrate inhibition previously was because the incorporation system was able to use ATP produced by substrate-level phosphorylation which was not inhibited by oligomycin. Thus ATP, and not non-phosphorylated intermediates, is apparently required for incorporation.

A final point of interest is the problem of whether oxidative phosphorylation only occurs in mitochondria. In general it is believed that extra-mitochondrial pathways for NADH and NADPH oxidation are not associated with any synthesis of ATP. However, there have been rather detailed studies of ATP synthesis in nuclei, apparently accompanying oxidative processes (e.g. Osawa, Allfrey and Mirsky, 1957). The nuclear phosphorylation differs in many of its properties from mitochondrial oxidative phosphorylation. The most important difference is that added ADP is not phosphorylated, but the nuclear ATP appears to be synthesized from a pool of endogenous adenine nucleotides. Also, certain inhibitors of oxidative phosphorylation, such as dicumarol, have no effect on nuclear phosphorylation. The matter is discussed in Chapter 2, and the safest interpretation of the present results is that although there is no doubt that isolated nuclei are able to synthesize

ATP, the exact relation to aerobic processes is uncertain. This is particularly so because of the absence of cytochrome oxidase (Hogeboom *et al.*, 1952) and succinoxidase (Roodyn, 1956) from nuclei.

G. ION TRANSPORT

Recent research has indicated that the transport of ions across the mitochondrial membrane is a reaction of fundamental importance and is intimately dependent on the structural and metabolic integrity of the mitochondrion. The reader is referred to a recent article by Harris and Judah (1966) for an up to date account of ion accumulation by mitochondria. Mitochondria from a wide range of sources have been shown to be able to take up a variety of ions, including Ca^{2+}, Sr^{2+}, Mg^{2+}, K^+ and inorganic phosphate from the medium. It is now becoming increasingly clear that a complex, balanced ion-translocation system exists in the mitochondrial membrane and that ingress of one ion is frequently accompanied by egress of another. For example, Carafoli, Rossi and Lehninger (1964) observed that Na^+ is taken up during uptake of Ca^{2+} and Mg^{2+} ions, and at the same time there is a loss of mitochondrial K^+ into the medium. Another aspect of great interest is that, particularly in the case of Ca^{2+} uptake it has been possible to show that in certain circumstances there is a stimulation of respiration during ion uptake, and this stimulation bears a stoichiometric relationship to the amount of ion taken up. An excellent demonstration of this is shown in Fig. 17, which is from Rossi and Lehninger (1964). It can be seen that the

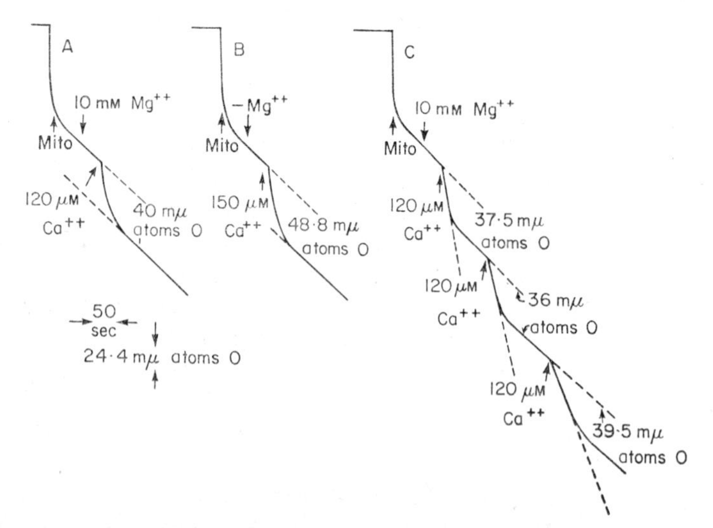

FIG. 17. Repeated stimulation of oxygen uptake by tightly coupled rat-liver mitochondria on addition of low concentrations of Ca^{2+} ions. Oxygen uptake indicated by a fall in the trace (from Rossi and Lehninger, 1964).

addition of low concentrations of Ca^{2+} results in a burst of oxygen uptake. This process can be repeated several times. A similar result may be obtained with Sr^{2+} uptake (Carafoli, 1965).

The relationship between ion uptake, phosphate uptake and the dense granules of mitochondria is discussed further in Section III. C. below.

H. ANABOLIC REACTIONS

There is increasing realization that the mitochondrion has considerable synthetic ability. Considering lipid synthesis first, there have been reports of the incorporation of radioactive precursors into the phospholipids of isolated mitochondria. For example, energy dependent uptake of phosphate into phospholipids was observed by Garbus, DeLuca, Loomans and Strong (1963). Using guinea pig liver mitochondria, Hajra, Seiffert and Agranoff (1965) showed that ATP labelled in the γ position with ^{32}P was incorporated into phosphatidyl inositol phosphate. The incorporation observed with free $^{32}P_i$ is probably via ATP. Hulsmann (1962) demonstrated the incorporation of acetate into saturated and unsaturated fatty acids by isolated heart sarcosomes. He also observed incorporation of malonate into fatty acids, and concluded that synthesis of fatty acids proceeds via the malonyl-CoA pathway. The synthesis of malonyl-CoA from oxalosuccinate plus acetyl-CoA was reported in a preparation of sonically disrupted rat-heart sarcosomes. (Hülsmann, 1963). Also, Hülsmann and Dow (1964) obtained results that indicated that mitochondria from lactating mammary gland could also synthesize fatty acids from malonyl CoA. The role of mitochondria in lipid synthesis has recently been discussed by Hülsmann *et al.* (1966). Apart from the enzymes described by Hülsmann and his co-workers, there have been reports of other systems for lipid synthesis. For example, Webster, Gerowin and Rakita (1965) purified an enzyme from bovine-heart mitochondria that catalysed the synthesis of butyryl-CoA. The enzyme also showed activity towards proprionate, valerate and caproate. Also, the incorporation of ^{14}C-palmitate into triglycerides, choline and ethanolamine phosphatides by isolated rat-liver mitochondria has been described by Johnson and Kerur (1963). This is a far from complete account of the role of mitochondria in the synthesis of lipids, and there is little doubt these reactions are of considerable importance, particularly in relation to problems of mitochondrial replication.

There have been an increasing number of reports in the last two years of RNA synthesis in isolated mitochondria. The incorporation of ^{14}C-UTP, GTP, and ATP into the RNA of isolated rat-liver mitochondria has been demonstrated by Neubert and Helge (1965). In intact mitochondria, the incorporation did not require added nucleotides, and was not sensitive to actinomycin C, however, swollen mitochondria that had enhanced permeability showed dependence on added nucleotides and inhibition by actinomycin C. The behaviour of the intact mitochondria was used to exclude the possibility that the observed activity was due to contamination with nuclear RNA polymerase (2.7.7.6.). The inhibition by actinomycin strongly suggested the presence of a DNA-dependent RNA polymerase in mitochondria. Additional evidence has been obtained recently for the presence of this enzyme in mitochondria isolated from a variety of biological materials [eg. yeast: Wintersberger (1964), a variety of warm-blooded animals: Neubert, Helge and Merker, (1965), lamb-heart: Kalf (1964) and *Neurospora crassa*: Luck and Reich (1964)]. Kalf (1964) also observed that amino acid incorporation into protein was inhibited by actinomycin D, and concluded that mitochondrial protein synthesis is dependent on a continually functioning RNA polymerase. A similar conclusion was reached by Kroon (1965). The subject is developing very rapidly at the present moment, and there is little doubt that the next few years will see an intensive investigation of the mechanism of RNA synthesis in mitochondria.

Although there have been several reports of the labelling of mitochondrial DNA *in vivo* by ^{3}H-thymidine (for example in *Tetrahymena pyriformis*: Parsons, 1965) the author is not aware of any reports of the demonstration of DNA synthesis in isolated mitochondria.

Perhaps the best established example of anabolic activity of the mitochondrion is the ability of isolated mitochondria to incorporate radioactive amino acids into protein. Incorporation of a range of labelled amino acids has been demonstrated in mitochondrial fractions from very diverse sources (See Roodyn, 1965b for a list of these). There is little doubt that this activity is a general property of the mitochondrion. The properties of the incorporation system are similar in all mitochondrial fractions examined, and differ from microsomal incorporation in being insensitive to rather massive amounts of added ribonuclease (2.7.7.16,17.), showing complete dependence on oxidative phosphorylation, having more prolonged progress curves (i.e. incorporation can continue for 1–3 hr, rather than 10–30 min) and being inhibited by chloramphenicol. Although some of the early reported values of cts/min/mg protein were rather low, recent improved methods have given activities that are at least comparable to those obtained with microsomal systems, particularly if disrupted mitochondria are used (Kroon, 1965). The essential properties of the incorporation system in rat-liver were summarized recently (Roodyn 1966) and are given in Fig. 18.

In the case of rat-liver, at least, it is reasonably certain that the incorporation is not due to bacterial, nuclear, lysosomal or microsomal contamination. Bacterial and microsomal contamination are also very unlikely in the case of muscle mitochondria. The incorporation is dependent on the maintenance of the structural integrity of the mitochondrial membrane system and its ability to carry out oxidative phosphorylation. With well-washed rat-liver mitochondria, there is a clear requirement for added Mg^{2+} ions, P_i, adenine nucleotides and one of a range of oxidizable substrates. Not all preparations, however, have been reported to show a requirement for added substrates, and presumably these use endogeneous substrates. The major site of incorporation is into insoluble proteins associated with the mitochondrial

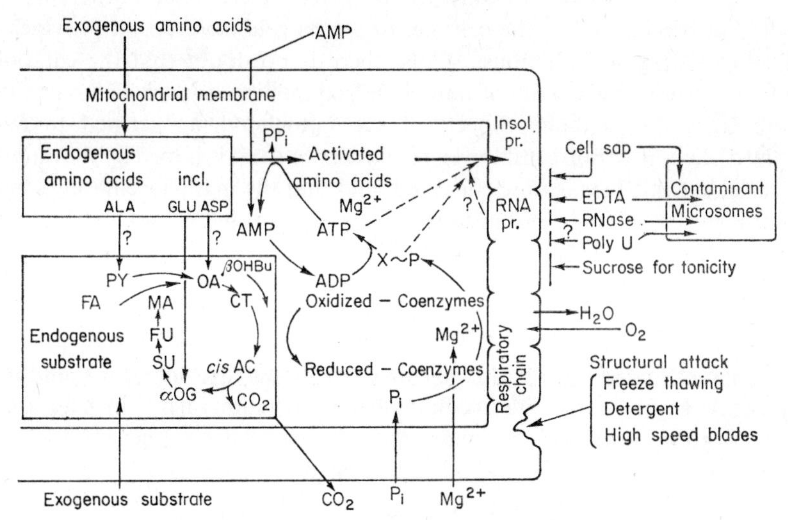

FIG. 18. Summary of factors affecting amino acid incorporation into protein by isolated rat-liver mitochondria. Thick lines represent direction of incorporation of labelled amino acids, thin continuous lines ancillary reactions involved in this, and broken lines (and question marks) areas of uncertainty. ALA: alanine, ASP: aspartate, *cis*AC: *cis*-aconitate, CT: citrate, FA: fatty acid, FU: fumarate, GLU: glutamic acid, MA: malate, αOG: α-oxoglutarate, β-(OH)Bu: β-hydroxybutyrate, OA: oxaloacetate, PP: pyrophosphate, PolyU: polyuridylic acid, PY: pyruvate (from Roodyn, 1966).

membrane. In liver mitochondria it has been shown that the labelled material shows some resemblance to the insoluble structural protein described by Criddle, Bock, Green and Tisdale (1962) (see Section III. F. below). The insoluble proteins were closely associated with the mitochondrial membrane and were extremely difficult to purify by conventional methods of protein fractionation (Roodyn, 1962). It was further shown that the enzymes cytochrome *c*, malate dehydrogenase and catalase, present in the mitochondrial fraction were not labelled *in vitro* (Roodyn *et al*., 1962). The physiological state of the animal affects the incorporation. In particular, thyroid hormones given *in vivo* stimulate the incorporation observed *in vitro* (Roodyn, Freeman and Tata, 1965).

There are reports of other synthetic reactions in isolated mitochondria, for example the synthesis of para-aminohippuric acid (Kielley and Schneider 1950). The formation of protoporphyrin and haem has been reported by Rimington and Tooth (1961) and related to this is the report of the synthesis of δ-amino-laevulinic acid by isolated mitochondria (Granick and Urata 1963). Also, Nishida and Labbe (1959) have demonstrated the conversion of protoporphyrin to haem by using ^{59}Fe as a tracer. Space does not allow a complete description of such reactions. However, the point to be stressed is that it is now clearly not correct to regard the prime function of the mitochondrion as the "power-house" of the cell, i.e. to see its role merely as a provider of ATP for endergonic reactions. Whilst there is no doubt that the catabolic and ATP-producing aspects of mitochondrial metabolism are of the greatest importance, the anabolic aspects of the mitochondrion seemed to have received scant attention until the last five years or so. It is hoped by the author, at least, that this bias is now corrected, and that the many fascinating implications of examining the mitochondrion as an integrated biosynthetic organelle will now be fully explored.

I. OTHER REACTIONS

One could include under this heading a truly massive literature, and it is impossible to provide a full account of these reactions here. Instead a few interesting examples will be taken, that are particularly relevant to other aspects of mitochondrial metabolism discussed in the article.

1. Amino Acid Metabolism

There are many enzymes in mitochondria involved in the various aspects of amino acid metabolism. Several amino-transferases have been reported in mitochondria, but it is interesting that as with certain of the enzymes of the tricarboxylic acid cycle, the activity is not uniquely localized in the mitochondrial fraction, but also appears in the soluble fraction. There are good indications that the soluble enzymes are different proteins, and not simply released from the mtiochondria during isolation. For example the soluble and mitochondrial aspartate aminotransferases (2.6.1.1.) have many different properties (for example different pH-activity curves: Sheid, Morris and Roth, 1965). The soluble and mitochondrial alanine aminotransferases (2.6.1.2,12) also appear to be different proteins. The mitochondrial enzyme is more unstable and has a lower K_m than the soluble enzyme (Swick, Barnstein and Stange, 1965). Morino, Kagamiyama and Wada (1964) extended their previous observation that crystalline mitochondrial and soluble beef-liver aspartate amino-transferases differed in physico-chemical and kinetic properties by showing that rabbit antisera prepared against the mitochondrial enzyme did not react with the soluble enzyme and vice versa. They also made the interesting observation that mitochondrial enzymes from a variety of different tissues were inhibited by the anti-mitochondrial enzyme serum, and the corresponding soluble enzymes were inhibited by the anti-soluble enzyme serum. Thus there was greater immunological similarity between enzymes from the same organelle in different tissues than between enzymes from different structures in the same tissue.

Another interesting aspect of the aminotransferases is that the enzymes attacking the various amino acids are by no means equally distributed between mitochondria and the soluble fraction. For example, the tyrosine amino-transferase activity (2.6.1.5.) of rat-liver mitochondria is far less than the aspartate or alanine aminotransferase activities, (Litwack, Sears and Diamondstone, 1963; Kenney, 1962).

Glutamate dehydrogenase is a key enzyme of mitochondrial amino acid metabolism. It has been known for many years that there is an active glutamate dehydrogenase in liver mitochondria (Hogeboom and Schneider, 1953). More refined fractionation methods have established without doubt that this enzyme is clearly associated with mitochondria, and in fact may be regarded as a "marker" enzyme for mitochondria (see Beaufay *et al.*, 1959). Because of the presence of glutamate-linked aminotransferases in mitochondria, it became generally accepted that they were the site of what may be called the "classical" transamination cycle. This is described fully in Braunstein (1957) and is shown in Fig. 19a. It is this cycle which is responsible for the production of ammonia during the oxidation of amino acids.

Recent work, however, particularly of Tager, Papa, Quagliariello and their associates has emphasized the importance of another mechanism that does not give rise to ammonia. This is shown in Fig. 19b. Although transamination

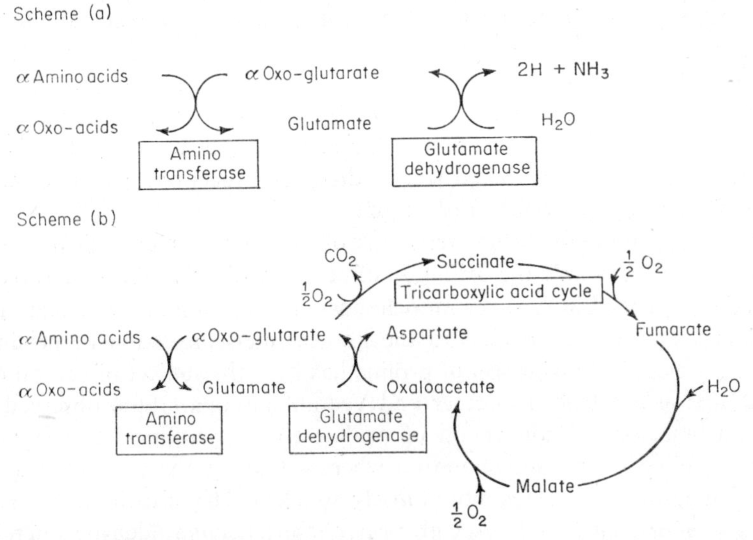

FIG. 19. (a) "Classical" transamination cycle for de-amination of amino acids to form ammonia (see Braunstein, 1957). (b) Alternative cycle for transamination of amino acids to form aspartate (see Quagliariello and Papa, 1964).

still occurs, the glutamate is not reconverted to α-oxoglutarate by the glutamate dehydrogenase reaction, but reacts by means of aspartate aminotransferase with oxaloacetate. This has been formed by oxidation of another molecule of α-oxoglutarate in the tricarboxylic acid cycle. The result is the formation of aspartate, and the regeneration of glutamate. This cycle is really an extension of the reactions studied by Krebs and Bellamy (1960). The effect of the "transamination cycle" is therefore to transfer the α-amino groups of amino acids to aspartate, which may then be used in synthetic reactions, or for the synthesis of urea in the urea cycle. Quagliariello, Papa, Saccone, Palmieri and Francavilla (1965) have recently shown that both the "classical" and "transaminase" routes for glutamate metabolism occur in rat liver mitochondria. However, ammonia production decreases with time, and the transamination route is more important in prolonged incubations.

One well-established observation that is relevant to this question is that in intact mitochondria the glutamate dehydrogenase activity is very low. In order to activate the enzyme fully, the mitochondria have to be disrupted by drastic procedures, such as by forcing under high pressure through a small orifice (Hogeboom and Schneider, 1953) or by exposure to high-speed blades, osmotic shock or detergents (Bendall and Duve, 1960). It may be noted that the phenomenon of latency of mitochondrial enzymes is not confined to this example, but has been observed with several other enzymes. For example the atency of rhodanese (2.8.1.1.) was the subject of a careful study by Greville and Chappell (1959).

Another interesting aspect of glutamate dehydrogenase activity that has emerged recently is that although the purified enzyme can react equally with NAD or NADP, and it had previously been thought that the enzyme was in fact NAD-linked (Hogeboom and Schneider, 1953), it is now thought likely that most of the oxidation of glutamate in intact mitochondria results in the formation of NADPH in the first instance. This is then used to reduce NAD^+ by means of an energy-linked transhydrogenase reaction. This is discussed more fully in Quagliariello *et al.* (1965).

There are numerous other reactions of amino acid metabolism in mitochondria. For example the oxidation of proline to Δ^1-pyrroline-5-carboxylate was demonstrated in rat-liver mitochondria by Johnson and Strecker (1962) and the reaction was related to the known inter-conversion of proline to glutamic acid. The oxidation of proline has been the subject of several other studies. For example, Brosemer and Veerabhadrappa (1965) observed that the enzyme for the conversion of proline to Δ^1-pyrroline-5-carboxylate is tightly bound to the mitochondria, whereas that for the conversion of this compound to glutamate is more loosely attached. The mitochondrion is also capable of oxidizing methylated glycines, choline, betaine aldehyde and related compounds (see for example Williams, 1952a, b; Hoskins and MacKenzie, 1961; Hoskins and Bjur, 1964; Bianchi and Azzone, 1964). It is therefore well established that the mitochondrion is the site of complex interactions between amino acids and their various derivatives.

2. Hydrolases

This is another group of enzymes of considerable importance in mitochondria. However, because of the serious and technically difficult problem of lysosomal contamination of mitochondrial fractions, it is difficult to be certain which hydrolases are truly mitochondrial. Perhaps the best established case, however, is that of ATPase. Considerable ATPase activity has been demonstrated in mitochondria from a very wide range of tissues, and there is little doubt that it is a fundamental mitochondrial enzyme. The ATPase activity is generally latent in intact mitochondria but is greatly increased if the mitochondria are damaged by a wide variety of treatments, or by additions of agents that uncouple oxidative phosphorylation, such as 2,4-dinitrophenol. In the last 10 years there has been extensive research into the precise relation between ATPase activity, the structural state of the mitochondrion and the reactions of oxidative phosphorylation.

It is quite likely that at least part of the ATPase activity observed in damaged mitochondria is due to reversal of the terminal reactions of oxidative phosphorylation, for example by hydrolysis of $X \sim P$. Thus instead of the reaction:

$$X \sim P + ADP \rightarrow X + ATP$$

hydrolysis of $X \sim P$ causes reversal of this reaction by removal of one of the reactants i.e.

$$X \sim P + H_2O \rightarrow X + P_i$$

$$ATP + X \rightarrow ADP + X \sim P$$

Sum: $$ATP + H_2O \rightarrow ADP + P_i$$

Net hydrolysis of ATP results. The effect could also be due to "discharge" of one of the non-phosphorylated energy rich intermediates, with resultant failure to form $X \sim P$. It could also be caused by the physical disorganization of the $\sim P$ acceptor system, so that water is used as an acceptor, instead of ADP. The agent 2,4-dinitrophenol probably acts by the discharge of energy-rich intermediates, while activation of ATPase by such treatments as repeated freezing and thawing probably acts by the latter mechanism.

An important finding in this respect is that oligomycin not only inhibits oxidative phosphorylation, but it also inhibits the dinitrophenol-stimulated ATPase (Lardy, Johnson and McMurray, 1958). As a result, the combined use of these two agents has been of great help in examining the terminal stages of oxidative phosphorylation, and of the so-called "partial reactions" (e.g. ATP/P_i exchange, ATP/ADP exchange and H_2O/ATP ^{18}O exchange).

The relation between these reactions, ATPase and the sites of action of oligomycin and dinitrophenol is given in Fig. 20 (from Lehninger, 1965). The interesting theory of the role of ATPase in oxidative phosphorylation presented by Mitchell has been discussed above (see Section II. F.).

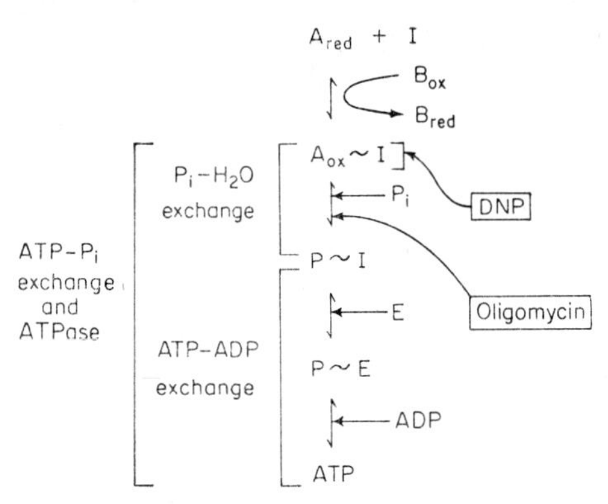

FIG. 20. "Partial reactions" of oxidative phosphorylation and sites of action of 2,4-dinitrophenol and oligomycin (from Lehninger, 1965).

In their attempts to fractionate the components of the respiratory chain phosphorylation system Penefsky, Pullman, Datta and Racker (1960) purified a soluble ATPase that had the ability to restore phosphorylation activity to suitably treated respiratory particles. It is an example of one of the several "coupling factors" described by this and other groups. The enzyme has the interesting and unusual property of being more unstable at 0° that at room temperature (Pullman, Penefsky, Datta and Racker, 1960). There have been many other investigations into the relationship between ATPase and the reactions of oxidative phosphorylation. An interesting example is from work done several years ago by Myers and Slater (1957). They studied the effect of pH on mitochondrial ATPase acitvity and obtained a complex pH-activity curve. They suggested that this was due to the super-imposed pH-activity curves of four different ATPases. Three of these, with pH optima at pH 6·3,7·4, and 8·5 were stimulated by dinitrophenol and it was suggested that they were specific ATPases acting at the three sites of ATP production in the respiratory chain.

Another important aspect of mitochondrial ATPase activity is its relationship to contraction of the membrane, and the possibility that there is a membrane bound actomyosin-like ATPase in mitochondria. In many ways there are similarities between the properties of the mitochondrial and muscular contraction systems (see Lehninger, 1965 for a discussion of this) and indeed there have been reports of the isolation of actomyosin-like "contractile proteins" from mitochondria (Neifakh and Kazakova 1963, and Ohnishi and

Ohnishi, 1962). Ohnishi, Kawamura, Takeo and Watanabe (1964) have extended their studies to the isolation of two proteins from rabbit liver mitochondria, one resembling actin, and the other resembling myosin. It may be concluded at this point that mitochondrial ATPase is clearly closely involved in the function and structure of the mitochondrial membrane system.

There have been reports of many other hydrolases in mitochondria and, as stated above, it is often difficult to assess the contribution of lysosomal contamination to these activities. One enzyme, however, is certainly not due to lyososomes present in the mitochondrial fraction. This is the alkaline DNase (3.1.4.5.). Unlike the acid DNase, which is a typical lysosomal enzyme, this enzyme clearly follows mitochondrial marker enzymes, such as cytochrome oxidase, in analytical gradient centrifuging (Beaufay *et al.*, 1959). The role of mitochondrial DNase is far from clear, but it may be associated in some way with the metabolism of the recently discovered mitochondrial DNA (see Section III. B.).

There is some doubt as to whether there are proteolytic enzymes in mitochondria. The classical "four-fraction" scheme of analysis (i.e. the isolation of nuclear, mitochondrial, microsomal and soluble fractions) revealed a wide range of proteolytic enzymes and amidases in the mitochondrial fraction. (For example, see Maver and Greco, 1951). More sophisticated schemes of fractionation, however, revealed that catheptic activity was concentrated in the lysosome-rich mitochondrial fraction (de Duve, Pressman, Gianetto, Wattiaux and Applemans, 1955) and "cathepsin" (a somewhat loose term) is now regarded as a typical lysosomal enzyme. However, some proteolytic activity has been ascribed to mitochondria. For example a CoA, ATP activated enzyme hydrolysing serum albumin was reported in rat-liver mitochondria by Penn (1960). Alberti and Bartley (1963) ascribed the increase in concentration of free amino acids that occurs on incubation of mitochondria *in vitro* to the action of soluble intramitochondrial proteolytic enzymes, but it is not certain that they could exclude the possibility of lysosomal contamination. A similar conclusion was reached by Baird (1964). The level of free amino acids in isolated mitochondria is surprisingly high, even in mitochondria that have been thoroughly washed (Roodyn *et al.*, 1961). This may be partly due to the action of proteolytic enzymes in the mitochondria, acting during the course of isolation.

The production of amino acids by isolated rat-liver mitochondria has recently been examined in more detail by Alberti and Bartley (1965) and the results correlated with the metabolic state of the mitochondrion. One interesting conclusion of these workers was that release of amino acids during incubation of mitochondria in the presence of radioactive amino acids can cause dilution of the label, and an apparent fall in incorporation into protein. Thus when Alberti and Bartley (1965) corrected the incorporation curve obtained by Roodyn *et al.* (1961) for this effect, a straight line was obtained over a 2 hr period. Thus proteolytic activity in mitochondrial preparations may have an important effect on the results of metabolic studies, particularly if long incubation periods are used. A most interesting aspect of the amino acid release by mitochondria is that it is inhibited by chloramphenicol (Alberti and Bartley, 1965) which is also a powerful inhibitor of mitochondrial protein synthesis.

3. Other Enzymes

It is impossible to deal here with the many other enzymes that have been detected in mitochondria from various sources. Simply giving two random examples, one may mention the enzyme rhodanese which is clearly associated with liver mitochondria and exhibits interesting activation and latency properties (Greville and Chappell, 1959). Secondly, the enzyme phosphoenol pyruvate carboxykinase (4.1.1.32.) has been demonstrated in mitochondria from various tissues. The amount varies somewhat with different mitochondria. For example, rabbit liver mitochondria are far more active than rat-liver mitochondria (Nordlie and Lardy, 1963). The enzyme is relatively soluble and there is considerable enzyme activity in the soluble fraction of the cell. It is interesting that there is evidence that the mitochondrial and soluble enzymes have certain differences in their properties, and may well be two different enzymes (Holten and Nordlie, 1965).

J. SUMMARY AND CONCLUSIONS

The author hopes that the reader will accept the many ommissions and oversimplifications in the above account of the metabolic activity of the mitochondrion. A most salutary method of proving the extraordinary wealth of reactions that have been described in isolated mitochondria is to follow the entries against "mitochondria" in the Chemical Abstracts. Any system of indexing or literature abstraction soon breaks down under the assault of the volume of material. Perhaps it is now no longer reasonable to attempt such a survey, any more than it is reasonable to survey the properties of cells. However, if the reader is left with the impression that the mitochondrion is an organelle possessing a wide range of interacting metabolic activities, this survey has not been completely without value. However, the reader is recommended to consult the appropriate reviews and monographs cited in the text for fine details of the above reactions.

III. THE STRUCTURAL BASIS OF MITOCHONDRIAL FUNCTION

Surveying the sub-cellular distribution of enzymes (see Roodyn, 1965a) is made relatively easy by the fortunate circumstance that the "four-fraction" scheme of Hogeboom, Schneider, Palade and co-workers developed in the late 1940's and early 1950's was generally adopted by large numbers of workers. It is thus possible to list activities in the nuclear, mitochondrial, microsomal and soluble fractions with relative ease, and to compare the results of different workers. Unfortunately, no one scheme for the sub-fractionation of mitochondria appears to have been generally adopted, and it is frequently difficult to compare the results obtained with different "sub-mitochondrial" particles. Also, centrifugal analysis of disrupted mitochondria does not result in such clear cut fractions as is the case with the fractionation of disrupted cells. There is a broad division between "soluble" and "insoluble" mitochondrial components, but the "insoluble" components actually include a range of sub-mitochondrial fragments (Watson and Siekevitz, 1956). Although there have been many studies of "compartmentation" in mito-chondria, it is difficult to establish the structural basis of these compartments with a great degree of certainty. For these reasons, sub-mitochondrial structure will only be considered under the three main headings of mitochondrial sap, mitochondrial inclusions and the mitochondrial membrane system.

A. MITOCHONDRIAL SAP

It is well established that isolated mitochondria, even after several washes, contain many soluble components. These include a variety of nucleotides and phosphate esters, a considerable amount of free amino acids, many different cations, and many soluble enzymes. Work on mitochondrial fluid spaces has indicated that there are two separate spaces, one more accessible to added solutes than the other. It is generally believed that the easily accessible space lies between the inner and outer membranes (and thus corresponds to the "outer compartment" or "cristae space"), whereas the less accessible space is the true "matrix" of the mitochondrion and is bounded only by the inner membrane (i.e., it may be called the "inner compartment".) Unfortunately, it is not possible in all cases to allocate in a quantitative fashion the various soluble components of the mitochondrion to the inner or outer compartments. However, as a result of combined enzymic, chemical and electron-microscope studies, Klingenberg and co-workers have recently been able to go some way towards demonstrating more precisely the location of some mitochondrial components. A model suggested by these workers (Heldt, 1966; Klingenberg and Pfaff, 1966) is given in Fig. 21 which shows the position of several enzyme systems and nucleotides, together with the supposed site of permeability barriers to these substances.

Other workers have frequently assigned activities to "inner membranes" or "outer membranes", but generally the correlation that is obtained with the actual mitochondrial membranes is not as well established as in the above work. An example of a rather complex chart of this kind is that given

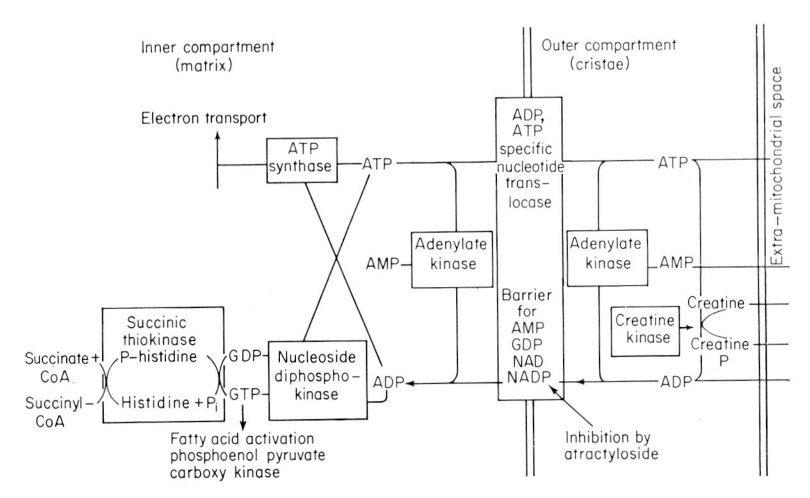

FIG. 21. "Compartmentation" of adenine nucleotide linked reactions of mitochondria (adapted from Heldt, 1966 and Klingenberg and Pfaff, 1966).

by Chappell (1961) to describe the suggested location of certain co-factors and enzymes involved in integrated oxidations in isolated mitochondria. This is given in Fig. 22. Several other similar diagrams have been presented to illustrate different aspects of mitochondrial metabolism (cf. Figs. 7 and 8).

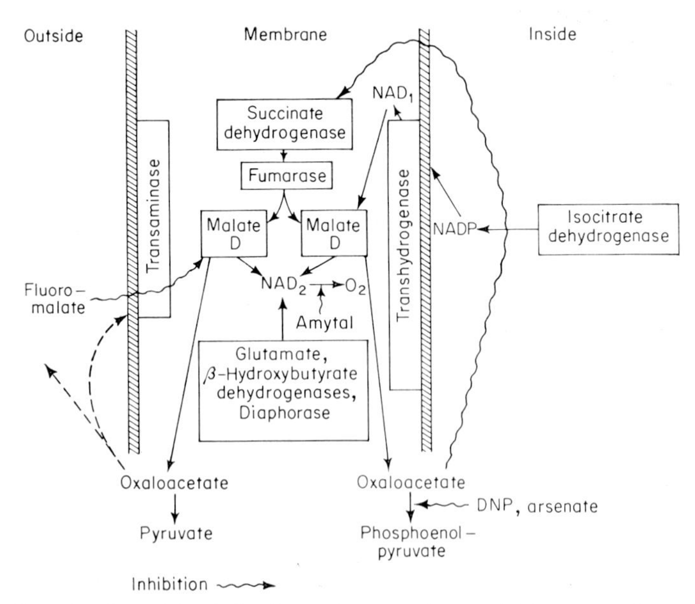

FIG. 22. Postulated spatial relationships of some mitochondrial enzymes (DPN_1, DPN_2 are bound and free forms of NAD; D dehydrogenase; TPN = NADP) (from Chappell, 1961).

In spite of the attraction of such diagrams, the precise location of submitochondrial components is certainly not as well known as generally thought. In the particular case of soluble proteins, definition of the term "soluble" is in fact rather difficult. (See Chapter 8 for a general discussion of this question.) One cannot simply state that a protein is either "soluble" or "insoluble" without more precise specification of the method of extraction of the protein. For example, it is well established that many proteins can be removed from mitochondria simply by washing them in salt media. For example, Dallam (1958) was able to extract 70% of the dry weight of rat-liver mitochondria by repeated washing with 0·16 M NaCl. At the other extreme, some proteins are only removed after total disruption of the mitochondria, for example by high concentrations of detergents. In a careful study of the mechanism of disruption of mitochondria by deoxycholate, Watson and Siekevitz (1956) showed that this does not occur by simple lysis of an all-or-none nature, with a sudden release of the soluble contents. Disruption is not simply bursting of a bag. By layering deoxycholate over a mitochondrial pellet and examining sections at different levels of the pellet, they were able to show that a series of changes occur under the influence of the disrupting agent. The mitochondria first swell greatly, and then they contract to about twice their original diameter. There follows dilution of the contents of the mitochondrion, followed by severe emptying and finally the formation of membranous elements. Thus the agent produces a series of changes, with progressive loss of mitochondrial material, rather than sudden lysis. These changes are shown in Fig. 23.

Support for these results comes from studies of the disruption of mitochondria with the Triton series of alkylphenoxypolyethoxyethanol detergents (Roodyn, 1962). Firstly, it was found that the ability to disrupt mitochondria depended greatly on the length of the side-chain of the detergent molecule. For example, concentrations of Triton-X-100 (with 9-10 ethylene oxide groups in the side chain) and Triton-X-305 (30 ethylene oxide groups) which had approximately the same surface active effect (i.e. would be expected to produce the same fall in surface tension) had very different effects on the mitochondrion. 0·1% Triton-X-100 produced complete lysis, whereas 0·2% Triton-X-305 had no effect. One could conclude that the action of the detergent depended on specific interaction with binding sites on the membrane that had to be of suitable size. This interaction results in loosening of membrane structure. Therefore increase in detergent concentration should result in more binding and a progressive loosening of structure. This was then confirmed by so-called "titration" experiments, in which mitochondria were exposed to increasing concentrations of detergent, and the release of various components followed. It was found that there were a series of "extraction curves" for the various components studied (malate dehydrogenase, total protein, succinoxidase, phospholipid, RNA and protein labelled after

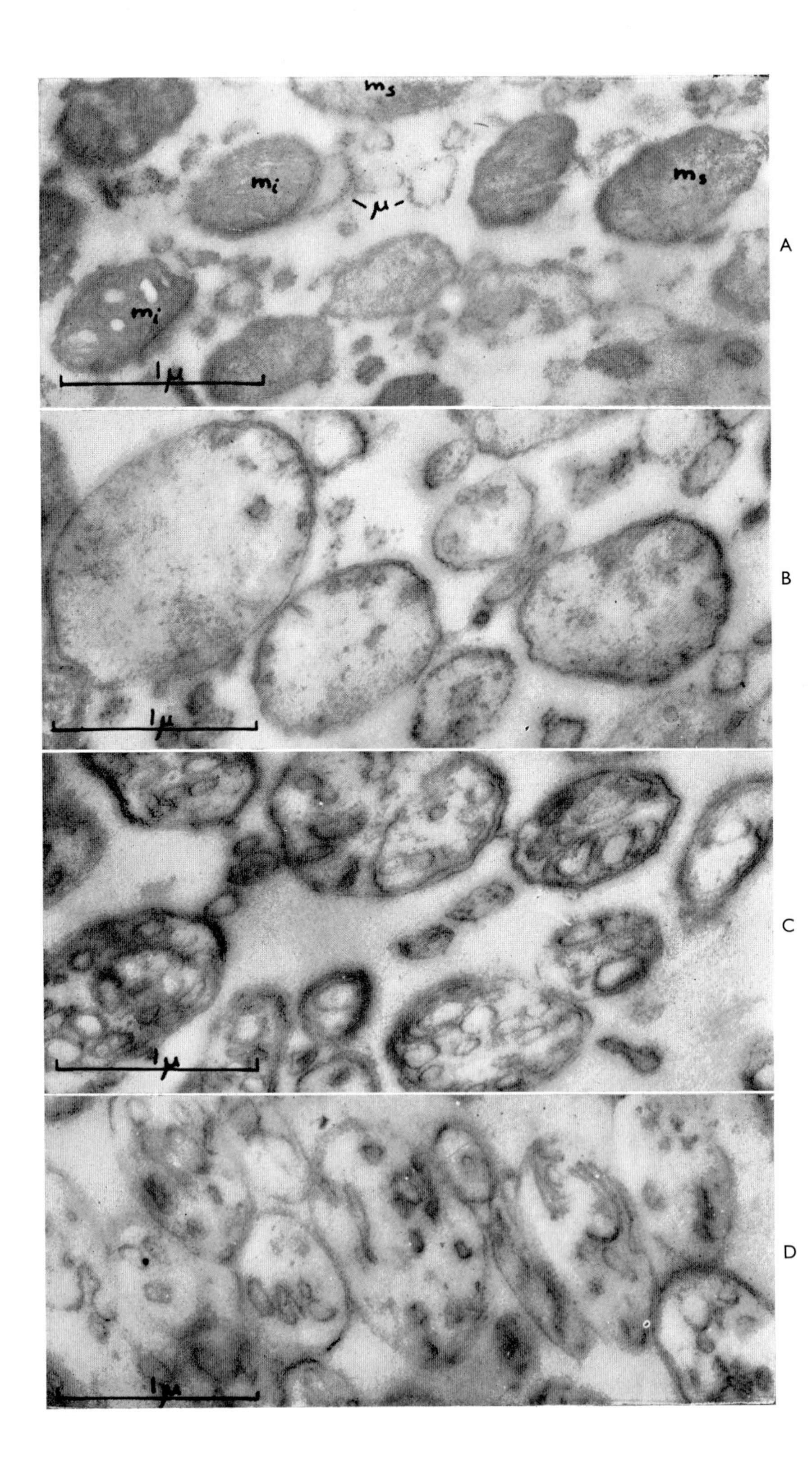

ms
mi
μ
ms
A
mi
1μ
B
1μ
C
1μ
D
1μ

incubation of mitochondria *in vitro* with ^{14}C-valine.) These curves are shown in Fig. 24, and it is clear that there is no sudden jump from "intact" to "disrupted" mitochondria, but rather a series of sequential extractions of the various components.

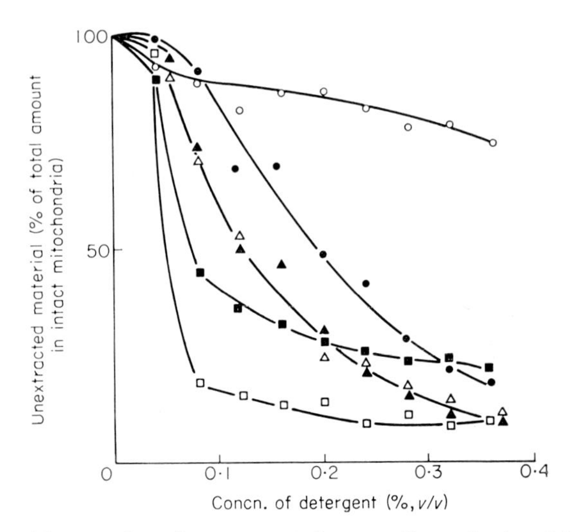

FIG. 24. Sequential extraction of components from rat-liver mitochondria by treatment with increasing concentrations of the neutral detergent, Triton-X-100. □ malate dehydrogenase, ■ total protein, △ phospholipid, ▲ succinate oxidase, ● radioactive protein after incubation *in vitro* with ^{14}C-valine, ○ RNA (from Roodyn, 1962).

It is therefore reasonable to imagine that the mitochondrion is not a simple bag containing a "solution" of enzymes of equal solubility, but rather contains a spectrum of components with a wide range of affinity for the most insoluble of the mitochondrial structures (presumably the structural proteins of the membrane—see Section III. F. below). There therefore may be little meaning in simply listing the "soluble" enzymes of the mitochondrion. If we define the term soluble as meaning that the enzyme is released into solution after disruption of the mitochondria, such a list would be quite large, and would include the enzymes involved in the β–oxidation of fats, several aminotransferases, several dehydrogenases, including malate dehydrogenase, isocitrate dehydrogenase, and glutamate dehydrogenase, fumarase, rhodanese, amino acid activating enzymes, certain phosphatases and phosphotransferases,

◀———

FIG. 23. Changes during disruption of mitochondria by 0·3% deoxycholate. The deoxycholate was layered over a pellet of intact mitochondria and sections taken at different levels of the pellet. (A) bottom of pellet shows intact mitochondria (m_1), and some beginning to swell (m_s); there is some microsomal contamination (μ). (B) middle of pellet, showing greatly swollen mitochondria. (C) top of pellet: mitochondria have shrunken somewhat and the matrix has become greatly diluted. (D) surface of pellet, showing free membranes with complete loss of matrix. $\times$ 30,000 (from Watson and Siekevitz, 1956).

the ATP/ADP exchange enzyme, cytochrome *c*, choline oxidase (1.1.99.1), and other enzymes. However, the procedures used to solubilize these various enzymes are different in many cases, and in order to be accurate one would have to specify the precise conditions of extraction. In order to emphasize that extractability is a very arbitrary definition of solubility, it may be pointed out that the most insoluble of all mitochondrial proteins, the structural protein, may in fact be extracted by a concentrated mixture of suitable detergents (Criddle *et al.*, 1962).

In a recent study of this problem, Klingenberg and Pfaff (1966) divided mitochondrial enzymes into three groups: 1. "easily associated", 2. "tightly enclosed" and 3. "structurally bound". These corresponded most probably to enzymes in the outer compartment, matrix and membrane, respectively. Enzymes in group 1 are easily extracted with salt media and include phosphotransferases, adenylate kinase (2.7.4.3.), creatine kinase (2.7.3.2.), the ATP/ADP exchange enzyme and cytochrome *c*. Enzymes in group 2 may be dissociated from the mitochondrial structure by mechanical means, such as

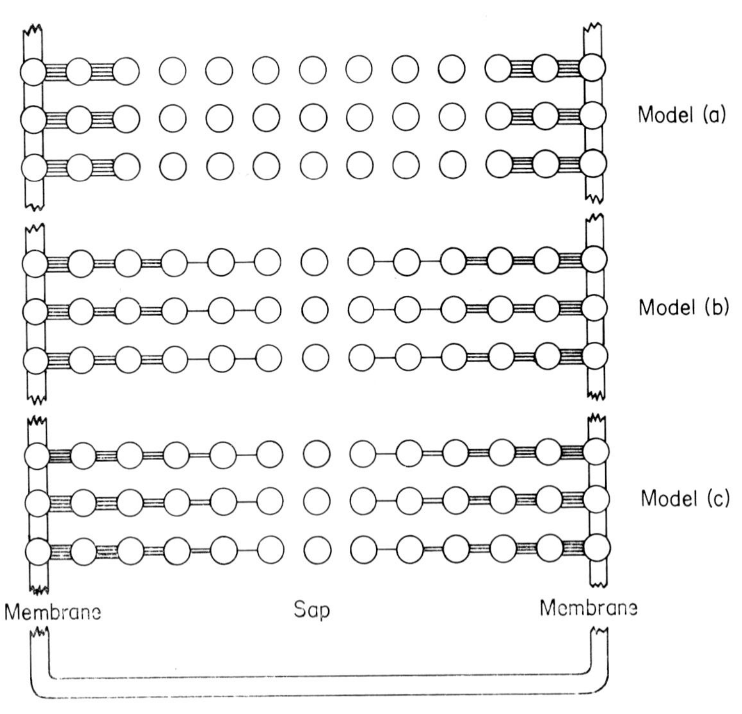

FIG. 25. Three models of mitochondrial structure showing possible binding of soluble proteins. (a) Proteins are either free or membrane-bound. (b) Intermediate situation in which proteins fall into several groups, including no binding, loose binding, and tight binding, and (c) Proteins have a continuous spectrum of binding. (Strength of binding is indicated by number of lines between molecules).

sonic vibrations, and include many NAD and NADP linked dehydrogenases (such as malate dehydrogenase, isocitrate dehydrogenase, glutamate dehydrogenase and lipoyl dehydrogenase (1.6.4.3.)), and a proportion of the aminotransferases, phosphotransferase, cytochrome *c* and ATPase. Finally the "structurally-bound" enzymes in group 3 are only released by detergent and include cytochromes a, a_3, c_1 and b, succinate dehydrogenase, NADH dehydrogenase, and some of the ATPase.

Thus the mitochondrion contains a number of components of varying solubility that can be released by more or less violent treatments. It is preferable not to talk simply of "soluble" or "insoluble" but to specify the treatment needed to detach the component from the least soluble structural elements of the mitochondrion. Three possible models of mitochondrial structure that illustrate this question are given in Fig. 25. In the first (Fig. 25a) the mitochondrion is depicted as consisting of a bag of soluble enzymes. Insoluble enzymes are attached to the inner surface of the bag, and represent the membrane-bound enzymes. In the second model (Fig. 25b) there are some completely soluble elements within the bag, but the walls of the bag are in fact multilayered with a range of more or less tightly bound components. In the third model (Fig. 25c) there are no "soluble" or free components at all, but rather a continous spectrum of binding from extremely tight binding to extremely weak binding. (These models ignore the question of compartments within the mitochondrion, and clearly may be drawn in a more complicated fashion). Of the three models that represent the intermediate situation, Fig. 25b is probably the most accurate on present evidence.

B. MITOCHONDRIAL INCLUSIONS: DNA

The term "inclusion" includes any large ordered structure lying in the mitochondrial sap. The density of the matrix is generally rather high in intact mitochondria, and it is difficult to make out any fine structure within it. For this reason there is considerable uncertainty as to whether there are intramitochondrial ribosomes or not. There has been a tendency to regard the matrix as a rather homogeneous material. However, this may be false, and more refined electron microscopy techniques may reveal structured elements hitherto not noticed. For example, sections of mitochondria have been examined for many years, and yet it is only recently that the presence of DNA fibrils has been well established.

DNA is the most important of the "inclusions". The history of the discovery of mitochondrial DNA is an interesting example of how an oversimplified view can easily develope into a false dogma that can then retard the development of further research. The false dogma in this case is that of the exclusively nuclear localization of DNA. Although there has been considerable genetic evidence for the existence of systems of cytoplasmic inheritance for many years and although there were several well-authenticated examples of cytoplasmic DNA (see Gahan and Chayen, 1965), until very recently, most biochemists, at least, held the view that in "normal" cells, such as rat-liver, all the DNA was in the nucleus. Since this would have made DNA a very useful chemical marker for nuclei, the author was rather surprised to find in surveying the evidence for this view that in fact there were very few experiments in which it had been rigorously established that 100% of the DNA was in the nuclear fraction (Roodyn, 1959). Thus, in most fractionations in which the bulk of the DNA (90–95% of the total) had been recovered in the nuclear fraction, no analyses had in fact been carried out on the cytoplasmic fractions. Although there are clearly great technical difficulties in establishing, for example, that 1% of the total DNA of the cell found in the mitochondrial fraction is not entirely due to contamination with nuclear fragments, it is an extraordinary illustration of the power of the dogma of the exclusive nuclear localization of DNA, that of the thousands of studies on mitochondria in the fifties and early sixties, there were extremely few reports of DNA assays in the mitochondrial fraction. Yet the discovery of mitochondrial DNA may well lead to a revolution in our thinking about mitochondrial function.

Chèvremont, Chèvremont-Comhaire and Baeckeland (1959) studied tissue-culture fibroblasts by autoradiography and cytochemical means, and reported the presence of DNA in mitochondria in these cells. Later they found that treatment of these cells with DNase resulted in greatly increased DNA contents in the mitochondria, as judged by the Feulgen stain. (Chèvremont, Baeckland and Chèvremont-Comhaire, 1960). However, it was not until the demonstration by Nass and Nass (1963) that fibrous inclusions frequently observed in the matrix of mitochondria (e.g. Nass and Nass, 1962) were in fact DNA fibrils that serious attention was focused on the possible existence of mitochondrial DNA. The DNA fibrils, which were present either in a clumped or dispersed state, were observed in mitochondria from a wide range of biological material (Nass *et al*, 1965) and there can be little doubt that they are present in all mitochondria. A typical example of these fibrils (in mitochondria in a gastropod) is shown in Fig. 26. The conclusion that DNA is present in all mitochondria appears to be confirmed by the isolation of DNA from mitochondria from a wide range of organisms. Thus preparations of DNA that are distinct from nuclear DNA have been isolated from mitochondria from rat, rabbit, pigeon, chick-embryo liver, chick-embryo and beef-heart, a variety of plant tissues, *Neurospora crassa*, *Tetrahymena pyriformis*, *Paramecium aurelia*, and *Saccharomyces cerevisiae*. A full survey of these findings is given elsewhere (Roodyn and Wilkie, 1966a)

FIG. 26. DNA-fibres in mitochondria of larval ectoderm of the gastropod *Littorina littorea*. Fixed in 2% OsO_4 in sea water. The fibres have clumped to form straight stubby rods (arrows) connected to the cristae by finer fibres × 64,000 (from Nass *et al*., 1965).

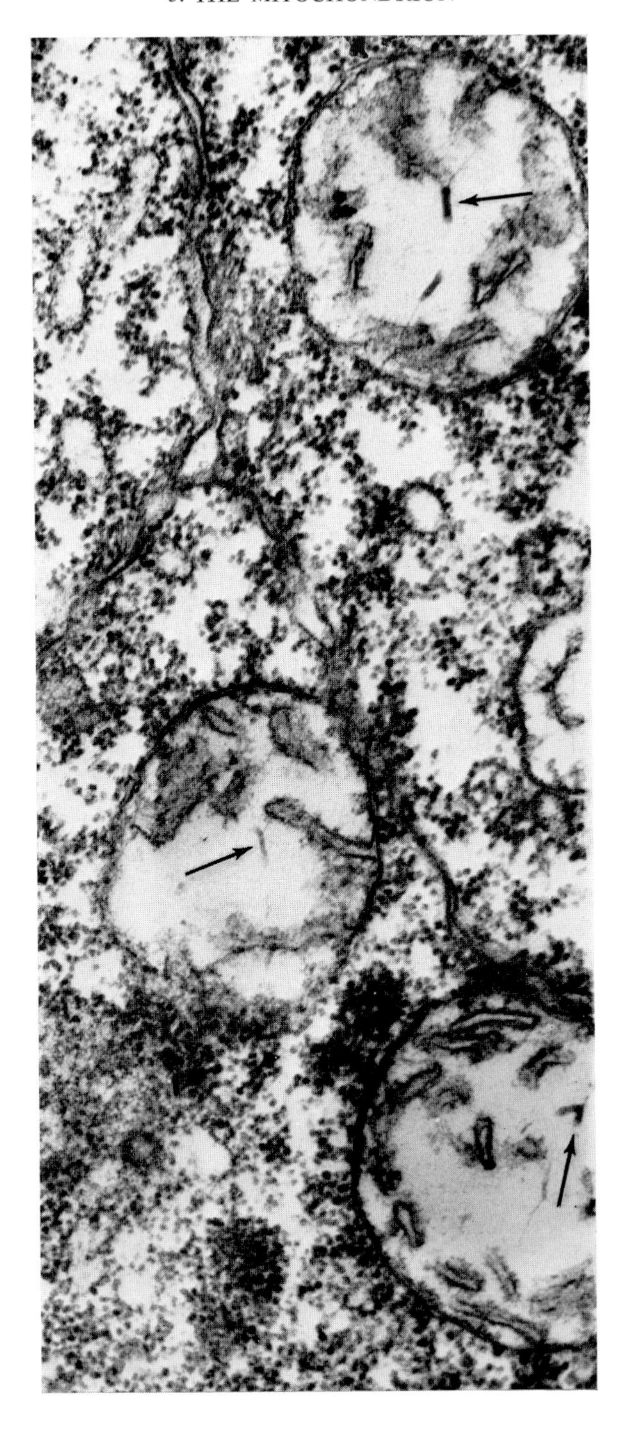

and it is clear that interest in mitochondrial DNA is growing very rapidly. The mitochondrial DNA is double-stranded and generally, but not always, has a lower specific gravity than nuclear DNA from the same organism. It has a different melting temperature and base composition. Caesium chloride density gradients of mitochondrial and nuclear DNA from chick embryo liver are shown in Fig. 27. These results are from Rabinowitz, Sinclair,

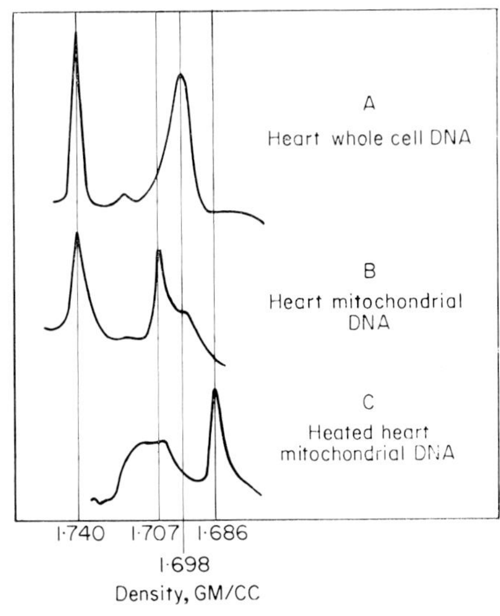

FIG. 27. Sedimentation profiles in caesium chloride gradients of DNA from nuclear and mitochondrial fractions of chick-embryo liver. The band at $\rho = 1{\cdot}740$ is a marker DNA (from Rabinowitz *et al.*, 1965).

DeSalle, Haselkorn and Swift (1965). Mitochondrial DNA of considerable size has been isolated from *Neurospora crassa* by Luck and Reich (1964). It appeared under the electron-microscope as a thread of length $6{\cdot}6\,\mu$, corresponding to a molecular weight of about 13×10^6. This rather striking thread is shown in Fig. 28. Tewari, Jayaraman and Mahler (1965) have recently made a careful study of DNA from yeast mitochondria, and have shown that mitochondrial DNA is very different in its properties (e.g. in its thermal transition properties) from DNA associated with purified yeast lactate dehydrogenase (cytochrome b_2) (1.1.2.3.) that is also associated with mitochondria. This DNA is a single strand of only 33 deoxyribonucleotides (Mahler and Pereira, 1962).

In the slime mould *Didymium nigripes*, the DNA in the mitochondria is located in a characteristic "core" that appears to consist of tightly coiled fibrils, and certainly has all the appearance of an organized structure (Schuster, 1965). It is clear from this work and that of Nass and Nass that the DNA fibrils do not lie in a randomly mixed "bundle" within the mitochondrion.

Mitochondria from certain cells, particularly in the nervous system, have been found to contain definite fibrous elements in the matrix. A particularly striking example of this was found by Mugnaini (1964) in certain glial cells (classified as astrocytes). In some mitochondria, filaments with a very clear helical

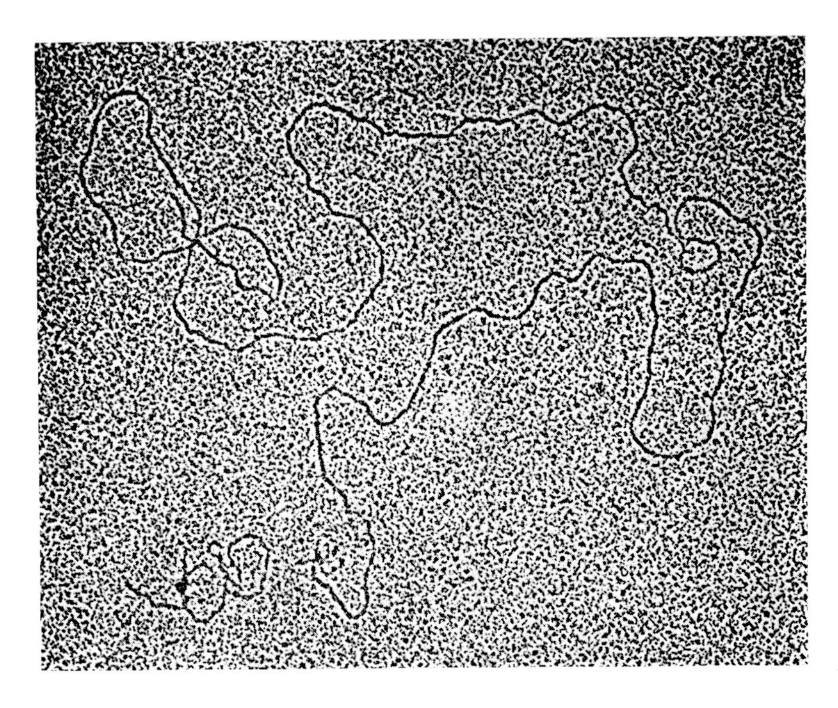

FIG. 28. Single molecule of DNA from mitochondria of *Neurospora crassa*. Length of molecule is 6·6 μ. $\times$ 55,000 (from Luck and Reich, 1964).

structure almost entirely filled the intracristal space. A model of such a spiral filament is shown in Fig. 29. It is not yet certain whether these structures are related to DNA-containing fibrils or not.

The relationship of DNA to the mitochondrial membrane is a question of some interest. Nass *et al.* (1965) observed that although the main mass of DNA fibrils appeared to lie in the matrix, it was frequently connected to the mitochondrial membrane or to the cristae by several fine fibrils. However it was not certain whether this was the result of fixation. Kroon (1965) has recently found that sub-mitochondrial particles prepared by digitonin lysis of beef-heart mitochondria were considerably enriched in DNA, relative to the intact mitochondria. This would suggest a rather close relationship between DNA and the mitochondrial membrane. As it appears that the membrane is the chief site of protein synthesis in the mitochondria (see Section

III. E.) and that DNA has a role to play in this synthesis, elucidation of the precise intra-mitochondrial localization of DNA is clearly of the greatest importance.

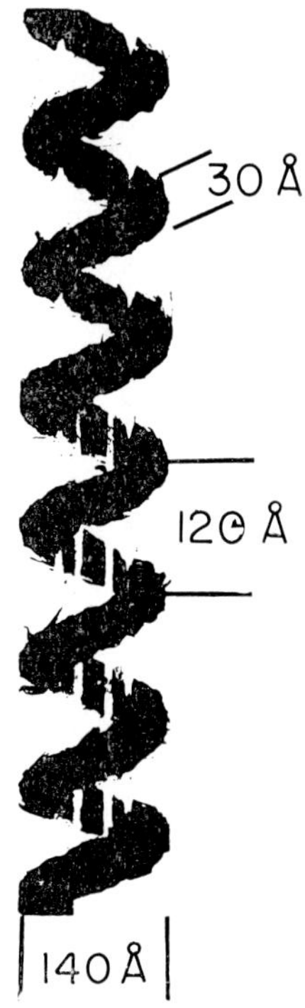

FIG. 29. Model of a helical filament observed in mitochondria from astrocytes in corpus striatum of the rat. The filament is about 30 Å thick, the total diameter of the helix is about 140 Å and the pitch is 120 Å (from Mugnaini, 1964).

C. MITOCHONDRIAL INCLUSIONS: DENSE GRANULES

These granules are very obvious in electron-micrographs of mitochondria (see Fig. 3 for example). However, their nature was unknown for many years. It has been shown recently that they are insoluble inorganic salts. Probably they normally consist mainly of the insoluble calcium phosphate salt hydroxyapatite, which has the composition $(Ca_3(PO_4)_2)_3.Ca(OH)_2$. However, the precise composition of the granules depends on the salts to which the mitochondria have been exposed. For example, Peachey (1964) incubated smooth muscle cells from toad bladder in calcium-free Ringer's solution containing 2 mM barium acetate. He found the appearance of dense intra-mitochondrial granules, with a 40 Å fine structure, and a characteristic hollow appearance. These were probably granules of barium phosphate (Fig. 30). Isolation of granules from Ca^{2+} loaded mitochondria has recently been achieved by Weinbach and Von Brand (1965). The major constituents of granules isolated from mitochondria disrupted by deoxycholate were found to be Ca^{2+} 30·0%, P_i 18·5%, Mg^{2+} 3·9% and CO_3^{2-} 4·0%. (These values are the % values of residual weight after ashing). The granules appeared

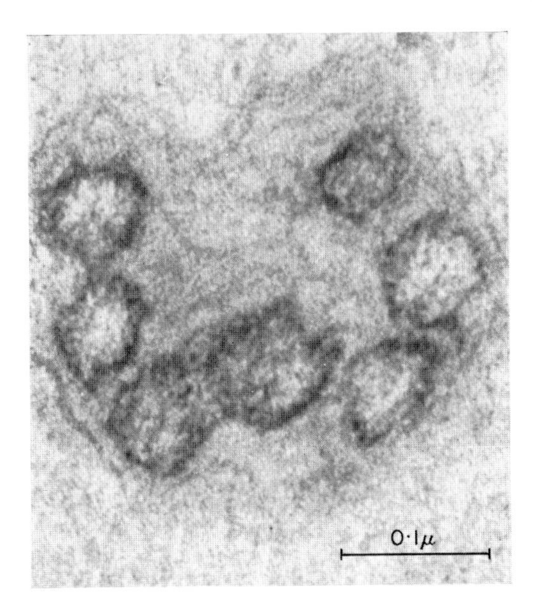

FIG. 30. Intra-mitochondrial granules, probably of barium phosphate, from smooth muscle cells of toad urinary bladder incubated in calcium-free Ringer's solution containing 2 mM barium acetate. Lead acetate stain. ×210,000 (from Peachey, 1964).

to contain some organic material, and their composition varied somewhat with the method of disruption of the mitochondria.

It is possible that the granules act as stores for Ca^{2+}, Mg^{2+} or even phosphate ions. It may be noted that H^+ ions would be liberated during the formation of calcium, magnesium, or even barium and strontium phosphates. For example, Brierley, Bachmann and Green (1962) formulated the uptake of Mg^{2+} and phosphate ions by beef-heart mitochondria by the reaction:

$$MgCl_2 + 2\,MgHPO_4 \leftrightharpoons (Mg)_3(PO_4)_2 + 2\,HCl$$

with one mole of acid being produced per mole of phosphate bound. These reactions have been expressed in a general way by Chappell and Crofts (1965a) as one of two possibilities:

$$3M^{2+} + 2HPO_4^{2-} \rightarrow M_3(PO_4)_2 + 2H^+ \text{ or}$$
$$3M^{2+} + 2H_2PO_4 \rightarrow M_3(PO_4)_2 + 4H^+$$

(M is a general symbol for a divalent cation).

Thus the H^+ production per divalent cation taken up will depend on the pH. At pH 7·2, near to the second acid dissociation constant of orthophosphate, the ratio will be approximately 1 H^+ per M^{2+} taken up. Chappell and Crofts (1965a) have exploited this fact to perform some ingenious experiments demonstrating the connection between ion transport, mitochondrial swelling and respiration. An interesting finding is that if excess Ca^{2+} is added, there is extensive swelling followed by release of Ca^{2+} into the medium. If this is then removed from the medium by a suitable chelating agent, the swelling is reversed (Crofts and Chappell, 1965). This reversal of swelling is dependent on respiration, and could be supported by electron transport through a limited portion of the respiratory chain. The contraction process is thought to be mediated by an intermediate acting at a site between X ~ I and the respiratory chain. The site of action of various inhibitors, and the supposed sites of interaction of the ion transport, contractile and ATP synthesizing systems are given in Fig. 31, from Crofts and Chappell (1965).

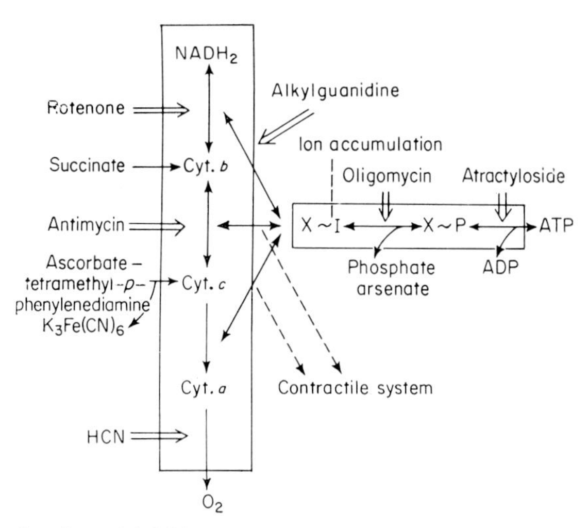

FIG. 31. Sites of action of inhibitors and associations of respiratory chain, contractile system, and ion accumulation as given by Crofts and Chappell (1965).

To conclude this brief account of mitochondrial inclusions reference should be made to protein inclusions, sometimes present as crystals of striking size, that are occassionally found in mitochondria. An elegant example of this is given by Ward (1962) in a study of mitochondria from oocytes of *Rana pipiens*, and a picture of this enormous inclusion is shown in Fig. 32. Oberling (1959) has described protein-storage deposits in rat-kidney mitochondria after injections of ovalbumin.

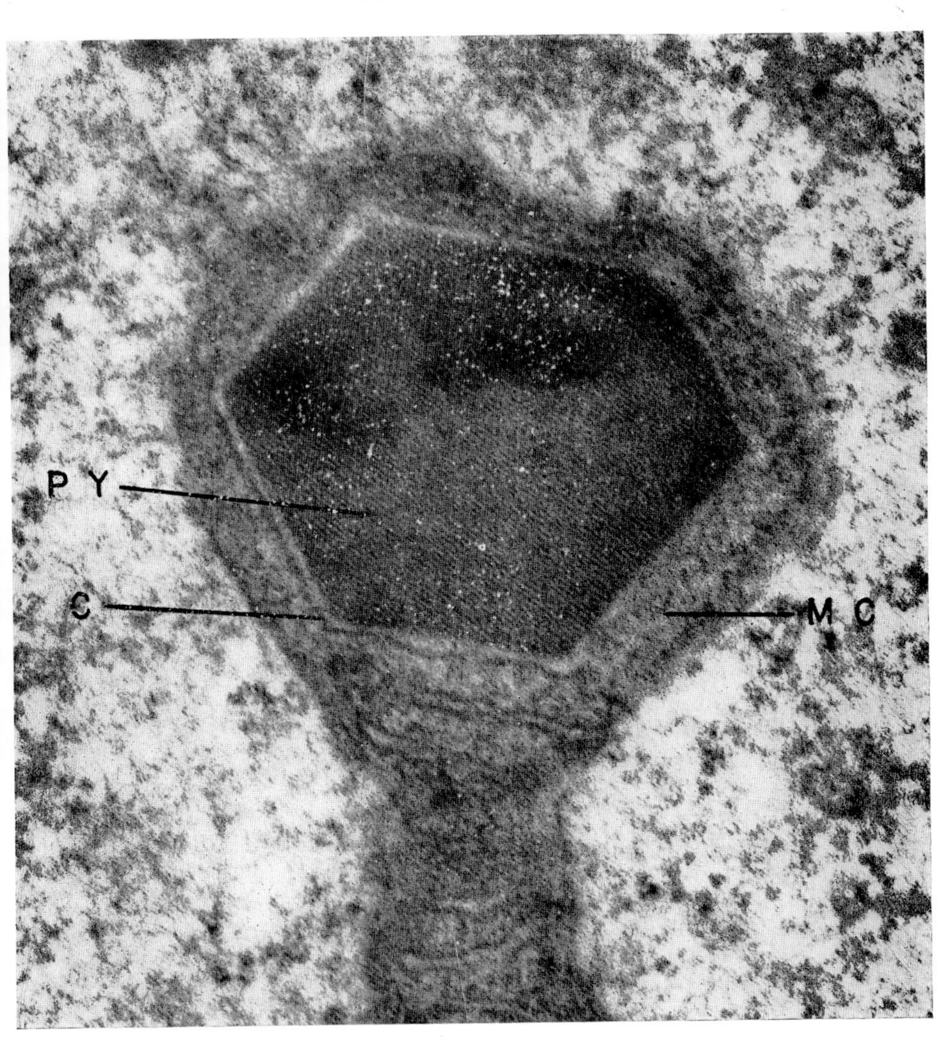

FIG. 32. Hexagonal crystal of yolk protein (yolk platelet PY) present in mitochondrial cortex (MC) and surrounded by a membrane (C) believed to be the limiting membrane of a greatly enlarged crista. $\times$100,000 (from Ward, 1962).

D. THE MITOCHONDRIAL MEMBRANE SYSTEM

Unfortunately, fractionation methods are not yet sufficiently refined to be able to separate the outer membranes, inner membranes and cristae membranes. Much of the evidence for localization in different parts of the membrane system is therefore essentially indirect. For example the greater number of cristae in mitochondria that have high respiratory activity per unit weight suggests that the respiratory system is located on the membranes of the cristae. The assignment of enzymes to specific parts of the mitochondrial membrane system is often based on indirect conclusions from metabolic studies and the effects of inhibitors (cf. Figs. 7, 8 and 22). Although some workers have presented rather detailed models of the structural organization of mitochondrial enzyme systems (see Section III. E. below) it should be stressed that at least some of these are still essentially conjectures or, at best, reasonable estimates, and are certainly not established with any great certainty. For these reasons it is best to consider the mitochondrial membranes as a whole, under the general heading of "membrane system".

E. ENZYMES OR ENZYME SYSTEMS ASSOCIATED WITH THE MEMBRANE

1. Respiratory Enzymes

Many of the activities described in Section II. above are associated with the mitochondrial membrane. The most well known example of this is the respiratory chain. There is little doubt that all the elements of the respiratory chain concerned with the oxidation of reduced NAD, NADP and FAD are bound to the mitochondrial membrane. It is also of considerable importance that the terminal enzyme of the chain, cytochrome oxidase, is clearly associated with the membrane (Siekevitz and Watson, 1956). In fact, it is generally believed that the respiratory enzymes are arranged in highly organized groups or complexes, usually called "respiratory assemblies", arranged along the surface of the mitochondrial membrane. There is still uncertainty as to the number, size and precise composition of these assemblies. However, it is generally agreed that in addition to the component enzymes of the respiratory chain (NADH dehydrogenase, cytochromes b, c_1, c, a and a_3) there are "auxiliary" respiratory enzymes associated with cellular oxidations, namely succinate dehydrogenase, NAD/NADP transhydrogenase (1.6.1.1.), fatty acyl CoA dehydrogenase (1.3.99.3.) and various other enzymes. Also there is little doubt that factors involved in oxidative phosphorylation are also associated with the respiratory assembly. (An up to date account of such "coupling factors" is given in Sanadi, 1965). An estimate of the composition and possible particle weight of the assembly is given in Lehninger (1965), and values for the chemical and enzymic constitution of the "unit of electron transfer" or "elementary particle" are given by Blair, Oda, Green and Fernàndez-Moran (1963). From these values, these workers calculated that the weight of the particle was about $2{\cdot}1 \times 10^6$. This is considerably higher than the value of $1{\cdot}4 \times 10^6$ which is the combined particle weights of the complexes thought to be present in the particle (see below). It was suggested that this discrepancy was due to the presence of a protein "not required for integrated electron transfer activity", probably corresponding to the structural protein of Criddle *et al.* (1962).

Some estimates have been made of the number of respiratory assemblies or "units" per mitochondrion. The number is certainly of the order of tens of thousands. For example Green and Oda (1961) give an estimated number of 17000–18000 per heart mitochondrion. These authors also calculate that there are about 300 units per crista. The number of cytochrome *a* molecules per "average" liver mitochondrion has been estimated at about 17000 in an interesting paper by Estabrook and Holowinsky (1961). However, this is not equal to the number of units, since there are probably several cytochrome *a* molecules per unit. Thus the number of respiratory assemblies per liver mitochondrion is probably considerably less than the number per heart mitochondrion. This is quite consistent with the fact the cristae in liver mitochondria are much less densely packed than in heart mitochondria. Estabrook and Holowinsky also made the interesting calculation that there were about 130,000 molecules of "enzymically-reducible pigments" and about 500,000 molecules of pyridine nucleotide (NAD + NADP) in each liver mitochondrion. This gives us some idea of the potential complexity of the mitochondrial respiratory system, and also shows the magnitude of the task of determining the molecular structure of the mitochondrion.

A question of considerable practical and theoretical importance is whether it is certain that the various components of the assemblies are present in stoichiometric amounts i.e. whether there are integral numbers of enzymes in each assembly, or whether the assembly has a somewhat loose composition with considerable variation in the relative number of components. The "integral", "quantal" or "stoichiometric" philosophy is satisfying intellectually and has been the driving force in the prolific studies on the organization of the mitochondrion that have come from the Madison School. In a large number of publications, Green and his co-workers have examined the detailed enzymology of a number of sub-mitochondrial preparations in order to determine the precise composition of the respiratory assembly and its relation to mitochondrial structure. (The relationship of the so-called "stalked-bodies" and the respiratory assembly will be discussed below: Section III. G.). As a result of these studies, Green and his colleagues have resolved the respiratory system into four complexes, and have attempted to define the exact number of molecules of each type of respiratory enzyme in each complex. The four complexes are I, NADH-coenzyme Q reductase; II, succinate-coenzyme Q reductase; III, reduced coenzyme Q-cytochrome *c* reductase, and IV, reduced cytochrome *c* oxidase. The molecular composition of these complexes and their inter-relations are given in Fig. 33, which was derived from Green and Fleischer (1963). The particle weights of the complexes are given in Blair *et al.* (1963) as follows: Complex I 530,000, Complex II 230,000, Complex III 200,000 and Complex IV 430,000. If the complexes were combined in a simple 1:1:1:1 ratio the complete system would thus have a particle weight of $1{\cdot}39 \times 10^6$, which is less than the weight of isolated elementary particles (see above).

Although there is still some discussion about the fine details of the composition of the respiratory assembly, there can be little doubt that the formulation in Fig. 33 is a reasonable picture of the major aspects of organization

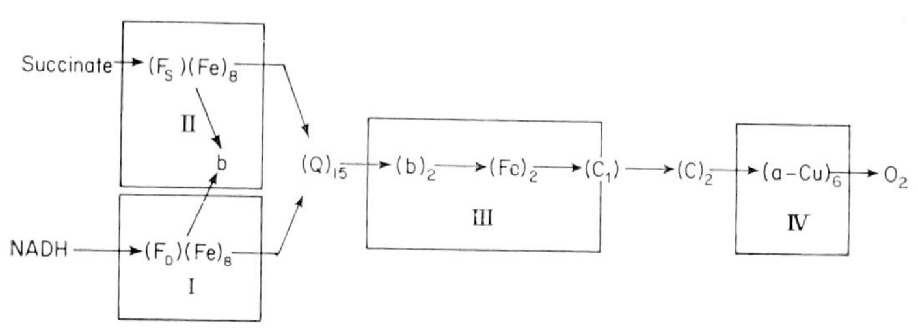

FIG. 33. Proposed stoichiometry of complexes of respiratory chain. The number of molecules or atoms of each component is given as a subscript. F_s: succinate dehydrogenase; FD: NADH (DPNH) dehydrogenase; *b*, c_1, *c*, *a* the various cytochromes; (Fe): non-haem iron; (Cu): copper; Q: coenzyme-Q. See text for names and molecular weights of the complexes I–IV (derived from Green and Fleischer, 1963).

of the assembly. Associated with these components are other organized enzyme systems concerned with feeding NADH to the respiratory chain. The most intensively studied of these "primary dehydrogenases", as they have been called, are the complexes involved in the oxidative decarboxylation of pyruvate and α-oxoglutarate. The detailed mechanism of these reactions has been described in Fig. 13 and space does not allow a full discussion of recent research into the structural basis of these reactions. (See Sanadi, 1963). An interesting example of these studies is a recent theoretical paper by Whitehead (1965) in which he proposes a model for the pyruvate oxidase complex given in Fig. 34. This complex had been studied in detail by Koike, Reed and Carroll (1963). It was isolated as a particle of molecular weight about 4×10^6. Koike and co-workers reversibly dissociated this into three sub-units which could be recombined with restoration of activity. The three sub-units were as follows:

1. A pyruvate-carboxylase that catalyses the reaction:

$$\text{pyruvate} + [\text{thiamine-PP}] \rightarrow [\text{acetaldehyde-thiamine-PP}] + CO_2$$

(The symbols [ ] indicate that the component is bound to the protein).

2. A lipoate reductase-transacetylase that catalyses the reaction:

$$[\text{acetaldehyde-thiamine-PP}] + \text{lip-}S_2 \rightarrow [\text{acetyl.S.lip.SH}] + [\text{thiaminePP}]$$

(where lip-S_2 is lipoic acid and acetyl-S-lip.SH is S-acetyl-dihydrolipoic acid) and the subsequent reaction:

$$[\text{acetyl-S-lip.SH}] + \text{CoA.SH} \rightarrow \text{acetyl.S.CoA} + [\text{lip.}SH_2]$$

(where lip.SH_2 is dihydrolipoic acid).

3. A dihydrolipoate dehydrogenase that catalyses the reaction:

$$[\text{lip.SH}_2] + \text{NAD}^+ \xrightarrow{\text{FAD}} [\text{lip.S}_2] + \text{NADH} + \text{H}^+$$

The spatial arrangement of these subunits given in Fig. 34 was derived by Whitehead on the basis of a general theory of the structure of multimolecular complexes. This system has been described in some detail in order to give an example of the sort of molecular organization that is possibly present in many regions of the mitochondrial membrane system.

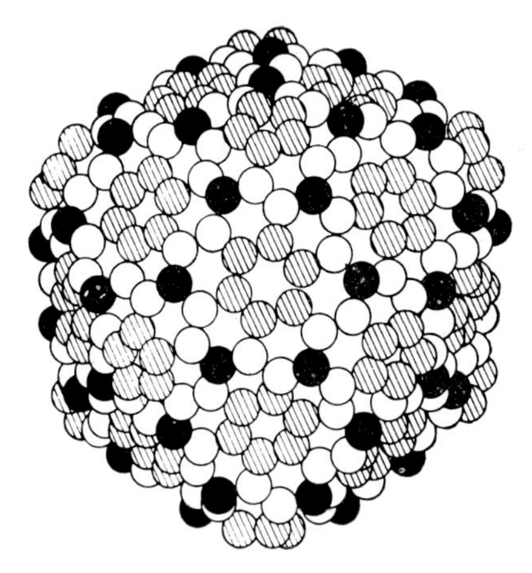

FIG. 34. Proposed structure of pyruvate decarboxylation complex. ●: dehydrogenase sub-units; ◍: carboxylase sub-units; ○: lipoic reductase-transacetylase sub-units (see text for activity of sub-units) (from Whitehead, 1965).

An interesting, and important, confirmation of the stoichiometric theory of respiratory chain structure has come from the work of Bücher, Pette and co-workers on the precise measurement of groups of mitochondrial enzymes (see for example, Pette, Klingenberg and Bücher, 1962; Klingenberg and Pette, 1962). It is found that if mitochondria from different sources are compared, some of the mitochondrial enzymes occur in groups of strikingly constant relative proportions, (the so-called "constant-proportion groups") For example, malate dehydrogense, succinate dehydrogenase, pyruvate oxidase and cytochrome *a* were always found in a certain ratio to each other, and also to the amount of cytochrome $c + c_1$ in the mitochondria. In addition to the constant proportion groups whose compositions are different in different types of mitochondria there are so called "specific proportion groups" that are characteristic of the type of mitochondrion. For example, insect flight muscle mitochondria have a high ratio of glycerol-1-phosphate dehydrogenase (1.1.99.5.) to glutamate dehydrogenase, whereas mammalain liver mitochondria have the converse. Such results suggest that the essential respiratory system common to all mitochondria consists of a number of rigidly determined units, which can vary in relative and absolute numbers in different mitochondria, but not in composition. "Ancillary" enzyme systems, however, may be present in different relative amounts, depending on the specialized function of the mitochondrion in question.

2. Enzymes Involved in Protein Synthesis

Another group of enzymes clearly associated with the mitochondrial membrane are those involved in amino acid incorporation into protein. If rat-liver mitochondria are incubated with radioactive amino acids, the most radioactive proteins formed are those associated with the mitochondrial membrane (Roodyn, 1962). The soluble mitochondrial enzymes malate dehydrogenase and cytochrome *c* do not appear to become labelled *in vitro* (Roodyn *et al.*, 1962). The radioactive product has many similar properties to the structural protein described by Criddle *et al.* (1962). The labelling of insoluble proteins in the membrane, rather than soluble proteins in the sap, may have some bearing on theories of the mechanism of mitochondrial replication, and it is interesting that similar labelling of insoluble proteins has been observed recently in the phylogenetically widely separated organisms, yeast (Wintersberger, 1965) and insect flight muscle (Bücher, 1965). It is therefore possible that labelling of membrane proteins by isolated mitochondria is a general phenomenon.

Studies of amino acid incorporation with sub-mitochondrial particles has confirmed that membrane-linked enzymes are actively involved in mitochondrial protein synthesis (Truman, 1963; Kroon, 1965). In fact, the amino acid incorporation activity observed by Kroon in sub-mitochondrial particles was comparable to that observed by other workers with isolated microsomal systems. As a result of inhibitor studies with chloramphenicol and actinomycin D, Kroon (1965) has suggested that mitochondrial protein synthesis is dependent on the continuous synthesis of RNA by mitochondrial DNA-dependent RNA polymerase. If this is so, the protein synthesis machinery in the mitochondrion would include DNA, RNA polymerase, some form of messenger RNA and enzymes involved in synthesis of peptide bonds. If all these are bound to the membrane, they would be in close proximity to the energy generating system of the mitochondrion. There is no doubt that there is a very close relationship between the protein synthesis and oxidative phosphorylation systems of mitochondria (Roodyn *et al.*, 1965; Roodyn, 1965b), and it would be most exciting if it could be shown that the mitochondrial membrane has at least some capacity for self-replication. Some of the advantages of such a limited self-replicating system are discussed in Roodyn and Wilkie (1966a).

3. Transport Systems

Another system of enzymes undoubtedly associated with the mitochondrial membrane are those concerned with the transport of metabolites and ions across the membrane. Interest in this field is rapidly expanding, and has been stimulated by investigations into the mode of action of the inhibitor of oxidative phosphorylation, the glycoside atractyloside. The drug is also called potassium atractylate and is extracted from the rhizome of a Mediterranean member of the Compositae called *Atractylis gummifera* (Bruni, Contessa and Luciani 1962). Inhibition of oxidative phosphorylation was reversed competitively by addition of adenine nucleotides. However, inhibition by oligomycin was not reversed by this. It appears, therefore, that unlike oligomycin which reacts directly with some intermediate in oxidative phosphorylation, these agents prevent the transfer of adenine nucleotides across the mitochondrial membrane. Chappell and Crofts (1965b) have made a detailed comparison of the actions of oligomycin and atractyloside. The effect of the drug on adenine nucleotide transport has been fully confirmed by direct measurement of these nucleotides in mitochondrial "spaces" (Klingenberg and Pfaff, 1966). The site of action of atractyloside is shown in Figs. 8 and 21. The drug may therefore be of great use in delineating the enzyme systems involved in the transport of adenine nucleotides. Carnitine-linked transfer systems in fat metabolism have also been described in the membrane, as shown above in Figs. 7 and 8. It is still too early to assign such "permeases" to the inner or outer membranes with complete certainty, but there is little doubt that we will have a much clearer picture of their location in the very near future.

There is little evidence to suggest at the moment that the transport of ions across the mitochondrial membrane is controlled by enzymes or "permeases" specific for each type of ion. However, it is certain that active ion transport is closely associated with the respiratory and energy-transfer systems, and is dependent on the physical integrity of the mitochondrial membrane. Also, there is little doubt that non-phosphorylated energy rich intermediates can provide the energy for ion translocation and accumulation (see for example Brierley, Murer, Bachmann and Green, 1963). [It should be mentioned here that it is important to distinguish between ion transport of relatively low concentrations of ions, and the rather massive uptake of Ca^{2+}, Mg^{2+} and phosphate ions that is associated with the formation of dense granules. Rossi and Lehninger (1964) discuss the differences in properties of these two systems.]

In a recent study, Chance (1965) suggests that Ca^{2+} ions can interact with the three energy-conservation sites in the respiratory chain, and the reaction is probably with the $X \sim I$ type of compound, although he does not completely rule out the possibility of reaction with $X{\sim}P$ to form P_i and $Ca^{2+} \sim X$. The reactions proposed by Chance for the overall mechanism of calcium binding and uptake by mitochondria are shown in Fig. 35, and it can be seen that the essential postulate is the formation of an intermediate of the type

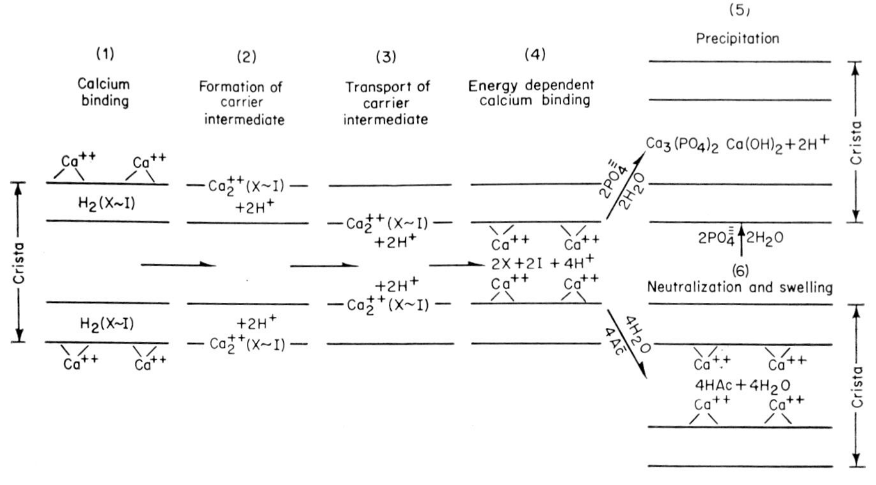

FIG. 35. Hypothesis for the accumulation of cations in mitochondrial membranes (from Chance, 1965).

$Ca^{2+}(X \sim I)$. The subject is in a state of rapid development at the moment, but it would not be unreasonable to imagine that the close connection between ion transport, oxidative phosphorylation, and the swelling-contraction cycles of the mitochondrial membrane is a result of the fact that all these activities are dependent on the maintenance of the identical ordered structure. They cannot be separated by classical fractionation methods, as would be used for enzyme purification, because they represent different aspects of the behaviour of a single entity. This view of the mitochondrial membrane as an integrated structure that is dependent on the maintenance of relative positions of its component parts for correct functioning is discussed more fully in Section III. H. below. Some of these questions are dealt with in an informative article by Lehninger (1962) on the relationship between water uptake and extrusion by mitochondria and the structure and function of the mitochondrial membrane.

The above account of the metabolic activity of the mitochondrial membrane system is far from complete, and is intended merely to present a picture of a series of complex systems that can interact both functionally and spatially.

F. CHEMICAL COMPOSITION OF THE MEMBRANE

Examination of the chemical composition of the membrane is essential to a correct understanding of its molecular organization. The most striking feature of the membrane is its high lipid content. The amount of lipid in isolated membrane fragments depends very much on the method of isolation used. However, in some cases it may be as high as 50% of the total weight.

Green and Fleischer (1963) have stressed the fundamental importance of lipids, and particularly phospholipids, in the maintenance of the integrity of the electron-transfer and oxidative phosphorylation systems. Indeed there is overwhelming evidence to support this view. Many respiratory enzymes have been shown to form complexes with phospholipids (e.g. cytochrome *c*: Das, Haak and Crane 1965). Also a variety of agents that either remove or destroy mitochondrial lipids have been shown to damage or to disrupt completely the respiratory and phosphorylation systems.

A good example of this is the observation of Sekuzu, Jurtshuk and Green (1961) that the tightly bound D(-)3-hydroxybutyrate dehydrogenase of beef heart mitochondria shows an absolute requirement for lecithin at all stages of purification after the initial extraction from the mitochondria. Failure to add lecithin either to the crude or purified enzyme resulted in complete loss of activity. (This type of phenomenon may well explain the difficulty frequently experienced in the purification of enzymes that are tightly bound to the membrane.) A more striking case is the experiment of Fleischer, Brierley, Klouwen and Slautterback (1962) in which intact mitochondria were treated with aqueous acetone. This resulted in the removal of more than 80% of the total phospholipid, and was accompanied by almost total loss of respiratory activity, as measured by succinate-cytochrome *c* reductase for example. However, respiratory activity could be restored completely by addition of phospholipid. The reactivation of succinoxidase activity in rat-liver mitochondria that had been treated by lecithinase was demonstrated by Casu and Modena (1964), who used lipid extracts of rat-liver as re-activating agents. It is interesting in this connection that there is an exact relationship between the loss of succinoxidase activity and the extraction of phospholipid during "titration" of mitochondria with Triton-X-100 (Roodyn, 1962, and Fig. 24 above).

The changes produced by lipid-damage may sometimes be of a subtle nature. An interesting example is given by Machinist and Singer (1965a,b). Sub-mitochondrial particles (the so-called "electron-transfer" particles or "ETP") are able to reduce added coenzyme Q homologues by NADH (Complex I, Fig. 33 above). As in intact mitochondria, the oxidation of NADH was inhibited by rotenone and Amytal. However, if these preparations were treated for a very short time with purified phospholipase A, the enzyme activity was lost. However, it could be restored by the addition of suitable phospholipids. If the phospholipase treatment was at all prolonged, however, it was no longer possible to restore activity by adding phospholipids. In addition, it is possible to obtain a preparation of purified NADH-coenzyme Q reductase from ETP. However, this preparation has a much lower phospholipid content than ETP, and in addition has lost the sensitivity to rotenone and Amytal.

Thus subtle changes in metabolic activity, or even in response to inhibitors, can result from alterations to the mitochondrial lipids. It is not surprising, therefore, that there are frequent differences between the results obtained by different workers with various sub-mitochondrial fragments or purified enzyme systems. Most, if not all, methods of mitochondrial fragmentation and sub-fractionation would depend on the rupture of lipid-protein bonds of one type or another, and translocation or even complete extraction of lipids would not be unexpected. The same considerations apply to the purification of bound enzymes. It is quite likely that the reason why such a large number of compounds can uncouple oxidative phosphorylation is that any agent which disturbs the precise inter-relationship of lipid and protein in the membrane will result in disturbance of the energy-transfer system. Thus any surface active or lipophilic agent is likely to act as an uncoupler. When one considers the complexity of enzyme systems present in mitochondria and the probability that lipid arrays play an important part in maintaining the correct spatial relationship of these systems, one realizes the magnitude of the task of obtaining a precise picture of the molecular organization of the membrane. This is particularly so because our methods of analysis undoubtedly disrupt these arrays.

Another important factor in the maintenance of membrane structure is undoubtedly the presence of insoluble proteins. The structural framework of the mitochondrion does not depend only on the presence of lipids, since Fleischer *et al*, (1962) have demonstrated the retention of reasonable fine structure in beef-heart mitochondria even after removal of most of the phospholipid by aqueous acetone (Fig. 36). It has been known for many years that most sub-cellular components contain insoluble "residual" proteins, which are not readily brought into solution, for example by repeated extraction with concentrated salt solutions. Levin and Thomas (1961) discuss the subject and describe the properties of insoluble lipoproteins in liver cell fractions. Interest in insoluble proteins responsible for maintaining the structure of cell components was stimulated recently by the isolation of the so-called "structural protein" of mitochondria (Criddle *et al.*, 1962). This protein is extremely insoluble in water at pH 7 but dissolved in anionic detergents, or extremes of pH. In solution structural protein has a marked tendency to polymerize, thus forming insoluble aggregates. The monomer has the rather low molecular weight of 20,000 to 30,000. Its most interesting property is its ability to form complexes with specific respiratory enzymes, e.g. cytochromes b,c_1 and a. It does not interact with cytochrome c except in the presence of phospholipid. It also has the property of combining spontaneously with phospholipid micelles at neutral pH (Richardson, Hultin and Fleischer, 1964). As a result of these properties, it was suggested that structural protein has a central role in the binding together of mitochondrial enzymes. A diagram showing the supposed interaction of lipids, cytochromes and structural protein is presented in Fig. 37 (from Criddle *et al.*, 1962). It is

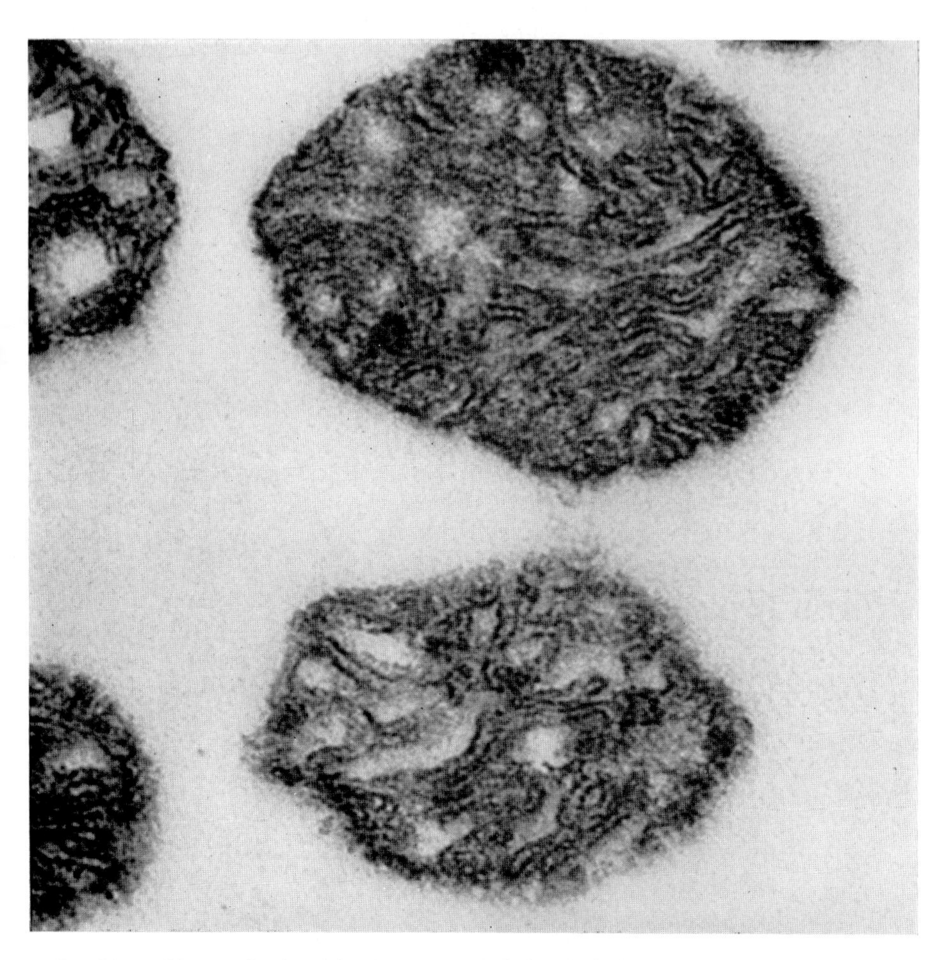

FIG. 36. Beef-heart mitochondria after removal of phospholipid by treatment with aqueous acetone. Note the retention of fine structure. ×86,000 (from Fleischer *et al.*, 1962).

thought that the chief interactions in these complexes are between non-polar regions of the molecules. It has also been suggested that similar proteins occur in all cell membranes and that the binding of phospholipids to structural protein is a phenomenon of general importance. For example, proteins resembling structural protein of mitochondria, and exhibiting binding properties with phospholipid have been isolated from liver microsomal fractions, erythrocyte stroma and leaf chloroplast lamellae (Richardson, Hultin and Green 1963).

The properties of structural protein would confer on it the possibility of being an "organizer" for mitochondrial assembly. One may speculate that the assembly of the mitochondrion may be controlled genetically by synthesis of

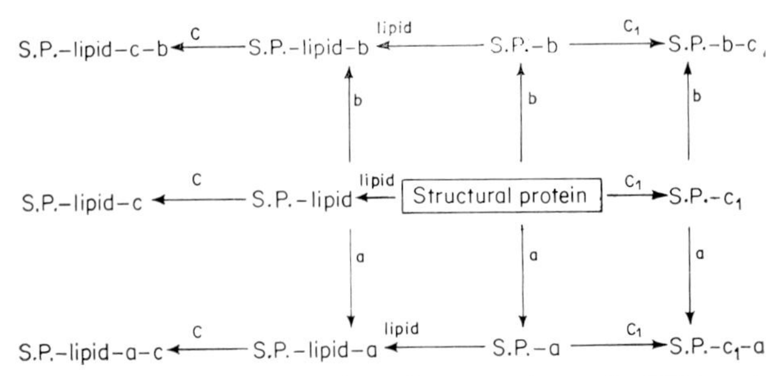

FIG. 37. Schematic summary of interaction of structural protein (S.P.) and cytochromes a, b, c, and c_1 (from Criddle *et al.*, 1962).

a specific mitochondrial "organizer protein" which would then form aggregates of the correct molecular organization by only complexing with those enzymes that had specific binding sites for the organizer protein. Thus if the protein has a region with affinity for a specific region of the cytochrome c molecule, and another region with affinity for part of cytochrome c_1, it would be able to form an aggregate of cytochrome c and c_1. The nature of this aggregate may be determined by complimentary sequences of amino acids in the peptide chains of the organizer and bound proteins. If the amino acid sequence of the organizer protein is itself determined by the nucleotide sequence in a specific type of "organizer" messenger RNA, which is itself encoded in mitochondrial DNA, one would have a genetically controlled system of reasonable simplicity for the ordered synthesis of the complex organization of the mitochondrial membrane.

An intriguing aspect of such a hypothesis is the finding that most of the amino acid incorporation in isolated mitochondrial is into insoluble membrane-bound protein, similar in many ways to the structural protein of Criddle *et al.* (1962) (see Roodyn, 1962). The matter is also discussed in Sections II H. and III E. An interesting experiment has been reported very recently which possibly argues against a role of structural protein in the control of mitochondrial replication, however. Katoh and Sanukida (1965) have shown that structural proteins isolated from mitochondria of wild-type and respiratory deficient yeast have the same sedimentation coefficients, amino acid compositions and serological properties. They also appear to be present in identical amounts in the two kinds of mitochondria. Since there is a gross disturbance of mitochondrial function and organization in respiratory deficient yeast, it seems that structural protein cannot have an important role in causing this. (The significance of this observation is discussed in further detail in Roodyn and Wilkie, 1966a). However, it is difficult at the moment to assess work with structural proteins, because the techniques of isolating them and studying their properties are still rather primitive. Current improvements in this field (for example the observation of Mac-

Lennan, Tzagoloff and Rieske (1965) that succinylation of structural protein results in an increase in its solubility) may well lead to important discoveries, not only about mitochondrial structure but also about the molecular mechanisms of mitochondrial formation.

The observations that RNA appears to be associated with the mitochondrial membrane appear to be relevant to this question (Roodyn, 1962; Kroon, 1966). Although there has been some doubt as to whether or not mitochondria contain RNA, it now seems reasonably certain that the small amounts of RNA found in the mitochondrial fraction are indeed due to mitochondria, and not to microsomal contamination (See Roodyn, 1966 for some recent views on mitochondrial RNA). Sub-fractionation of mitochondria shows that there is very little RNA in the soluble fraction, and most of it remains associated with membranous material. It is very resistant to the action of ribonuclease, even after disruption of the mitochondria to allow access of the enzyme. Resistance to ribonuclease is a diagnostic property of mitochondrial amino acid incorporation into protein. It is interesting that RNA has been found in other "smooth-membranes" (see for example Chauveau, Moulé, Rouiller and Schneebeli, 1962) although the amount per mg protein is far less than in ribosomes. Also, some "post-microsomal" membrane fractions have been shown to have amino acid incorporation activity that is not inhibited by ribonuclease (for an account of these see Hird, McLean and Munro, 1964). This may imply that there are non-ribosomal protein synthesis systems associated with certain membranes. It may be that a special type of "membrane RNA" exists in cell membranes and is concerned in the control of their replication. The question of membrane oriented protein synthesis is fully discussed in an interesting recent article by Hendler (1965).

In conclusion, reference should be made to an interesting aspect of the chemistry of the mitochondrial membrane. Patterson and Touster (1962) reported the presence of sialic acid in mitochondria. However, the microsomal fraction was rich in this material, and the possibility of cross-contamination should not be ruled out. Nevertheless the possible presence of such materials in the membrane certainly deserves further detailed study.

G. STALKED BODIES AND ELEMENTARY PARTICLES

The current state of knowledge of this subject has been discussed recently by Whittaker (1966), from which the following account has mainly been drawn. The application of the technique of negative staining to studies of mitochondrial sub-structure resulted in the discovery of rather striking stalked particles along the surface of the cristae. They were arranged in ordered rows along the membrane and appeared to be attached to it by a definite "neck" or "stalk". This is shown in Fig. 38, taken from Fernández-Moran (1962). Particle studded membranes were observed by many other

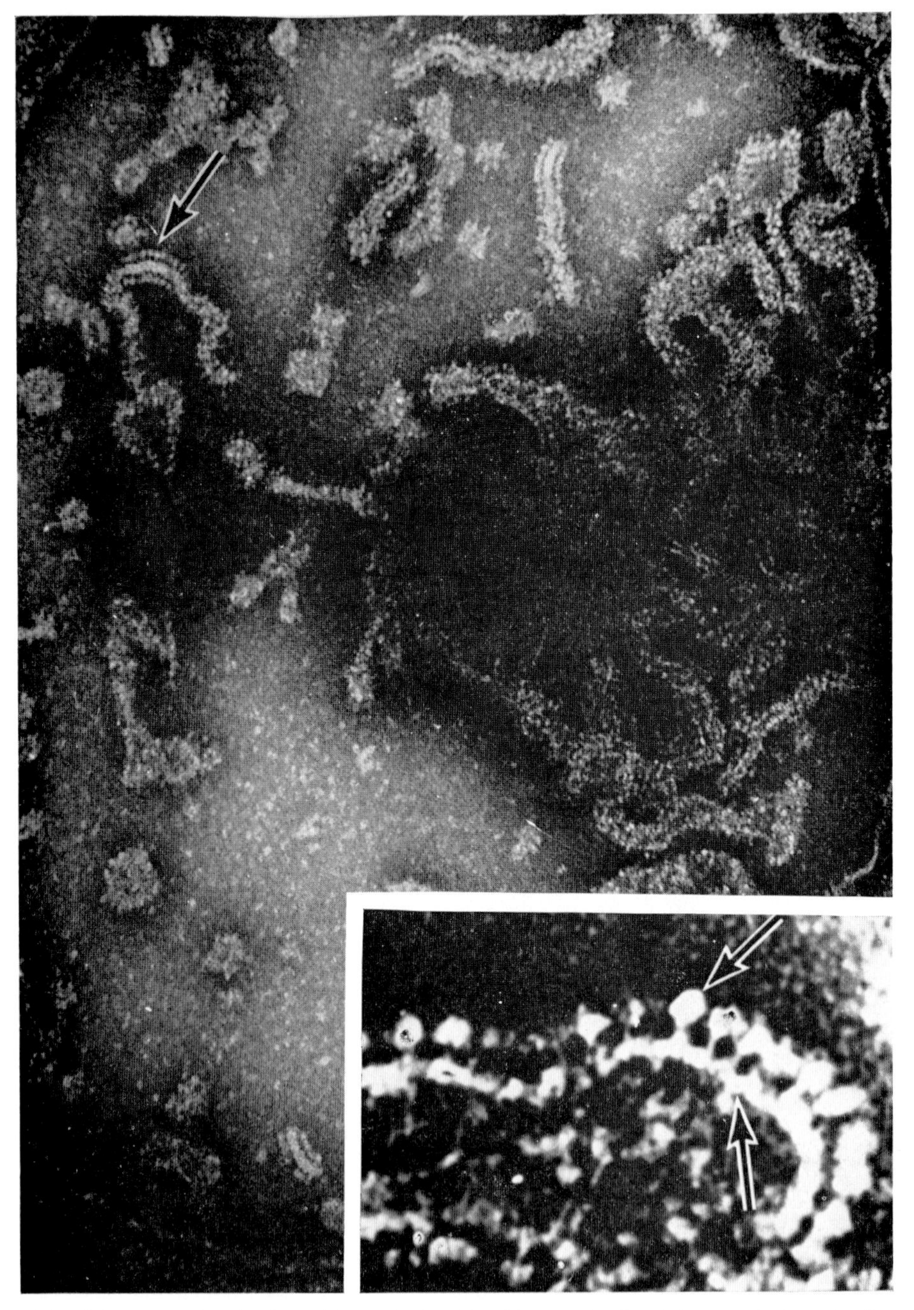

FIG. 38. Arrays of stalked particles in isolated beef-heart mitochondria, negatively stained in phosphotungstic acid. ×130,000. Insert shows fine structure of particles, with head piece, stalk and base piece. ×650,000 (from Fernández-Moran *et al.*, 1964).

workers in negatively stained mitochondria from diverse sources. A particularly clear example found in plant mitochondria by Nadakavukaren (1964) is shown in Fig. 39. Perhaps rather prematurely, these particles were identified by Blair *et al.* (1963) with the fundamental unit or "elementary particle" of the electron-transfer system. This was particularly premature because the isolated elementary particle had no attached stalk (Fernández-Moran, Oda, Blair and Green (1964)). More careful analysis, however, suggested that the particles were too small to contain the entire respiratory assembly. Thus the minimum particle weight for the assembly is about $1{\cdot}4 \times 10^6$ (See Section III. E.). As the spherical head of the stalked body is only about 90 Å in diameter, it can be calculated that the particle weight would only be about 230,000 if the particles had a density near to unity. Even if they had a density of 1·3, and the dimensions had been underestimated by 12%, the value would still only be about 420,000 (Whittaker, 1966). Stoeckenius (1963) in fact suggested a maximum particle weight of 500,000. It has therefore been suggested by Fernández-Moran *et al.* (1964) that the knob, stalk and the base-piece embedded in the membrane all take part in formation of the respiratory assembly, as their combined weight is $1{\cdot}3 \times 10^6$.

However, there is further evidence that the stalked bodies are not respiratory assemblies. For example, mitochondria that had been stripped of all inner membrane structure still retained their cytochromes (Chance, Parsons and Williams, 1964). Stasny and Crane (1964) observed that fractions of sonically disrupted beef-heart mitochondria that were richest in succinoxidase and cytochromes consisted of membranes that were free of attached stalked bodies. On the other hand, the sub-mitochondrial fraction that had the least respiratory activity was in fact richest in these sub-units. Chance and Parsons (1963) observed that mitochondria from *Ascaris lumbricoides* lack cytochromes c_1, a and a_3. Nevertheless, negative staining still revealed the presence of 80 Å diameter particles along the surface of the membrane. Ogawa and Barnett (1965) used histochemical staining and electron microscopy to examine the localization of succinate dehydrogenase. Although clear deposits of formazans could be observed after staining, these did not correspond in position to the stalked particles.

It has been suggested that the particles may be the F_1 coupling factor of oxidative phosphorylation. Racker, Chance and Parsons (1964) observed that the isolated F_1 factor consisted of 90 Å diameter particles. Non-phosphorylating sub-mitochondrial preparations were found to have about six particles per μ of membrane. Addition of F_1 restored the capacity for phosphorylation, and at the same time increased the number of attached particles to $30/\mu$. (The number in beef-heart mitochondria is $60/\mu$). The reader is referred to Racker (1962) for a description of the coupling factors. Apart from these suggestions, there is the more worrying possibility that the particles are in

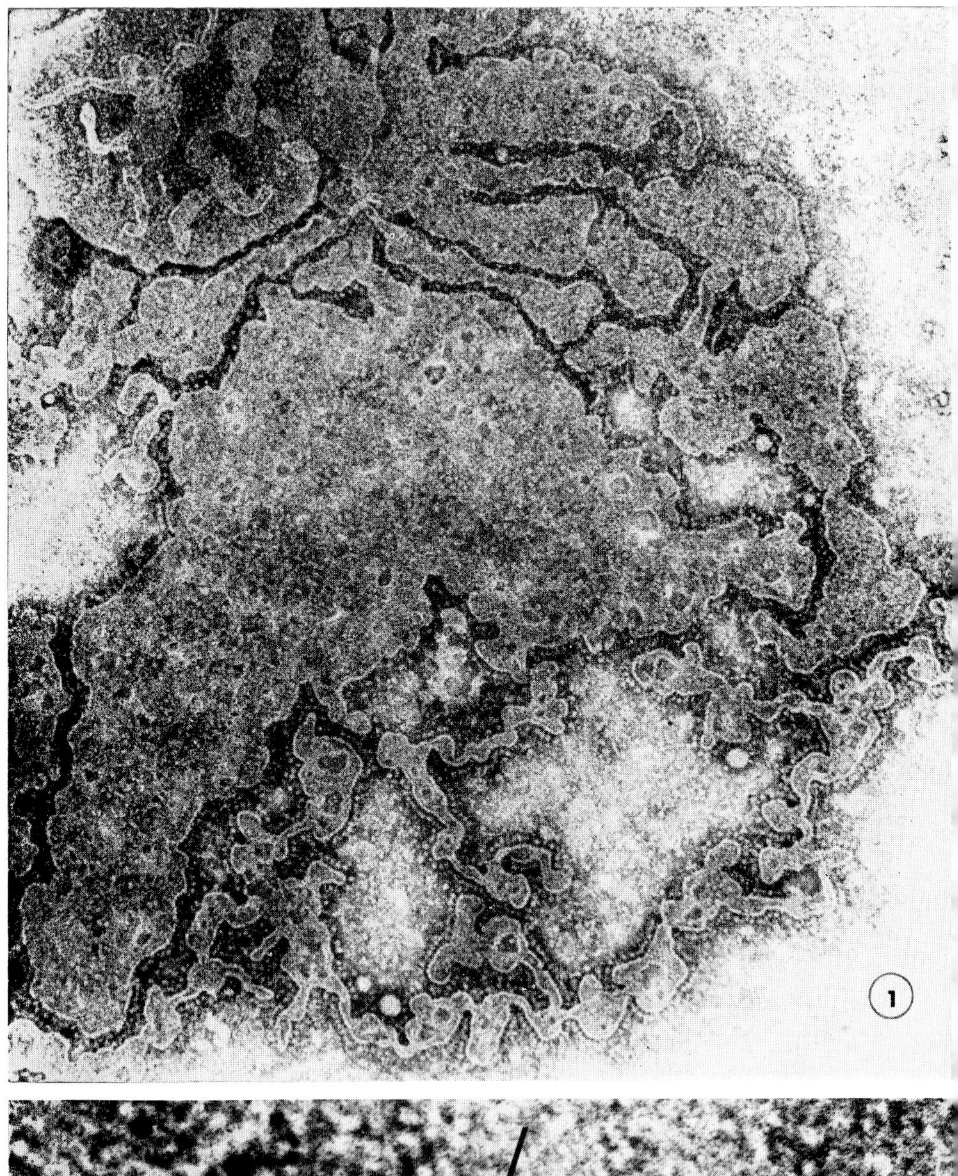
1

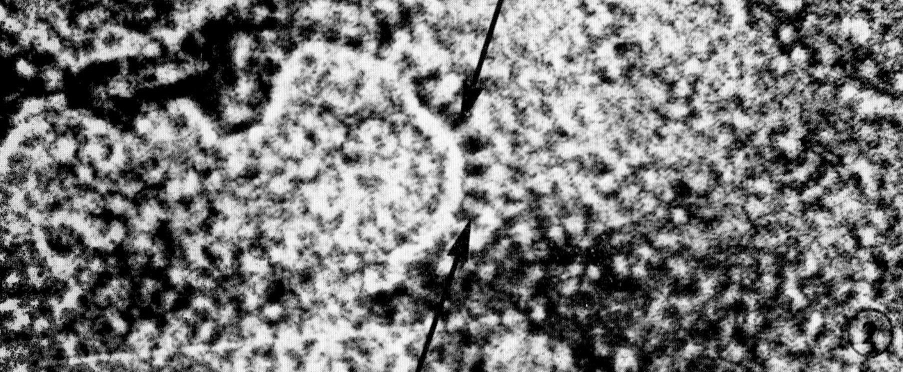

fact artifacts of the negative staining method. Whittaker (1966) discusses this problem and shows a very interesting electron-micrograph from Bangham and Horne (1964) in which artificial lamellae of ovolecithin that had been treated with lysolecithin showed arrays of globular structures 70–80 Å in diameter (Fig. 40). Although this does not establish conclusively that the stalked bodies are artifacts, it demonstrates the extreme caution that is needed in the interpretation of such structures.

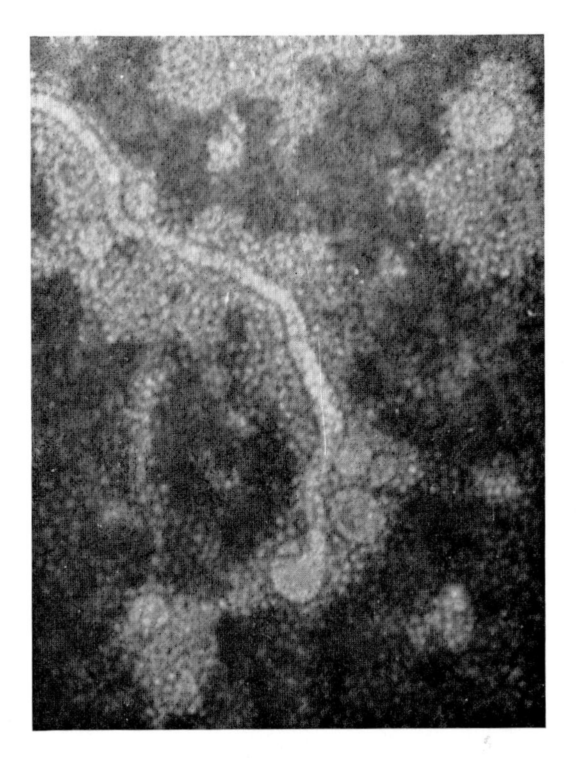

FIG. 40. Dispersion of purified ovolecithin after treatment with lysolecithin and negative staining in phosphotungstate. The particles are estimated to be 70–80 Å in diameter. × 210,000 (from Bangham and Horne, 1964).

H. THE MITOCHONDRIAL MEMBRANE AS AN INTEGRATED STRUCTURE

The above account demonstrates that the mitochondrial membrane system is the site of complex oxidation and energy transfer reactions, protein synthesis, ion translocation, water transport and probably the active transport of many metabolites into and out of the mitochondrion. There is a marked

FIG. 39. Stalked bodies on the cristae of mitochondria from Castor beans (*Ricinus communis*) negatively stained in phosphotungstic acid. 1. General view of preparation; 2. larger magnification showing stalks (arrows) (from Nadakavukaren, 1964).

inter-dependence of many of these reactions on one another. Sometimes the relationship is so close that it is difficult, if not impossible, to separate the two systems.

Perhaps an example from the author's experience will show the difficulty that is often encountered in separating effects on interacting mitochondrial systems. It was found by Tata *et al.* (1963) that treatment of thyroidectomized rats with thyroid hormone resulted in enhanced respiration and phosphorylation activity per mg mitochondrial protein. As there was also a stimulation in amino acid incorporation in microsomal protein, it was suggested that the hormone had produced a stimulation of protein synthesis, and a net increase in the number of respiratory assemblies per mitochondrion. This thesis was confirmed in a most striking fashion by the electron microscope studies of Gustafsson, Tata, Lindberg and Ernster (1965) in which it was shown that there was a clear increase in the mitochondrial population and also in the ratio of cristae to matrix. A photomicrograph from this paper is presented in Fig. 41 in order to illustrate the striking mitochondrial hypertrophy that can

FIG. 41. Hypertrophy of rat skeletal muscle mitochondria in rat rendered hypermetabolic by treatment with L-thyroxine for three weeks. The mitochondrion marked with arrows measures about three sarcomere lengths. The normal mitochondrion in rat skeletal muscle measures only $\frac{1}{4}$–$\frac{1}{2}$ sarcomere length (from Gustafsson *et al.*, 1965).

occur. (This is an excellent example of the dynamic changes that can take place *in vivo* in the morphology of mitochondria). Because of these observations, it was therefore not surprising to find that similar treatment resulted in a stimulation of amino acid incorporation into protein observed in isolated mitochondria (Freeman, Roodyn and Tata, 1963). However, when the effect was studied in more detail it was not possible to establish in a conclusive fashion whether the increase in incorporation was due to specific stimulation of mitochondrial protein synthesis, or was simply because the enhanced ATP synthesis made more energy available for incorporation (Roodyn *et al.*, 1965). The problem could not be resolved because amino acid incorporation was seriously affected by any process that damaged the oxidative phosphorylation system and it was not possible to supply energy for incorporation from added ATP. Thus the two systems could not be separated, and it was not possible to localize the effect of thyroid hormones.

There are many other examples of such unsuccessful attempts to separate closely associated processes in the mitochondrion. Attempts to sub-fractionate or "purify" a complex system frequently fail when a critical amount of damage has been inflicted on the mitochondrial organization. Are we therefore advancing into a blind-end by the conventional analytical approach? In many ways the philosophy of biochemistry is totally alien to the concept that the whole can only be understood as a whole, and not deduced from the properties of its parts. Such a view would be condemned as "vitalism", a most sinful term. Admittedly, it is a most depressing prospect if it can be shown that many of the integrated processes occurring in the mitochondrion may only be detected in intact membranes, and are absent from membrane fragments, simply because they can only occur in a structure that has the correct spatial organization. Thus it may be just as hopeless to look for integrated processes in sub-mitochondrial fragments below a certain size as it would be to look for catalytic activity in small fragments of an enzyme (for example in dipeptides or free amino acids). The catalytic activity of the enzyme is entirely dependent on the spatial configuration of its component parts, and clearly could never be "restored" by mixing suitable proportions of amino acids.

These problems are discussed in a fascinating essay by Mitchell entitled "Metabolism, Transport and Morphogenesis: Which drives Which?" (Mitchell, 1962). His chemi-osmotic theory has been discussed above (Section II. E.) and there is still considerable discussion as to its significance. However, it seems to the author at least that the point of supreme issue is not whether this or that mechanism for oxidative phosphorylation is correct, but whether attempts to "sub-fractionate" many membrane systems are based on false premises and are inherently doomed to failure. One may argue from the rich harvest of results in the field of protein synthesis that the sub-fractionation approach is valid. Yet even here at least some have expressed doubts at the validity of ignoring the role of the intact membrane (Hendler, 1965). Whatever the final outcome, there is little doubt that the elucidation of mitochondrial function in molecular terms is a central challenge to modern biology and will require the most gentle dissection of the mitochondrial membrane by the most subtle research weapons. The days of the hammer are over.

IV. Addendum: The Structure and Function of Peroxisomes

For many years the sub-cellular distribution of the enzymes catalase and urate oxidase (1.7.3.3.) were difficult to elucidate. The enzymes were at first thought to be in mitochondria but more refined fractionation methods suggested that this was incorrect. However, the enzymes did not show precisely the same sedimentation properties as lysomal enzymes. More detailed analysis strongly suggested that the enzymes catalase, urate oxidase and another enzyme, D-amino acid oxidase (1.4.3.3.), in fact belonged to a separate class of particle. In a detailed study, Beaufay *et al.* (1964) resolved rat-liver mitochondrial fractions into three distinct groups of particles, mitochondria, lysosomes and the particles containing urate oxidase, catalase and D-amino acid oxidase. There was some difficulty in separating these particles from ysosomes, however. Fortunately, it was found by Wattiaux, Wibo and Baudhuin (1963) that injection of suitable doses of the neutral detergent Triton WR 1339 produced massive swelling of the lysosomes, but was without effect on the urate oxidase particles. By this means it was possible to obtain a reasonably pure preparation of these particles (Baudhuin, Beaufay and de Duve, 1965). Electron microscopy revealed that the liver particles had the characteristic morphology of the so-called "microbodies" previously described by Rouiller and Bernhard (1956). They were slightly smaller than mitochondria, had a single bounding membrane and contained a characteristic central core. There was a clear correlation between the number of microbodies in the preparation and the specific activities of urate oxidase, catalase and D-amino acid oxidase. (Because they were more numerous in regenerating liver, Rouiller and Bernhard (1956) had suggested that microbodies might be precursors of mitochondria. However, this now seems to be very unlikely (de Duve and Baudhuin, 1966)).

A comparison of some properties of liver mitochondria and microbodies is given in Table I, taken from de Duve (1965). Most of these values were obtained from results of density gradient centrifuging experiments, using cytochrome oxidase as a marker for mitochondria, and urate oxidase as a marker for microbodies.

Very recent developments in this field are discussed extensively in the

TABLE I.

	Mitochondria	Microbodies
Dry wt (μg)	10^{-7}	$2 \cdot 4 \times 0^{-8}$
Osmotically active solutes		
(milli-osmoles/g dry wt)	0·157	0
Water compartments (cm^3/g dry wt)		
Hydration	0·430	0·214
Sucrose space	0·905	2·51
Osmotic space	0·595	0
Total in 0·25 M sucrose	1·930	2·724
Sedimentation coefficient in		
0·25 M sucrose (Svedberg Units)	10^3	$4 \cdot 4 \times 10^4$
Diameter in 0·25 M sucrose (μ)	0·8	0·54
Density in 0·25 M sucrose	1·099	1·095

(Data computed from results of density gradient centrifuging using cytochrome oxidase as marker for mitochondria and urtea oxidase as marker for microbodies. From de Duve, 1965).

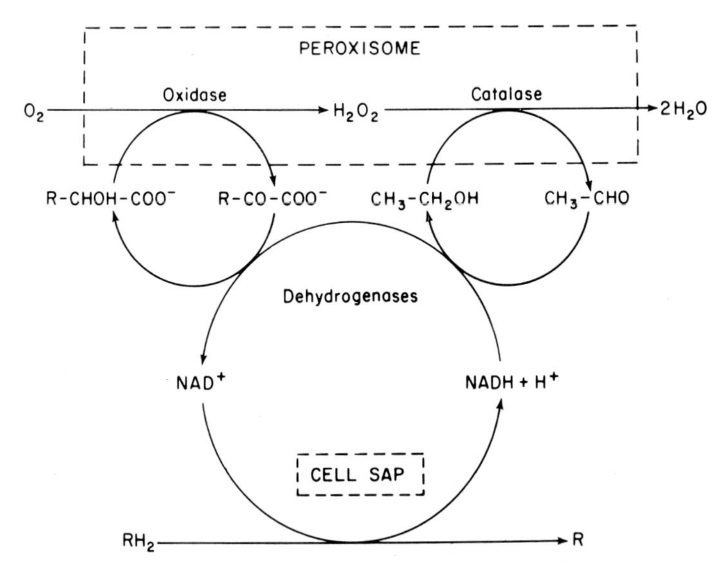

FIG. 42. Possible role of peroxisomes in extra-mitochondrial oxidation of NADH. The α-hydroxy acid oxidase produces H_2O_2 (e.g. by lactate oxidation) and this is used by catalase, in high concentration in the peroxisome, to oxidize a range of compounds, for example alcohol. Reduction of the product is achieved by NAD-linked dehydrogenases, and serves as a means of removing NADH produced by enzymes in the cell sap (e.g. triosephosphate dehydrogenase) (from de Duve and Baudhuin, 1966).

review by de Duve and Baudhuin (1966). The enzyme L-α-hydroxy-acid oxidase has been added to the list of enzymes found in microbodies. In addition, particles with very similar enzymic properties to liver microbodies have been isolated from rat-kidney and *Tetrahymena pyriformis*. Vaes and Jacques (1965) found that catalase sedimented in a different way to mitochondria from homogenates of rat bone, and from the activation properties of catalase Vaes (1965) concluded that the enzyme was present in particles similar to rat-liver microbodies. Since it is not certain that all these particles have identical morphology to rat-liver microbodies, the general name of "peroxisome" has been proposed (see Duve and Baudhuin, 1966). It is suggested that peroxisomes have an important role in extra-mitochondrial oxidations, by the mechanisms described in Fig. 42. The fascinating possibility that the particles are really "fossil organelles" of some primitive pre-mitochondrial oxidative system is discussed by de Duve and Baudhuin.

The fine structure of the liver microbody is now being actively examined. The pronounced central core has a regular sub-structure, and there now seems little doubt the core is the site of location of urate oxidase (Baudhuin

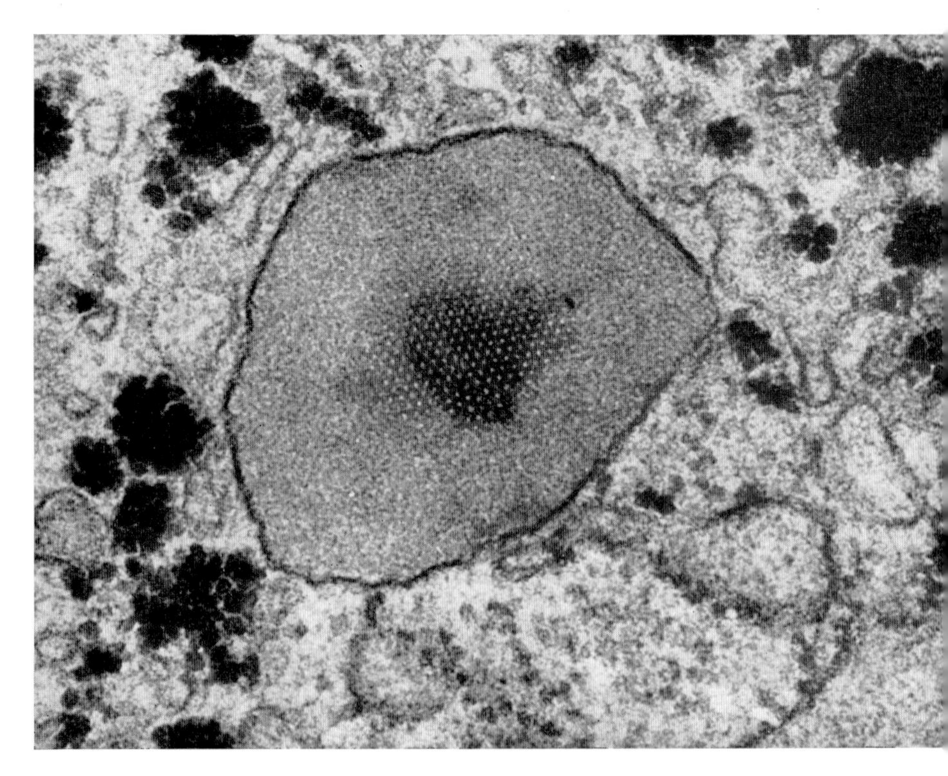

FIG. 43. Guinea-pig liver microbody showing hexagonal lattice in core. The spacing between the elements of the lattice is about 110 Å. × 126,000 (from Baudhuin *et al.*, 1965).

et al., 1965). The outer "sap" contains catalase and the other peroxisomal enzymes, and it may be noted that the catalase concentration appears to be very high. An electron-micrograph of a microbody from guinea-pig liver is shown in Fig. 43. One can clearly see the single membrane the sap, and the core, which has a hexagonal lattice structure.

The origins, formation and biological function of these extremely interesting structures is under active investigation at the present moment, and there is little doubt that many fascinating results will soon be obtained.

REFERENCES

Alberti, K. G. M. M. and Bartley, W. (1963). *Biochem. J.* **87**, 104.
Alberti, K. G. M. M. and Bartley, W. (1965). *Biochem. J.* **95**, 641.
Baird, G. D. (1964). *Biochim. biophys. Acta* **93**, 293.
Bangham, A. D. and Horne, R. W. (1964). *J. mol. Biol.* **8**, 660.
Baudhuin, P., Beaufay, H. and de Duve, C. (1965). *J. Cell Biol.* **26**, 219.
Baudhuin, P., Beaufay, H., Rahman-Li, Y., Sellinger, O. Z., Wattiaux, R., Jacques, P. and de Duve, C. (1964). *Biochem. J.* **92**, 179.
Beaufay, H., Bendall, D. S., Baudhuin, P., Wattiaux, R. and de Duve, C. (1959). *Biochem. J.* **73**, 628.
Beaufay, H., Jacques, P., Bauduin, P., Sellinger, O. Z., Berthet, J., and de Duve, C. (1964) *Biochem. J.* **92**, 184.
Bendall, D. S. and de Duve, C. (1960). *Biochem. J.* **74**, 444.
Bianchi, G and Azzone, G. F. (1964). *J. biol. Chem.* **239**, 3947.
Blair, P. V., Oda, T., Green, D. E. and Fernandez-Moran, H. (1963). *Biochemistry* **2**, 756.
Boyer, P. D. (1965). *Nature, Lond.* **207**, 409.
Braunstein, A. E. (1957). *Adv. Enzymol.* **19**, 335.
Bremer, J. (1963). *J. biol. Chem.* **238**, 2774.
Brierley, G. P., Bachmann, E. and Green, D. E. (1962). *Proc. natn. Acad. Sci. U.S.A.* **48**, 1928.
Brierley, G., Murer, E., Bachmann, E. and Green, D. E. (1963). *J. biol. Chem.* **238**, 3482.
Bronk, J. R. (1963). *Proc. natn. Acad. Sci. U.S.A.* **50**, 524.
Brosemer, R. W. and Veerabhadrappa, P. S. (1965). *Biochim. biophys. Acta* **110**, 102.
Bruni, A., Contessa, A. R. and Luciani, S. (1962). *Biochim. biophys. Acta* **60**, 301.
Bruni, A., Luciani, S. and Bortignon, C. (1965). *Biochim. biophys. Acta* **97**, 434.
Bücher, N. L. R., Overath, P. and Lynen, F. (1960). *Biochim. biophys. Acta* **40**, 491.
Bücher, Th. (1965). *Biochem. Soc. Symp.* **25**, 15.
Carafoli, E. (1965). *Biochim. biophys. Acta* **97**, 107.
Carafoli, E., Rossi, C. S. and Lehninger, A. L. (1964). *J. biol. Chem.* **239**, 3055.
Casu, A. and Modena, B. (1964). *Ital. J. Biochem.* **13**, 197.
Chance, B. ed. (1963). "Energy-linked Functions of Mitochondria", Academic Press, New York and London.
Chance, B. (1965). *J. biol. Chem.* **240**, 2729.
Chance, B. and Hollunger, G. (1961). *J. biol. Chem.* **236**, 1534.
Chance, B. and Hollunger, G. (1963). *J. biol. Chem.* **278**, 418.
Chance, B. and Parsons, D. F. (1963). *Science, N.Y.* **142**, 1176.

Chance, B. and Williams, G. R. (1956). *Adv. Enzymol.* **17**, 65.
Chance, B., Parsons, D. F. and Williams, G. R. (1964). *Science, N.Y.* **143**, 136.
Chappell, J. B. (1961). *In* "Biological Structure and Function" (T. W. Goodwin and O. Lindberg, eds.), Vol. II, p. 71. Academic Press, London and New York.
Chappell, J. B. and Crofts, A. R. (1965a). *Biochem. J.* **95**, 378.
Chappell, J. B. and Crofts, A. R. (1965b). *Biochem. J.* **95**, 707.
Chauveau, J., Moulé, Y., Rouiller, C. and Schneebeli, J. (1962). *J. Cell Biol.* **12**, 17.
Chèvremont, M., Chèvremont-Comhaire, S. and Baeckeland, E. (1959). *Arch. biol. Liegè* **70**, 811.
Chèvremont, M., Baeckeland, E. and Chèvremont-Comhaire, S. (1960). *Biochem. Pharmac.* **4**, 67.
Criddle, R. S., Bock, R. M., Green, D. E. and Tisdale, H. (1962). *Biochemistry* **1**, 827.
Crofts, A. R. and Chappell, J. B. (1965). *Biochem. J.* **95**, 387.
Dallam, R. D. (1958). *Archs Biochem. Biophys.* **77**, 395.
Das, M. L., Haak, E. D. and Crane, F. L. (1965). *Biochemistry* **4**, 859.
de Bernard, B. (1957). *Biochem. biophys. Acta* **23**, 510.
de Duve, C. (1965). *In* "The Harvey Lectures" Series 59, p. 49, Academic Press, New York and London.
de Duve, C. and Baudhuin, P. (1966). *Physiol. Rev.* **46**, 323.
de Duve, C., Pressman, B. C., Gianetto, R., Wattiaux, R. and Applemans, F. (1955). *Biochem. J.* **60**, 604.
de Duve, C., Wattiaux, R. and Baudhuin, P. (1962). *Adv. Enzymol.* **24**, 291.
Deshpande, P. D., Hickmann, D. D. and Von Korff, R. W. (1961). *J. biophys. biochem. Cytol.* **11**, 77.
Ernster, L. and Lee, C. P., (1964). *Annu. Rev. Biochem.* **33**, 729.
Ernster, L. and Lindberg, O. (1958). *Annu. Rev. Physiol.* **20**, 13.
Estabrook, R. W. and Holowinsky. A. (1961). *J. biophys. biochem. Cytol.* **9**, 19.
Fernandez-Moran, H. (1962). *Circulation* **26**, 1039.
Fernandez-Moran, H., Oda., T., Blair, P. V. and Green, D. E. (1964). *J. Cell Biol.* **22**, 63.
Fleischer, S., Brierley, G. Klouwen, H. and Slautterback, D. B. (1962). *J. biol. Chem.* **237**, 3264.
Freeman, K. B., Roodyn. D. B. and Tata, J. R. (1963). *Biochim. biophys. Acta* **72**, 129.
Fritz, I. B. and Marquis, N. R. (1965). *Proc. natn. Acad. Sci. U.S.A.* **54**, 1226.
Gahan, P. B. and Chayen, J. (1965). *Int. Rev. Cytol.* **18**, 223.
Garbus, J., DeLuca, H. F., Loomans, M. E. and Strong, F. M. (1963). *J. biol. Chem.* **238**, 59.
Granick, S. and Urata, G. (1963). *J. biol. Chem.* **238**, 821.
Green, D. E. and Fleischer, S. (1960). *In* "Metabolic Pathways" (D. M. Greenberg, ed.), Vol. 1, p. 41. Academic Press, London and New York.
Green, D. E. and Fleischer, S. (1963). *Biochim. biophys. Acta* **70**, 554.
Green, D. E. and Oda, T. (1961). *J. Biochem. Japan* **49**, 742.
Greville, G. D. (1966). *In* "Regulation of Metabolic Processes in Mitochondria" (J. M. Tager, S. Papa, E. Quagliariello and E. C. Slater, eds.). Elsevier (BBA Library No. 7), Amsterdam. p. 86.
Greville, G. D. and Chappell, J. B. (1959). *Biochim. biophys. Acta* **33**, 267.
Gustafsson, R., Tata, J. R., Lindberg, O. and Ernster, L. (1965). *J. Cell Biol.* **26**, 555.
Hajra, A. K., Sieffert, U. B. and Agranoff, B. W. (1965). *Biochem. biophys. Res. Comm.* **20**, 199.

Harris, E. J. and Judah, J. D. (1966). *In* "Current Topics in Bioenergetics" (D. R. Sanadi, ed.), Vol. 1, Academic Press, London and New York. (In press.)

Heldt, H. W. (1966). *In* "Regulation of Metabolic Processes in Mitochondria" (J. M. Tager, S. Papa, E. Quagliariello and E. C. Slater, eds.), p. 51, Elsevier (BBA Library No. 7), Amsterdam.

Hendler, P. (1965). *Nature, Lond.* **207**, 1053.

Hird, H. J., McLean, E. J. T. and Munro, H. N. (1964). *Biochim. biophys. Acta* **87**, 219.

Hogeboom, G. H. and Schneider, W. C. (1950). *J. biol. Chem.* **186**, 417.

Hogeboom, G. H. and Schneider, W. C. (1953). *J. biol. Chem.* **204**, 233.

Hogeboom, G. H., Schneider, W. C. and Striebich, M. J. (1952). *J. biol. Chem.* **196**, 111.

Holten, D. D. and Nordlie, R. C. (1965). *Biochemistry* **4**, 723.

Hoskins, D. D. and Bjur, R. A. (1964). *J. biol. Chem.* **239**, 1856.

Hoskins, D. D. and Mackenzie, C. G. (1961). *J. biol. Chem.* **236**, 177.

Hülsmann, W. C. (1962). *Biochim. biophys. Acta* **58**, 417.

Hülsmann, W. C. (1963). *Biochim. biophys. Acta* **77**, 502.

Hülsmann, W. C., Wit-Peeters, E. M. and Benckhuysen, C. (1966). *In* "Regulation of Metabolic Processes in Mitochondria" (J. M. Tager, S. Papa, E. Quagliariello and E. C. Slater, eds.), p. 460. Elsevier (BBA Library Vol. 7), Amsterdam.

Hülsmann, W. C. and Dow, D. S. (1964). *Biochim. biophys. Acta* **84**, 486.

Johnson, R. M. and Kerur, L. (1963). *Biochim. biophys. Acta* **70**, 152.

Johnson, A. B. and Strecker, H. J. (1962). *J. biol. Chem.* **237**, 1876.

Kalf, G. F. (1964). *Biochemistry* **3**, 1702.

Katoh, T. and Sanukida, S. (1965). *Biochem. biophys. Res. Commun.* **21**, 373.

Kenney, F. T. (1962). *J. biol. Chem.* **237**, 1610.

Kielley, R. K. and Schneider, W. C. (1950). *J. biol. Chem.* **185**, 869.

Klingenberg, M. and Pette, D. (1962). *Biochem. biophys. Res. Commun.* **7**, 430.

Klingenberg, M. and Pfaff, E. (1966). *In* "Regulation of Metabolic Processes in Mitochondria" (J. M. Tager, S. Papa, E. Quagliariello and E. C. Slater, eds.), Vol. 7. Elsevier (BBA Library), Amsterdam. p. 180.

Klingenberg, M. and Schollmeyer, P. (1960). *Biochem. Z.* **333**, 335.

Koike, M., Reed, L. J. and Carroll, W. R. (1963). *J. biol. Chem.* **238**, 30.

Krebs, H. A. and Bellamy, D. (1960). *Biochem. J.* **75**, 523.

Krebs, H. A. and Lowenstein, J. M. (1960). *In* "Metabolic Pathways" (D. M. Greenberg, ed.), Vol. 1, p. 129. Academic Press, London and New York.

Kroon, A. M. (1964). *Biochim. biophys. Acta* **91**, 145.

Kroon, A. M. (1965). *Biochim. biophys. Acta* **108**, 275.

Kroon, A. M. (1966). Ph.D. Thesis, University of Amsterdam.

Lardy, H. A., Connelly, J. L. and Johnson. D. (1964). *Biochemistry* **3**, 1961.

Lardy, H. A., Johnson, D. and McMurray, W. C. (1958). *Archs Biochem. Biophys.* **78**, 587.

Lehninger, A. L. (1962). *Physiol. Revs.* **42**, 467.

Lehninger, A. L. (1965). "The Mitochondrion", Benjamin Press, New York.

Levin, E. and Thomas, L. E. (1961). *Expl Cell Res.* **22**, 363.

Litwack, G., Sears, M. L. and Diamondstone, T. I. (1963). *J. biol. Chem.* **238**, 302.

Luck, D. J. and Reich, E. (1964). *Proc. natn. Acad. Sci. U.S.A.* **52**, 931.

Lynen. F. (1955). *Annu. Rev. Biochem.* **24**, 653.

Machinist, J. M. and Singer, T. P. (1965a). *Proc. natn. Acad. Sci. U.S.A.* **53**, 467.

Machinist, J. M. and Singer, T. P. (1965b). *J. biol. Chem.* **240**, 3182.

MacLennan, D. H., Tzagoloff, A. and Rieske, J. G. (1965). *Archs Biochem. Biophys.* **109**, 383.
Mahler, H. R. and Pereira, A. da S. (1962). *J. mol. Biol.* **5**, 325.
Massey, V. (1963). *In* "The Enzymes" (P. D. Boyer, H. Lardy and K. Mÿrback, eds.), Vol. 7, p. 275. Academic Press, New York.
Maver, M. E. and Greco, A. E. (1951). *J. natn. Cancer Inst.* **12**, 37.
Mitchell, P. (1961). *Nature, Lond.* **191**, 144.
Mitchell, P. (1962). *J. gen. Microbiol.* **29**, 25.
Mitchell, P. (1966a). *In* "Regulation of Metabolic Processes in Mitochondria" (J. M. Tager, S. Papa, E. Quagliariello and E. C. Slater, eds.), Vol. 7. Elsevier (BBA Library), Amsterdam. p. 65.
Mitchell, P. (1966b). *Biol. Revs.* **41**, 445.
Morino, Y., Kagamiyama, H. and Wada, H. (1964). *J. biol. Chem.* **239**, PC 943.
Mugnaini, E. (1964). *J. Cell Biol.* **23**, 173.
Myers, D. K. and Slater, E. C. (1957). *Biochem. J.* **67**, 558.
Nadakavukaren, M. J. (1964). *J. Cell Biol.* **23**, 193.
Nass, M. M. K. and Nass, S. (1962). *Expl Cell Res.* **26**, 424.
Nass, M. M. K. and Nass, S. (1963). *J. Cell Biol.* **19**, 593.
Nass, M. M. K., Nass, S. and Afzelius, B. A. (1965). *Expl Cell Res.* **37**, 516.
Neifakh. S. A. and Kazakova, T. B. (1963). *Nature, Lond.* **197**, 1106.
Neubert, D. and Helge, H. (1965). *Biochem. biophys. Res. Commun.* **18**, 600.
Nuebert, D., Helge, H. and Merker, H. J. (1965). *Biochem. Z.* **343**, 44.
Nishida, G. and Labbe, R. F. (1959). *Biochim. biophys. Acta* **31**, 519.
Nordlie, R. C. and Lardy, H. A. (1963). *J. biol. Chem.* **238**, 2259.
Norum, K. R. (1964). *Biochim. biophys. Acta* **89**, 95.
Novikoff, A. B. (1961). *In* "The Cell" (J. Brachet and A. E. Mirsky, eds.), Vol. II, p. 299. Academic Press, New York and London.
Oberling, C. (1959). *Int. Rev. Cytol.* **8**, 1.
Ogawa, K. and Barnett, R. J. (1965). *J. Ultrastr. Res.* **12**, 488.
Ohnishi, T., Kawamura, H., Takeo, K. and Watanabe, S. (1964). *J. Biochem., Japan* **56**, 273.
Ohnishi, T. and Ohnishi, T. (1962). *J. Biochem., Japan* **51**, 380.
Osawa, S., Allfrey, V. G. and Mirsky, A. E. (1957). *J. gen. Physiol.* **40**, 491.
Packer, L. (1961). *In* "Biological Structure and Function" (T. W. Goodwin and O. Lindberg, eds.), Vol. II, p. 85. Academic Press, London and New York.
Palade, G. E. (1953). *J. Cell Biol.* **1**, 188.
Parsons, J. A. (1965). *J. Cell Biol.* **25**, 641.
Patterson, M. K. and Touster, O. (1962). *Biochim. biophys. Acta* **56**, 626.
Peachey, L. D. (1964). *J. Cell Biol.* **20**, 95.
Penefsky, H. S., Pullman, E., Datta, A. and Racker, E. (1960). *J. biol. Chem.* **235**, 3330.
Penn, N. W. (1960). *Biochim. biophys. Acta* **37**, 55.
Pette, D., Klingenberg, M. and Bücher, Th. (1962). *Biochem. biophys. Res. Commun.* **7**, 425.
Pharo, R. L. and Sanadi, D. R. (1964). *Biochim. biophys. Acta* **85**, 346.
Pullman, M. E., Penefsky, H. S., Datta, A. and Racker, E. (1960). *J. biol. Chem.* **235**, 3322.
Quagliariello, E. and Papa, S. (1964). "Atti del Seminario di Studi Biologici, Bari", p. 351.
Quagliariello, E., Papa, S., Saccone, C., Palmieri, F. and Francavilla, A. (1965). *Biochem. J.* **95**, 742.

Rabinowitz, M., Sinclair, J., DeSalle, L., Haselkorn, R. and Swift, H. (1965). *Proc. natn. Acad. Sci. U.S.A.* **53**, 1126.
Racker, E. (1961). *Adv. Enzymol.* **23**, 323.
Racker, E. (1962). *Proc. natn. Acad. Sci. U.S.A.* **48**, 1659.
Racker, E., Chance, B. and Parsons, D. F. (1964). *Fed. Proc.* **23**, 431.
Richardson, S. H., Hultin, H. O. and Fleischer, S. (1964). *Archs Biochem. Biophys.* **105**, 254.
Richardson, S. H., Hultin, H. O. and Green, D. E. (1963). *Proc. natn. Acad. Sci. U.S.A.* **50**, 821.
Rimington, C. and Tooth, B. E. (1961). *J. Biochem., Japan* **49**, 456.
Roodyn, D. B. (1956). *Biochem. J.* **64**, 361.
Roodyn, D. B. (1959). *Int. Rev. Cytol.* **8**, 279.
Roodyn, D. B. (1962). *Biochem. J.* **85**, 177.
Roodyn. D. B. (1965a). *Int. Rev. Cytol.* **18**, 99.
Roodyn. D. B. (1965b). *Biochem. J.* **97**, 782.
Roodyn, D. B. (1966). *In* "Regulation of Metabolic Processes in Mitochondria" (J. M. Tager, S. Papa, E. Quagliariello and E. C. Slater, eds.), Vol. 7. Elsevier (BBA Library), Amsterdam. p. 383.
Roodyn, D. B. and Wilkie, D. (1966a). "The Biogenesis of Mitochondria" (In preparation).
Roodyn, D. B. and Wilkie, D. (1966b). (In press.)
Roodyn, D. B., Freeman, K. B. and Tata, J. R. (1965). *Biochem. J.* **94**, 628.
Roodyn, D. B., Reis, P. J. and Work, T. S. (1961). *Biochem. J.* **80**, 9.
Roodyn, D. B., Suttie, J. W. and Work, T. S. (1962). *Biochem. J.* **83**, 29.
Rossi, C. S. and Lehninger, A. L. (1964). *J. biol. Chem.* **239**, 3971.
Rossi, C., Hauber, J. and Singer, T. P. (1964). *Nature, Lond.* **204**, 167.
Rouiller, C. and Bernhard, W. (1956). *J. biophys. biochem. Cytol.* **2**, 355.
Sanadi, D. R. (1963). *In* "The Enzymes" (P. D. Boyer, H. Lardy and K. Mÿrback, eds), Vol. 7, p. 307. Academic Press, New York and London.
Sanadi, D. R. (1965). *Annu. Rev. Biochem.* **34**, 21.
Schneider, W. C. (1946). *J. biol. Chem.* **165**, 585.
Schneider, W. C. and Hogeboom, G. H. (1950). *J. natn. Cancer Inst.* **10**, 969.
Schneider, W. C. and Hogeboom, G. H. (1956). *Annu. Rev. Biochem.* **25**, 201.
Schuster, F. L. (1965). *Expl Cell Res.* **39**, 329.
Sekuzu, L., Jurtshuk, P. and Green, D. E. (1961). *Biochem. biophys. Res. Commun.* **6**, 71.
Sheid, B., Morris, H. P. and Roth, J. S. (1965). *J. biol. Chem.* **240**, 3016.
Siegel, L. and Englard, S. (1962). *Biochim. biophys. Acta* **64**, 101.
Siekevitz, P. and Watson, M. (1956). *J. biophys. biochem. Cytol.* **2**, 653.
Slater, E. C. (1965). *Nature, Lond.* **207**, 411.
Slater, E. C. (1966). *In* "Comprehensive Biochemistry" (M. Florkin and E. H. Stotz, eds.), Vol. 14. Elsevier, Amsterdam. (In press.)
Slater, E. C. and Tager, J. M. (1963). *Biochim. biophys. Acta* **77**, 276.
Sperti, S., Pinna, L. A., Lorini, M., Moret, V. and Siliprandi, N. (1964). *Biochim. biophys. Acta* **93**, 284.
Stasny, J. T. and Crane, F. L. (1964). *J. Cell Biol.* **22**, 49.
Stoeckenius, W. (1963). *J. Cell Biol.* **17**, 443.
Swick, R. W., Barnstein, P. L. and Stange, J. L. (1965). *J. biol. Chem.* **240**, 3334.
Tager, J. M. and Slater, E. C. (1963). *Biochim. biophys. Acta* **77**, 227.
Tager, J. M., Papa, S., Quagliariello, E. and Slater, E. C. eds. (1966). "Regulation of Metabolic Processes in Mitochondria", Vol. 7. Elsevier (BBA Library), Amsterdam.

Tata, J. R., Ernster, L., Lindberg, O., Arrhenius, E., Pedersen, S. and Hedman, R. (1963). *Biochem. J.* **86**, 408.
Tewari, K. K., Jayaraman, J. and Mahler, H. R. (1965). *Biochem. biophys. Res. Commun.* **21**, 141.
Truman, D. E. S. (1963). *Expl Cell Res.* **31**, 313.
Vaes, G. (1965). *Biochem. J.* **97**, 393.
Vaes, G. and Jacques, P. (1965). *Biochem. J.* **97**, 389.
Ward, R. T. (1962). *J. Cell Biol.* **14**, 309.
Watson, M. L. and Siekevitz, P. (1956). *J. biophys. biochem. Cytol.* **2**, 639.
Wattiaux, R., Wibo, M. and Baudhuin, P. (1963). *In* "Lysosomes" Ciba Foundation Symposium (A. V. S. de Reuck and M. P. Cameron, eds.), p. 176. Churchill, London.
Webster, L. T., Gerowin, L. D. and Rakita, L. (1965). *J. biol. Chem.* **240**, 29.
Weinbach, E. C. and von Brand, T. (1965). *Biochem. biophys. Res. Commun.* **19**, 133.
Whitehead, E. P. (1965). *J. theoret. Biol.* **8**, 276.
Whittaker, V. P. (1966). *In* "Regulation of Metabolic Processes in Mitochondria" (J. M. Tager, S. Papa, E. Quagliariello and E. C. Salter, eds.), Vol. 7. Elsevier (BBA Library), Amsterdam. p. 1.
Williams, J. N. (1952a). *J. biol. Chem.* **194**, 139.
Williams, J. N. (1952b). *J. biol. Chem.* **195**, 37.
Wintersberger, E. (1964). *Z. Physiol. Chem.* **336**, 385.
Wintersberger, E. (1965). *Biochem. Z.* **341**, 409.

Chapter 4

THE CHLOROPLAST

D. O. HALL and F. R. WHATLEY

King's College
University of London, London, England

I. Introduction

The chloroplast occupies a particularly important place in the economy of the plant cell, being the site of the whole complex of reactions leading to the capture of light energy and its eventual storage in chemical form as carbohydrates. Isolated chloroplasts have proved to be convenient experimental objects for the study of many of the partial reactions which together lead to their overall metabolism.

It was early found that certain of these processes were associated with the water-insoluble fraction of the chloroplasts and the enzymes catalysing them could be readily localized within the chloroplast. However, the detailed *characterization* of the water-insoluble enzymes has proved more difficult.

On the other hand many of the enzymes thought to be originally present within the chloroplast are easily extracted by water or dilute salt solutions. These have frequently been characterized in some detail as soluble enzymes but their definitive *localization* within the chloroplast has proved to be correspondingly more difficult. The technique of isolating the chloroplasts in non-aqueous solvents has provided a promising approach to proving the original distribution of many of these easily water-soluble enzymes.

In this chapter we shall describe the structure of the chloroplasts (and other plastids) and discuss some of the evidence for the localization of a considerable number of enzymic components within chloroplasts.

II. Chloroplasts and Other Plastids

Plastids may be generally divided into two groups. (a) the chromoplasts or coloured plastids and (b) the leucoplasts or colourless plastids The chromoplasts may be conveniently divided into (i) chloroplasts and other photosynthetically active chromoplasts, which contain chlorophyll *a* plus varying accessory pigments (ii) chromatophores which contain the far-red absorbing bacteriochlorophyll (photosynthetically active) and (iii) non-photosynthetic chromoplasts which usually lack chlorophylls but contain carotenoids (Granick, 1961). This classification is convenient but by no means ideal since a particular plastid may occur as a colourless, green, red or yellow plastid at different periods of its existence.

A. Chloroplasts

In this category we shall include the green plastids of green algae and higher plants where chlorophyll *b* is the accessory pigment; the brown or yellow plastids of the brown algae and diatoms in which fucoxanthin is the accessory pigment; the red plastids of the red algae where phycoerythrin is the accessory pigment; and the green plastids of the blue-green algae which have phycocyanin and phycoerythrin as the accessory pigments (Haxo and ÓhEocha, 1960) This review will be primarily concerned with the chloroplasts of green algae and higher plants since most of the research in photosynthesis has been done with these organisms.

Chloroplasts in higher plants are saucer shaped organelles 1 to 5μ in diameter. In algae the chloroplast may be of much more complicated shape, varying from the simple cup-shaped chloroplast of *Chlorella* to the stellate chloroplast of *Zygnema* or the spirally wound, ribbon-shaped chloroplast of *Spirogyra*. Chloroplasts possess a lamellar structure. Discrete chloroplasts are lacking in blue-green algae but a lamellar structure is diffused throughout the cell. In the light microscope chloroplasts can often be seen to contain grana, within which the pigment appears to be concentrated. Electron microscope studies have shown further details of the structure. As shown in Fig. 1a, representing an ultra-thin section of bean chloroplast, the organelle is bounded by a double membrane within which is contained the ground substance, the

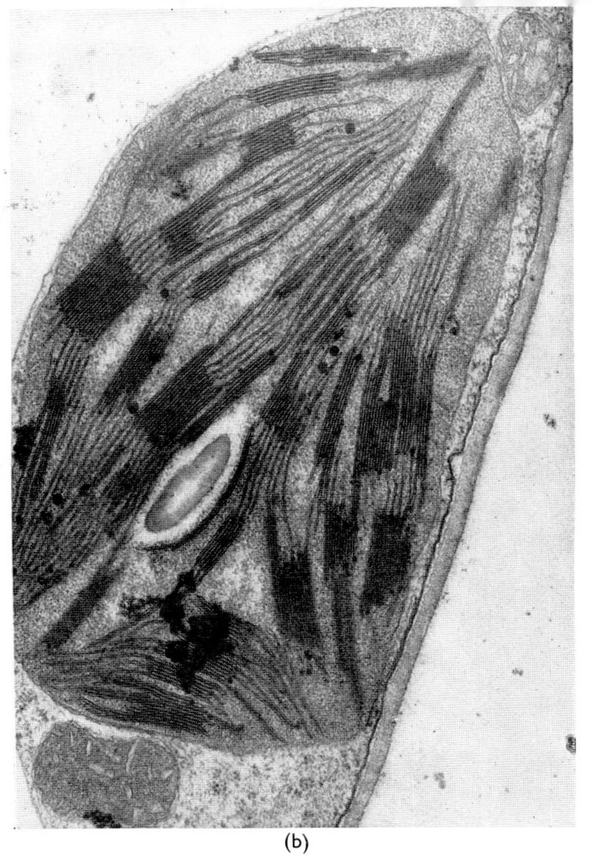

(b)

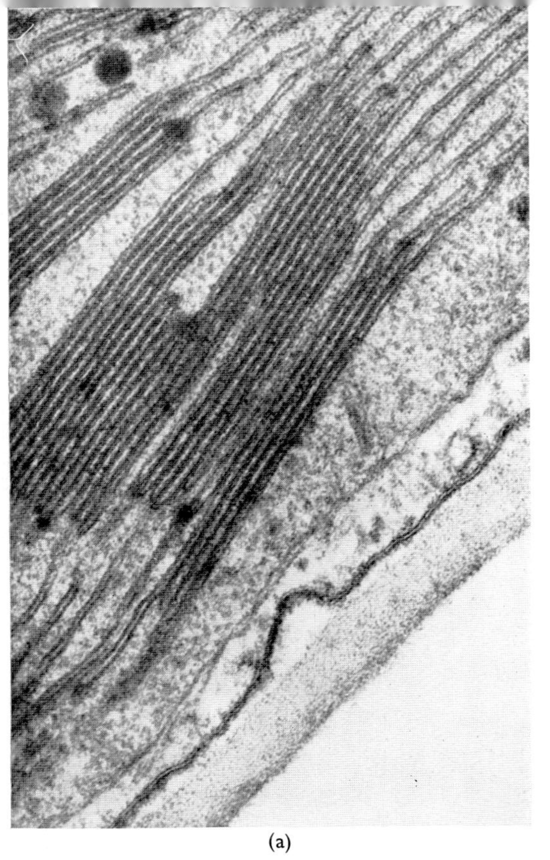

(a)

FIG. 1(a). Ultra-thin section of chloroplast in bean leaf cell fixed in glutaraldehyde, pH7, and Dalton's medium. Note double membrane surrounding the chloroplast, ground substance (stroma), lipid droplets, lamellar structure including grana and starch inclusion. At either end of the chloroplast a mitochondrion is seen. (Courtesy of T. E. Weier). (b). Two fold magnification (approx.) of portion of Fig. 1(a) to show greater detail of lamellar structure.

stroma, and various apparently lipid droplets visible only under certain fixation regimes. Embedded in the stroma are the lamellar structures, consisting of flattened discs, which in some cases are concentrated to form distinct "stacks" corresponding to the grana in the light microscope. Within the grana certain of the lamellae appear to end, but it will be observed (Fig. 1b) that a considerable number extend beyond the limits of the grana, becoming stroma lamellae. The distinction between stroma and grana lamellae may or may not be significant. The highly magnified section shows the detailed lamellar structure to be quite complicated (see Weier, 1961; Weier and Thomson, 1962, for discussion). The discrete structure which can be observed along the double lamellae may be related to the elementary photosynthetic particles, the quantasomes. In Fig. 2a is shown a single lamella from a spinach

(a)

FIG. 2(a). Chromium shadowed spinach chloroplast lamella (surface view) showing regular arrangement of quantasomes (Park and Biggins, 1964).

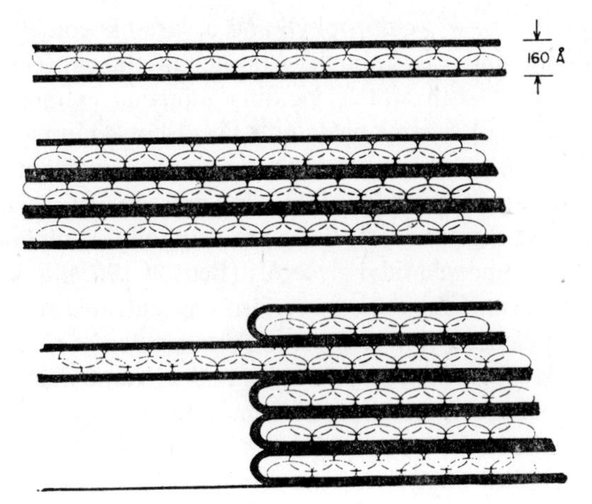

FIG. 2 (b). Diagram of typical lamellar structure from blue green algae, *Euglena* and higher plants. (Park and Pon, 1963).

grana fraction in surface view. The surface has become mechanically disrupted so that the regular orientation of particles underneath can be clearly seen. It is thought that these particles correspond to the quantasomes. Park and Pon (1963) have suggested a possible arrangement of the quantasomes along the lamellae (Fig. 2b) which may correspond to the particles observable in the very highly magnified section in Fig. 1b. Figure 2b also suggests the evolution of the lamellar structure from that found in the blue-green algae (no distinct chloroplast, but well developed lamellae), through that found in green algae and many higher plants (in which the lamellae become stacked), to the condition in the chloroplasts of most higher plants (in which grana formation occurs).

Chloroplasts contain ribonucleic acid, a component which is normally associated with ribosomes. Lyttleton (1962) has isolated a fraction corresponding to ribosomes by sonic disruption of isolated chloroplasts and subsequently centrifuging in a sucrose gradient, indicating the existence of these protein-synthesizing particles within the chloroplast.

When starch is deposited within the chloroplast it is laid down between the grana lamellae, eventually pushing them apart and leading to a great disorganization of the structure—a phenomenon which may represent a mechanism for controlling the overall extent of starch synthesis.

The chemical structure of chloroplasts of various higher plants has been determined (see Thomas, 1960; also Menke, 1962, 1963, for references). Chloroplasts contain 75% to 80% water and between 30% and 60% of the total protein in the leaf is associated with them. Zucker and Stinson (1962) exceptionally found that *Oenothera* chloroplasts contain 75% of the leaf protein. On a dry weight basis chloroplasts contain 25% to 35% lipid, 45%

to 60% protein, 4% to 7% chlorophyll and a variable content of the carotenoids, β-carotene and xanthophyll. About half the protein is readily extracted by treatment with water, yielding aqueous extracts containing a considerable number of enzyme proteins, as is discussed in more detail in this chapter. The lipids in the chloroplasts contain only a small amount of neutral triglycerides, but are characterized by the presence of galactolipids (e.g. β-galacto-pyranosyl diglycerides), the plant sulpholipid sulphodeoxyglycosyl diglyceride, and the phosphatidyl glycerols (Benson 1963). Small amounts of metabolically important substances are also concentrated within the chloroplasts. These include catalysts of electron flow, e.g. ferredoxin, cytochrome *f*, cytochrome b_6, flavoproteins, plastocyanin and pyridine nucleotides, various metals, e.g. iron, copper, manganese, magnesium and zinc, and a number of quinones, e.g. vitamin K, tocopherylquinone and the plastoquinones.

A detailed chemical analysis of the composition of the purified lamellar fraction has been published by Park and Biggins (1964). The large amounts of several of the quinones and pigments suggests a structural role for these substances in the chloroplasts. The proposed model of Frey-Wyssling (1953) for the chloroplast lamellae has lipoprotein structures which is in accordance with this analysis. The chloroplast is the site of a number of important synthetic activities of the cell including photophosphorylation, CO_2 fixation, carbohydrate interconversions as well as a number of important ancillary reactions. These activities are discussed in some detail in succeeding sections.

B. CHROMATOPHORES

Chromatophores are particles approximately 600Å in diameter which can be isolated from photosynthetic bacteria by sonication or by mechanical disruption followed by differential centrifuging. This is an operational definition. It is now thought that the chromatophores isolated from the cells of purple bacteria broken by such methods represent fragments derived from an initially continuous cytoplasmic membrane and therefore, if we may quote, "chromatophores . . . are structural artefacts produced by the disruption of the bacterial cytoplasmic membrane" (Cohen-Bazire and Kunisawa, 1963). Thus a membrane bound vesicular system within the cytoplasm constitutes the intracellular site of the photosynthetic apparatus in these photosynthetic bacteria. The chromatophores when isolated contain all the photosynthetic pigments of the bacterial cells including the bacteriochlorophyll which resembles chlorophyll *a* in plants although there is no accessory pigment similar to chlorophyll *b* or one of the phycobilins found in higher plants. Large amounts of carotenoids and phospholipoproteins are found, it having been observed by Newton and Newton (1957) that these constituents occur in the ratio bacteriochlorophyll:carotenoid:cytochrome in the proportion 10 : 5 : 1. Protein constitutes approximately 40% on a weight basis and 21% is lipid, principally phospholipid. Chromatophores contain considerable quantities of non-haem iron. They also contain pyridine nucleotides, flavin, a considerable amount of a bacterial cytochrome which resembles cytochrome *f* of plants in many respects and a haem protein called Rhodospirillum haem protein. Isolated chromatophores are able to carry out photophosphorylation and the photoreduction of pyridine nucleotide and these reactions are presumably catalysed by the membrane system within the bacteria from which they are derived. The review by Gest and Kamen (1960) may be consulted for further details of the metabolism of photosynthetic bacteria.

C. NON-PHOTOSYNTHETIC CHROMOPLASTS

These non-photosynthetic plastids lack chlorophyll but often contain considerable amounts of carotenoids. The plastids contain pigments which are red or yellow and impart colours to some flowers, fruits, roots, fungi, etc. They are usually spindle- or needle-shaped bodies which occur either singly or grouped in bundles. They may also consist of droplets containing the carotenoids, each droplet representing a degenerated plastid. The colour may change from green due to the disappearance of chlorophyll and the unmasking of the carotenoid pigments; when this happens in existing chloroplasts the red or yellow coloured chromoplasts are obviously derived from original chloroplasts.

Carotenoids commonly occurring in red and yellow coloured chromoplasts are α- and β-carotene and the xanthophylls lutein, violaxanthin and neaxanthin (Goodwin, 1960). The yellow chromoplasts of carrot roots develop from leucoplasts, and as the carotene is synthesized the starch contained by the leucoplast disappears. Older chromoplasts appear as large flat plates or as crystalline needles. Carotene may make up to 20% to 56% of the plastid on a dry weight basis. When the carotene content of the plastid is about 30% the other lipids may amount to 30% to 40% and the protein to some 15% (Granick, 1961).

D. LEUCOPLASTS

This term is applied to all colourless plastids, the term proplastid designating immature colourless plastids occurring in developing tissues (Granick, 1955; Steffen, 1955). Leucoplasts are predominantly storage plastids. According to the product principally stored the leucoplasts are named amyloplast if starch predominates, elaioplast if oil predominates and aleuroplast when protein granules predominate.

1. Amyloplasts

These are mature leucoplasts filled with starch and generally found in storage tissues in tubers, cotyledons and endosperms. Reserve starch is laid down as large starch grains usually made up of a series of concentric layers successively deposited about a centre—the amyloplast membrane and accompanying stroma becomes greatly distended and finally becomes so thin as to be scarcely perceptible as a pellicle enveloping the starch grain. Each leucoplast apparently produces only a single starch grain. These vary greatly in size (1–150μ in diameter) and in number in different cells. Their shape is characteristic for a given species.

2. Elaioplasts

These are plastids which develop a preponderance of oil which may appear in old chromoplasts which lose their pigment. Oil droplets are however formed independently of plastids. The oils are usually considered to be a reserve storage product.

3. Aleuroplasts

These are colourless plastids which contain protein. The proteins are normally "globular" and are water soluble in contrast to fibrous proteins. They tend to crystallize if solvent is withdrawn. The size of the aleuroplast varies between 3 and 20μ. They are found in many fruits and seeds. The protein bodies in developing wheat endosperm contained 68% protein, 23% lipid (most phospholipid), 6% nucleic acid present as ribosomes and 4% of phytic acid. The protein bodies are enclosed within a distinctive lipoprotein membrane and the whole structure is considered to be an aleuroplastid or proteoplast (Morton, Palk and Raison 1964) i.e., the plastids consist of a lipoprotein membrane enclosing the protein body and a region rich in ribosomes. The aleuroplast contains the enzymes and nucleic acid components necessary for the incorporation of amino acid into the protein. The protein body within the plastid is formed by the accumulation of storage protein within the plastid membrane.

III. Methods for Isolation of Chloroplasts

Chloroplasts may be isolated from leaves in aqueous or non-aqueous media. If the intracellular distribution of water-soluble enzymes is to be studied, it may be desirable to isolate the particles in non-aqueous solvents. However, chloroplasts isolated in this way may lack a number of the overall photochemical processes normally associated with chloroplasts which have been isolated in dilute salt or sugar solution. Protocols are given in Figs. 3, 6 and 7 below of methods which have been commonly used for the isolation of chloroplasts.

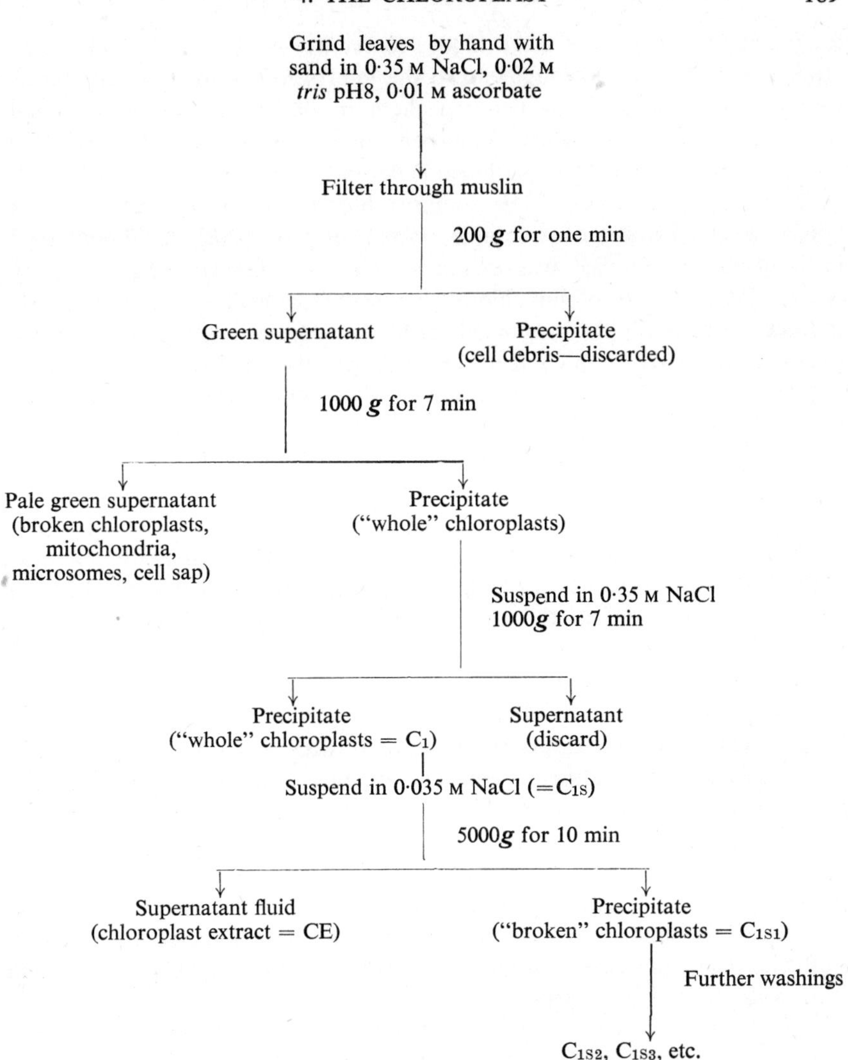

FIG. 3. Isolation of chloroplasts in aqueous medium (Whatley and Arnon, 1963).

A. ISOLATION IN AQUEOUS MEDIUM

Most of the work which has been carried out on the enzymology of chloroplasts have been done with chloroplasts isolated in 0·35M sodium chloride and frequently broken subsequently by suspension in much lower salt concentration. In the method described by Whatley and Arnon, 1963 (Fig. 3) spinach leaves were gently ground in buffered 0·35M sodium chloride with sand, using a pestle and mortar; other species have been used in a similar way. The resultant slurry was filtered through cheese cloth and submitted to a brief centrifuging at slow speed (200*g*) to remove coarse debris and nuclei. Higher speeds (1000*g*) brought down the unbroken ("whole") chloroplasts, which were washed by suspending them in 0·35M sodium chloride and centrifuging again at 1000*g*. A suspension of "whole" chloroplasts (C_1) was obtained by taking up the green pellet in 0·35M sodium chloride. If the pellet was taken up in 0·035M sodium chloride the chloroplasts became "broken" (C_{1s}) and a considerable amount of protein originally contained in the whole chloroplast was extracted (CE). Further washings of C_{1s} by suspending in 0·035M sodium chloride and centrifuging down the chloroplasts to remove water-soluble components yielded suspensions of "broken" chloroplasts labelled C_{1s1}, C_{1s2}, C_{1s3}, etc., according to the number of washings.

Kahn and von Wettstein (1961) have isolated chloroplasts in 0·35M NaCl (C_1) and examined them in the light and electron microscopes. Their electron micrographs shown in Figs. 4a and b represent those with smooth outlines and no clearly visible grana when viewed with the light microscope. These chloroplasts retain the normal structural features of those in the leaf and possess a lamellar system, stroma and surrounding membrane. Other chloroplasts isolated in 0·35M NaCl lack the smooth outlines when viewed in the light microscope and in the electron microscope can be seen to consist solely of the lamellar system and lacking a surrounding membrane and stroma (Fig. 4c). Despite the absence of the confining structures the lamellar systems of such chloroplasts appear identical to those of the chloroplasts in the leaf. When these NaCl chloroplasts are treated with water (or dilute NaCl: fraction C_{1s}) the surrounding membrane together with the stroma is lost and the regular arrangement of the lamellar discs is greatly distorted (Fig. 4d). These distorted, "broken" chloroplasts are however still capable of carrying out many biochemical reaction sequences.

Other workers e.g. James and Das (1957), Park and Pon (1961), Greenwood, Leech and Williams (1963), and Leech (1964), have preferred the use of buffered 0·3–0·5M sucrose solutions for the isolation of chloroplasts, but otherwise follow similar procedures.

Chloroplasts isolated in buffered sucrose have been examined under the electron microscope by Greenwood *et al.* (1963). An unwashed pellet (C_1) prepared from bean leaves contained chloroplasts together with some chloroplast fragments and mitochondria. These particles are all shown in Fig. 5a. The chloroplasts were of two easily distinguishable types, the first retaining the lamellar structure and stroma within a surrounding membrane (more opaque), the second lacking the membrane and stroma and corresponding to the naked lamellar structures (Fig. 4c) of Kahn and von Wettstein in salt-isolated chloroplasts. If the unwashed chloroplast fraction (C_1) was subjected to a sucrose-glycerol gradient differential centrifuging (James and Das, 1957) the naked lamellar structures were selected without contamination—see Fig. 5c. On subjecting C_1 to a sucrose gradient differential

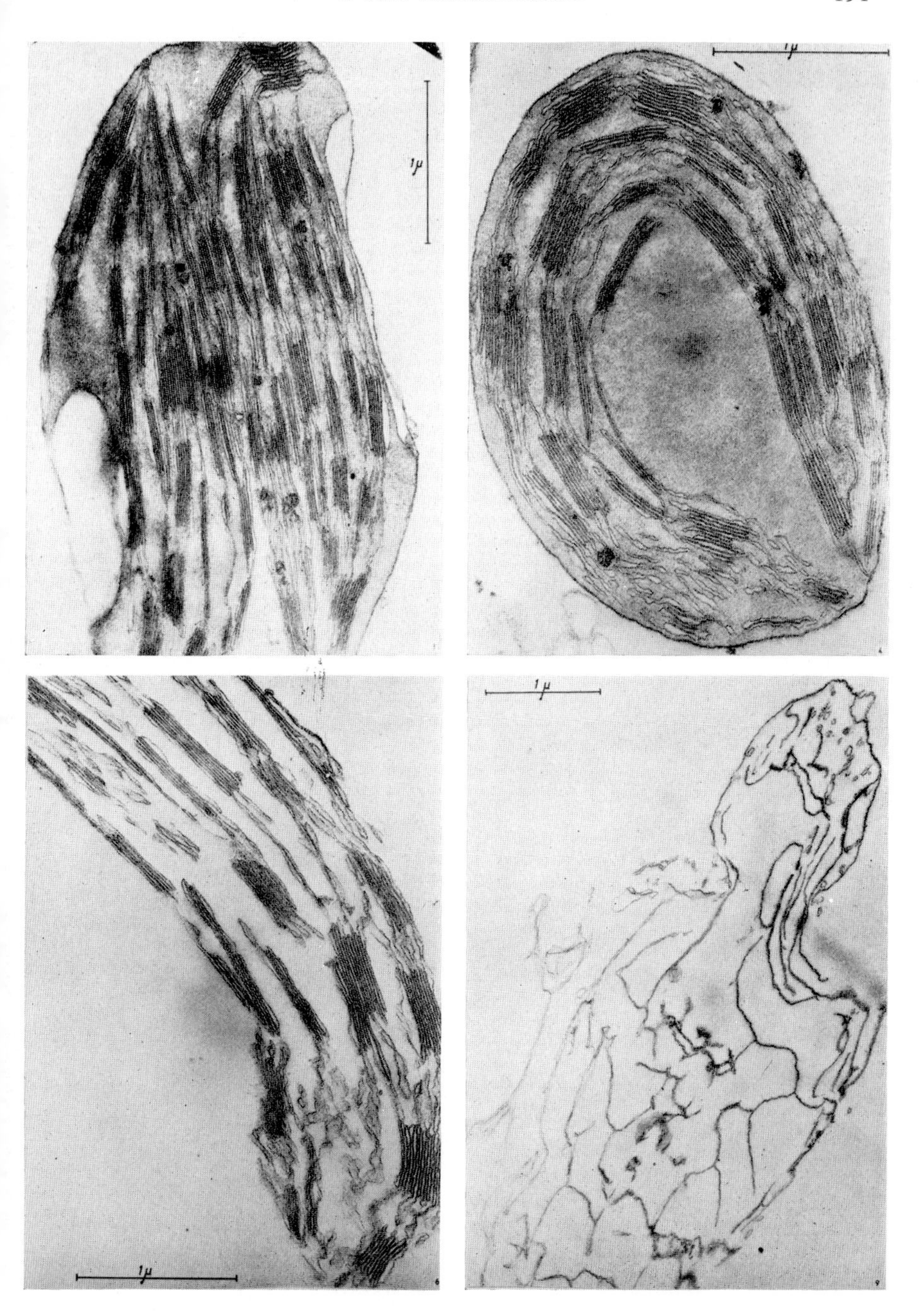

FIG. 4. Sections of spinach chloroplasts isolated in salt media, fixed in 2% $KMnO_4$. For descriptive details see text. (Kahn and von Wettstein, 1961).

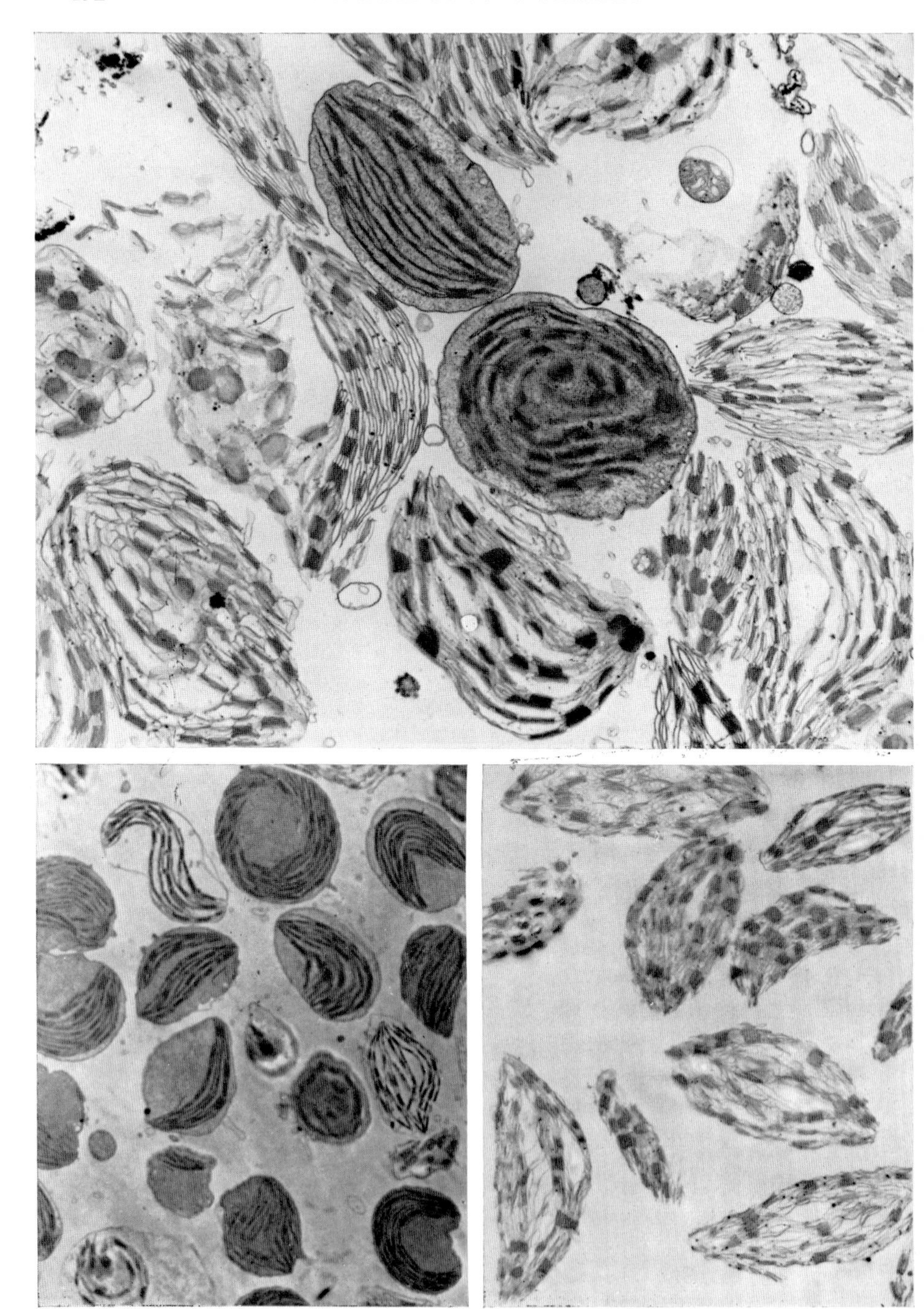

centrifuging (Leech, 1964) intact chloroplasts were selected with minimal contamination—see Fig. 5b.

B. ISOLATION IN NON-AQUEOUS MEDIUM

Chloroplasts have also been prepared in non-aqueous solvents by several workers using essentially the methods of either Stocking, 1959 (Fig. 6) or Thalacker and Behrens, 1959 (Fig. 7). In Stocking's method spinach leaves

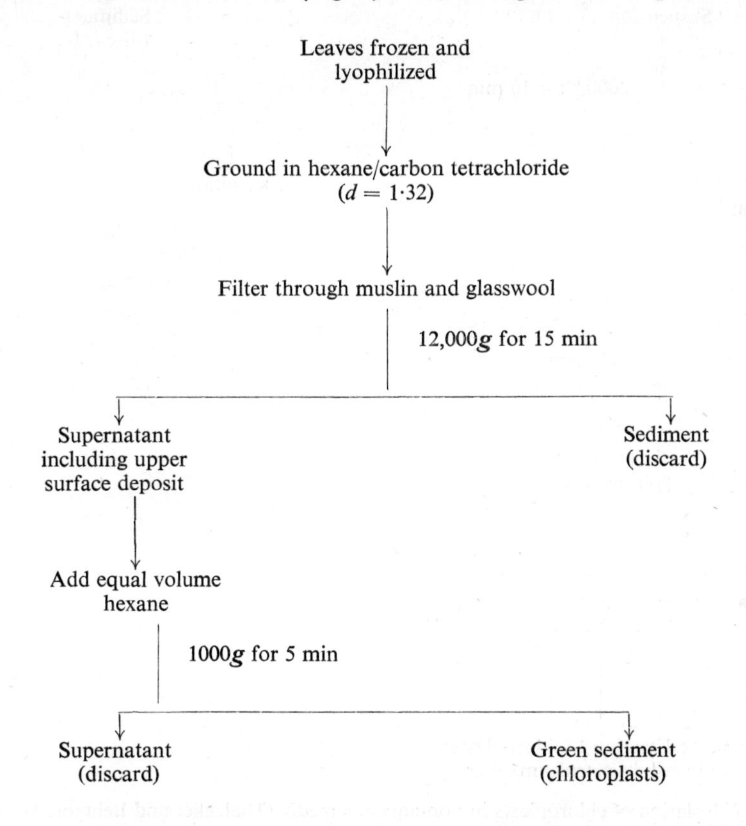

FIG. 6. Preparation of chloroplasts in non-aqueous media (Stocking, 1959)

were freeze-dried and ground in a mixture of hexane and carbon tetrachloride adjusted to give a density of 1·32. After filtering through cheesecloth the suspension was centrifuged at 12,000g for 15 minutes and the precipitate (density greater than 1·32) was rejected. To the supernatant an equal volume

FIG. 5. Sections of bean chloroplasts isolated in buffered sucrose media, fixed in OsO_4. For descriptive details see text. Magnification approximately (a) top: ×8000 (b) bottom left: ×3200 (c) bottom right: ×5500. (Courtesy of A. D. Greenwood.)

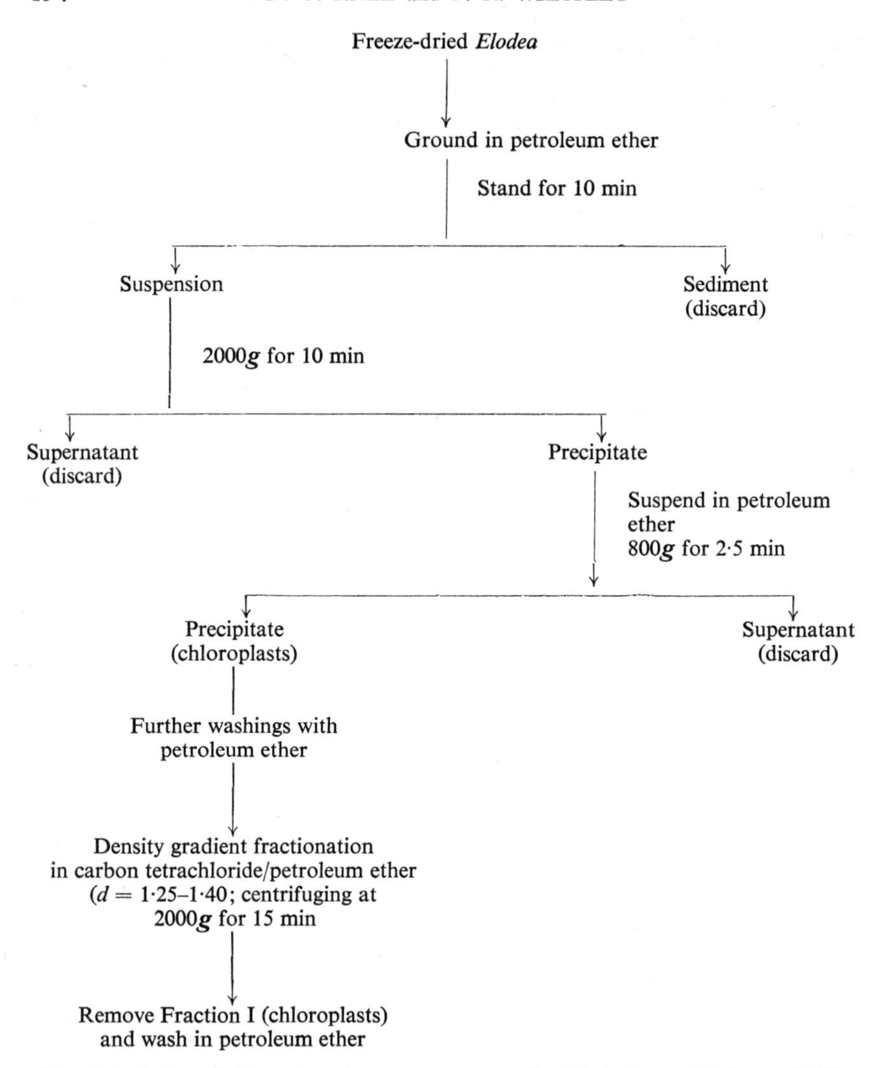

FIG. 7. Isolation of chloroplasts in non-aqueous media (Thalacker and Behrens, 1959)

of hexane was added and the mixture was centrifuged at 1000*g* for 5 minutes. The green precipitate (density less than 1·32) consisted almost entirely of chloroplast material.

In the method of Thalacker and Behrens (Fig. 7) leaves of *Elodea* were freeze-dried, ground in petroleum ether and allowed to stand for 10 minutes to allow the larger debris to settle out. The supernatant was then centrifuged for 10 minutes at 2000*g* and the precipitate of chloroplast material collected. The chloroplasts were washed by further suspending them in petroleum ether and centrifuging at 800*g*. Finally a density gradient fractionation between the density limits of 1·25 to 1·40 was carried out and fraction 1 (less than 1·32) was collected. The chloroplast fraction was finally washed once in the centrifuge with petroleum ether. Other investigations (e.g. Heber, Pon and Heber, 1963) with chloroplasts prepared in non-aqueous solvent systems have used essentially one or other of these methods.

C. COMMENTS ON THE LOCALIZATION OF ENZYME ACTIVITY

The difficulties of localizing the distribution of enzymes within chloroplasts have been discussed in some detail by Smillie (1963). When chloroplasts are isolated by grinding leaves in sodium chloride or sucrose solutions a common deficiency is that soluble proteins are lost. The loss of the protein may exceed 50% of the total protein content of the chloroplast and has been shown to vary with such factors as the method of breaking the cells, the medium used in the isolation of the chloroplast, the plant tissue used and the condition and physiological age of the material. If the chloroplasts contain starch they are often disrupted during centrifuging since the starch is denser than the chloroplast material. The loss of protein from the chloroplasts which occurs during the isolation has been repeatedly demonstrated and attention will be drawn to this phenomenon in subsequent discussions. Where the particular enzymic component under study is insoluble or relatively insoluble in water the chloroplasts isolated in aqueous media are often the most appropriate to use. Such chloroplasts exhibit the maximum range of activity of photochemical reactions, chloroplasts isolated in non-aqueous media being normally unable to catalyse such basic reactions as the Hill reaction (see below). The rates of reactions carried out in the light by chloroplasts, which depend on the presence of soluble proteins have, however, been found to be much lower than the rates of comparable preparations of leaf material. Two rather important examples of the loss of key proteins from chloroplasts during isolation in aqueous solvents include the loss of carboxydismutase from both "whole" and "broken" chloroplasts and the loss of the ferredoxin on breaking the chloroplast. Although very little of the cellular ferredoxin is found in chloroplasts isolated in dilute sodium chloride solutions ferredoxin is required for the reduction of NADP by enzymes which are bound to the chloroplasts and in the intact cell the ferredoxin is undoubtedly localized within the chloroplast.

The loss of materials from isolated chloroplasts has also been demonstrated by electron microscopy (Kahn and von Wettstein, 1961). As mentioned above two types of plastids were found in the population of spinach chloroplasts isolated in a sodium chloride medium. One type retains an apparently continuous outer membrane and contains the interlamellar material similar to that seen in chloroplasts in these sections. The remainder of the plastids are damaged to a greater or lesser extent and their outer membranes, and in more extreme cases interlamellar material, are missing. On treatment of the chloroplasts with dilute sodium chloride or water only the latter broken plastid type was observed. It is apparent that some tissues when treated by a procedure which gives a high percentage of intact chloroplasts from spinach yield in the main broken chloroplasts. This is an additional complication which may be of importance in evaluating the results of experiments on chloroplasts isolated in aqueous medium.

Smillie (1963) goes on to point out that in cases where the isolation of the chloroplast in aqueous media causes the loss of much protein from their original site, fractionation in non-aqueous media may be a useful alternative. In this procedure described in Figs 6 and 7, freeze-dried tissues are used and the entire isolation is carried out in media in which proteins are insoluble. The proteins are then preserved in their original location and are not leached out of the subcellular structures (cf. isolation of nuclei—Chapter 2). It has been pointed out that the contamination of the chloroplasts by co-precipitation of proteins originating from other portions of the cell might take place during the grinding and subsequent isolation procedures. This has not so far turned out to be a serious problem in the isolation of chloroplasts. Most enzymes survive the isolation procedure well. Enzymes which can be demonstrated in aqueous extracts of cells can also be detected after non-aqueous isolation of the chloroplasts.

Some difficulty may be experienced with the isolation of chloroplasts in non-aqueous media if they contain starch. The average density of the chloroplast is increased when it has accumulated starch and the fractionation will not proceed in the same way in the case of chloroplasts containing starch as occurs with chloroplasts that do not contain starch. It has been particularly noted that in the gradient centrifuging method of the type described by Thalacker and Behrens (1959, Fig. 7) chloroplasts with high starch content are not adequately separated from other cell components. Because of the difficulty inherent in the fractionation of the leaf material by the non-aqueous method Smillie has chosen to use the unicellular alga *Euglena gracilis*. Although the organism contains storage polysaccharide granules these are located outside the chloroplasts. The organism can be grown under well controlled conditions, by contrast with the frequently variable conditions under which leaves can be grown. Leaves, however, have several types of cell instead of being a uniform material. Moreover *Euglena* can be lyophilized very rapidly and the lyophilized material can be broken down more readily than can leaf material.

The fractionation and isolation of chloroplast material from *Euglena* has demonstrated very clearly that those fractions which contain chlorophyll, corresponding to chloroplast material, are characteristically rich in those enzymes which are confined to the chloroplasts in the intact cell, and a distribution of the enzyme activity identical with the distribution of chlorophyll has been found to occur for a number of enzymes. Equally clearly other enzymes are not localized in the chloroplasts and are presumably linked to glycolysis and respiratory processes.

The basic method which is used is to carry out a fractionation as cleanly as possible and then to assay for some "endogenous marker" component, e.g. chlorophyll in the case of chloroplasts or succinic dehydrogenase and cytochrome oxidase in the case of the mitochondria. By comparison of the distribution of the enzymic activity which is under study and the distribution of the component characteristic of the particular organelle in the various fractions it is possible to arrive at a conclusion about which particular organelle is the original site of localization of the enzyme activity. This technique has been very successfully applied to animal tissues by de Duve, Wattiaux and Baudhuin, (1962) and to plant tissues by Smillie (1955, 1963) and by other workers. (See Chapter 1 for a general discussion of this approach.)

IV. Enzymes of Chloroplasts

A. Enzymes for Assimilatory Power

In order to reduce CO_2 to the level of carbohydrates the chloroplast must first synthesize the components of the assimilatory power, viz. $NADPH_2$ and ATP. These products are formed by the processes of cyclic and noncyclic photophosphorylation.

Cyclic photophosphorylation which yields only ATP as a net product, can be represented by the equation:

$$\text{ADP} + \text{orthophosphate} \xrightarrow[\text{chloroplast}]{\text{light}} \text{ATP}$$

The process was first observed in isolated spinach chloroplasts by Arnon, Allen and Whatley (1954). Subsequent work showed that under appropriate experimental conditions this process is capable of high rates of ATP formation commensurate with the rates of photosynthesis in intact leaves (see review by Arnon, 1961).

The process of noncyclic photophosphorylation on the other hand yields both ATP and $NADPH_2$ according to the equation:

$$\text{NADP} + \text{ADP} + \text{orthophosphate} + H_2O \xrightarrow[\text{chloroplast}]{\text{light}} \text{NADPH}_2 + \text{ATP} + \tfrac{1}{2}O_2$$

The reaction was first observed with isolated chloroplasts by Arnon, Whatley and Allen (1958).

1. Noncyclic Photophosphorylation

A scheme to represent the electron flow from H_2O to NADP coupled with the formation of ATP is given in Fig. 8. It incorporates many of the current ideas of the mechanism of the process (see, for example, Whatley and Losada, 1964 for references). In a light reaction catalysed by a pigment system in which chlorophyll *b* or other accessory pigment predominates, electrons are transferred from H_2O ($E'_0 = +0{\cdot}81V$) to plastoquinone ($E'_0 = ca\ 0$ V). Oxygen is released in this process as a waste product of the

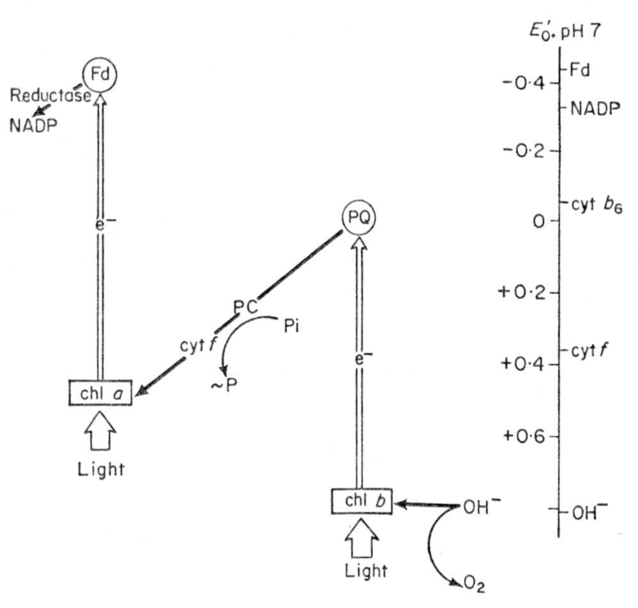

FIG. 8. Scheme for noncyclic photophosphorylation. Details in text. Fd: Ferredoxin, chl *a*: chlorophyll *a*, chl *b*: chlorophyll *b*., *PQ*: plastoquinone.

photo-oxidation of H_2O. In a subsequent light reaction catalysed by a pigment system in which chlorophyll *a* predominates electrons are transferred from cytochrome *f* ($E'_0 = +0{\cdot}37V$) to ferredoxin ($E'_0 = -0{\cdot}42V$). This reduced ferredoxin then donates its electrons to NADP ($E'_0 = -0{\cdot}34V$) in a dark reaction. The two photochemical reactions are joined by a series of dark reactions involving the transfer of electrons from plastoquinone to cytochrome *f* via plastocyanin in the direction of the thermochemical gradient. Part of the energy released during the passage of electrons is coupled to the formation of ATP. The participation and characterization of several of the enzymic intermediates is discussed below.

(*a*) *Photo-oxidation of water.* The enzymic apparatus necessary for the photo-oxidation of water, with the accompanying evolution of oxygen as a waste product, is located in the water-insoluble portion of the chloroplast, i.e., washed chloroplast fragments ($C_{1s}1$, $C_{1s}2$, $C_{1s}3$—see Fig. 3). Photo-oxidation of water can also be carried out in the very small chloroplast sub-units representing a small number of quantasomes (Park and Pon, 1963;

Becker, Shefner and Gross, 1965). The ability of chloroplast fragments to carry out the photo-oxidation of water in the light can be determined by measuring the rate of oxygen evolution accompanying the photo-reduction of various artificial electron acceptors such as ferric oxalate, ferricyanide, quinones and various dyes. The continuing evolution of oxygen from water (the electron donor) by illuminated chloroplasts in the presence of artificial electron acceptors is called the Hill reaction after its discoverer. The detailed mechanism of that part of the Hill reaction which leads to the evolution of oxygen has not so far proved susceptible to experimental analysis.

The photo-oxidation of water requires the presence of chloride as was first shown by Warburg in 1948. Chloride could be replaced by bromide but not by other anions and was considered to be essential for the evolution of oxygen. Later Bové, Bové, Whatley and Arnon (1963), confirmed Warburg's conclusion that chloride is essential only for those photosynthetic reactions in which oxygen is liberated. Chloride was not required for the cyclic photo-phosphorylation carried out by chloroplasts. A concentration of approximately 0·01M chloride is necessary to saturate the oxygen evolving reaction.

Manganese is also shown to be related to oxygen evolution. Experiments summarized by Pirson (1960) and Kessler (1960) using the green alga *Ankistrodesmus* which could be adapted to use hydrogen as the reductant instead of water and which can be grown under manganese deficient conditions, localized the site of action of manganese at the oxygen evolving site in photosynthesis. Spencer and Possingham (1961) working with chloroplasts isolated from manganese-deficient spinach leaves, came to the conclusion that manganese was required for oxygen evolution and noncyclic photophosphorylation but was not required for the cyclic electron flow.

Plastoquinone was shown to be necessary for the overall Hill reaction by Bishop (1959). The role of plastoquinone in the photoreduction of NADP by water, the physiological counterpart of the Hill reaction, has now been investigated. Since the noncyclic electron flow in chloroplasts has been shown to be composed of two partial reactions (i) the photo-oxidation of water leading to oxygen evolution and (ii) the subsequent photoreduction of NADP usually coupled with ATP formation, it has been possible to identify the site of action of plastiquinone with more certainty. Plastoquinone was found to be needed only for the photo-oxidation of water and not for the subsequent reduction of NADP. Plastoquinone may very well be the initial electron acceptor for the auxiliary light reaction catalysed by the accessory pigment (photoreaction B). However, from his work on fluorescent compounds, Duysens (1964) proposes that an unidentified substance termed "quencher" is the initial electron acceptor located between photoreaction B and plastoquinone.

A number of substances have been found to exert a powerful inhibitory effect on the oxygen evolving reaction. These include phenylurethane and the substituted phenyl ureas such as the herbicide CMU (3-(4-chlorophenyl)-1,1-dimethylurea), *o*-phenanthroline, hydroxylamine and azide (Losada and Arnon, 1963). Duysens (1964) has suggested that CMU interrupts the electron flow immediately prior to the point at which plastoquinone accepts electrons. The site of action of the other inhibitors of oxygen evolution are quite unknown.

The photo-oxidation of water (photoreaction B) is driven by light absorbed by the accessory pigment system, as indicated in Fig. 8. In this accessory pigment system, chlorophyll b_{650} is the primary light absorber and is accompanied by small amount of chlorophyll *a* (chlorophyll a_{673}) which appears to act as the sink of the energy transfer chain. For the photo-reduction of NADP (photoreaction A) chlorophyll *a* with an absorption maximum at 683mμ (chlorophyll a_{683}) is the primary light absorber; it is accompanied by a pigment P700 (absorption maximum at or above 700mμ) which acts as the energy sink for chlorophyll a_{683} (French, 1964).

(*b*) *Plastocyanin.* Katoh (1960) and Katoh, Suga, Shiratori and Takamiya, (1961) isolated a blue copper-containing protein called plastocyanin from the leaves of higher plants and from various algae. This protein has a molecular weight of approximately 21,000 and in the oxidized form possesses an absorption spectra with a broad maximum at 597mμ. The redox potential was found to be +0·370 volts. Plastocyanin contains two atoms of copper per mole and approximately 50% of the copper contained in the chloroplast is present in plastocyanin. One mole of plastocyanin ($2Cu^{2+}$ atoms) is present per 3–400 moles of chlorophyll in the spinach chloroplast (Katoh *et al.*, 1961). Katoh and Takamiya (1963) found that added plastocyanin could become photoreduced on illumination with chloroplast fragments, i.e., it could act as a Hill reagent.

In studying the photoreduction of NADP with water as the electron donor, Trebst (1963) found that the overall reaction was inhibited by the copper chelating agent salicylaldoxime. If ascorbate, in the presence of the dye dichlorophenolindophenol, was substituted for water as the electron donor salicylaldoxime was no longer an effective inhibitor. It was suggested that plastocyanin might have been the copper-containing compound upon which salicylaldoxime acted. This was later substantiated by the experiments of Trebst and Pistorius (1965) who studied the role of plastocyanin in digitonin-extracted chloroplasts. On treatment with digitonin the plastocyanin was lost and at the same time the ability to carry out the Hill reaction was lost. The use of ascorbate as an alternative electron donor showed that electrons were readily donated to the electron transport chain through the mediation of the dye which could also be replaced by plastocyanin. This indicated that reduced plastocyanin is capable of reacting readily with the components of the chain and provides some evidence for the functional role of plastocyanin. These experiments are related to those of Nieman, Nakamura and Vennesland,

(1959) who used digitonin-treated chloroplasts and showed that they were capable of carrying out a photo-oxidation of reduced cytochrome *c*. On fractionation of the chloroplast preparation a supernatant factor was found which was necessary to permit the reaction of the cytochrome *c* with the electron transport chain. This was later shown by Katoh and Takamiya (1963) to be identical with plastocyanin. Further evidence of the role of copper-containing compounds in the electron transport chain comes from the work of Spencer and Possingham (1960) on copper deficient tomato plants. Chloroplasts isolated from leaves of copper deficient plants showed only a feeble Hill activity which they showed to be due to an effect on the reductive portion of the Hill reaction and not related to the oxygen evolving system, in contrast to manganese.

The fact that purified plastocyanin is photo-reduced by chloroplasts and can also be photo-oxidized if the chloroplasts are pre-treated with digitonin has led Kok, Rurainski and Karmon (1964) to suggest that plastocyanin could mediate the transfer of electrons between the two light reactions of photosynthesis. Evidence for this functional role has been obtained by de Kouchkovsky and Fork (1964) who analysed the light induced absorption changes in the spectral region where the difference between the oxidized and reduced form of the copper protein is maximum, namely 591mμ. Another compound which markedly inhibits the Hill reaction at high concentrations is KCN. This probably functions in a manner similar to that of salicylald-oxime by reacting with the copper containing compound.

(*c*) *Cytochrome* f. Special cytochrome components have been discovered in photosynthetic cells of green plants and several algae and appear to be concentrated in the chloroplasts (Hill and Scarisbrick, 1951; Davenport and Hill, 1952). Cytochrome *f* was isolated, as a haem protein of molecular weight 110,000, from photosynthetic tissues and shown to be present in chloroplasts in the ratio of 1 mole of cytochrome *f* to 400 moles of chlorophyll. Cytochrome *f* has an unusually oxidizing redox potential (E'_0, pH7 $= +0{\cdot}365$ V). When isolated from the chloroplasts it occurs in the reduced form, characterized by a very sharp α band at 554·5mμ. Katoh (1959) has described the isolation and properties of a cytochrome of the *c*-type from various algae of the family Rhodophyceae, Phaeophyceae, Chlorophyceae, and Cyanophyceae. The cytochromes from these widely different algae had practically identical redox potentials and absorption spectra, with E'_0, pH 7 $= +0{\cdot}30$ to $+0{\cdot}34$ V and an α band at 553mμ. They appear to be the algal equivalent of cytochrome *f*, which itself is a cytochrome of the *c*-type. Cytochrome *f* has been isolated and purified from parsley, spinach and from the alga, *Euglena*.

The presence of cytochrome *f* in chloroplasts can be shown directly since its spectrum is visible not only in leaves but also in isolated chloroplasts after acetone treatment to remove the chlorophyll. In the living state cytochrome *f* may also be seen in pale varieties of plants or mutants of algae. Cytochrome *f* exists in the unilluminated leaf in the reduced form. It is not auto-oxidizable and no cytochrome *f* oxidase has ever been detected in plants. Cytochrome *f* is not oxidized by cytochrome *c* oxidase and thus cytochrome *f* does not react with oxygen. However, on illumination of leaves or algae the reduced cytochrome *f* becomes oxidized (see Duysens, 1964, for discussion and references). A clear cut demonstration of the photo-oxidation of cytochrome *f* was obtained by Duysens (1955) with the red alga *Porphyridium cruentum*. In this case the α band of cytochrome *f* was not completely masked by the chlorophylls and spectral changes around 550mμ were clearly seen. Similar changes were also observed by Chance and Sager (1957) with a *Chlamydomonas* mutant that had a low content of chlorophyll and carotenoids. In normal leaves the changes in spectrum at 550mμ could not be observed because of masking by chlorophyll and evidence demonstrating the photo-oxidation of cytochrome *f* could only be inferred from changes at 420mμ. On turning off the light the cytochrome returned to the reduced level. Similar cytochromes have also been detected in photosynthetic bacteria. Smith and Baltscheffsky (1959) made the interesting observation that in chromatophores cytochromes could become photo-oxidized in the light but that this only occurred in the presence of a phosphate accepting system, adenosine diphosphate + inorganic phosphate. The coupling of photo-oxidation of the cytochrome with phosphorylation was thus shown in the bacterial system and it is inferred by analogy that it takes place in the chloroplast. In Fig. 8 we represent the pigment system accepting an electron from cytochrome *f* and passing the electron on to ferredoxin in a reaction catalysed by the chlorophyll *a* system. The special pigment component of this system, P700 first observed by Kok in 1956 (see Kok, Cooper and Yang, 1963 for discussion) is believed to participate as a catalyst within the system.

(*d*) *Ferredoxin*. Spinach ferredoxin (Tagawa and Arnon, 1962) is a water-soluble protein of molecular weight of approximately 14,000, characterized by possessing two iron atoms per molecule in a non-haem combination and two labile sulphur atoms which are released on acidification as H_2S. In addition the molecule contains six sulfhydryl groups in cysteine. The protein had been studied extensively under a series of different names before the identity of the various synonyms was recognized; it was earlier studied as the methaemoglobin reducing factor, later as photosynthetic pyridine nucleotide reductase, as the NADP reducing factor, as the red enzyme, and finally as ferredoxin. One mole of ferredoxin is found per approximately 400 moles of chlorophyll both in leaves and in isolated whole chloroplasts. On breaking the chloroplasts the ferredoxin is released into the solution. The identity of the various substances mentioned above has been shown on the basis of complete amino acid analysis (Davenport and Hill 1960, Fry and San Pietro, 1963) by a comparison of the activity of the purified compounds, and by a demonstration of their interchangeability in a number of reactions. The redox potential of spinach ferredoxin, (E'_0 pH 7·5) is – 0·430 V. The role assigned to ferredoxin in metabolism of the chloroplast is as an intermediate in electron transport at a very reduced potential. In the presence of illuminated chloroplasts ferredoxin will accept electrons by a photochemical reaction and can subsequently pass these electrons on to acceptors such as NADP in the presence of the appropriate enzyme, ferredoxin-NADP reductase. Its role as a carrier of electrons can be readily studied by observing changes in the absorption spectrum and it has been shown to become reduced by accepting one electron per mole in a photochemical reaction and to donate this electron via the flavoprotein enzyme to NADP in a subsequent dark reaction (Whatley, Tagawa and Arnon, 1963). Ferredoxin also seems to serve as a branching point in the electron transport systems that result either in cyclic or noncyclic photophosphorylation. When the photoreduced ferredoxin is reoxidized by NADP (with the aid of ferredoxin-NADP reductase) noncyclic photophosphorylation results; when oxidized NADP is unavailable as an acceptor, the photoreduced ferredoxin is reoxidized by a bound component of the grana [possibly cytochrome b_6—see Section IV. A. 2. e.] and cyclic photophosphorylation results. The expected stoichiometry between the number of moles of ferredoxin reduced and of oxygen evolved in noncyclic photophosphorylation has also been observed (Arnon, Tsujimoto and McSwain, 1964). Other reactions in addition to the reduction of NADP or the catalysis of cyclic photo phosphorylation have also been identified. Reduced ferredoxin is involved for example in the reduction of nitrite, in the photoreduction of hydrogen and in the reduction of various dyes. Many of these reactions have been summarized recently by San Pietro and Black (1965).

o-Phenanthroline, an inhibitor of photosynthesis, combines readily with ferredoxin, removing the iron atoms from the molecule and releasing H_2S. This action is however not the basic reason for the inhibition of photosynthesis by *o*-phenanthroline, which apparently acts directly on the oxygen evolving reaction. Mercury compounds also combine readily with ferredoxin (mercuric chloride and *p*-chloromercuribenzoate have been studied in some detail) and the action of mercuric compounds on the inhibition of cyclic and noncyclic photophosphorylation in which ferredoxin is involved can be attributed directly to the reaction between these compounds and ferredoxin.

(*e*) *Ferredoxin-NADP reductase* (1.6.99.4.). As already stated, ferredoxin-NADP reductase is a flavoprotein enzyme of the chloroplast which catalyses the transfer of the hydrogens from reduced ferredoxin to NADP. Shin, Tagawa and Arnon (1963) prepared the reductase in crystalline form and investigated a number of its properties. It has a much higher affinity for NADP than for NAD (400 times greater; K_m for NADP $= 9{\cdot}78 \times 10^{-6}$M; for NAD $= 3{\cdot}75 \times 10^{-3}$M). Ogren and Krogmann (1963) showed that more than 90% of the NADP but only 50% of the NAD in the leaf of spinach is located in the chloroplast. (These values were determined on chloroplast material isolated in non-aqueous solvent since chloroplasts isolated in aqueous solvents contain virtually no pyridine nucleotide). In view of the marked affinity of the reductase for NADP it seems reasonable, therefore, to believe that NADP is of predominant biosynthetic significance in the chloroplast.

In addition to its main physiological function as a ferredoxin-NADP reductase the enzyme was also found by Shin *et al.* (1963) to have other activities from which were derived an ability to catalyse the oxidation of $NADPH_2$. They observed (i) transfer of hydrogens from $NADPH_2$ to a number of hydrogen or electron acceptors such as FMN, FAD, ferricyanide and various dyes; (ii) NADP reductase was also found to have transhydrogenase activity. It catalysed a transfer of hydrogens from $NADPH_2$ to NAD though not $NADH_2$ to NADP; (iii) it acted as a $NADPH_2$-cytochrome *c* reductase in the presence of one of several electron acceptors mentioned above such as vitamin K_3, FMN, FAD or ferredoxin. In all these cases the role of the ferredoxin-NADP reductase was to transfer electrons from $NADPH_2$ to the respective electron acceptor which could then reduce cytochrome *c* non-enzymically.

These additional activities of the ferredoxin-NADP reductase, in addition to other evidence, indicate that the enzyme is the same substance as the chloroplast $NADPH_2$—diaphorase isolated and purified by Avron and Jagendorf (1956) and the chloroplast transhydrogenase isolated and purified by Keister, San Pietro and Stolzenbach (1960). The latter investigators have already discussed the similarities and possible identity of transhydrogenase (1.6.1.1.) and $NADPH_2$ diaphorase. The spectrum of the enzyme which was crystallized by Shin *et al.* (1963) indicates that it contains flavin. Avron and Jagendorf, studying the enzyme under the name of diaphorase, found it to contain FAD. The enzyme is shown to be localized in the chloroplast since it is unnecessary to add the enzyme initially in studying the reduction of NADP by illuminated chloroplasts in the presence of ferredoxin. However, the extraction of chloroplasts by dilute *tris* buffer causes the solubilzation of the reductase, and treated chloroplasts cannot then reduce NADP via ferredoxin. On adding the purified enzyme or simply on adding the purified extract this ability reappears. Shin and Arnon (1964) have shown that the action of the enzyme is to become reversibly reduced by diaphorase and to be reoxidized by NADP, a phenomenon readily observed through changes in the absorption spectrum. Ferredoxin-NADP reductase is inhibited by adenosine-2′-monophosphate, a competitive inhibitor which was previously shown also to inhibit transhydrogenase and diaphorase activity. Mercuric chloride, *p*-chloromercuribenzoate, and iodoacetate also inhibit. The diaphorase activity was not inhibited by cyanide, *o*-phenanthroline, ethylenediamine tetraacetate (EDTA), 3-(4-chlorophenyl)-1,1-dimethylurea, (CMU) or azide. Since ferredoxin reacts with *o*-phenanthroline one cannot test the sensitivity of the ferredoxin-NADP reductase activity directly. The molecular weight of the protein appears to be approximately 35,000 and it contains 1 mole of FAD per mole enzyme.

(*f*) *Factors affecting coupled phosphorylation.* The scheme for electron transport (Fig. 8) indicates that ATP formation is coupled to the electron flow. The overall reaction is represented by the equation:

$$\text{NADP} + \text{H}_2\text{O} + \text{ATP} + \text{P} \xrightarrow{\text{light}} \text{NADPH}_2 + \text{ATP} + \tfrac{1}{2}\text{O}_2$$

This stoichiometry was first observed by Arnon *et al.* (1958) and has been confirmed by several investigators. Replacement of NADP by various artificial electron acceptors, e.g. ferricyanide or quinones, still leads to the same ATP/2e$^-$ ratio. Evidence on this point is summarized by Jagendorf (1962). In the absence of ADP and phosphate reduction of the acceptor is normally greatly diminished, provided no other component is rate limiting, but the rate can be restored to its original value by the addition of ADP and phosphate.

On the addition of a number of reagents which can act as uncouplers, e.g. NH_4^+, carbonyl cyanide *m*-chlorophenylhydrazone (see Losada and Arnon, 1963 for further examples), the rate of electron transport remains either unchanged at the "ADP + P-stimulated" rate or may be slightly increased, whereas the ATP formation ceases. Compounds such as NH_4^+, primary amines and several anions at high concentration, atebrine at 5×10^{-5}M, and the very potent carbonyl cyanide phenylhydrazone inhibitors cause an uncoupling which can be reversed by simply washing them out of the chloroplasts. The electron transport can be irreversibly uncoupled by many fold dilution so as to reduce the cation concentration below a critical level. If chloroplast suspensions are frozen they usually become uncoupled to a considerable extent.

Similar work on the uncoupling agents of oxidative phosphorylation by mitochondria have suggested the involvement of proteins in the mitochondria which couple phosphorylation and electron transport. The isolation of these coupling agents from mitochondria has been attempted with some success (Lehninger, 1964). Although it is not exactly clear how these protein compounds function, their participation in the coupling in oxidative phosphorylation is indicated (see Chapter 3). In photosynthetic phosphorylation in chloroplasts the only experiments on isolation of the coupling factors so far reported are those of Avron (1963) who found a 20% or somewhat greater stimulation on the re-addition of a heat labile, non-dialysable factor to depleted chloroplasts. Further experiments may demonstrate the obligatory nature of protein coupling factors in photosynthetic phosphorylation which may function similarly to those isolated from mitochondria.

2. Cyclic Photophosphorylation

In this process the net product is ATP. It is thought to result from the passage of electrons from a photochemically produced reductant to a photochemically produced oxidant in an electron flow which is cyclic in nature (Arnon, 1959). In Figs. 9a,b schemes are suggested for two examples of this process. Figure 9a represents a system probably closely related to the system

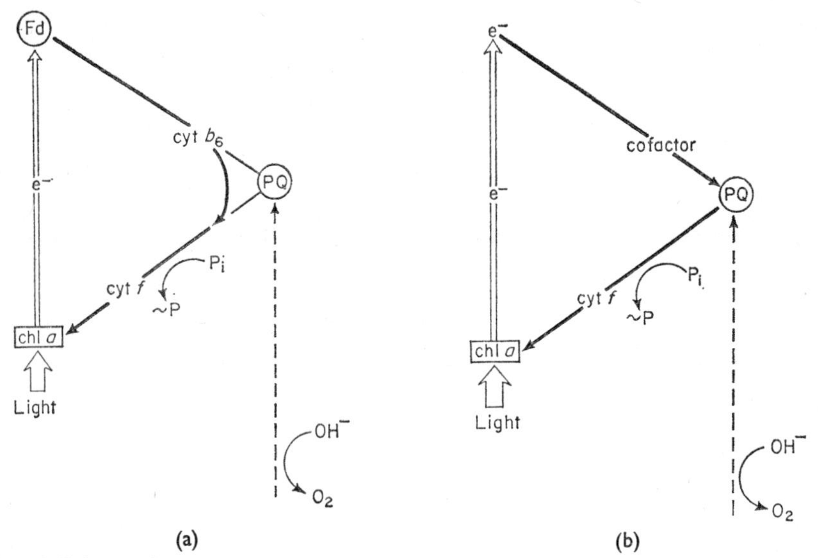

FIG. 9. Schemes for cyclic photophosphorylation. Details in text. Fd: Ferredoxin, chl *a*: chlorophyll *a*, PQ: plastoquinone.

occurring *in vivo* in which electrons are received by ferredoxin from the photochemical reaction and returned by a series of dark reactions via cytochrome b_6 and cytochrome *f*. The energy released by this transfer of electrons is coupled to ATP formation. The photo-oxidation of H_2O, an integral part of noncyclic photophosphorylation, is not involved in the cyclic electron flow.

Figure 9b represents a system in which washed chloroplast fragments depleted of ferredoxin (C_{1s1}, see Fig. 3) catalyse the formation of ATP only on the addition of an artificial cofactor, e.g. phenazine methosulphate, vitamin K_3 and flavin mononucleotide. These artificial cofactors substitute for ferredoxin and cytochrome b_6, which are themselves components of the physiological electron flow system.

Plastoquinone is included in Fig. 9b as an intermediate in the cyclic electron flow on the basis of the experiments of Krogmann and Olivero (1962) and Whatley and Horton (1963), showing the depression of cyclic photophosphorylation when plastoquinone was removed and its restoration when plastoquinone was added back. There is no direct evidence available on the participation of plastoquinone in the cyclic photophosphorylation catalysed by ferredoxin.

The participation of several of the intermediates shown in Figures 9a and 9b is discussed below.

(*a*) *Plastocyanin.* Direct evidence is lacking but Trebst (1963) has shown that salicylaldoxime at 10^{-2}M inhibits cyclic photophosphorylation with vitamin K_3 as the co-factor to about the same extent as it inhibits noncyclic photophosphorylation. On the basis of this inhibition by salicylaldoxime, taken in conjunction with the removal and re-addition of plastocyanin to digitonin treated chloroplasts, Trebst concluded that plastocyanin participates in the electron transport system of noncyclic photophosphorylation. It seems reasonable then to assume that plastocyanin also participates in the cyclic electron transport system. Details of plastocyanin are given in the section (IV.A.1.*d*) above.

(*b*) *Cytochrome* f. No direct evidence for its role in cyclic photophosphorylation is available. We place cytochrome *f* in the scheme adjacent to chlorophyll *a* by analogy with the position assigned to it in non-cyclic electron flow—see Section (IV.A.1.*c*) above. It should be pointed out that in green plants technical problems make it extremely difficult to observe any changes in the redox state of cytochrome *f* upon illumination. Although changes indicating the oxidation of cytochrome *f* have been seen with golden varieties of leaves or chlorophyll deficient mutants of *Scenedesmus* it has been pointed out by Hill and Bonner (1961) that these varieties may not necessarily reflect conditions in the normal plant.

(*c*) *Ferredoxin.* Tagawa, Tsujimoto and Arnon (1963) have demonstrated a functional role for ferredoxin as a catalyst in cyclic photophosphorylation under anaerobic conditions in the presence of CMU to stop the photo-oxidation of water. Prior to this several investigators had detected ATP formation dependent upon ferredoxin ("photosynthetic pyridine nucleotide reductase"), under conditions where oxygen could reoxidize the reduced co-factor. Although this appeared to be a cyclic electron flow it actually closely resembled a non-cyclic system described above and was inhibited by CMU. This oxygen-dependent type of cyclic photophosphorylation has been termed pseudocyclic photophosphorylation. (See Arnon, 1965 for details and references).

The cyclic photophosphorylation catalysed by ferredoxin and described by Tagawa *et al.* (1963) is sensitive to antimycin A, thereby differentiating it from cyclic photophosphorylation catalysed by artificial catalysts and non-cyclic photophosphorylation which are not sensitive to antimycin A. Reasoning by analogy with the sensitivity of mitochondrial electron flow to antimycin A at the level of cytochrome *b*, the sensitivity of the ferredoxin catalysed photophosphorylation to antimycin A suggests that ferredoxin may react with the cytochrome b_6 found in chloroplasts and this in turn with cytochrome *f*. Urbach and Simonis (1964) studying inhibition of the endogenous photophosphorylation in the alga *Ankistrodesmus* have found this to be sensitive to antimycin A, suggesting that *in vivo* the cyclic electron flow catalysed by ferredoxin is in fact operative.

(*d*) *Ferredoxin-NADP reductase.* There is no evidence to implicate the reductase in the electron transport of cyclic photophosphorylation catalysed by ferredoxin. Its role appears to be confined to the catalysis of NADP reduction in noncyclic photophosphorylation. However, the possibility cannot be excluded that the flavoprotein may act as a diaphorase in those artificial cyclic photophosphorylation systems which are catalysed by vitamin K_3, FMN and various dyes.

(*e*) *Cytochrome* b_6 (General reference Losada and Arnon, 1964). In addition to cytochrome *f*, Hill (1954) found another cytochrome (in the *b* group) which he called cytochrome b_6, to be present in chloroplasts. It is auto oxidizable, has an E'_0 at pH 7 of $-0{\cdot}06$ V and a sharply defined absorption band at 563mμ. The molar ratio of b_6 to *f* was determined by Hill and Bonner (1961) to be 1·3. It is clear from the work of Hill and Bonner (1961) and of Lundegårdh (1962) that chloroplasts of higher plants contain at least two characteristic cytochrome components, *f* and b_6, and that these compounds are present in relatively high amounts to the extent of one mole of each cytochrome per 400 moles of chlorophyll. Cytochrome b_6 has not been isolated from plants since it appears to be bound firmly to the insoluble portion of the particle. It is however relatively easy to observe in acetone extracted chloroplasts.

On the basis of its redox potential and the sensitivity of the ferredoxin catalysed cyclic photophosphorylation to antimycin A it seems reasonable to assign cytochrome b_6 to an electron transporting role between ferredoxin and cytochrome *f*. There is no direct evidence to warrant the inclusion or exclusion of plastocyanin in the ferredoxin catalysed cyclic photophosphorylation, even though there is, as mentioned above, evidence of the salicylaldoxime sensitivity of the vitamin K_3 catalysed cyclic photophosphorylation.

(*f*) *Factors affecting coupled photophosphorylation.* The same observations and comments apply to cyclic photophosphorylation as to the non-cyclic photophosphorylation described above (Section IV.A.1.*f*) The only direct evidence which bears on the subject comes from the work of Avron (1963) who has shown some degree of restoration of activity with protein components.

B. ENZYMES FOR CO_2 FIXATION CYCLE

Isolated chloroplasts can fix CO_2 and reduce it to the level of carbohydrate in the light as was first shown by Arnon, Allen and Whatley (1954)—see Allen, Arnon, Capindale, Whatley and Durham (1955) for further details.

The incorporation of CO_2 in photosynthetic tissues takes place according to a scheme which may be termed the reductive pentose cycle or Calvin cycle. Evidence for the operation of the reductive pentose cycle in algae and leaves comes from the work summarized by Bassham *et al.* (1954). Subsequent work by Losada, Trebst and Arnon (1960) indicates that the reductive pentose cycle also operates in an isolated chloroplast system.

In the condensed diagram (Fig. 10) the cycle is indicated as consisting of three phases. In the carboxylative phase ribulose diphosphate accepts a

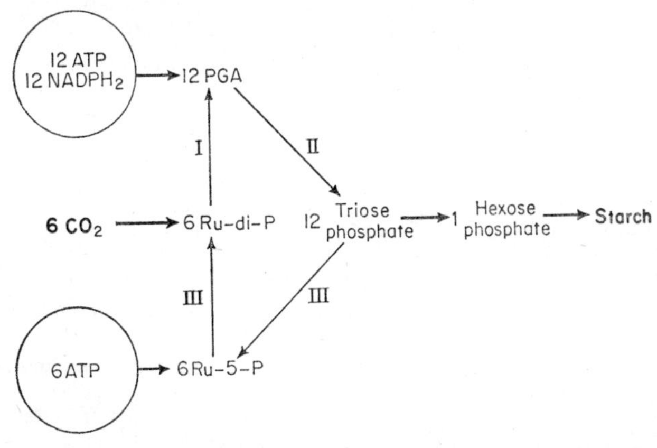

FIG. 10. Condensed diagram of the reductive carbohydrate cycle in chloroplasts. The cycle consists of three phases. In the carboxylative phase (I), ribulose diphosphate (Ru-di-P) accepts a molecule of CO_2 and is cleaved to 2 molecules of phosphoglyceric acid (PGA); in the reductive phase (II) PGA is reduced to triose phosphate; in the regenerative phase (III) triose phosphate is partly converted into Ru-di-P and partly into hexose phosphate and starch. All the reactions of the cycle occur in the dark. The reactions of the carboxylative and reductive phases are driven by ATP and $NADPH_2$ formed in the light. One complete turn of the cycle results in the assimilation of 1 mole of CO_2 at the expense of 3 moles of ATP and 2 moles of $NADPH_2$. (Losada *et al*, 1960.)

molecule of CO_2 and yields two molecules of phosphoglyceric acid. In the reductive phase phosphogylceric acid is reduced to triosephosphate. In the regenerative phase triosephosphate is partly converted into ribulose diphosphate and partly into hexose phosphate and starch. All of these reactions occur in the dark. The energy which they require is supplied by ATP and $NADPH_2$ formed in the light (assimilatory power). One complete turn of the cycle results in the assimilation of one mole of CO_2 at the expense of 3 moles of ATP and 2 moles of $NADPH_2$. In Fig. 11 the carbon compounds occurring as intermediates in the cycle are summarized and the enzymes believed to be involved in their transformation are listed. These enzymes and their distribution within the cell are described in the following sections.

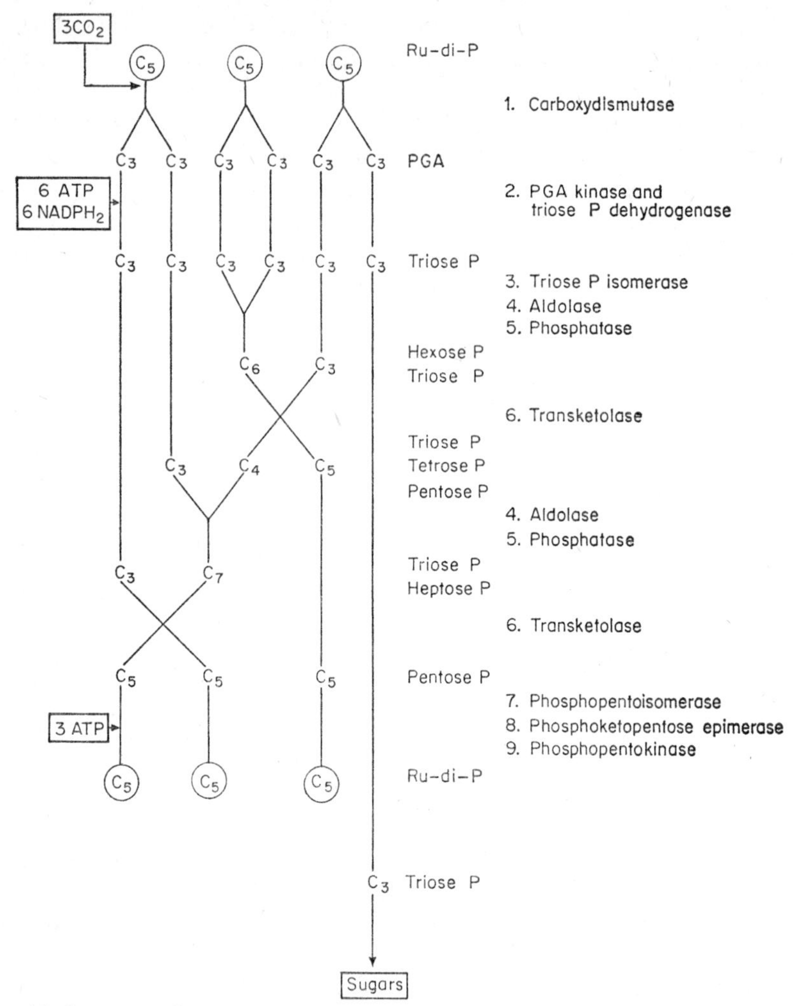

FIG. 11. Summary of reactants and enzymes of the reductive pentose cycle. PGA: phosphoglycerinc acid, Ru-di-P: Ribulose diphosphate.

1. Carboxylative Phase

Ribulose-1,5-diphosphate carboxylase (carboxydismutase: 4.1.1.39.,1). This enzyme catalyses the reaction:

$$CO_2 + \text{D-ribulose-1,5-diphosphate} + H_2O \rightarrow 2\ \text{D-3-phosphoglycerate}$$

This enzyme has been isolated and purified from spinach leaves by Weissbach *et al.* (1956). The enzyme has also been identified in other leaves and in algae (see references in Thomas, 1960; Calvin and Bassham, 1962). Carboxydismutase is localized in the chloroplast. Water extracts (CE in Fig. 3) of chloroplasts isolated in 0·35M sodium chloride contain the enzyme

(Losada *et al.*, 1960). Jacobi and Perner (1961) showed the loss of the carboxydismutase from whole chloroplasts on washing in 0·35M sodium chloride, indicating its original location in the chloroplasts. This loss of carboxydismutase may well acount for the loss of CO_2 fixing ability during the isolation and washing of spinach chloroplasts (Smillie and Fuller, 1959). When chloroplasts were isolated in non-aqueous media Smillie and Fuller found that the activity of the carboxydismutase paralleled the chlorophyll content of the fractions of frozen dried leaves. Heber *et al.* (1963) further confirmed the location of carboxydismutase in chloroplasts from several species, isolating the chloroplasts in non-aqueous media. Smillie (1963) has also shown that carboxydismutase is localized in the chloroplast of light grown *Euglena* by carrying out the isolation procedure in non-aqueous media.

The major component of the soluble leaf protein fraction from green leaves is a high molecular weight substance which sediments as a single protein in the ultra-centrifuge. It accounts for a remarkably high proportion (up to 40%–50%) of the total leaf protein (Dorner, Kahn and Wildman, 1957). This protein, originally termed Fraction I protein (Wildman and Bonner, 1947) is localized in the chloroplast (Lyttleton and Ts'o, 1958) and has since been shown to exhibit carboxydismutase activity (van Noort and Wildman, 1964).

Although the enzyme in green plants is confined to the chloroplasts it has also been shown by Trudinger (1956) and by Aubert, Milhaud and Millet (1957) to be present in the autotrophic, CO_2 fixing, non-photosynthetic sulphur bacterium *Thiobacillus denitrificans* (in which, of course, no chloro plasts or chromatophores are present).

The enzyme purified from leaves (Weisbach, Horecker and Hurwitz, 1956; Jakoby, Brummond and Ochoa, 1956) has a molecular weight of about 300,000, is unstable below pH6, is specific for ribulose-1,5-diphosphate and has a K_m = approximately $2{\cdot}3 \times 10^{-4}$M for ribulosediphosphate and about $1{\cdot}5 \times 10^{-2}$M for HCO_3^- at pH8. The latter is an unexpectedly high value for the principal CO_2 fixing enzyme which in nature depends on an external CO_2 concentration of 0·03% CO_2. In studying the stoichiometry of the reaction catalysed by the isolated enzyme it was found that in the presence of excess CO_2, one mole of ribulose-1,5-diphosphate gave 2 moles of 3-phosphoglyceric acid. The reaction was apparently irreversible. The purified enzyme was inactive in the absence of divalent metal ions; magnesium, nickel and cobalt are effective activators. The enzyme is also activated by sulfhydryl compounds (cysteine and glutathione) and is inhibited by sulfhydryl binding agents. Mercuric chloride at a concentration of 2×10^{-4}M inhibited the enzyme completely and *p*-chloromercuribenzoate inhibited 50% at this concentration. Arsenite did not inhibit, indicating that dithiol groups were not involved.

Cyanide was found by Trebst, Losada and Arnon (1960) to be an effective inhibitor of the carboxylating enzyme, a concentration of 10^{-4}M giving 60% inhibition. For further discussion of the inhibitors of CO_2 fixation in general see the review by Losada and Arnon (1963).

2. *Reductive Phase*

According to the ideas expressed by Calvin in his description of the photosynthetic CO_2 fixation cycle the reductive phase involves the conversion of phosphoglyceric acid into phosphoglyceraldehyde, a reaction sequence catalysed by two enzymes (i) phosphoglycerate kinase (2.7.2.3) and (ii) the NADP dependent triosephosphate dehydrogenase (1.2.1.13). The reactions catalysed by these enzymes are:

(i) ATP + D-3-phosphoglycerate → ADP + D-1,3-diphosphoglycerate
(ii) D-1,3-diphosphoglycerate + $NADPH_2$ →
NADP + D-glyceraldehyde-3-phosphate + orthophosphate.

The action of the dehydrogenase has been measured both in the forward and reverse directions. Starting with glyceraldehyde-3-phosphate as a substrate, Jacobi and Perner (1961) added arsenate to render the reaction irreversible and were able to measure the activity of the dehydrogenase alone. Most other workers have chosen to assay the reactions starting with phosphoglyceric acid and have thereby obtained a composite measure of both the enzymes involved in the reductive phase.

The enzymes have been isolated from the green tissues of several algae and higher plants (see Thomas, 1960 and Calvin and Bassham, 1962 for references). Both enzymes have been detected in water extracts of the chloroplasts isolated in 0·35M sodium chloride by Losada *et al.* (1960) who claimed that this provided evidence that the enzymes are present in the chloroplasts of intact cells. However, Smillie (1963) has commented that since the percentage of the total cellular activity was not stated he considered that the data were insufficient to justify this claim. The activity found in the chloroplast extract was low compared with the values for the intact leaves indicating that the bulk of the activity was present in non-chloroplast fractions.

Jacobi and Perner (1961) have already observed that "whole" chloroplasts isolated in 0·35M sodium chloride lose triosephosphate dehydrogenase activity during washing at the same time as they lose their water soluble proteins by leakage. This implies an original localization for the NADP-dependent dehydrogenase in the chloroplasts. The ready leakage of chloroplast proteins probably accounts for the low activity reported by Losada *et al.* (1960) since the "whole" chloroplasts from which the cytoplasmic extract was made had already undergone preliminary washing. A direct demonstration of the original distribution of the enzyme within the cell has been obtained by Smillie (1963) for pea leaves and for *Euglena*, and by Heber *et al.* (1963)

for spinach and other leaves. Using chloroplasts isolated in non-aqueous media they showed that all the NADP-dependent triosephosphate dehydrogenase was localized within the chloroplast. The phosphoglyceric acid kinase is however distributed between the chloroplast and the rest of the cytoplasm an observation consistent with its participation both in photosynthesis and in glycolysis.

In addition to the NADP-dependent dehydrogenase which is closely associated with photosynthetic reactions, green leaves also contain a NAD-dependent dehydrogenase (Gibbs, 1952; Arnon, 1952). This has been detected in water extracts of isolated chloroplasts by Losada *et al.* (1960) and by Smillie and Fuller (1960) and is also found in non-green tissues of several plants where it presumably participates in glycolysis. Further experiments using chloroplasts prepared in non-aqueous solvents have shown that the enzyme is found in both chloroplast and non-chloroplast fractions, 20%–40% of the total amount of the NAD-dependent triosephosphate dehydrogenase activity being found in the chloroplast.

It is worth noting that Rosenberg and Arnon (1955) have found a third enzyme which oxidizes triosephosphate but which acts in the absence of phosphate or arsenate and catalyses a reversible conversion of phosphoglyceraldehyde to phosphoglyceric acid. It requires NADP but does not require cooperation of the phosphoglyerate kinase. The dehydrogenase is present only in green tissue but its role in photosynthetic reactions is unknown.

The triosephosphate dehydrogenases are sensitive to iodoacetate and addition of this inhibitor would be expected to result in the prevention of the reductive steps of photosynthetic CO_2 fixation. Work on the inhibition of photosynthesis by iodoacetate using intact *Elodea* leaves (Simonis and Weichart, 1958), and with *Chlorella* (Kandler, Liesenkotter and Oaks, 1961), has shown the accumulation of phosphoglyceric acid and phosphoenolpyruvate and a decrease in sugar phosphates, a result in agreement with this expectation. However, other workers have found that in isolated systems the enzyme phosphoribulokinase (see below) is sensitive to iodoacetate and this is probably the site of action of iodoacetate with isolated chloroplasts.

Investigations have also been made on the possible competitive inhibition by various sugar phosphates of the triosephosphate dehydrogenases. D-sedoheptulose-7-phosphate was shown by Gibbs (1963) to inhibit CO_2 fixation by isolated chloroplasts and in addition to inhibit both the NADP- and NAD-linked dehydrogenases. Work with threose-2,4-diphosphate (a specific inhibitor of muscle and yeast triosephosphate dehydrogenase) has however, given equivocal results and the use of this inhibitor has so far proved to have been of limited value (see Gibbs, 1963).

Gibbs and Turner (1964) have summarized additional details of all three dehydrogenases and of the phosphoglycerate kinase from leaf material.

3. Regeneration of the CO_2 Acceptors

(*a*) *Triosephosphate isomerase* (5.3.1.1.). This enzyme catalyses the reaction:

D-Glyceraldehyde-3-phosphate→Dihydroxyacetone phosphate.

The isomerase has been found in pea seeds (Tewfik and Stumpf, 1951), in spinach leaves (Peterkovsky and Racker, 1961) and in algae (Richter, 1959). On the basis of ^{14}C labelling experiments with spinach chloroplasts, Trebst and Fiedler (1962) and Losada *et al.* (1960) have concluded that the photosynthetic scheme according to Calvin operates in the isolated chloroplast system. This implies the presence of the enzyme triosephosphate isomerase in the systems. Recently Smillie (1963) has reported on the intracellular distribution of this enzyme in both pea leaves and *Euglena*; by isolating subcellular fractions in non-aqueous media about 40% of the total isomerase activity was shown to be in the chloroplast fraction in both organisms. This is in accordance with the expected participation of triosephosphate and isomerase in both photosynthesis and glycolysis. Gibbs and Turner (1964) have summarized additional information on this enzyme.

(*b*) *Aldolase* (4.1.2.7.). Aldolase catalyses the reaction:

Dihydroxyacetone phosphate + D-Glyceraldehyde-3-phosphate

→D-Fructose-1,6-phosphate.

and is believed also to catalyse the second similar aldolase type conversion:

Dihydroxyacetone phosphate + D-Erythrose-4-phosphate

→D-Sedoheptulose-1,7-diphosphate.

The enzyme has been demonstrated in leaves and algae (references are given in Thomas, 1960). The enzyme has been shown to be present in water extracts of isolated spinach chloroplasts (Losada *et al.*, 1960) as well as "whole" chloroplasts by Jacobi and Perner (1961) who also showed that the enzyme leaks out of the chloroplasts during washing. After these results had indicated a partial localization of aldolase within the chloroplast, the work of Smillie (1963) with subcellular fractions of *Euglena* prepared in non-aqueous media showed that some 60% of the aldolase activity within the cell was present in the chloroplast fraction. These data are in agreement with the participation of aldolase in both photosynthetic reactions and glycolysis. Further details of this enzyme are given in the review of Gibbs and Turner (1964).

(*c*) D-*fructose*-1,6-*diphosphatase* (3.1.3.11.). This enzyme catalyses the reaction:

D-Fructose-1,6-diphosphate + H_2O →

D-Fructose-6-phosphate + orthophosphate

The enzyme was shown to be present in leaves (see Thomas, 1960 and Calvin and Bassham, 1962 for references) and in algae (Peterkovsky and Racker, 1961, Smillie, 1964). The enzyme has been shown to be present in the aqueous extract of isolated spinach chloroplasts (Losada *et al.*, 1960). Smillie (1964) has clearly distinguished between the distribution of alkaline and acid phosphatases within different parts of the pea plant and has shown that the alkaline phosphatase is confined to the green tissues. In *Euglena* it has been shown to be completely confined to the chloroplasts by fractionation in non-aqueous media, whereas the acid phosphatases are associated with the non-chloroplast fractions. This indicates the importance of the alkaline fructose-diphosphatase in the photosynthetic organelle. The enzyme is highly specific for D-fructose-1,6-diphosphate; it requires Mg^{2+} for its activity and has an optimum pH between 8·5 and 9·5 (see Gibbs and Turner, 1964 for additional details).

(*d*) *Transketolase* (2.2.1.1.). This enzyme catalyses the reactions:

D-Glyceraldehyde-3-phosphate + D-Fructose-6-phosphate →

D-Erythrose-4-phosphate + D-Xylulose-5-phosphate

and

D-Glyceraldehyde-3-phosphate + D-Sedoheptulose-7-phosphate →

D-Ribose-5-phosphate + D-Xylulose-5-phosphate.

Transketolase has been found in green leaves by numerous investigators (see Thomas, 1960 and Calvin and Bassham, 1962 for references). It has also been found in algae by Peterkovsky and Racker (1961) and Smillie (1963). On the basis of experiments involving the fractionation of subcellular particles in non-aqueous media, Smillie (1963) has shown that transketolase is present in the chloroplasts of *Euglena* but is also found in other parts of the cell indicating a dual metabolic role for this enzyme. The evidence for the occurrence of the enzyme in isolated chloroplasts rests upon ^{14}C labelling experiments which indicate that the complete photosynthetic reductive cycle operates *in vitro* (Losada *et al.*, 1960; Trebst and Fiedler, 1962). The presence of transketolase may be inferred from these experiments. This enzyme is described in further detail in the review of Waygood and Rohringer (1964).

(*e*) *Ribosephosphate isomerase* (phosphopento-isomerase) (5.3.1.6.). This enzyme catalyses the reaction:

D-Ribose-5-phosphate → D-Ribulose-5-phosphate

The enzyme has been isolated and partially purified from spinach leaves by Weissbach, Smyrniotis and Horecker (1954). Peterkovsky and Racker (1961) have confirmed the presence of ribosephosphate isomerase in spinach leaves and shown it also to be present in *Euglena* and *Chlorella*. Smillie (1963) has shown that the enzyme is present in the chloroplasts of *Euglena* but is also found in other subcellular fractions. The ^{14}C labelling experiments of Losada *et al.* (1960) and Trebst and Fiedler (1962) indicate the presence of this enzyme in isolated spinach chloroplasts. Further details of the enzyme are given in the review of Waygood and Rohringer (1964).

(*f*) *Ribulosephosphate*-3-*epimerase* (phosphoketopentose epimerase) (5.1. 3.1.). This enzyme catalyses the reaction:

$$\text{D-Xylulose-5-phosphate} \rightarrow \text{D-Ribulose-5-phosphate}$$

The enzyme has been demonstrated to occur in spinach leaves (Horecker, Hurwitz and Smyrniotis, 1956; Peterkovsky and Racker, 1961) and it is present also in *Euglena* and *Chlorella* (Peterkovsky and Racker, 1961). Its presence in the isolated chloroplasts is inferred from the ^{14}C labelling experiments. Waygood and Rohringer (1964) have summarized further details of the enzyme.

(*g*) *Phosphoribulokinase* (phosphopentokinase) (2.7.1.19.). This enzyme catalyses the reaction:

$$\text{ATP} + \text{D-Ribulose-5-phosphate} \rightarrow \text{ADP} + \text{D-Ribulose-1,5-diphosphate}$$

The enzyme has been purified from spinach leaves by Hurwitz, Weissbach, Horecker and Smyrniotis (1956) (see Thomas, 1960 for other references). Losada *et al.* (1960) indicated that the enzyme is present in water extracts of isolated spinach chloroplasts. Peterkovsky and Racker (1961) showed that the enzyme was present in cell free extracts of *Euglena* and *Chlorella*, while Smillie (1963) localized the phosphoribulokinase as occurring solely in the chloroplasts of *Euglena*, after isolating the chloroplasts in non-aqueous media.

The properties of the purified enzyme were studied by Hurwitz *et al.* (1956) who found that it was specific for ribulose-5-phosphate and ATP. The Michaelis constants were $2{\cdot}5 \times 10^{-4}$M for ribulose-5-phosphate and $2{\cdot}8 \times 10^{-4}$M for ATP. The phosphorylation reaction proceeds at a maximum velocity at pH 7·9 and requires Mg^{2+} at a concentration of 5×10^{-3}M for maximum velocity. Phosphoribulokinase appears to be a sulphydryl enzyme. It is inhibited by the addition of the heavy metals Cu^{2+} and Hg^{2+} and by *p*-chloromercuribenzoate. The inhibition by metal and by *p*-chloromercuribenzoate was completely reversed by the addition of cysteine. Trebst *et al.* (1960) and Calo and Gibbs (1960) found that iodoacetate inhibited the carboxylative phase of photosynthetic CO_2 fixation in isolated chloroplasts, and the latter workers showed in addition that iodoacetamide inhibits the phosphoribulokinase, thus establishing the site of action of this inhibitor in photosynthesis.

Table I shows a summary of the known intracellular distribution of the enzymes involved in photosynthetic CO_2 fixation. The enzymes have been

TABLE I. *Distribution of Enzymes of the Reductive Pentose Cycle in Several Plants. Numbers refer to listing of enzymes in Fig.* 11. (For references, see text).

	Spinach	Pea	*Euglena*
1. Carboxydismutase	++	++	++
2. Triose phosphate dehydrogenase (NADP)	++	++	++
2. Phosphogylcerate kinase	+	+	+
3. Triose phosphate isomerase	+	+	+
4. Aldolase	+	−	+
5. FDP phosphatase (alkaline)	+	++	++
6. Transketolase	+	−	+
7. Phosphopentoisomerase	+	−	+
8. Phosphoketopentoepimerase	+	−	+
9. Phosphopentokinase	+	−	++

\+ = Present in chloroplast
++ = Confined to chloroplast
− = Unknown.

discussed individually in the preceding sections where the evidence to support their placing in the table is given.

C. ADDITIONAL ENZYME REACTIONS

1. *Carbohydrate Metabolism*

(*a*) *Formation of glucose*-1-*phosphate from fructose*-6-*phosphate*. The conversion of fructose-6-phosphate (a nett product of the reductive CO_2 fixation cycle) to glucose-1-phosphate (precursor of starch) requires the presence of two enzymes. These are

(i) Glucosephosphate isomerase (5.3.1.9.) which catalyses the reaction:

D-Fructose-6-phosphate → D-Glucose-6-phosphate

and

(ii) Phosphoglucomutase (2.7.5.1.) which catalyses the reaction:

D-Glucose-6-phosphate + D-Glucose-1,6-diphosphate →

D-Glucose-1,6-diphosphate + D-Glucose-1-phosphate

The conversion of [^{14}C]-glucose-1-phosphate into [^{14}C]-fructose-6-phosphate has been shown in intact leaves and leaf homogenates of sugar beet by Burma and Mortimer (1956) indicating the presence of both enzymes in leaf tissues. The glucosephosphate isomerase has been isolated and partially purified from seeds of *Phaseolus radiatus* by Ramasamna and Giri (1956). The isomerase is widely distributed in plants and in animals. The phosphoglucomutase has been reported by Sissakian (1958) to be localized in the chloroplast. Recent evidence by Bird, Porter and Stocking (1965) using chloroplasts from tobacco leaves isolated in non-aqueous media has confirmed the presence of both glucose phosphate isomerase and phosphoglucomutase within the chloroplast. In the special case of *Euglena* which forms starch outside the chloroplast, the isomerase and the mutase are found in the non-chloroplast fraction (Smillie 1963).

Photosynthetic experiments using $^{14}CO_2$ *in vivo* (see Bassham and Calvin, 1960 for references) have shown the occurrence of the intermediates fructose-6-phosphate and glucose-1-phosphate as early products of CO_2 fixation, suggesting that glucose-1-phosphate is formed from fructose-6-phosphate by the action of the isomerase and the mutase. Since sucrose has been shown by Stocking, Williams and Ongun (1963) to be formed within the chloroplast and presumably requires the prior formation of glucose-1-phosphate as an intermediate, the presence of the isomerase and mutase in the chloroplast had already been presumed.

(*b*) *Starch synthesis*. Starch consists of two components, amylose and amylopectin. In amylose glucose molecules are joined by α-1,4-linkages in an unbranched chain. In amylopectin there is a high degree of branching with relatively short α-1,4-linked glucose chains connected by α-1,6-linkages to other α-1,4-linked glucose chains (Feingold, Neufeld and Hassid, 1964). Two enzymes are needed to synthesize starch. These are phosphorylase and the Q enzyme or branching enzyme.

(i) Phosphorylase (2.4.1.1.) catalyses the reaction:

$$\alpha\text{-D-Glucose-1-phosphate} + (\alpha\text{-1,4-glucosyl})_n \rightarrow (\alpha\text{-1,4-glucosyl})_{n+1} + \text{Orthophosphate}$$

The enzyme is widely distributed in plants both in green and non-green parts. In leaves of algae it has been found in both the chloroplasts and in the cytoplasm (see Sissakian, 1958 and Thomas, 1960 for references). There is no absolute agreement between various workers concerning the distribution of phosphorylase. Some workers emphasize that it is predominantly localized in the chloroplasts and others believe in an equal distribution of the enzyme between chloroplast and cytoplasm. Stocking (1959) isolated chloroplasts from tobacco leaves in non-aqueous media and detected at least 54% of the total phosphorylase activity in the chloroplast fraction. Since starch is in any case synthesized in higher plants only within the chloroplasts or leucoplasts and is an insoluble product, it seems reasonable to expect that the enzymes required for its synthesis will be localized within the chloroplast or leucoplast.

There is an alternative method known for the synthesis of α-1,4-linkages; it utilizes UDP-D-glucose as the substrate (see Section IV.C.1.*d*). This enzyme system has been found in bean cotyledons by Rongine de Fekete, Leloir and Cardini (1960). An interconversion of sucrose to starch via UDP-glucose was also envisaged by them. The pathway via UDP-glucose would be energetically more favourable than the phosphorylase pathway. Unfortunately, however, no information is available concerning the distribution of this synthetase enzyme within the cell.

(i) The Q (or branching) enzyme (α glucan-branching glycosyl transferase: 2.4.1.18.) catalyses the formation of α-1,6-linkages in amylopectin. The reaction occurs by the transfer of a terminal segment of an α-1,4-linked glucose chain to the primary hydroxyl of a non-terminal D-glucose residue, i.e., to position 6. We know nothing of its intracellular location but its presence in chloroplasts and in leucoplasts must be inferred from the ability of these organelles to deposit starch internally.

(*c*) *Starch breakdown.* In view of the equilibrium constant of the synthesis of starch from glucose-1-phosphate (which is directly and markedly affected by the pH) it has been postulated that the action of phosphorylase is the primary mechanism responsible for the breakdown of starch in plants.

An alternative enzyme system responsible for starch breakdown has been provided by the enzyme α-amylase (3.2.1.1.) which hydrolyses the α-1,4-linkages of starch in a random fashion. A similar enzyme β-amylase (3.2.1.2.) removes maltose units (glucosyl-1,4-α-glucose) hydrolytically from starch. Although the enzymes are widely distributed they are especially active in germinating seeds where rapid starch breakdown is required. The amylases are presumably located in the plastids, as this is where starch is stored. The enzymes have been detected in leaves and algae (see Sissakian, 1958 and Thomas, 1960 for references).

(*d*) *Sucrose synthesis.* Two related mechanisms have been identified for sucrose synthesis in plants. These are catalysed by the enzymes:

(i) UDP glucose-fructose glucosyltransferase (2.4.1.13.) which catalyses the reaction:

$$\text{UDP-D-Glucose} + \text{D-Fructose} \rightarrow \text{UDP} + \text{Sucrose}$$

(ii) UDPglucose-fructose-phosphate glucosyltransferase (2.4.1.14.) which catalyses the reaction:

$$\text{UDP-D-Glucose} + \text{D-Fructose-6-P} \rightarrow \text{UDP} + \text{Sucrose-6-phosphate}$$

The involvement of a phosphatase to split off the phosphate from the sucrose-6-phosphate is presumed since no sucrose phosphate accumulates in the plant. The operation of such a phosphatase, which has been identified in some tissues, would render the second reaction irreversible. Bird *et al.* (1965) have shown sucrose-6-phosphatase to be present in tobacco chloroplasts isolated in non-aqueous media.

The substrate for the transferases, UDPG, is synthesized by the enzyme:

(iii) UDPG pyrophosphorylase (2.7.7.9.) which catalyses the reaction:

$$\text{D-Glucose-1-phosphate} + \text{UTP} \rightarrow \text{UDPG} + \text{Pyrophosphate}$$

This enzyme was also found in tobacco chloroplasts isolated in non-aqueous media by Bird *et al.* (1965). The enzyme is absent from the cytoplasm.

The two transferases were first isolated from wheat germ by Leloir and Cardini (1955). Burma and Mortimer (1956) have reported the synthesis of sucrose by such mechanisms using excised sugar beet leaves and leaf homogenate, indicating the presence of the two enzymes in leaf material. Evidence indicating the occurrence of both transferases in chloroplasts has been given by Bird *et al.* (1965), as noted above.

Radioactive sucrose is formed during the incorporation of $^{14}CO_2$ in the light by *Chlorella*, *Scenedesmus* and soyabean leaves and is the first free sugar to become labelled to any extent (see Calvin and Bassham, 1962). The intracellular distribution of the [^{14}C]-sucrose formed in photosynthesis was studied by Stocking *et al.* (1963) in tobacco leaves. After treatment with $^{14}CO_2$, leaves were freeze-dried and separated in non-aqueous media to isolate various subcellular fractions. After a 20 sec treatment with $^{14}CO_2$, sucrose appears in the chloroplast fraction only. Appreciable sucrose did not appear in the non-cellular fractions up to 1 to 2 min after $^{14}CO_2$ feeding had taken place. This suggests that the primary site of sucrose synthesis in tobacco leaves is in the chloroplast and implies therefore that there is glucosyl transferase in the chloroplast. This is in agreement with the identification in isolated chloroplasts of the transferases, pyrophosphorylase and phosphatase, required for the synthesis of sucrose.

The properties of these enzymes are described in detail in the review by Feingold *et al.* (1964).

2. *Carboxylic Acids*

(*a*) *Entry of PGA into the Krebs cycle.* The enzymes needed to transform phosphoglyceric acid to pyruvate include:

(i) Phosphoglycerate mutase (5.4.2.1.) [possibly (2.7.5.3.)] which catalyses the reaction:

$$\text{3-Phospho-D-glycerate} \rightarrow \text{2-Phospho-D-glycerate}$$

(ii) 2-phospho-D-glycerate hydrolyase (enolase) (4.2.1.11.) which catalyses the reaction:

$$\text{D-2-Phosphoglycerate} \rightarrow \text{Phosphoenolpyruvate} + H_2O$$

(iii) Pyruvate kinase (2.7.1.40) which catalyses the reaction:

$$\text{Phosphoenolpyruvate} + \text{ADP} \rightarrow \text{Pyruvate} + \text{ATP}$$

These three enzymes have been found in the leaves of various plants (see Gibbs and Turner, 1964 for references). Evidence for the occurrence of the enzymes in spinach chloroplasts comes from experiments on aqueous chloroplast extracts (Rosenberg, Capindale and Whatley, 1958). Although the evidence is insufficient to determine the extent to which the enzymes are localized within the chloroplast of spinach, Smillie (1963) found approximately 10% of the enolase activity to be concentrated in the chloroplasts in pea leaf fractions isolated in non-aqueous media.

Pyruvate would normally be expected to enter the Krebs cycle via CoA formed by way of the pyruvate oxidase system (i.e. pyruvate dehydrogenase, 1.2.4.1; lipoate acetyl transferase, 2.3.1.12; and lipoamide dehydrogenase, 1.6.4.3.). Although the system has been characterized in detail in bacterial and animal cells, little is known about it in plant systems, apart from the observation that certain plant mitochondrial systems can oxidize pyruvate. In all probability the pyruvate oxidase system is not present in chloroplasts.

The enzymes which convert phosphoenolpyruvate to malate, aspartate and oxaloacetate have been detected both in green leaves and in water extracts of isolated spinach chloroplasts (see Rosenberg *et al.*, 1958 for references; also Zelitch and Barber, 1960b for malate dehydrogenase in washed spinach chloroplasts). These enzymes include:

(iv) Phosphopyruvate carboxylase (4.1.1.31.) which catalyses the reaction:

$$\text{phosphoenolpyruvate} + CO_2 + H_2O \rightarrow \text{oxalacetate} + \text{orthophosphate}$$

(v) Malate dehydrogenase (1.1.1.37.) which catalyses the reaction:

$$\text{Oxalacetate} + NADH_2 \rightarrow \text{L-Malate} + NAD$$

(vi) Aspartate aminotransferase (2.6.1.1) which catalyses the transformation:

$$\text{Oxalacetate} + \text{L-Glutamate} \rightarrow \text{L-Aspartate} + \text{2-Oxoglutarate}$$

For further details of phosphoenolpyruvate carboxylase and malate dehydrogenase the review by Davies and Ellis (1964) may be consulted. Sanwal, Zink and Din, (1964) have given further details of the enzyme aspartate transaminase in their review article.

(*b*) *Krebs cycle enzymes.* Losada and Arnon (1964) have drawn attention to the occurrence of a number of enzymes participating in the Krebs cycle which were found in water extracts of isolated spinach chloroplasts. Tests for individual enzymes have demonstrated the requisite enzymic equipment for the conversion of pyruvate to acetate and oxoglutarate via citrate and isocitrate, the continued formation of oxalacetate necessary for the operation of the condensing enzyme presumably coming via the carboxylation of phosphoenol-pyruvate to oxalacetate. The closely related malate and fumarate presumably arise when excess reduced pyridine nucleotide is available (the phosphoenolpyruvate carboxylase and the malate dehydrogenase are discussed above in section (IV c. 2.*a*)).

Those enzymes normally associated with the Kreb's cycle which have been identified in chloroplast extract (CE) are the acetate activating enzyme (acetyl CoA synthetase, 6.2.1.1), condensing enzyme (citrate synthase, 4.1.3.7.), aconitate hydratase (4.2.1.3.), isocitrate dehydrogenase (1.1.1.42.), malate dehydrogenase (1.1.1.37.) and fumarase (fumarate hydratase, 4.2.1.2.) —these enzymes catalyse the conversion of oxalacetate to fumarate. The enzyme succinate dehydrogenase (1.3.99.1.) has not been reported in isolated chloroplasts.

The available data are insufficient to admit us to claim the localization of these enzymes within the chloroplasts or to specify the extent to which their distribution indicates an original association with the chloroplasts. It is of interest to point out that Zelitch and Barber (1960a) have obtained mitochondrial preparations from spinach leaves which can carry out the oxidation of succinate, citrate, isocitrate, oxoglutarate, pyruvate and $NADH_2$ accompanied by ATP formation (oxidative phosphorylation). The enzymes required for this complex of oxidations are obviously located within the mitochondria, and might conceivably represent a source of contamination in the chloroplast extract.

(*c*) *Two carbon acids.* The enzyme glycollate oxidase (1.1.3.1.) catalyses the reaction:

$$\text{Glycollate} + O_2 \rightarrow \text{Glyoxylate} + H_2O_2$$

The enzyme has been identified in chloroplasts of various leaves (see Zelitch, 1964 and Davies and Ellis, 1964, for references and details of the properties of the enzyme). Pierpoint (1962) using a sucrose density gradient centrifuging method to fractionate tobacco leaf homogenates, states that very little of the glycollate oxidase appeared to be associated with the chloroplast fraction but is instead principally a soluble enzyme. However some fraction of the glycollate oxidase activity always remains associated with chloroplasts.

The function of the enzyme in chloroplasts is unclear although it has been suggested that it forms part of the terminal oxidase system in conjunction with glyoxalate reductase (Zelitch, 1964).

The glycollate oxidase activity is weak or absent in extracts of etiolated plants but on illumination the activity increases markedly. Tolbert and Cohan (1953) have shown that this response to light is indirect. Etiolated plants sprayed with glycollate in the dark showed an increase in the glycollate oxidase activity and it is suggested that glycollate oxidase is an adaptive enzyme, appearing in response to the formation of glycollate produced during photosynthesis. It is also interesting that the oxidase activity is induced in cell free extracts of etiolated plants on the addition of glycollate. Although at first sight the involvement of light in the inducing of glycollate oxidase appears to indicate the localization of the enzyme within the chloroplast, the fact that the effect is indirect, depending eventually on the production of glycollate upon greening, make such a conclusion unreliable.

The glycollate pathway in photosynthesizing cells is further discussed in the paper of Tolbert (1963).

The enzyme glyoxylate reductase (1.1.1.26) catalyses the reaction:

$$\text{Glyoxylate} + \text{NADH}_2 \rightarrow \text{Glycollate} + \text{NAD}$$

The enzyme is present in spinach leaves and occurs in washed suspensions of spinach chloroplasts (Zelitch and Barber, 1960b). The enzyme has been crystallized from tobacco leaves (see Davies and Ellis 1964, for details of the enzyme). Crude extracts of tobacco and spinach leaves have activity with $NADH_2$ for glyoxylate reduction. Zelitch and Gotto (1962) have partially separated the NADP-and NAD-dependent enzymes and shown them to be distinct entities.

It has been suggested that the function of the enzyme is as a terminal oxidase system in conjunction with glycollate oxidase. Butt and Peel (1963) have suggested that the glycollate-glyoxylate cycle participates in the re-generation of ATP in illuminated *Chlorella* by removing excess $NADH_2$. There is no reported evidence to indicate the intracellular localization of the NADP dependent glyoxylate reductase.

In tissues rich in fats, particularly storage tissues, e.g. castor bean seeds, glyoxylate is involved in the operation of the glyoxylate cycle, whereby acetate is converted to malate and other intermediates of the Krebs cycle. Of the two special enzymes involved, malate synthase (4.1.3.2.) and iso-citratase (isocitrate lyase) (4.1.3.1.), only the former has been detected in leaf tissue (see Davies and Ellis, 1964 for further discussion of the glyoxylate cycle in plants).

3. *Amino Acid Synthesis*

A number of enzymes concerned with amino acid synthesis have been identified both in leaves and algae and in water extracts of isolated chloro-plasts. These enzymes are (*a*) glutamate dehydrogenase, catalysing the con-version of ammonia to α-amino groups and (*b*) transaminases, catalysing the transfer of α-amino groups to newly synthesized α-keto acids.

(*a*) *Glutamate dehydrogenase* (1.4.1.2). This enzyme catalyses the reaction:

$$\text{2-Oxoglutarate} + \text{NH}_3 + \text{NADH}_2 \rightarrow \text{L-Glutamate} + \text{H}_2\text{O} + \text{NAD}$$

Glutamate dehydrogenase has been found in leaves (see Thomas, 1960 for references). Although the enzyme appears to be principally localized in the mitochondria (Sanwal and Lota, 1964) it has been detected in water extracts of isolated chloroplasts (see Losada and Arnon, 1964).

(*b*) *Transaminases* (2.6.1.-.). These enzymes catalyse the general reaction:

$$\alpha\text{-Keto-acid A} + \alpha\text{-Amino-acid B} \rightarrow$$
$$\alpha\text{-Amino-acid A} + \alpha\text{-Keto-acid B}$$

Transaminases specific for glutamate, aspartate and alanine have been identified in leaves and seedlings of a number of plants.

Aspartate transaminase was already mentioned in section (IV.C.2.*a*) as being present in aqueous extracts of chloroplasts isolated from spinach leaves. It is also present in extracts of castor bean endosperm and wheat germ, and in mitochondria, indicating a partial localization in both chloroplasts and mitochondria.

In experiments with whole *Chlorella* cells a high proportion of the $^{14}CO_2$ fixed in the light may appear very rapidly in amino acids, especially alanine, glutamate, aspartate and serine. The synthesis of these amino acids can account for 60% of the ^{14}C fixed by the algae. The available evidence indicates that there are at least two pools of these predominant amino acids and that one of them becomes labelled with such rapidity as to indicate "that the site of their synthesis must be freely accessible to their photosynthetically formed precursors, viz., phosphoenolpyruvate and phosphoglyceric acid" (Calvin and Bassham, 1962). These authors concluded that in *Chlorella* the more rapidly labelled pools of the amino acids are located at the site of photosynthetic crabon reduction, probably in the chloroplast. This supports the conclusion that the transaminases (e.g. aspartate transaminase, (2.6.1.1.) and alanine transaminase, (2.6.1.2 and 2.6.1.12.) are located within the chloroplast.

4. *Protein Synthesis*

App and Jagendorf (1964) have pointed out the difficulties in obtaining substantive evidence for protein synthesis in isolated chloroplasts. However, recent work of Spencer and Wildman (1964), Eisenstadt and Brawerman (1964) and Goffeau and Brachet (1965) indicates that protein synthesis (the incorporation of the amino acids into proteins) may occur in isolated chloroplasts of tobacco, *Euglena* and *Acetabularia.*

Chloroplasts contain ribosomes which when extracted from the chloroplasts are able to incorporate amino acids into proteins (see Goffeau and Brachet, 1965, for references). Isolated chloroplasts may contain DNA (see Gibor and Granick, 1964 and Kirk, 1963 for references) and also RNA (Eisenstadt and Brawerman, 1964).

If the chloroplasts are to synthesize proteins they must be capable of activating the amino acids. Amino acid activating enzymes have been demonstrated in isolated chloroplasts (Bové and Raacke, 1959; Marcus, 1959). These enzymes (6.1.1.-) catalyse the general reaction:

$$\text{Amino acid} + \text{sRNA} + \text{ATP} \rightarrow$$
$$\text{Amino acyl-sRNA} + \text{AMP} + \text{Pyrophosphate}$$

Bové and Raacke (1959) found that water extracts of isolated chloroplasts contained the enzymes needed to activate tyrosine and to a lesser extent methionine and leucine. Washed "whole" chloroplasts were found by Marcus (1959) to activate leucine, isoleucine and valine and to a lesser extent histidine, tyrosine, asparagine, cysteine, proline and alanine. Amino acid activation was also detectable in the cytoplasm, but when the amino acids listed above were tested with cytoplasmic extracts the relative rates of their activation were different from those found with the chloroplast extract. This indicates that the amino acid activation observed in the chloroplasts was probably originally associated with and located in the chloroplast and did not represent simply contamination from extra-chloroplastic sources.

In the experiments of Spencer and Wildman (1964) with tobacco leaf fractions the product of amino acid activation, viz. aminoacyl-sRNA, as well as the enzymes themselves, were found in the chloroplast fraction. In experiments on the incorporation of ^{14}C labelled amino acid, 67% of the ^{14}C incorporated in one min. was found in the fraction containing amino acyl-sRNA, but the incorporation into this fraction dropped to 25% of the total ^{14}C incorporated after 30 min. This indicates that the amino acyl-sRNA is an early intermediate of amino acid incorporation, the fraction containing it attaining a maximum radioactivity after 5 min. This labelling was maintained during the subsequent period of amino acid incorporation, so that the percentage of ^{14}C in the amino acyl-sRNA fraction decreases with time after 5 min.

5. *Lipid Synthesis*

The ability of isolated spinach and lettuce chloroplasts to incorporate acetate in the light into lipids has been shown by a number of investigators (see Stumpf *et al.*, 1963, for references). Stumpf and James (1962) and Stumpf, Bové and Goffeau (1963) found that the system for the light-stimulated incorporation of acetate into oleate and palmitate required the presence of CoA, HCO_3^-, Mg^{2+} and Mn^{2+}, and was stimulated when NADP, ADP and orthophosphate were present. They interpreted their results to show that the acetate incorporation into fatty acid in the chloroplast depended upon the occurrence of noncyclic photophosphorylation (see Fig. 8) which results in the formation of $NADPH_2$, ATP and oxygen. They also concluded that in fatty acid synthesis by illuminated chloroplasts light has another, as yet unknown, role besides simply promoting the formation of $NADPH_2$, ATP and O_2 during noncyclic photophosphorylation. The individual enzymes concerned with the biosynthesis of long-chained fatty acids from acetate have not been studied in isolated chloroplasts and one can only assume that all the enzymes required for the biosynthesis of these fatty acids are present in the chloroplasts.

6. *Pigment Metabolism*

(*a*) *Chlorophyll.* The steps believed to be responsible for the biosynthesis of chlorophyll have been well outlined in the review by Smith and French

(1963). The initial steps include the condensation of succinyl CoA and glycine to δ-amino-laevulinic acid (δ-ALA), the condensation of which leads to the tetrapyrrole structure which is subsequently transformed via Mg-protoporphyrin to chlorophyll *a* or chlorophyll *b*. The enzymes required for all the transformation steps have not been identified in plant material, although it is clear that chlorophyll is accumulated and presumably synthesized within the grana of the chloroplasts. The enzyme chlorophyllase (3.1.1.14) which may catalyse the phytylation of chlorophyllide has, however, been identified in chloroplasts (see Sissakian, 1958). A few of the other enzymes needed in chlorophyll synthesis have been identified in photosynthetic bacteria and in *Chlorella*. The study of chlorophyll synthesis, using labelled intermediates and mutants of *Chlorella* has established the probable biosynthetic pathway in the algae.

The early precursors, succinyl-CoA and glycine, are probably both synthesized in the chloroplasts. Glycine can be formed by transamination from glyoxylate, which is itself an early product of CO_2 fixation within the chloroplast (Zelitch, 1964). The enzymes required for the formation of α-oxoglutarate from phosphoenolpyruvate are present in chloroplasts. (see Section IV.C.2.*b*). The oxidation of α-oxoglutarate (a reaction not thus far demonstrated in chloroplasts) would then be expected to yield succinyl-CoA, which could be fed directly into the reaction sequence leading to the synthesis of chlorophyll. It has also been shown that succinic dehydrogenase is significantly absent from the chloroplast, thus preventing the operation of the Krebs cycle.

(*b*) *Carotenoids.* These pigments (carotenes and xanthophylls) are found to vary in the extent of their occurrence in plastids, depending on the physiological state of the plastid. In experiments with $^{14}CO_2$ there was rapid incorporation of ^{14}C into β-carotene in illuminated seedlings and leaves. Illumination of etiolated plants (from which carotene is absent) resulted in the formation of functional chloroplasts in which the carotenoids and xanthophylls were synthesized together (see Goodwin, 1960 for references). It has been deduced that β-carotene produced on illumination is formed *de novo* in newly formed chloroplasts from photosynthetically fixed CO_2.

Biosynthesis of the carotenoids (tetra-terpenes) is believed to involve the synthesis of mevalonic acid from acetate or leucine and its transformation to isopentenylpyrophosphate, which is the unit which becomes polymerized to form the terpene unit. β-Carotene and the xanthophylls would appear to be synthesized in the chloroplast. However, not all of the carotenoids present in the leaf are found within the chloroplast, some being localized in chromoplasts and being characteristic of non-green tissues.

7. *Nitrate Metabolism*

The reduction of nitrate to ammonia by leaf homogenates can be accomplished in the dark using reduced pyridine nucleotides or in the light photochemically. Non-green parts of the plant (and leaves in the dark) may carry out nitrate reduction using the pyridine nucleotide dependent nitrate reductase (1.6.6.2) and subsequent enzymes to the level of ammonia (NH_4^+) via nitrite reductase (1.6.6.4), but in the chloroplast it appears that a somewhat different mechanism may occur. Ramirez, Del Campo, Paneque and Losada (1964) have shown that flavin nucleotides can be photochemically reduced by isolated chloroplasts and subsequently used for the reduction of nitrate to nitrite using the enzyme nitrate reductase, which they isolated from spinach leaves. The further reduction of nitrite to NH_4^+ was shown by Paneque, Ramirez, Del Campo and Losada (1964) to proceed by way of photoreduced ferredoxin as the electron carrier. The enzyme complex nitrite reductase needed for this last reaction sequence was also isolated from spinach leaves.

It is presumed that in the photochemical nitrate and nitrite reducing systems, in contrast to the dark system where pyridine nucleotides are the reducing agents, the reductase enzymes are present inside the chloroplasts. This is rendered more probable since the protein electron carrier, ferredoxin, is known to be localized within the chloroplast.

A number of the enzymes involved in the dark and photochemical reduction of nitrate to ammonia have been identified and partly purified from higher plants, but no attempt has been made to localize them within a particular cell organelle (see Kessler, 1964 for references).

8. Adenosine Triphosphatases (ATPases) (3.6.1.-)

These enzymes catalyse the reaction:

$$ATP + H_2O \rightarrow ADP + \text{Orthophosphate}$$

Three types of ATPases have been reported in isolated chloroplasts, namely (i) dark, (ii) light-dependent, and (iii) light-triggered ATPases (see Bennun and Avron, 1964 for references). In the dark, chloroplasts hydrolyse ATP at a very low rate. This rate is stimulated several-fold in the light due to the "triggering" of an ATPase, which is detectable only in the presence of Mg^{2+} and a large quantity of cysteine. By contrast the light-dependent ATPase is stimulated by Ca^{2+}; in the dark no ATPase is detectable in the presence of Ca^{2+}. It should be noted that the maximum rates of ATPase activity associated with the chloroplasts are 1/50 of the maximum rates of photophosphorylation (Avron, 1962), indicating that the ATPases probably do not contribute markedly to the loss of ATP during photophosphorylation.

9. Adenylate Kinase (2.7.4.3)

This enzyme catalyses the reaction:

$$ATP + AMP \rightarrow ADP + ADP$$

The enzyme has been shown to be present in isolated spinach chloroplasts (Mazelis, 1956) and is easily removable by washing "whole" chloroplasts.

When this enzyme is present AMP can act as the phosphate acceptor in a phosphorylation experiment, and under these conditions is as effective in photosynthetic phosphorylation as ADP (Allen, Whatley and Arnon, 1958). Smillie (1963), using a non-aqueous solvent technique, has shown that the enzyme is present both in the chloroplast and in the non-chloroplast fractions isolated from *Euglena.*

10. Various Redox Enzymes

(*a*) *Glucose*-6-*phosphate dehydrogenase* (1.1.1.49). This enzyme catalyses the reaction:

$$\text{Glucose-6-phosphate} + \text{NADP} \rightarrow \text{Glucono-}\beta\text{-lactone-6-phosphate} + \text{NADPH}_2$$

This dehydrogenase has been extracted and purified from acetone treated spinach, wheat and tobacco leaves and from algae (see Thomas, 1960 and Waygood and Rohringer, 1964 for references).

Evidence for the occurrence of the enzyme in chloroplasts comes from unpublished observations of Whatley with water extracts of isolated spinach chloroplasts, where the reduction of NADP was followed with glucose-6-phosphate as the substrate. Losada *et al.* (1960) have also shown that glucose-6-phosphate greatly stimulates $^{14}CO_2$ uptake in an isolated chloroplast system. It has been shown by Smillie (1963) that the enzyme is largely concentrated outside the chloroplast in *Euglena.*

(*b*) *Phosphogluconate dehydrogenase* (1.1.1.44). The reaction catalysed by this enzyme is:

$$\text{6-Phospho-D-gluconate} + \text{NADP} \rightarrow \text{D-Ribulose-5-phosphate} + CO_2 + \text{NADPH}_2$$

The enzyme is widely distributed in all parts of the plant, including the leaves (Waygood and Rohringer, 1964). Evidence for its occurrence in chloroplasts comes from the experiments of Losada *et al.* (1960) where 6-phosphogluconate stimulated the photosynthetic $^{14}CO_2$ uptake of isolated spinach chloroplasts some fifteen-fold. The enzyme is largely concentrated outside the chloroplasts in *Euglena* (Smillie, 1963).

(*c*) *Catalase* (1.11.1.6). This enzyme catalyses the reaction:

$$H_2O_2 + H_2O_2 \longrightarrow O_2 + 2H_2O$$

Krogmann (1960) has shown that about 30% of the catalase activity of the whole leaf homogenate was associated with the unwashed chloroplast, which retained this activity for one washing and gradually lost the catalase activity on subsequent washing. The enzyme was also found in broken chloroplasts isolated from spinach leaves by Trebst and Eck (1961) who studied the effect of various catalase inhibitors on H_2O_2 accumulation in photosynthetic phosphorylation.

(*d*) *Polyphenol oxidase* (*o*-diphenol oxidase) (1.10.3.1).
This enzyme catalyses the reaction:

$$o\text{-Diphenol} + \tfrac{1}{2}O_2 \rightarrow o\text{-Quinone} + H_2O$$

Polyphenol oxidase is present in many leaves, but is not universally distributed (see Thomas, 1960, for references). Arnon (1949) demonstrated that in sugar beet the enzyme was associated with broken, washed chloroplasts. Trebst and Wagner (1962) found that freshly isolated spinach chloroplasts showed little if any polyphenol oxidase activity but on breaking the lipid structure of the chloroplasts the latent polyphenol oxidase activity was revealed.

(*e*) *Diaphorase* (or transhydrogenase). As noted in section (IV.A.1.*e*), diaphorase, transhydrogenase, and ferredoxin-NADP reductase are probably one and the same enzyme and occur in isolated chloroplasts. The diaphorase is further discussed in section (IV.A.1.*e*) under the name ferredoxin-NADP reductase.

(*f*) *Cytochrome* c *photo-oxidase system.* No cytochrome *c* oxidase (1.9.3.1) has been found in chloroplasts although it is present in plant mitochondria (Hackett, 1955). However, in the light reduced cytochrome *c* is oxidized by oxygen or NADP using isolated chloroplasts which have been pretreated with digitonin. This cytochrome *c* photo-oxidase activity, which is cyanide insensitive, is a manifestation of the activity of a terminal portion of the photosynthetic electron transport chain. It should be noted that cytochrome *f*, which occurs naturally within the chloroplast in the reduced form, becomes oxidized on illumination and reduced again in the dark (see review by Duysens, 1964), another manifestation of this type of activity.

11. Miscellaneous Enzymes

(*a*) *Hexokinase* (2.7.1.1.) Catalyses the reaction:

$$\text{ATP} + \text{D-Hexose} \rightarrow \text{ADP} + \text{D-Hexose-6-phosphate}$$

The enzyme is widespread in the plant kingdom (see Waygood and Rohringer, 1964, for references). However, hexokinase has not been directly demonstrated in isolated chloroplasts. The only evidence suggesting its occurrence in chloroplasts comes from the experiments of Losada *et al.* (1960), where the addition of glucose or fructose to the isolated chloroplast system assimilating ^{14}C in the light resulted in a 10–15 fold increase in CO_2 uptake. It seems justifiable to conclude that the "chloroplasts have the enzyme systems necessary for catalysing the rearrangements of the added primary substances to yield the pentosemonophosphate which is needed for the carboxylation phase of CO_2 assimilation." The enzymes would include hexokinase.

(*b*) *Phospholipase C* (3.1.4.3) *and Phospholipase D* (3.1.4.4). Both these enzymes hydrolyse phospholipids, and have been detected in plastids isolated from a number of leaves, although the enzymes are also present in the non-chloroplast fractions of leaves (see Barron, 1964 for references).

D. INDUCED ENZYME FORMATION

The localization of some enzymes within the chloroplast may be deduced from the behaviour of etiolated plants on illumination. The rapid greening which occurs in the light is accompanied by the development of the lamellar structure of the chloroplast and at the same time there is the development of a number of chloroplast enzymes. Increases in the total protein content of the plastid fraction have been noted and the development of a particular enzyme has been studied. On illumination, increases occur in the activity of the NADP-dependent triosephosphate dehydrogenase, ferredoxin, ferredoxin-NADP reductase (transhydrogenase), ribulose-1, 5-diphosphate carboxylase, alkaline fructose-1, 6-diphosphatase and glycollate oxidase. It has been shown that the glycollate oxidase is induced by the prior formation of glycollate in the light as an early product of photosynthesis (see Section IV.C.2*c*). These enzymes appear to be quite definitely located within the chloroplast.

In addition Smillie (1962) has pointed out that certain enzyme activities in cells containing developing chloroplasts undergo changes which are more closely connected with the changes in photosynthetic rates (initially low and increasing) than with the changes in respiration rates (initially high and decreasing). This was true of ribulose-1,5-diphosphate carboxylase and of ferredoxin. However, other enzymes e.g. enolase and 6-phosphogluconate dehydrogenase, showed changes related more closely to the changes in the respiration rate and were presumably associated with the non-chloroplast fraction. The changes in the activity of the enzyme transketolase were intermediate, suggesting its participation in both photosynthesis and respiration and therefore its partial localization in the chloroplast in addition to its occurrence in the cytoplasm. Both of these approaches afford useful techniques for studying the intracellular localization of enzymes.

V. Comparison of Oxidative and Photosynthetic Phosphorylation

The electron transport and accompanying ATP formation carried out by mitochondria in oxidative phosphorylation and by chloroplasts in noncyclic photophosphorylation are shown in simplified form in Fig. 12. In mitochondria the starting point is $NADH_2$, electrons being transferred in a series of identifiable steps through flavoprotein and ubiquinone to cytochrome *c* and eventually to molecular oxygen yielding water as the end product (see Chapter 3). Each of the steps represents the passage of an electron from a more reduced to a more oxidized potential, and each step is coupled to the formation of a molecule of ATP. For purposes of comparison the second step, ubiquinone to cytochrome *c*, is emphasized in the figure. In chloroplasts the starting point is water, electrons being transferred in a series of identifiable steps through plastoquinone and cytochrome *f* to NADP, yielding $NADPH_2$ as the end product. The step plastoquinone to cytochrome *f* represents the "fall" of an electron to a more oxidized state and is accompanied by the formation of a molecule of ATP.

Although the overall electron flow pathways of oxidative and photosynthetic phosphorylation appear to be direct opposites the figure indicates that this is not true of all the component partial reactions. The actual steps resulting in ATP formation, a product common to both reactions, accompany the release of chemical energy during a redox reaction (compare especially the heavy lines in Fig. 12). The electrons are, however, brought to the more

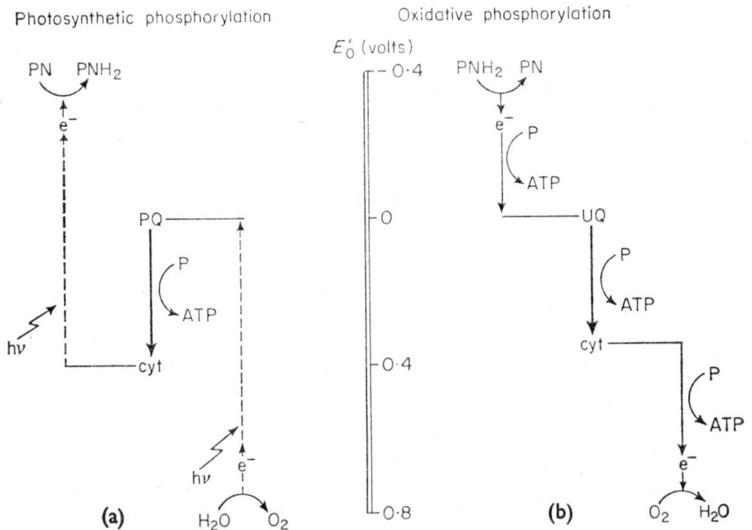

FIG. 12. Diagramatic comparison of electron transport and ATP formation in photosynthetic and oxidative phosphorylation. PN, PNH_2: oxidized and reduced pyridine nucleotides, PQ: plastoquinone, UQ: ubiquinone, cyt: cytochrome.

reducing potential by very different means in the two processes. In oxidative phosphorylation the reducing electrons in $NADH_2$ arise by a thermochemical reaction from substrate molecules e.g. pyruvate; by contrast, in noncyclic photophosphorylation the reducing electrons result from a photochemical reaction in which water acts as the electron donor, and plastoquinone is the initial electron acceptor. Following the reaction sequence leading to ATP formation (heavy line in Fig. 12) a second photochemical reaction leads to the reduction of NADP. Neither photochemical reaction is accompanied by ATP formation. In noncyclic photophosphorylation the ATP formation is a dark reaction only secondarily dependent on the photochemical reactions, and closely resembles the dark production of ATP in the mitochondrial system. In cyclic photophosphorylation (Fig. 9) the ATP formation is also a dark reaction dependent on a prior photochemical step.

ATP formation is sensitive to a number of inhibitors which act either as inhibitors of electron transport or as uncoupling agents. The same inhibitor does not always produce the same effect in both mitochondrial and chloroplast systems (Avron and Shavit, 1963; Losada and Arnon, 1964). Although the formation of ATP is in fact a similar thermochemical reaction in both oxidative and photosynthetic phosphorylation, the differential effects of inhibitors indicate that the details of the two processes may be appreciably different. This is true of the intermediates in the two electron transport sequences and also of the ATP formation with its associated coupling factors.

The metabolism of mitochondria and chloroplasts may be contrasted by emphasizing that mitochondria appear to be organelles adapted to yield ATP, using respiratory substrates as the energy source and oxygen as terminal electron acceptor. They contain the necessary complement of Krebs' cycle enzymes as ancillary equipment to enable them to degrade the substrates by way of reduced pyridine nucleotide. Mitochondria are thus largely concerned with the oxidation of the substrates, and any synthetic reducing activities they show are associated with the $NADH_2$ and $NADPH_2$ formed from the substrates. By contrast chloroplasts yield ATP together with $NADPH_2$ discarding oxygen as a waste product. They are largely concerned with reactions needing a large supply of reductants. The principal synthetic reactions characteristic of the chloroplasts are CO_2 fixation leading to sucrose and starch synthesis, and the synthesis of fats, pigments and proteins.

References

Allen, M. B., Arnon, D. I., Capindale, J. B., Whatley, F. R. and Durham, L. J. (1955). *J. Am. chem. Soc.* **77**, 4149.

Allen, M. B., Whatley, F. R. and Arnon, D. I. (1958). *Biochim. biophys. Acta* **27**, 16.

App, A. A., and Jagendorf, A. T. (1964). *Pl. Physiol.* **39**, 772.

Arnon, D. I. (1949). *Pl. Physiol.* **24**, 1.

Arnon, D. I. (1952). *Science N.Y.* **116**, 635.

Arnon, D. I. (1959). *Nature, Lond.* **184**, 10.

Arnon, D. I. (1961). *Bull. Torrey bot. Club* **88**, 215.

Arnon, D. I. (1965). *Science N.Y.* **149**, 1460.

Arnon, D. I. and Horton, A. A. (1963). *Acta chem. scand.* **17**, Suppl. 1, 135.

Arnon, D. I., Allen, M. B. and Whatley, F. R. (1954). *Nature, Lond.* **174**, 394.

Arnon, D. I., Whatley, F. R. and Allen, M. B. (1958). *Science N.Y.* **127**, 1026.

Arnon, D. I., Tsujimoto, H. Y. and McSwain, B. D. (1964). *Proc. natn. Acad. Sci. U.S.A.* **51**, 1274.

Aubert, J. P., Milhaud, G. and Millet, J. (1957). *Annls Inst. Pasteur, Paris* **92**, 515.

Avron, M. and Jagendorf, A. T. (1956). *Archs Biochem. Biophys.* **65**, 475.
Avron, M. (1962). *J. biol. Chem.* **237**, 2011.
Avron, M. (1963). *Biochim. biophys. Acta* **77**, 699.
Avron, M. and Shavit, M. (1963). *In* "Photosynthetic Mechanisms of Green Plants" p. 611, Publication 1145, National Academy of Science—National Research Council, Washington, D.C.
Barron, E. J. (1964). *In* "Modern Methods of Plant Analysis" (Paech, K. and Tracey, M. V., eds.) Vol. VII, p. 454. Springer-Verlag, Berlin.
Bassham, J. A. and Calvin, M. (1960). *In* "Encyclopedia of Plant Physiology" (Ruhland, W., ed.) Vol. V, Pt. 1, p. 884. Springer-Verlag, Berlin.
Bassham, J. A., Benson, A. A., Kay, L. D., Harris, A. Z., Wilson, A. T. and Calvin, M. (1954). *J. Am. chem. Soc.* **76**, 1760.
Becker, M. J., Shefner, A. M. and Gross, J. A. (1965). *Pl. Physiol.* **40**, 243.
Bennun, A. and Avron, M. (1964). *Biochim. biophys. Acta* **79**, 746.
Benson, A. A. (1963). *In* "Photosynthetic Mechanisms of Green Plants", p. 571, Publication 1145, National Academy of Science—National Research Council, Washington, D.C.
Bird, J. F., Porter, H. K. and Stocking, C. R. (1965). *Biochim. biophys. Acta* **100**, 366.
Bishop, N. I. (1959). *Proc. natn. Acad. Sci.* (*U.S.A.*) **45**, 1696.
Bové, J. and Raacke, I. D. (1959). *Biochim. biophys. Acta* **85**, 521.
Bové, J. M., Bové, C., Whatley, F. R. and Arnon, D. I. (1963). *Z. Naturf.* **18b**, 683.
Burma, D. P. and Mortimer, D. C. (1956). *Archs Biochem. Biophys.* **62**, 16.
Butt, V. S. and Peel, M. (1963). *Biochem. J.* **88**, 31p.
Calo, N. and Gibbs, M. (1960). *Z. Naturf.* **15b**, 287.
Calvin, M. and Bassham, J. A. (1962). "The Photosynthesis of Carbon Compounds". W. J. Benjamin, Inc., New York.
Chance, B. and Sager, R. (1957). *Pl. Physiol.* **32**, 548.
Cohen-Bazire, G. and Kunisawa, R. (1963). *J. Cell Biol.* **16**, 401.
Davenport, H. E. and Hill, R. (1952). *Proc. R. Soc.* B**139**, 327.
Davenport, H. E. and Hill, R. (1960). *Biochem. J.* **74**, 493.
Davies, D. D. and Ellis, R. J. (1964). *In* "Modern Methods of Plant Analysis" (Paech, K. and Tracey, M. V. eds.) Vol. VII, p. 616, Springer-Verlag, Berlin.
de Duve, C., Wattiaux, R. and Baudhuin, P. (1962). *Adv. Enzymol.* **24**, 291.
de Kouchovsky, Y. and Fork, D. C. (1964). *Proc. natn. Acad. Sci. U.S.A.* **52**, 232.
Dorner, R. W., Kahn, A. and Wildman, S. G. (1957). *J. biol. Chem.* **229**, 945.
Duysens, L. N. M. (1955). *Science, N.Y.* **121**, 120.
Duysens, L. N. M. (1964). *Prog. Biophys. biophys. Chem.* **14**, 1.
Eisenstadt, J. M. and Brawerman, G. (1964). *J. mol. Biol.* **10**, 392.
Feingold, D. S., Neufeld, E. F. and Hassid, W. Z. (1964). *In* "Modern Methods of Plant Analysis" (Paech, K. and Tracey, M V., eds.) Vol. VII, p. 474, Springer-Verlag, Berlin.
French, C. S. (1964). Carnegie Institute of Washington Yearbook **63**, 414.
Frey-Wyssling, A. (1953). "The Submicroscopic Morphology of Protoplasm and its Derivatives", Elsevier, Amsterdam.
Fry, K. T. and San Pietro, A. (1963). *In* "Photosynthetic Mechanisms of Green Plants" p. 252, Publication 1145, National Academy of Science—National Research Council, Washington, D.C.
Gest, H. and Kamen, M. D. (1960). "Encyclopedia of Plant Physiology" (Ruhland, W., ed.) Vol. V, pt. 2, p. 568. Springer-Verlag, Berlin.

Gibbs, M. (1952). *Nature, Lond.* **170**, 164.
Gibbs, M. (1963). "Photosynthetic Mechanisms of Green Plants", p. 663, Publication 1145, National Academy of Science—National Research Council, Washington, D.C.
Gibbs, M. and Turner, J. F. (1964). *In* "Modern Methods of Plant Analysis" (Paech, K. and Tracey, M. V., eds.) Vol. VII, p. 520.
Gibor, A. and Granick, S. (1964). *Science, N.Y.* **145**, 890.
Goffeau, A. and Brachet, J. (1965). *Biochim. biophys. Acta* **95**, 302.
Goodwin, T. W. (1960). *In* "Encyclopedia of Plant Physiology" (Ruhland, W. ed.) Vol. V, pt. 1, p. 394. Springer-Verlag, Berlin.
Granick, S. (1955). *In* "Encyclopedia of Plant Physiology" (Ruhland, W., ed.) Vol. I, p. 507. Springer-Verlag, Berlin.
Granick, S. (1961). *In* "The Cell" (Brachet, J. and Mirsky, A. E., eds.) Vol. II, p. 489, Academic Press, New York and London.
Greenwood, A. D., Leech, R. M. and Williams, J. P. (1963). *Biochem. biophys. Acta* **78**, 148.
Hackett, D. P. (1955). *Int. Rev. Cytol.* **4**, 143.
Haxo, F. T. and ÓhEocha, C. (1960). *In* "Encyclopedia of Plant Physiology" (Ruhland, W., ed.) Vol. V, pt. 1, p. 497. Springer-Verlag, Berlin.
Heber, U., Pon, N. G. and Heber, M. (1963). *Pl. Physiol.* **38**, 355.
Hill, R. (1954). *Nature, Lond.* **174**, 501.
Hill, R. and Bonner, W. D. Jnr. (1961). *In* "Light and Life". (McElroy, W. D. and Glass, B., eds.) p. 424. Johns Hopkins Press, Baltimore, Maryland, U.S.A.
Hill, R. and Scarisbrick, R. (1951). *New Phytol.* **50**, 98.
Horecker, B. L., Hurwitz, J. and Smyrniotis, P. Z. (1956). *J. Am. chem. Soc.* **78**, 692.
Hurwitz, J., Weissbach, A., Horecker, B. L. and Smyrniotis, P. Z. (1956). *J. biol. Chem.* **218**, 769.
Jacobi, H. and Perner, E. (1961). *Fiora* **150**, 209.
Jagendorf, A. T. (1962). *Surv. Biol. Prog.* **4**, 181.
Jakoby, W. B., Brummond, D. O. and Ochoa, S. (1956). *J. biol. Chem.* **218**, 811.
James, W. O. and Das. V. S. R. (1957). *New Phytol.* **56**, 325.
Kahn, A. and Wettstein, D. von (1961). *J. Ultrastruct. Res.* **5**, 557.
Kandler, O., Liesenkotter, J. and Oaks, B. A. (1961). *Z. Naturf.* **16b**, 50.
Katoh, S. (1959). *J. Biochem.* (*Tokyo*) **46**, 629.
Katoh, S. (1960). *Nature, Lond.* **186**, 533.
Katoh, S., Suga, J., Shiratori, J. and Takamiya, A. (1961). *Archs Biochem. Biophys.* **94**, 136.
Katoh, S. and Takamiya, A. (1963). *In* "Photosynthetic Mechanisms of Green Plants" p. 262. Publication 1145, National Academy of Science—National Research Council, Washington, D.C.
Keister, D. L., San Pietro, A. and Stolzenbach, F. E. (1960). *J. biol. Chem.* **235**, 2989.
Kessler, E. (1960). *In* "Encyclopedia of Plant Physiology" (Ruhland, W., ed.) Vol. V, pt. 1, p. 951. Springer-Verlag, Berlin.
Kessler, E. (1964). *Annu. Rev. Pl. Physiol.* **15**, 57.
Kirk, J. T. O. (1963). *Biochim. biophys. Acta* **76**, 417.
Kok, B., Cooper, B. and Yang, L. (1963). *In* "Studies on Microalgae and Photosynthetic Bacteria" (Edited by Japanese Society of Plant Physiologists) pp. 373.
Kok, B., Rurainski, H. J. and Karmon, E. A. (1964). Plant Physiol. 39, 513.
Krogmann, D. W. (1960). *J. biol. Chem.* **235**, 3630.

Krogmann, D. W. and Olivero, E. (1962). *J. biol. Chem.* **237**, 3292.
Leech, R. M. (1964). *Biochim. biophys. Acta* **79**, 637.
Lehninger, A. L. (1964). "The Mitochondrion", W. A. Benjamin Inc., New York.
Leloir, L. F. and Cardini, C. E. (1955). *J. biol. Chem.* **214**, 157.
Losada, M. and Arnon, D. I. (1963). *In* "Metabolic Inhibitors" (Hochster, R. M. and Quastel, J. H., eds.) Vol. II, p. 559. Academic Press, New York and London.
Losada, M. and Arnon, D. I. (1964) *In* "Modern Methods of Plant Analysis" (Paech, K. and Tracey, M. V., eds.) Vol. VII, p. 569, Springer-Verlag, Berlin.
Losada, M., Trebst, A. V. and Arnon, D. I. (1960). *J. biol. Chem.* **235**, 832.
Lundegårdh, H. (1962). *Nature, Lond.* **192**, 243.
Lyttleton, J. W. and Ts'o, P. O. P. (1958). *Archs Biochem. Biophys.* **73**, 120.
Lyttleton, J. W. (1962). *Expl Cell Res.* **26**, 312.
Marcus, A. (1959). *J. biol. Chem.* 234, 1238.
Mazelis, M. (1956). *Pl. Physiol.* **31**, 37.
Menke, W. (1962). *Annu. Rev. Pl. Physiol.* **13**, 27.
Menke, W. (1963). *In* "Photosynthetic Mechanisms of Green Plants", p. 537. Publication 1145, National Academy of Science, National Research Council Washington, D.C.
Morton, R. K., Palk, B. A. and Raison, J. K. (1964). *Biochem. J.* **91**, 522.
Newton, J. W. and Newton, G. A. (1957). *Archs Biochem. Biophys.* **71**, 250.
Nieman, R. H., Nakamura, H. and Vennesland, B. (1959). *Pl. Physiol.* **34**, 262.
Ogren, W. L. and Krogmann, D. W. (1963). *In* "Photosynthetic Mechanisms of Green Plants", p. 684. Publication 1145, National Academy of Science—National Research Council, Washington, D.C.
Paneque, A., Ramirez, J. M., Del Campo, F. E. and Losada, M. (1964). *J. biol. Chem.* 239, 1737.
Park, R. B. and Biggins, J. (1964). *Science, N.Y.* **144**, 1009.
Park, R. B. and Pon, N. G. (1961). *J. mol. Biol.* **3**, 1.
Park, R. B. and Pon, N. G. (1963). *J. mol. Biol.* **6**, 105.
Peterkovsky, A. and Racker, E. (1961). *Pl. Physiol.* **34**, 409.
Pierpoint, W. S. (1962). *Biochem. J.* **82**, 143.
Pirson, A. (1960). *In* "Encyclopedia of Plant Physiology" (Ruhland, W., ed.) Vol. V, pt. 2, p. 123. Springer-Verlag, Berlin.
Ramasamna, T. and Giri, K. V. (1956). *Archs Biochem. Biophys.* **62**, 91.
Ramirez, J. M., Del Campo, F. F., Paneque, A. and Losada, M. (1964). *Biochem. biophys. Res. Commun.* **15**, 297.
Redfearn, E. R. and Friend, J. (1961). *Nature, Lond.* **191**, 806.
Richter, G. (1959). Naturwissenschaften **46**, 604.
Rongine de Fekete, M.A., Leloir, L. F. and Cardini, C. S. (1960). *Nature, Lond.* **187**, 918.
Rosenberg, L. L. and Arnon, D. I. (1955). *J. biol. Chem.* **217**, 361.
Rosenberg, L. L., Capindale, J. B. and Whatley, F. R. (1958). *Nature, Lond.*, **181**, 632.
San Pietro, A. and Black, C. C. (1965). *Annu. Rev. Pl. Physiol.* **16**, 155.
Sanwal, B. D. and Lota, M. (1964). *In* "Modern Methods of Plant Analysis" (Paech, K. and Tracey, M. V., eds.) vol. VII, p. 290. Spring Verleg, Berlin.
Sanwal, B. D., Zink, M. W. and Din, G. (1964). *In* "Modern Methods of Plant Analysis" (Paech, K. and Tracey, M. V., eds.) Vol. VII, pp. 361–391, Springer-Verlag, Berlin.
Shin, M. and Arnon, D. I. (1964). *Fed. Proc.* **23**, 227.

Shin, M., Tagawa, K. and Arnon, D. I. (1963). *Biochem. Z.* **338**, 84.
Simonis, W. and Weichart, G. (1958). *Z. Naturf.* **13b**, 694.
Sissakian, N. M. (1958). *Adv. Enzymol.* **20**, 201.
Smillie, R. M. (1955). *Aust. J. Biol. Sci.* **8**, 186.
Smillie, R. M. (1962). *Pl. Physiol.* **37**, 716.
Smillie, R. M. (1963). *Can. J. Bot* **41**, 123.
Smillie, R. M. (1964). *In* "Fructose-1,6-diphosphatase and its Role in Gluconeogenesis" (McGilvery, R. W. and Pogell, B M., eds.) p. 31, Amer. Inst. Biol. Sci., Washington, D.C.
Smillie, R. M. and Fuller, R. C. (1959). *Pl. Physiol.* **34**, 651.
Smillie, R. M. and Fuller, R. C. (1960). *Biochem. biophys. Res. Commun.* **3**, 368.
Smith, L. and Baltscheffsky, M. (1959). *J. biol. Chem.* **234**, 1575.
Smith, J. H. C. and French, C. S. (1963). *Annu. Rev. Pl. Physiol.* **14**, 181.
Spencer, D. and Possingham, J. V. (1960). *Aust. J. Biol. Sci.* **13**, 441.
Spencer, D. and Possingham, J. V. (1961). *Biochim. biophys. Acta* **52**, 379.
Spencer, D. and Wildman, S. G. (1964). *Biochemistry* **3**, 954.
Steffen, K. (1955). *In* "Encyclopedia of Plant Physiology" (Ruhland, W., ed.) Vol. I, 401. Springer-Verlag, Berlin.
Stocking, C. R. (1959). *Pl. Physiol.* **34**, 56.
Stocking, C. R., Williams, G. R. and Ongun, A. (1963). *Biochem. biophys. Res. Commun.* **10**, 416.
Stumpf, P. K., Bové, J. M. and Goffeau, A. (1963). *Biochim. biophys. Acta* **70**, 260.
Stumpf, P. K. and James, A. T. (1962). *Biochim. biophys. Acta* **57**, 400.
Tagawa, K. and Arnon, D. I. (1962). *Nature, Lond.* **195**, 537.
Tagawa, K., Tsujimoto, H. Y. and Arnon, D. I. (1963). *Proc. natn. Acad. Sci. U.S.A.* **49**, 567.
Tewfik, S. and Stumpf, P. K. (1951). *J. biol. Chem.* **192**, 519.
Thalacker, R. & Behrens, M. (1959). *Z. Naturf.* **14b**, 443.
Thomas, J. B. (1960). *In* "Encyclopedia of Plant Physiology" (Ruhland, W., ed.) Vol. V, pt. 1, p. 511. Springer-Verlag, Berlin.
Tolbert, N. E. (1963). *In* "Photosynthetic Mechanisms of Green Plants" p. 648. Publication 1145, National Academy of Science—National Research Council, Washington, D.C.
Tolbert, N. E. and Cohan, M. S. (1953). *J. biol. Chem.* **204**, 639.
Trebst, A. (1963). *Z. Naturf.* **18b**, 817.
Trebst, A. and Eck, H. (1961). *Z. Naturf.* **16b**, 455.
Trebst, A. and Fiedler, F. (1962). *Z. Naturf.* **17b**, 553.
Trebst, A. and Pistorius, E. (1965). *Z. Naturf.* **20b**, 885.
Trebst, A. V. and Wagner, S. (1962). *Z. Naturf.* **17b**, 396.
Trebst, A., Eck, H. and Wagner, S. (1963). *In* "Photosynthetic Mechanisms of Green Plants" p. 174. Publication 1145, National Academy of Science, National Research Council, Washington, D. C.
Trebst, A. V., Losada, M. and Arnon, D. I. (1960). *J. biol. Chem.* **235**, 840.
Trudinger, P. A. (1956). *Biochem. J.* **64**, 274.
Urbach, W. and Simonis, W. (1964). *Biochem. biophys. Res. Commun.* **17**, 39.
van Noort, G. and Wildman, S. G. (1964). *Biochim. biophys. Acta* **90**, 309.
Warburg, O. (1948). "Schermetalle als Wirkungsgruppe von Fermenten", Saenger, Berlin.
Waygood, E. R. and Rohringer, R. (1964). *In* "Modern Methods of Plant Analysis" (Paech, K. and Tracey, M. V., eds.) Vol. VII, p. 546, Springer-Verlag, Berlin.

Weier, T. E. (1961). *Am. J. Bot.* **48**, 615.
Weier, T. E. and Thomson, W. W. (1962). *J. Cell Biol.* **13**, 89.
Weissbach, A., Horecker, B. L. and Hurwitz, J. (1956). *J. biol. Chem.* **218**, 795.
Weissbach, A., Smyrniotis, P. Z. and Horecker, B. L. (1954). *J. Am. Chem. Soc.* **76**, 5572.
Whatley, F. R. and Arnon, D. I. (1963). *In* "Methods in Enzymology" (Colowick, S. P. and Kaplan, N. O., eds.) Vol. VI, p. 308. Academic Press, New York and London.
Whatley, F. R. and Horton, A. A. (1963). *Acta chem. scand.* **17**, Suppl. 1, 140.
Whatley, F. R. and Losada, M. (1964). *In* "Photophysiology" (Giese, A. C., ed.) p. 111. Academic Press, New York and London.
Whatley, F. R., Tagawa, K. and Arnon, D. I. (1963). *Proc. natn. Acad. Sci U.S.A.* **49**, 266.
Wildman, S. G. and Bonner, J. (1947). *Archs Biochem. Biophys.* **14**, 381.
Zelitch, I. (1964). *Annu. Rev. Pl. Physiol.* **15**, 121.
Zelitch, I. and Barber, G. A. (1960a). *Pl. Physiol.* **35**, 205.
Zelitch, I. and Barber, G. A. (1960b). *Pl. Physiol.* **35**, 626.
Zelitch, I. and Gotto, A. M. (1962). *Biochem. J.* **84**, 541.
Zucker, M. and Stinson, H. T. (1962). *Archs Biochem. Biophys.* **96**, 637.

Chapter 5

LYSOSOMES, PHAGOSOMES AND RELATED PARTICLES

W. STRAUS[1]

Division of Metabolic Research, Institute for Medical Research, The Chicago Medical School, Chicago, Illinois, U.S.A.

[1] This work was supported by research grants from the U.S. Public Health Service (No. GM 12123) and from the American Heart Association.

I. Introduction

The history of the study of lysosomes began with the discovery by de Duve, Pressman, Gianetto, Wattiaux and Applemans (1955) that a group of acid hydrolytic enzymes was concentrated in the "light" mitochondrial fraction isolated from liver homogenates of rats by differential centrifuging. The enzymes were thought to belong to a special type of cytoplasmic granule, 0·2 to 0·4 μ diameter, and to function in the degradation of various biological substrates; therefore, the name "lysosomes" was given to these granules (de Duve *et al.*, 1955). De Duve and co-workers also observed that the hydrolytic enzymes of the isolated lysosome fraction were active only if the granules were damaged by osmotic shock, freezing and thawing, or by detergents (de Duve, 1959a). De Duve concluded that the lysosomes were surrounded by a membrane of lipoprotein nature restricting the accessibility of external substrates to the internal enzymes.

Soon after the separation of lysosomes from liver cells by de Duve and coworkers, granules containing the same types of acid hydrolytic enzymes were isolated from kidney homogenates of normal rats (Straus, 1954, 1956). The size of the kidney lysosomes varied within a wide range, from 0·1 to 3 μ diameter. Figure 1 shows a purified kidney fraction with granules of intermediate size (0·5 to 1·5 μ diameter). The most interesting property of the kidney lysosomes was their ability to concentrate injected foreign proteins. After injection of egg white or horse-radish peroxidase, for example, the size of many granules increased, and the foreign proteins were concentrated together with the hydrolytic enzymes in the isolated lysosome fractions (Straus and Oliver, 1955; Straus, 1957a, b). These experiments suggested for the first time that the lysosomes may function in the digestion or detoxification of foreign material ingested by pinocytosis or phagocytosis, and that the well-known property of the reticulo-endothelial system to segregate ingested foreign material in intracellular granules may be closely related to the activity of lysosomes.

Progress during the early phase of lysosome research was based mainly on the biochemical analysis of isolated fractions. In recent years, however, cytochemical methods and electron microscopy have contributed much to

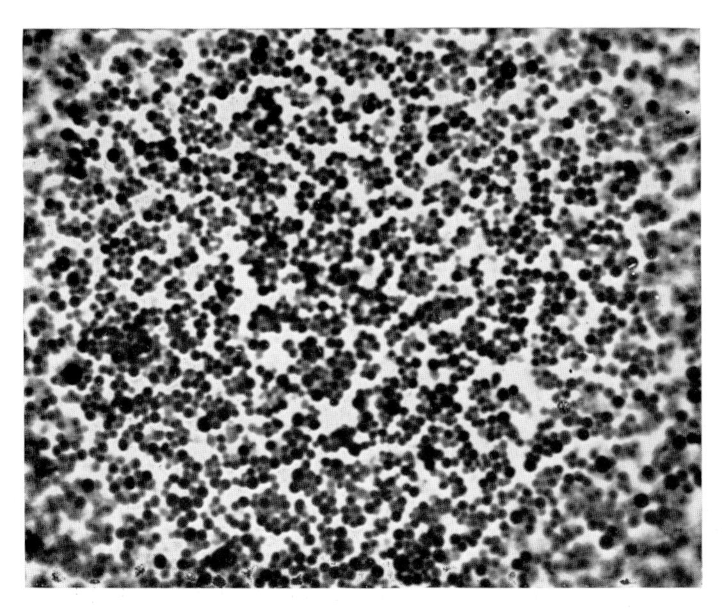

FIG. 1. Lysosomes of intermediate size (0·5–1·5 μ diameter) isolated from rat kidney. ×400. (Reproduced with permission from the *J. biol. Chem.* **207**, 750 (1954.)

the advance of knowledge in this field. In particular, the cytochemical reaction for one lysosomal enzyme, acid phosphatase (3.1.3.2.), was very useful in identifying lysosomes in various tissues. Several years after the early cytochemical detection of acid phosphatase in kidney droplets (Holt, 1954), this cytochemical method was extended to the identification of lysosomes in other tissues (Novikoff, 1961; and others). Of special interest was the demonstration by Barka, Schaffner and Popper (1961) that acid-phosphatase-positive granules were numerous in the cells of the reticulo-endothelial system. Early observations by electron microscopy led to a misunderstanding of the relationship between lysosomes and mitochondria. Several electron microscopists had come to the conclusion that the lysosomes of kidney and certain other cells were derived from mitochondria. This error, which can now be understood by the existence of "autolytic vacuoles" containing mitochondria (Section VI) was corrected only in 1960 by Miller in his work on the reabsorption of injected haemoglobin by the tubule cells of the kidney (Miller 1960). In the case of the liver, fortunately, no such confusion between lysosomes and mitochondria occurred, and the tentative identification by electron microscopy of lysosomes with pericanalicular bodies (Novikoff, Beaufay, and de Duve, 1956) was confirmed by later work. The identification of polymorphous bodies as lysosomes was much facilitated during recent years by the application of the cytochemical acid phosphatase method of Gomori (1941, 1952), at the electron microscope level (Holt and Hicks, 1961b; Daems, 1962; Miller, 1962; Novikoff, 1963; Miller and Palade, 1964; Ericsson, 1964).

As may be seen from this brief review on the development of lysosome research, the study of these granules is now accessible to three main approaches: biochemical analysis of lysosomal enzymes in isolated fractions combined with tests for "latency", cytochemical staining reactions for lysosomal enzymes, and histochemical electron microscopy for acid phosphatase. In view of the changing properties of lysosomes during their life cycle, a correlation of biochemical with cytochemical methods seems to be essential for the study of these cell organelles.

Since excellent reviews on lysosomes have been written in recent years by de Duve (1959a) and by Novikoff (1961), and since various aspects of the problem were discussed at a recent symposium on lysosomes (De Reuck and Cameron, 1963), it is not appropriate to repeat in detail points already covered there.[1] In the following pages the main emphasis will be on the relationship between lysosomes and pinocytosis and phagocytosis, a subject with which the writer has some experience from his own work.

A comment on terminology: The reading of the literature on lysosomes is made more difficult by the proliferation of new terms for the same type of particles. In part, new names were given to bodies before their function and origin were well understood. In order to avoid confusion of terms in this review, it seems appropriate to comment on terminology at this stage.

"Lysosomes" were defined by de Duve and co-workers (1955) as cytoplasmic granules containing high concentrations of several acid hydrolytic enzymes and showing "latency" of these enzymes. On the basis of more recent knowledge of lysosomes, de Duve (1963a) distinguished four subtypes of lysosomes: a storage granule (for lysosomal enzymes), a digestive vacuole, an autophagic vacuole, and a residual body. The properties on which these terms are based will be discussed in detail in the following Sections.

The term "phagosome" was suggested originally to indicate the segregating ability of granules which also contained the lysosomal enzymes (Straus, 1958, 1959). Later, when the relationship between lysosomes and phago-

[1] While this article was in press, several comprehensive reviews on lysosomes appeared. Berthet (1965), Weissmann (1965), and Wattiaux and de Duve (1966) discussed the general properties and function of lysosomes, and Woessner (1965a) reported on lysosomes and lysosomal enzymes with special reference to connective tissue.

somes was understood better, the term "phagosome" was suggested to designate phagocytic or pinocytic vacuoles or vesicles alone, before their fusion with lysosomes (de Duve, 1963a; Straus, 1963, 1964a, b). The term "phago-lysosome" or "lyso-phagosome" was used for the combined granules (Straus, 1963, 1964a, b). This terminology simplified the description of experiments in which the fusion of phagosomes with lysosomes was deduced from double staining for acid phosphatase and injected plant peroxidase in the same tissue sections (Straus, 1963, 1964a, b).

Enlarged lysosomes which also contained mitochondria or other cell components were termed "cytolysomes" by Novikoff (1960). Since the original idea that these bodies function in cytolysis was not confirmed, the terms "areas of focal degradation" (Hruban, Spargo, Swift, Wissler and Kleinfeld 1963) or "autophagic vacuoles" (de Duve, 1963b) seem to be more appropriate. The term "segresome" was proposed by Tanaka (1962) for granules in which vital stains were segregated. The term "cytosegresomes" was used by Ericsson and Trump (1964) for areas of focal degradation. The neutral term "composite bodies" was preferred by Straus (1961) for much enlarged phago-lysosomes which seemed to contain also cell fragments (mitochondria, nuclei, brush border material). Thus, the terms cytolysome, area of focal degradation, autophagic (autolytic) vacuole, cytosegresome, or composite body all refer to similar types of structures. Gordon, Miller and Bensch (1965) recently suggested a unifying terminology. The terms "proto-lysosomes" were introduced for "young" lysosomes, "telo-lysosomes" for "old" lysosomes (phago-lysosomes), and "auto-lysosomes" for bodies functioning in the digestion of endogenous cell material (autophagic vacuoles). The diagram from the paper of Gordon, Miller, and Bensch (1965) showing the interrelationship of these particles, is reproduced in Fig. 2.

The term "cytosome" was used by Lindner (1957) and by Schulz (1958) and later by other electron microscopists to designate dense and polymorphous bodies in electron micrographs which probably were lysosomes. "Cytosis" was suggested by Novikoff (1961) as a general term for ingestion processes by pinocytosis or phagocytosis. "Endocytosis" for intake and "exocytosis" for extrusion were proposed by de Duve (see, de Reuck and Cameron, 1963, p. 126). Zucker-Franklin and Hirsch (1964) suggested the term "exoplasmosis" for the release of enzymes from cells or into phagocytic vacuoles. The terms used by earlier investigators in relation to ingestion processes were reviewed by Chapman-Andresen (1962).

II. Properties of Lysosomes

A. Enzymes

1. Types of Lysosomal Enzymes

As mentioned above, the high concentration of certain hydrolytic enzymes and their "latency" in the intact granules were considered by de

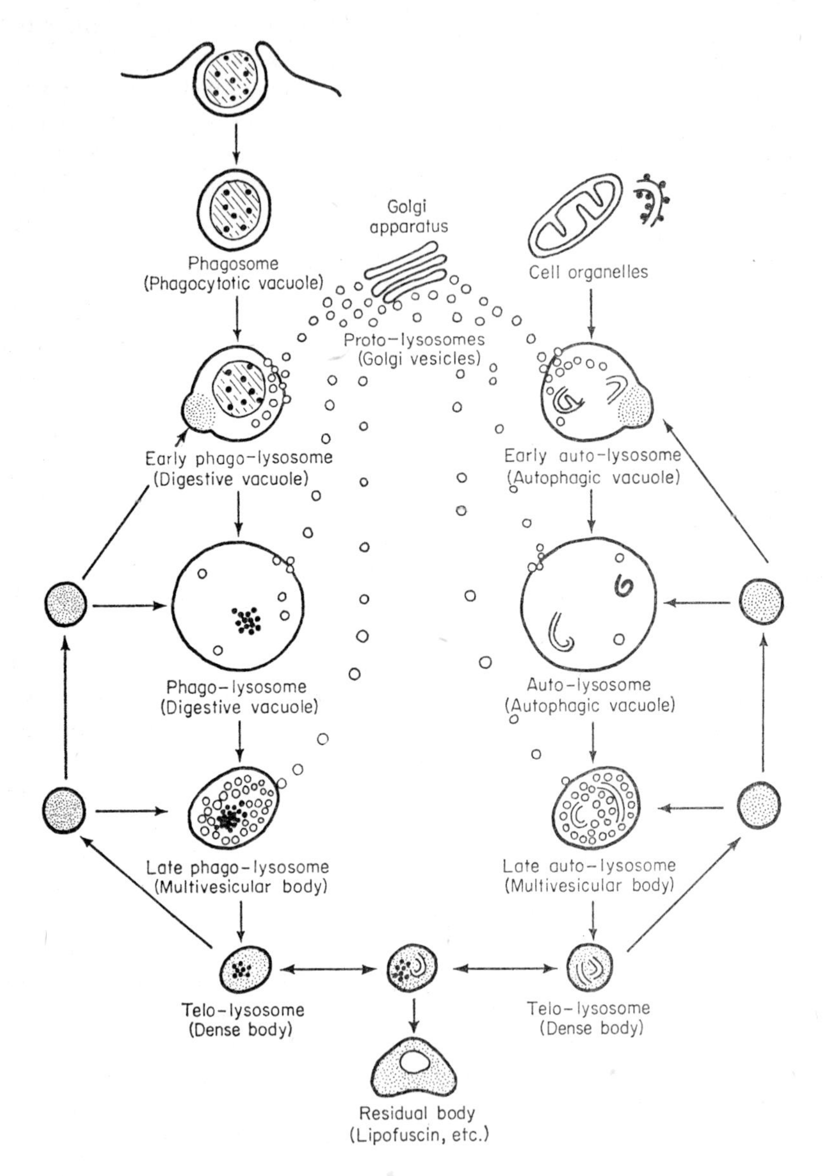

FIG. 2. Interrelationship of cytoplasmic granules, as suggested by Gordon, Miller and Bensch (1965), functioning in intracellular digestion in strain L cells. (Reproduced with the permission of *J. Cell Biol.*)

Duve (1963a) to be the most characteristic properties of lysosomes. The following five enzymes were found to be concentrated in rat-liver lysosomal fractions by de Duve and co-workers (1955) and de Duve (1959a): acid phosphatase, acid ribonuclease (2.7.7.16), acid deoxyribonuclease (3.1.4.6), β-glucuronidase (3.2.1.31), and cathepsin (3.4.4.9). The same enzymes were also found to be highly concentrated in "droplet" fractions isolated from rat kidney (Straus, 1954, 1956). More recently, the same acid hydrolases were demonstrated in partially purified fractions isolated from polymorphonuclear leucocytes (Cohn and Hirsch, 1960a), eosinophil leucocytes (Archer and Hirsch, 1963), macrophages (Cohn and Wiener, 1963a), and in fractions isolated from spleen (Bowers, 1964), intestine (Hsu and Tappel, 1964) and brain (Koenig, Gaines, McDonald, Gray and Scott, 1964).

Certain other hydrolytic enzymes were reported to be concentrated in lysosome fractions of liver: arylsulphatase (3.1.6.1) (Viala and Gianetto, 1955; Roy, 1958), β-galactosidase (3.2.1.23), β-N-acetylglucosamine hydrolase (3.2.1.–or 3.2.1.29), α-mannosidase (3.2.1.24) (Sellinger, Beaufay, Jacques, Doyen and de Duve, 1960), collagenase (3.4.4.19) (Frankland and Wynn, 1962; Schaub, 1964), α-glucosidase (3.2.1.20) (Lejeune, Thinès-Sempoux, and Hers, 1963), hyaluronidase (4.2.99.1) (Aronson and Davidson, 1965) and phosphatidate phosphatase (3.1.3.4) (Wilgram and Kennedy, 1963; Sedgwick and Hübscher, 1965). The presence of phosphoprotein phosphatase (3.1.3.16) in lysosomes, reported by Paigen and Griffiths (1959) was not confirmed by Tappel, Sawant and Shibko (1963) by analysis of highly purified preparations. The possibility of a high concentration of alkaline phosphatase (3.1.3.1) in lysosomes of leucocytes (Cohn and Hirsch, 1960a), and of lipase (3.1.1.3) in the lysosomes of macrophages (Cohn and Wiener, 1963a) requires confirmation since these enzymes were not present in high concentrations in lysosomes of other cells.

Uncertainties arose concerning the concentration of esterase (3.1.1.1) in lysosomes. The lysosomes show strong staining with cytochemical reactions for esterase (Holt and Withers, 1952; Shnitka and Seligman, 1961) but biochemical determination of esterase in tissue fractions indicated that the main esterase activity was in the microsomal fraction (Underhay, Holt, Beaufay and de Duve, 1956). However, the two enzymes seem to be different since the microsomal esterase was sensitive and the lysosomal esterase was resistant to inhibition by organo-phosphorous compounds (Shnitka and Seligman, 1961; Holt, 1963). Hess and Pearse (1958) suggested that the cytochemical esterase reaction might be due to cathepsin. This question was clarified to a certain extent by the work of Shibko and Tappel (1964) who observed esterase activity in highly purified lysosomes from rat liver. The activity was lower than that of the other hydrolytic enzymes and the enzyme was fully active in the intact granules. Shibko and Tappel suggested, therefore, that esterase was bound to the lysosomal membrane.

It is of considerable interest that lysosomes seem to contain an oxidoreductase. A $NADH_2$ cytochrome *c* reductase (1.6.99.3) was detected by Cagan and Karnovsky (1964) and by Rossi and Zatti (1964a) in the granules of leucocytes. The presence of the same enzyme was also reported in highly purified liver lysosomes by Tappel and Ragab (1964).

It should be noted that most of the hydrolytic enzymes mentioned were much more concentrated in the isolated lysosome fractions than in the original homogenates. However, lysosomes may contain several other enzymes in *low* concentrations. It is technically difficult to determine whether low concentrations of enzymes in particulate fractions belong to cell organelles *in vivo*, or whether they were adsorbed on to granules during their isolation. This complication is especially difficult to surmount in the case of lysosomes. As will be mentioned later, enzymes characteristic of other cell organelles may be incorporated into lysosomes by processes of fusion. For example, the fusion of lysosomes with phagosomes will introduce phagosomal enzymes, and ATPase activity (3.6.1.3) present in highly purified kidney lysosomes (Straus, 1954) may be derived from the phagosomes. Low concentrations of alkaline phosphatase observed by Straus (1954) and by Shibko and Tappel (1965) in highly purified lysosomal fractions of rat kidney may belong to the phagolysosomes, since contamination of the purified fractions by microsomes (rich in alkaline phosphatase) was very low in these fractions. The sequestration of lysosomes together with mitochondria and cell fragments (Sections VI. and VIII. A) may introduce low concentrations of non-lysosomal enzymes into these "composite bodies." These considerations show that biochemical and cytochemical observations have to be correlated in the analysis of lysosomal enzymes.

2. *Heterogeneity of Lysosomes and Lysosomal Enzymes*

It has already been mentioned that lysosomes vary greatly in size and density. A simple cytochemical staining reaction for acid phosphatase often shows at first sight striking differences in the size, number, and location of acid phosphatase-positive granules in the same tissue section. This difference can be seen between the proximal and distal tubule cells in kidney sections (Straus, 1964b) and between parenchymal and Kupffer cells in liver sections. The ultrastructure of the lysosomes as observed by electron microscopy, also reveals the difference in size and the pleomorphism of the bodies (dense bodies, vacuolated bodies, multivesicular bodies, lamellated bodies (Farquhar and Palade, 1962; Miller and Palade, 1964).

The heterogeneity of lysosomes is related to the dynamic changes which take place in the granules during their life cycle. As will be reviewed in greater detail in Section V.A.2, lysosomes often fuse with pinocytic or phagocytic vacuoles in which various ingested material is concentrated and gradually digested. Small areas of the cytoplasm may be sequestered together with lysosomes and may form "autolytic vacuoles" (Sections VI. and VIII.A). Since the changes connected with fusion or degradation are probably more pronounced in some lysosomes than in others of the same cell or tissue, the heterogeneity of the granules is accentuated.

It may be expected that the hydrolytic enzymes also undergo changes during the life cycle of the lysosomes. The concentration of hydrolytic enzymes probably is greater in "young" lysosomes than in older ones which have fused with phagosomes. If lysosomes are sequestered together with larger cell components (Sections VI. and VIII.A), the original enzyme concentration may be further diluted. Another cause of heterogeneity would be introduced if differences existed in the rates of synthesis or decay between the various lysosomal enzymes. Nothing definite is known about this point. It also is not known to what extent enzymes which are highly concentrated in most lysosomes, may be present in only low concentration in others, or may be completely lacking. In order to answer this question, it would be necessary to isolate "pure" lysosomes from homogenous cell types, and to correlate biochemical, cytochemical and electron microscopic observations. Lutzner (1964) attempted to obtain such a correlation on different types of blood cells and observed considerable heterogeneity. Shibko, Caldwell, Sawant and Tappel (1963) reported that muscle tissue showed a high concentration of cathepsin and acid RNase but a low concentration of acid phosphatase, aryl-sulphatase, β-glucuronidase, and β-galactosidase. It would be interesting to test whether purified lysosomes isolated from muscle tissue have the same properties and whether the enzymes were derived from lysosomes of muscle cells or blood cells.

In spite of these variations, the similarities in the enzyme content of different lysosomes are probably more marked than the differences. As was mentioned, the same hydrolytic enzymes were found to be concentrated in the lysosomes of liver, kidney, leucocytes, macrophages, spleen, intestine, and brain. It therefore seems justified to treat the lysosomes as a distinct group of cytoplasmic particles, and to sub-divide them into certain types as proposed by de Duve (1963a).

3. Distribution Patterns of Lysosomal Enzymes

The difference in size and density of the lysosomes makes it difficult or impossible to isolate a large proportion of these granules from a complex tissue in a highly purified state (Section IX. A.). Most investigators found it easier, therefore, to compare the distribution patterns of lysosomal enzymes in *partially* purified fractions obtained from tissue homogenates by differential centrifuging or by density gradient centrifuging.

Certain factors which alter the distribution of lysosomal enzymes *in vivo* and *in vitro* were not well understood until recently. With fractions isolated from kidney homogenates, for example, it was observed that the distribution pattern of lysosomal enzymes was altered after intraperitoneal injection of egg white (Straus, 1957a). Similar changes in the distribution of lysosomal enzymes were observed by Cohn and Hirsch (1960b) and by Cohn and Wiener (1963b) in granule fractions isolated from leucocytes and macrophages after phagocytosis of bacteria. In both cases, fusion of lysosomes with phagosomes (Section V.A.2) was responsible for the altered distribution pattern of the enzymes. Similar changes may be expected if lysosomes fuse with phagosomes containing unknown, *endogenous* material in untreated animals and during the subsequent degradation of this material. In view of these continuous changes during the life cycle of the granules, no constancy of their centrifugal behaviour can be expected.

The risk of artifacts of diffusion or adsorption of enzymes occurring during the fractionation of tissue homogenates has been pointed out by several investigators (see, for example, Schneider and Hogeboom, 1951). In the case of lysosomes, it is important to consider the variation in the fragility of the lysosomal membranes (Section II. C) in relation to the functional state of the granules. Artifacts of diffusion, therefore, occur with some types of lysosomes more readily than with others. Thus, "phago-lysosomes" newly formed after ingestion of bacteria or proteins (Cohn and Hirsch, 1960b, Straus, 1957a) may be especially fragile, and they may release their hydrolytic enzymes, in part, into the supernatant fluid during the homogenization and fractionation of the tissue.

The study of distribution patterns of lysosomal enzymes is further complicated by the existence in the same tissue of different enzymes which react with the same substrates. These enzymes may have similar or different intracellular localization. The possibility of different enzymes in lysosomal and supernatant fractions from rat liver was raised by Reid and Nodes (1963) in the case of acid ribonucleases, and by Shibko and Tappel (1963) in the case of acid phosphatases. The most comprehensive study of the multiplicity of hydrolases which react with the same substrate in guinea pig and mouse liver was made by Neil and Horner (1964a, b). These investigators demonstrated the existence of 4 different enzymes, localized in the large granule, lysosomal, microsomal, and supernatant fractions respectively, which all reacted with the same substrate (nitrophenylphosphate). The enzymes could be distinguished by specific activators and inhibitors, difference in heat stability, and pH activity curves. In their excellent discussion of the problem, Neil and Horner pointed out the importance of the following factors: (a) the relative affinity of each enzyme for the particular substrate used; (b) the relative amounts of the different enzymes present in each cell fraction; (c) the influence of the assay conditions on each enzyme. In the case of rat and guinea pig liver, for example, Neil and Horner observed glycerophosphatase activity to be mainly localized in the particulate (lysosomal) fraction. In the case of mouse liver, however, appreciable amounts of a different enzyme also splitting β-glycerophosphate were present in the supernatant fractions. (This enzyme was, however, more active with nitrophenylphosphate than with β-glycerophosphate). For cytochemical work, it is of interest that Neil and Horner observed much stronger inhibition by formaldehyde of the supernatant enzyme than of the lysosomal enzyme. This may explain why cytochemical staining for acid phosphatase, after formaldehyde fixation, mainly reveals lysosomal but not cytoplasmic acid phosphatase in certain tissues. Neil and Horner reported good agreement between the properties of nitrophenylphosphatases, separated by chromatography on DEAC-cellulose by Moore and Angeletti (1961), and of those studied by them. The agreement was less good in regard to fractions obtained by Barka (1961a, b) by electrophoresis on polyacryl gel and DEAC-cellulose chromatography. Allen and Gockerman (1964) separated two acid phosphatases by gel electrophoresis from the mitochondrial-lysosomal fraction of rat liver, one of which was bound more firmly to the granules than the other. Schuel and Anderson (1964) by using a new centrifugation procedure, zonal ultracentrifugation, also concluded that several nitrophenylphosphatases existed in rat liver. Rosenbaum and Rolon (1962) and Wachstein, Meisel, and Ortiz (1962) discussed the heterogeneity of acid phosphatases in liver and kidney on the basis of cytochemical observations.

A few specific questions relating to the presence of lysosomal enzymes in conventionally prepared nuclear, mitochondrial, microsomal, and supernatant fractions may be reviewed briefly. (i) "Nuclear" fraction: A considerable number of large lysosomes contaminate "nuclear" fractions separated from rat kidney homogenates by differential centrifugation (Straus, 1954, 1956, 1957a). It is not certain whether acid hydrolytic enzymes are contained in the nuclei themselves as suggested by Siebert (1963) and by Swingle and Cole (1964). The presence of acid hydrolytic enzymes was demonstrated in nuclear inclusion bodies by Leduc and Wilson (1959). (ii) "Mitochondrial" fraction: Mitochondrial fractions are often contaminated by lysosomes, especially in tissues relatively rich in lysosomes (kidney, liver) (Kuff and Schneider, 1954). The degree of contamination can be estimated by electron microscopy, or by assay for lysosomal enzymes. In certain types of lysosomes ("composite bodies", "autolytic vacuoles", Sections VI. and VIII. A.), mitochondria and lysosomes are enclosed within a common membrane. The separation of mitochondrial and lysosomal enzymes from these particles would not be possible, of course, without disintegration of the particles. (iii) "Microsomal" fractions: Ten to 20% of lysosomal enzymes of the total homogenates were present in the microsomal fractions of kidney and liver (de Duve and co-workers, 1955; Straus, 1956). It is of considerable interest that the microsomal activities were increased in regenerating liver, possibly indicating new formation of lysosomal enzymes (Walkinshaw and van Lancker, 1964). (iv) "Supernatant" fraction: The increase of lysosomal enzymes in the supernatant fluid, due to the release of the enzymes from fragile "phago-lysosomes" *in vitro*, was already mentioned. In pathological conditions, lysosomal enzymes may be released *in vivo* in areas of focal degradation or they may be extruded from the cells (Sections VI. and VIII.). Both processes may result in the appearance of these enzymes in the supernatant fluids after fractionation of homogenates.

B. NON-ENZYMIC COMPONENTS OF LYSOSOMES

Tappel, Sawant and Shibko (1963) and Sawant Shibko, Kumta and Tappel (1964) succeeded in isolating lysosomes from rat liver in good purity and in relatively good yield, and, therefore, were able to test for some non-enzymic components. They reported the presence of small amounts of lipids (lecithin, phosphatidylethanolamine, sphingomyelin, and cerebrosides). The amounts of phospholipids were sufficient to account for a unit membrane around the lysosomes. Small amounts of lipid phosphorus were also detected in highly purified kidney lysosomes (Straus, 1954).

Tappel, Sawant and Shibko (1963) noted that highly purified liver lysosomes showed the fluorescence spectra of flavin and pyridine nucleotides. Maunsbach (1964a) also observed fluorescence of highly purified kidney lysosomes, and Koenig (1963a) described the fluorescence of lysosomes *in situ*.

Lysosomes show a positive periodic acid-Schiff reaction (Davies, 1954; Novikoff, 1961). According to Koenig (1962) and Tessenow (1964) this reaction can be attributed to the presence of glycolipids. Fedorko and Morse (1965) isolated acid mucopolysaccharides, among them hyaluronic acid, from the granules of leucocytes. However, no carbohydrates of this class could be extracted from the lysosome-rich fraction of macrophages.

Cytoplasmic granules isolated from leucocytes were reported to contain bactericidal (Cohn and Hirsch, 1960a; Zeya and Spitznagel, 1963) and inflammation-producing substances (Janoff and Zweifach, 1964). It is not certain, however, whether or not these compounds are localized in the lysosomes or in accompanying granules. Leake and Myrvik (1964) concluded from differences in osmotic properties of isolated granules that lysozyme was contained in other types of granules, not in the lysosomes of alveolar macrophages, as suggested by Cohn and Hirsch.

Since lysosomes often fuse with phagosomes (Section V. A. 2.), they may contain various types of exogenous or endogenous ingested materials at different stages of digestion. Residues of poorly digestible matter, resembling myelin figures, can be seen in electron micrographs of some lysosomes. Crystalline inclusions were frequently observed by the writer in kidney lysosomes after injection of horse-radish peroxidase. Figures 3 and 4 show such crystal-containing bodies. These structures may be related to crystalline bodies demonstrated in kidney cells by electron microscopy by Maunsbach (1966). It is interesting to note that calcium seems to accumulate

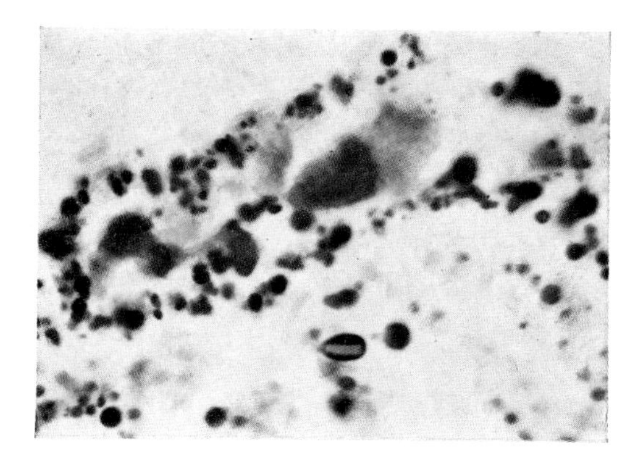

FIG. 3. Phago-lysosomes in kidney cells, 48 hr after intravenous injection of horse-radish peroxidase; stained with benzidine. Note that a large phagosome (phago-lysosome) in lower part of picture contains a crystal, and that most of the peroxidase in other phagosomes of this nephron was already digested. ×660.

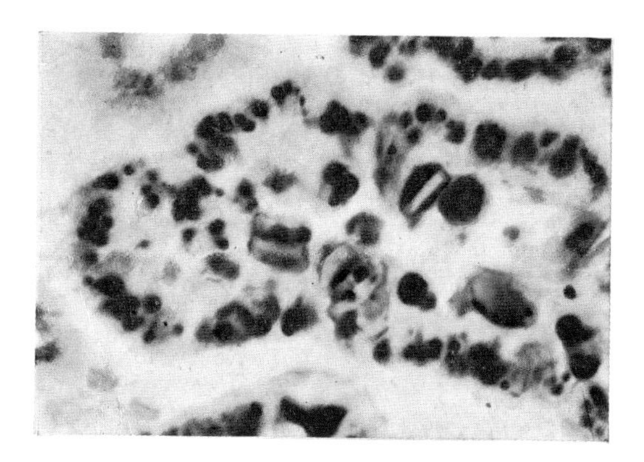

FIG. 4. Two phago-lysosomes in kidney cells showing crystals, 48 hr after intravenous injection of horse-radish peroxidase; stained for acid phosphatase by an azo-dye procedure. ×660.

in the lysosomes of the kidney medulla during magnesium deprivation (Schneeberger and Morrison, 1965; Battifora, Eisenstein, Laing, and McCreary, 1966).

C. LYSOSOMAL MEMBRANE AND ACTIVATION OF ENZYMES

The importance of the lysosomal membrane for the "activation" of the hydrolytic enzymes *in vitro* and *in vivo* was recognized by de Duve and co-workers early in their work. These investigators systematically studied various treatments which injured the membrane (freezing and thawing, mechanical disruption, osmotic shock, surface-active agents, proteolytic agents, and "autolysis") (Gianetto and de Duve, 1955; Wattiaux and de Duve, 1956; Beaufay and de Duve, 1959). In general, each treatment caused the release of the various hydrolytic enzymes in a similar fashion, although slight variations from one enzyme to the other were observed. These properties supported the hypothesis that the different enzymes belong to the same type of organelle. The "latency" of the enzymes could be explained, according to de Duve and co-workers, by envisaging a model in which the enzymes were contained in a sac of fluid enclosed by a semipermeable membrane. A different model was proposed by Koenig and Jibril (1962). According to these investigators, the hydrolytic enzymes are linked within the granules by ionic bonds to a matrix of a glycolipid nature. This hypothesis was based on the observation that cationic substances caused the release of the enzymes. Jacques, Ennis and de Duve (1964) did not confirm the releasing effect of some of the cationic compounds tested by Koenig and Jibril. Sawant, Desai and Tappel (1964a) discussed the possibility that lysosomal enzymes might be bound to the *membrane* of the granules by electrostatic bonds.

During recent years, the effects of the following agents which cause release of lysosomal enzymes *in vitro* have been studied: Vitamin A (Lucy, Dingle, and Fell, 1961; de Duve, Wattiaux and Wibo, 1962; Dingle, 1963; Weissmann and Thomas, 1963; Dingle, Sharman, and Moore, 1966), vitamin D, vitamin E, vitamin K, and sterols (de Duve *et al.*, 1962), endotoxin (Weissmann and Thomas, 1962), streptolysin (Weissmann, Keiser and Bernheimer, 1963; Hirsch, Bernheimer and Weissmann, 1963), ultraviolet irradiation (Weissmann and Dingle, 1961); Desai, Sawant and Tappel, 1964) and cations (Koenig and Jibril, 1962; Sawant, Desai and Tappel, 1964a; Jacques, Ennis and de Duve, 1964). Of practical interest is the observation by Sawant *et al.* (1964a) that the enzymes were released more rapidly in 0·25 M sucrose, often used for fractionation experiments, than in 0·7 M sucrose. Kidney lysosomes were found by Shibko and Tappel (1965) to be much less affected by changes in the osmolarity of the suspending medium than liver lysosomes. Shibko, Pangborn and Tappel (1965) observed that the enzymes were released from kidney lysosomes in two stages during incubation at 37°. In the first stage, increasing amounts of the enzymes became available to the external substrates but the granules did not release their enzymes and did not show any change in morphology. In the second stage, the enzymes were released at the same time as the granules disintegrated. Several of the agents which cause the release of lysosomal enzymes also labilize the membranes of mitochondria and erythrocytes (Dingle, 1963; Keiser, Weissmann and Bernheimer, 1964). It may be of interest to mention in this context that in kidney homogenates in 0·88 M sucrose at room temperature, swelling of mitochondria occurred much more readily, than swelling of lysomes (Straus, 1959). Only one agent, cortisone, seemed to make the lysosomal membrane more resistant toward the release of enzymes (Weissmann and Dingle, 1961; de Duve, *et al.*, 1962; Weissmann and Thomas, 1962, 1963). It would be of considerable interest if cortisone specifically affected the lysosomal membrane.

In all these studies, the possibility does not seem to have been considered sufficiently that the membrane properties may differ with the physiological state of the cytoplasmic granules. It would not be surprising if the membranes of "young" lysosomes, "old" lysosomes, lysosomes soon after their fusion with phagosomes, and lysosomes after pathological enlargement (Section VIII. A.) behave quite differently. Reports in the literature also indicate that the properties of lysosomal membranes are not uniform. Rahman (1964) showed that the membrane properties of lysosomes (as studied by release of acid phosphatase during incubation at pH 5·2 at 37°, or after treatment with phospholipase at pH 7·0) were quite different in homogenates of liver, spleen, and thymus. Bitensky (1963), by using Gomori's acid phosphatase reaction in a controlled temperature freezing and sectioning procedure, noted that the fragility of the lysosomes varied with the physiological state of the granules.

The factors causing the release of enzymes from lysosomes are little understood. Whereas de Duve and co-workers, in general, found that the release proceeded in a parallel fashion for different enzymes, Sawant, Desai and Tappel (1964a) noted that the various enzymes were released to a different extent. As was pointed out by de Duve *et al.* (1962), the releasing effect by biologically active agents *in vitro* does not prove that these agents also have this effect *in vivo*. Before the effect of an agent on the lysosomal membranes *in vivo* can be evaluated, it should be determined by light microscopy or electron microscopy whether or not the agent caused the transformation of one type of lysosome into another. This question of the release of lysosomal enzyme *in vivo* will also be discussed in Section VIII.

The fusion of lysosomes with phagosomes, to be discussed later (Section V. A. 2.), is probably closely related to properties of the membranes. This was foreseen in the hypothesis by Bennett (1956) on "membrane flow". The factors causing the fusion of membranes between lysosomes, phagosomes, and plasmalemma, are unknown.

III. Occurrence of Lysosomes

A quick estimation of the number and location of lysosomes in different types of cells can be obtained in many cases by the cytochemical staining for β-glycerophosphatase in formalin-fixed, frozen sections. Such a survey shows that lysosomes are numerous in a number of epithelial cells of organs of absorption, secretion and excretion (kidney, liver, prostate, epididymis, for example). Considerable numbers of lysosomes are also present in epithelial cells of the intestine, lung, uterus, and in nerve cells. Relatively few acid phosphatase-positive granules were seen in muscle cells and in the acinar cells of the pancreas. On the other hand, all phagocytic cells or reticulo-endothelial cells (macrophages) contain large numbers of lysosomes: for example, the phagocytic cells of the spleen, bone marrow, liver, and connective tissue. It should be noted that acid phosphatase may not always be a reliable marker for lysosomes and that a cytochemical reaction for acid phosphatase of a cell particle only points to its possible lysosomal nature but is not sufficient proof. This has to be emphasized since many investigators seem to rely on this criterion alone for the detection of lysosomes. In order to establish the lysosomal nature of a cell particle, it has to be shown that *several* acid hydrolytic enzymes are present in relatively high concentrations and that the enzymes show "latency" (see also de Duve, 1963a).

In the following paragraphs, a few representative studies from the literature have been selected in which pertinent data on lysosomes in various tissues may be found.

A. Lysosomes in Epithelial Cells

1. Liver

The fundamental biochemical work which established the presence of acid hydrolytic enzymes in a distinct class of cytoplasmic granules was done on liver by de Duve and co-workers (1955) and de Duve (1959a). Sawant, Shibko, Kumta and Tappel (1964) succeeded in improving considerably the procedure for the isolation of liver lysosomes.

By electron microscopy, Holt and Hicks (1961b) demonstrated the reaction product of acid phosphatase in liver lysosomes. This was also achieved by Daems (1962) who investigated the relationship between liver lysosomes and the vacuoles formed after injection of dextran and iron-dextran. Kent, Volini, Orfei, Minick and de la Huerga (1963), Kent, Minick, Volini, Orfei and de la Huerga (1963) in similar work, studied the uptake of iron-dextran by liver cells and its appearance in lysosomes; they noticed, by electron microscopy, increased uptake after injury associated with the fusion of granules. Increased uptake of a protein after injury and enlargement of phago-lysosomes in liver cells, were also reported by Straus (1963, 1964a) by double cytochemical staining for acid phosphatase and injected plant peroxidase (1.11.1.7).

2. Kidney

Biochemical analysis of 5 lysosomal enzymes in highly purified "droplet" fractions from kidney homogenates was reported by Straus (1954, 1956). Maunsbach (1964a) and Shibko and Tappel (1965) also obtained highly purified kidney lysosomes and studied some of their properties (see Section IX. A.).

In cytochemical work, Shnitka and Seligman (1961) stained kidney lysosomes by the reaction for esterase, and differentiated inhibitor resistant esterase in the lysosomes from an inhibitor-sensitive enzyme in the cytoplasm. Wachstein, Meisel and Ortiz (1962) compared by cytochemical staining for acid phosphatase lysosomes in kidneys of rat, mouse, guinea pig, and rabbit. Straus (1964a, b) investigated phago-lysosomes in tubule cells of rat kidney by double staining for acid phosphatase and injected peroxidase.

Acid phosphatase-positive granules in cells of the collecting tubules in potassium deficiency were investigated by histochemical electron microscopy by Morrison and Panner (1964). Ericsson and Trump (1964) and Ericsson (1964) studied by histochemical electron microscopy acid phosphatase-positive granules in the proximal tubule cells. Fisher (1964) using the same method, investigated the lysosomes during increased protein absorption induced by nucleoside nephrosis. Miller and Palade (1964) by histochemical electron microscopy, demonstrated the presence of acid phosphatase and re-absorbed haemoglobin or ferritin in the same granules.

3. Intestine

Biochemical studies on the intracellular distribution of lysosomal enzymes were reported by Hsu and Tappel (1964). Gitzelmann, Davidson and Osinshak (1964) also fractionated intestinal homogenates and studied the acid phosphatase reaction of the isolated fractions biochemically and in the electron microscope.

Sheldon, Zetterquist and Brandes (1955) were the first to demonstrate acid phosphatase-positive granules at the level of the electron microscope. More recently, Behnke (1963) observed enlarged lysosomes, by histochemical electron microscopy for acid phosphatase, in foetal rat duodenum during certain stages of differentiation, and Behnke and Moe (1964) investigated Paneth cells by the same method. Barka (1964) applied the Gomori reaction for acid phosphatase at the electron microscope level to identify lysosomes in intestinal epithelial cells and possible changes during fat absorption.

4. Brain

In biochemical work, Koenig *et al.* (1964) fractionated mitochondrial and microsomal fractions from rat brain into several sub-fractions by discontinuous density gradient centrifuging. They showed that 5 lysosomal enzymes were concentrated in some of the sub-fractions and also presented electron micrographs of the sub-fractions. By cytochemical staining for acid phosphatase, Becker, Goldfischer, Shin and Novikoff (1960), Barron and Sklar (1961), and Anderson and Song (1962) showed the presence of numerous lysosomes in nerve cells. Chouinard (1964) correlated the cytochemical reactions for several lysosomal enzymes with the periodic acid-Schiff reaction, in Purkinje cells.

Histochemical electron microscopy was applied to demonstrate the acid phosphatase reaction product in lysosomes of brain stem by Torack and Barrnett (1962) and in neurosecretory cells by Osinshak (1964).

5. Prostate

Bertini and Brandes (1964) reported (in an abstract) the biochemical analysis of acid phosphatase and cathepsin in fractions isolated from prostate homogenates by differential centrifuging. Brandes, Groth and Gyorkey (1962) and Brandes (1963, 1965) investigated lysosomes in prostatic epithelium by electron microscopy, and by cytochemical staining for acid phosphatase.

6. Epididymis

In biochemical work, Schneider and Kuff (1954) observed concentration of acid phosphatase activity in isolated Golgi material. By cytochemical staining for acid phosphatase, Allen and Slater (1958) demonstrated the presence of many lysosomes in the epithelial cells of the epididymis.

7. Urinary Bladder

Kanczak, Krall, Hayes, and Elliot (1965) isolated lysosomal fractions rich in acid phosphatase and β-glucuronidase from canine and bovine transitional epithelium. The granules were also identified *in situ* by their cytochemical reaction for acid phosphatase.

8. Uterus, Ovaries and Endometrium

Nilsson (1962c) and Fuxe and Nilsson (1963) observed lysosomes in the uterus of spayed mice by cytochemical staining for acid phosphatase and esterase, and noted the effect of oestrogen. Nilsson (1962a, b) investigated lysosomes by electron microscopy in the human uterus during the folicular and luteal phases. Banon, Brandes and Frost (1964) made a cytochemical study of lysosomes in rat ovaries and endometrium during the oestrus cycle. Woessner (1965b) analysed changes in the concentration of several lysosomal enzymes in homogenates and fractions from rat uterus in relation to pregnancy, post-partum involution, and collagen breakdown.

9. Thyroid, Adrenals

Hosoya (1963) fractionated thyroid tissue and found that acid phosphatase was concentrated in a fraction which also showed dense bodies under the electron microscope. Wollman, Spicer and Burstone (1964) demonstrated lysosomes in thyroid epithelial cells by cytochemical staining for acid phosphatase and esterase and studied the relationship to colloid droplets. Smith and Winkler (1966) fractionated homogenates from the bovine adrenal medulla by centrifuging in a discontinuous sucrose gradient. They separated the lysosomal fraction (in which 6 acid hydrolytic enzymes were concentrated) from the mitochondrial fraction and the chromaffine granule fraction.

B. LYSOSOMES IN RETICULO-ENDOTHELIAL CELLS AND BLOOD CELLS

The isolation of lysosomes and the biochemical analysis of their hydrolytic enzymes from polymorphonuclear leucocytes was reported by Cohn and Hirsch (1960a), from eosinophil leucocytes by Archer and Hirsch (1963), and from macrophages by Cohn and Wiener (1963a). Lutzner (1964) briefly reported the isolation, biochemical analysis and electron microscopy of lysosomes from various types of blood cells. Marcus, Zucker-Franklin, Safier, and Ullman (1966) fractionated homogenates from human blood platelets and obtained fractions rich in acid phosphatase, cathepsin, and β-glucuronidase.

In cytochemical work, acid phosphatase in lysosomes was demonstrated in reticulo-endothelial cells of spleen, liver, thymus, lymph nodes, and other organs, by Barka and co-workers (1961). Thorbecke, Old, Benacerraf and Clarke (1961) observed acid phosphatase-positive granules in Kupffer cells of liver during various treatments modifying the activity of the reticulo-endothelial system. Acid phosphatase- and esterase-positive granules in alveolar and peritoneal macrophages and in polymorphonuclear leucocytes were demonstrated by Dannenberg, Burstone, Walter and Kinsley (1963). Acid phosphatase-positive vacuoles in blood platelets were detected by histochemical electron microscopy by White, Krivit and Vernier (1964). The presence of lysosomes in lymphocytes and their increase after phytohaemagglutinin treatment was reported by Hirschhorn and co-workers (1965), Allison and Malucci (1964a), Diengdoh and Turk (1965), and Parker, Wakasa, and Lukes (1965) (see also Chapter 7). Straus (1964a, Fig. 8 and 1964d, Fig. 10) showed "phago-lysosomes" in Kupffer cells of liver and in macrophages of spleen by double staining for acid phosphatase and injected peroxidase in the same tissue sections. In experiments, illustrated in Figs. 5–7 of the present article, this procedure was applied to demonstrate phagolysosomes in macrophages of thymus, lung, and prostate.

Rahman (1962a, b) purified "dense bodies" probably derived from phagocytic cells of the thymus after X-ray irradiation and measured acid phosphatase and β-glucuronidase activities of the isolated fractions. Ruyter (1964) showed acid phosphatase-positive granules in osteoclasts cytochemically. Vaes and Jacques (1965) fractionated bone tissue homogenates and demon-

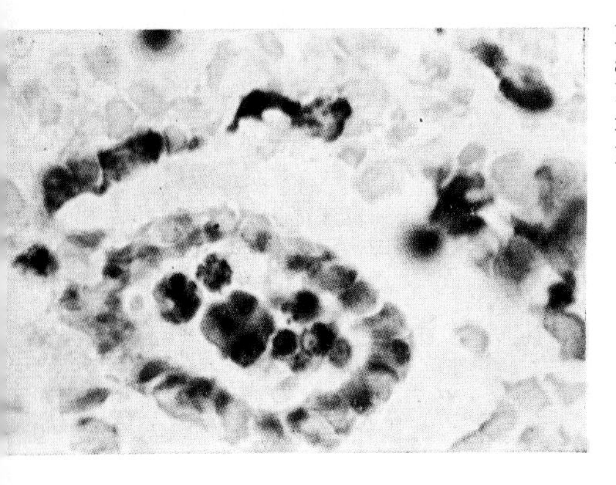

FIG. 5. Phago-lysosomes in macrophages of lung, 3 hr after intravenous injection of horse-radish peroxidase. Note purple (blue-red) color of phago-lysosomes in macrophages after double staining for acid phosphatase and peroxidase in the same section. ×660.

G. 6. Phago-lysosomes in macrophages of connective sue of prostate, 2 hr after intravenous injection of rse-radish peroxidase. Note purple (blue-red) color large phago-lysosomes in macrophages after double ining for acid phosphatase and peroxidase, and acid osphatase reaction (without peroxidase) in lyso- mes of prostatic epithelial cells. ×660.

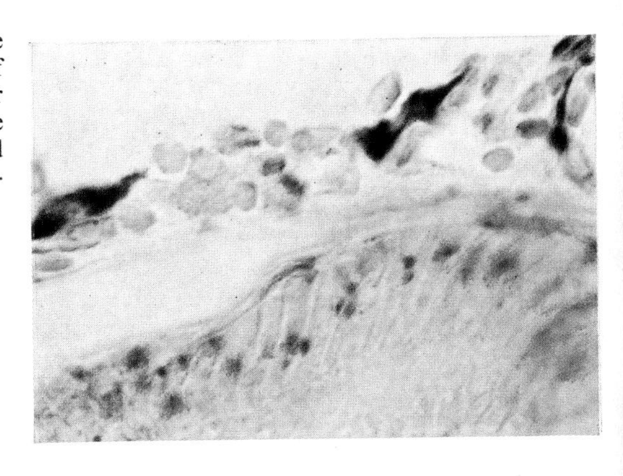

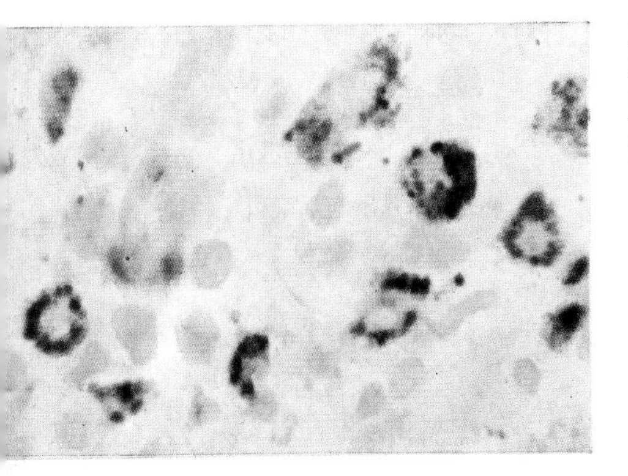

FIG. 7. Phago-lysosomes in macrophages of thymus, 24 hr after intravenous injection of horse-radish peroxidase; double staining for acid phosphatase and peroxidase. Note that the peroxidase in some phago-lysosomes seems to have been digested already. ×660.

strated the concentration of 8 lysosomal enzymes in the "light" mitochondrial (lysosomal) fraction.

C. LYSOSOMES IN DIFFERENT SPECIES

Acid phosphatase activity in fractions isolated from amoebae was studied in relation to metachromatic vacuoles by Quertier and Brachet (1959). Lagunoff (1964) observed no significant change in the acid protease and acid phosphatase activities of amoebae after pinocytosis of proteins. Biochemical investigation of 6 lysosomal enzymes in muscle, spleen, and liver of various animals (marine invertebrates, bees, frogs, turtles, pigeons, ox, sheep, and hogs) was carried out by Shibko, Caldwell, Sawant and Tappel (1963). Zeidenberg and Janoff (1964) analysed 3 lysosomal enzymes in the particulate fractions of kidneys from fishes.

Cytochemical staining for acid phosphatase and esterase in food vacuoles of protozoa (paramecium) was applied by Müller (1962); Müller and Törö (1962); Müller, Tóth and Törö (1962); Müller, Röhlich, Tóth and Törö (1963); Müller, Törö, Polgar and Druga (1963) and by Rosenbaum and Wittner (1962). Acid phosphatase-positive granules in amoeba were reported by Birns (1960), in ciliates by Seaman (1961) and by Klamer and Fennell (1963), and in planarians by Rosenbaum and Rolon (1960a). Rosenbaum and Ditzion (1963) made cytochemical observations on lysosomal enzymes in the digestive glands of a snail. Brandes, Buetow, Bertini and Malkoff (1964) studied the development of autophagic vacuoles after carbon starvation in *Euglena*, by histochemical electron microscopy for acid phosphatase. Wächtler and Pearse (1966) demonstrated lysosomes in the pars distalis of the amphibian pituitary by cytochemical staining for acid phosphatase, β-glucosaminidase, β-glucuronidase, sulphatase, and esterase.

D. LYSOSOMES IN TISSUE CULTURE CELLS

Wattiaux (1962) separated fractions enriched in several lysosomal enzymes, from HeLa cells, and Munro, Daniel and Dingle (1964) analysed fractions isolated from cultured fibroblasts. Cohn and Benson (1965b) measured the increase in 3 lysosomal enzymes in monocytes during differentiation in tissue culture.

Acid phosphatase-positive granules were demonstrated in monocytes and macrophages in tissue culture by Weiss and Fawcett (1953), in HeLa cells by Green and Verney (1956), in fibroblasts by Ogawa, Mizuno and Okamoto (1961), in cultures of chick cells by Mulnard (1961) and Flaxman and Mulnard (1961). Hirschhorn, Kaplan, Goldberg, Hirschhorn and Weissman (1965) observed a striking increase in acid phosphatase-positive granules in lymphocytes in tissue culture in the presence of haemagglutinin, and made correlated biochemical measurements of enzyme activities.

E. Lysosome-like Granules in Plant Cells

Lysosome-like granules also seem to exist in plant cells. Walek-Czernecka (1962, 1963, 1965) observed that the spherosomes of plant cells showed cytochemical reactions for acid phosphatase, esterase, β-glucuronidase, β-galactosidase, β-glucosidase, lipase, arylsulphatase, and desoxyribonuclease, and was the first to suggest that these granules might be analogous to animal cell lysosomes. Olszewska and Gabara (1964) reported that granules in plant cells showing the cytochemical reactions for esterase, acid and alkaline phosphatase, arylsulphatase, desoxyribonuclease and β-glucuronidase seemed to be involved in the formation of the cell plate after cell division, and pointed out the analogies between spherosomes and animal cell lysosomes. Harrington and Altschul (1963) measured acid phosphatase activities in particulate fractions isolated from the embryo and endosperm of germinating onion seed and concluded that lysosome-like particles were present. Matile, Balz, and Semadeni (1965) purified spherosomes from corn and tobacco seedlings by continuous density gradient centrifugation. Acid proteinase, acid phosphatase, esterase, and acid ribonuclease were concentrated in fractions which, by electron microscopy, showed the presence of spherosomes. The authors believe that the spherosomes of plant cells, in analogy with animal cell lysosomes, function in the autolytic degradation of cell material.

IV. Origin and Formation of Lysosomes and of Lysosomal Enzymes

The formation of lysosomes and of their enzymes is not yet understood completely. It was suggested that "young" lysosomes might be a type of secretory granule comparable to the zymogen granules of the pancreas (Straus, 1957a). The main difference between both types of secretory granules would be in the pH optima of their hydrolytic enzymes, and in the fact that secretion of enzymes from the cells into ducts occurs in the case of zymogen granules, and that utilization of enzymes for intracellular digestion in the same cells in which they are produced occurs in the case of the lysosomes.

It may be expected that the synthesis of the lysosomal enzymes takes place on the ribosomes of the endoplasmic reticulum where the synthesis of other proteins is known to occur. One lysosomal enzyme, acid phosphatase, appeared initially in the cisternae of the endoplasmic reticulum in ciliae, as demonstrated by Carasso, Favard and Goldfischer (1964) by electron microscopy. Later, the acid phosphatase in these cells was seen in small pinocytic vesicles which seemed to bud off from the food vacuole. Brandes (1965) demonstrated by electron microsopy that acid phosphatase-positive bodies were formed in the rough-surfaced endoplasmic reticulum of prostatic and seminal vesicle epithelial cells.

As shown by Caro and Palade (1964), secretory enzymes formed in the endoplasmic reticulum of the pancreas are assembled in the Golgi region and later discharged at the periphery of the cells. The suggestion of Novikoff (1961) that the lysosomal enzymes are also assembled in the Golgi zone therefore seems plausible. Novikoff observed a positive acid phosphatase reaction, by electron microscopy, in Golgi cisternae and in small vesicles close to them, in several types of cells. Smith (1963) also reported acid phosphatase activity within the inner Golgi cisternae and in small vesicles associated with them, in the adenohypophysis, and increased activity at both sites during periods of heightened secretion. Osinshak (1964) in a careful study by electron microscopy of neurosecretory cells, observed that some Golgi cisternae in these cells showed the reaction product for acid phosphatase. The activity was also seen in small vesicles, probably young secretory granules, which formed from the Golgi cisternae. Osinshak suggested that acid phosphatase was "carried over" and was transient, and was not related to the neurosecretory process. No evidence could be found by Osinshak for the origin of acid phosphatase in the large lysosomes of the neurosecretory cells. Gordon, Miller and Bensch (1965) clearly demonstrated a relationship between Golgi vesicles, probably containing acid phosphatase, and phago-lysosomes. According to the observations of these authors, the small vesicles of "multivesicular bodies" are enzyme carriers derived from the Golgi apparatus. Figure 8 reproduced from the article by Gordon, Miller, and Bensch (1965), shows such small vesicles together with exogenous gold-labelled DNA-protein complexes in a multivesicular body of a fibroblast in tissue culture. Bainton and Farquhar (1966) also demonstrated clearly that lysosomes are derived from the Golgi complex. These investigators examined developing leucocytes from rabbit bone marrow by electron microscopy and observed that two types of lysosomes, the "specific granules" and the "azurophil" granules, originated from different faces of the Golgi complex during different stages of granulocyte development. Interesting observations on the relationship between the Golgi zone and acid phosphatase-positive granules were reported by Mulnard (1961) and by Flaxman and Mulnard (1961) in cultured chick cells, using light microscopy, and by Brandes *et al.* (1964) on carbon-starved *Euglena*, using electron microscopy.

Other investigators, however, did not find any acid phosphatase activity, by electron microscopy, in the Golgi zone. Barka (1964) looked for this in vain in the intestinal epithelial cells, and Behnke and Moe (1964) in the Paneth cells (secretory cells of the intestinal epithelium). Miller and Palade (1964) observed a positive acid phosphatase reaction in the Golgi cisternae only in 7 out of 170 sections of kidney cells using electron microscopy. Thus, the question of the acid phosphatase activity in the Golgi apparatus is still controversial. Perhaps, acid phosphatase is only detected in the Golgi zone in cells which do not contain many "old" lysosomes and in which

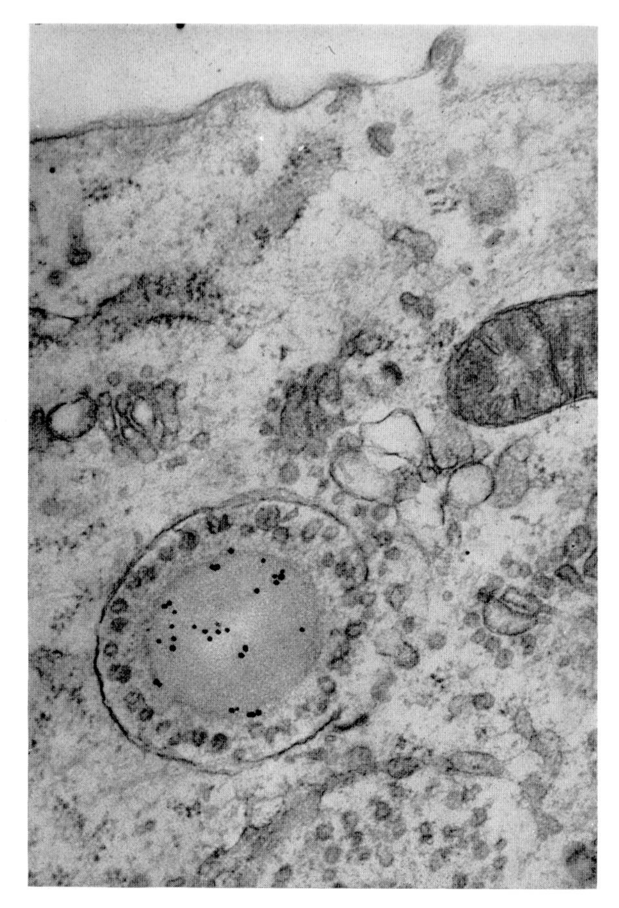

FIG. 8. Multivesicular body, reproduced from the paper by Gordon, Miller, and Bensch (1965), showing relationship of phagosome, containing ingested gold-DNA-protein complex, to vesicles derived from the Golgi apparatus. × 32,000. (Reproduced with the permission of *J. Cell Biol.*)

lysosomal enzymes (or lysosomes) are newly formed after stimulation of the enzyme-forming system by increased pinocytosis or autophagy.

Several investigators observed that lysosomal enzymes increased in activity after various treatments. However, in most cases, it was not certain whether there was a real increase of lysosomal enzymes in one type of cell, or whether an infiltration of macrophages, rich in lysosomes, into the tissue had taken place. Recently, however, several workers have demonstrated a real increase of lysosomes or of lysosomal enzymes in one type of cell: in the cells of the collecting tubules after K^+ deficiency (Morrison and Panner, 1964), in macrophages differentiating from monocytes in tissue culture (Cohn and Benson, 1965a), and in lymphocytes in tissue culture in the presence of hemagglutinin (Hirschhorn *et al.*, 1965). Cohn and Benson (1965b) observed that the formation of lysosomes and of hydrolytic enzymes in macrophages, cultured *in vitro*, was directly related to the uptake of proteins by pinocytosis.

Some investigators have expressed quite different ideas about the origin and development of lysosomes to those mentioned above. According to van Lancker (1964), acid hydrolytic enzymes, probably synthesized in the endoplasmic reticulum, would appear in areas of focal degradation (Sections VI. and VIII.). The hydrolytic enzymes, due to their greater resistance towards autolytic degradation, would accumulate there while the other portions of the sequestered areas undergo autolysis. In this way, enlarged areas of focal degradation would be transformed into smaller lysosomes of normal size. Most investigators who expressed these, or similar ideas (Ashford and Porter, 1962; van Lancker, 1964; Confer and Stenger, 1964) have mainly studied pathologically altered lysosomes. The hypothesis does not explain how the hydrolytic enzymes arrive at the areas of focal degradation in the many cases where the sequestered areas do not contain endoplasmic reticulum. It also ignores processes of fusion between lysosomes and phagosomes which often seem to be related to the development of areas of focal degradation (Straus, 1964a, b; see also Section VIII. A.). Another observation which speaks against the derivation of normal lysosomes from areas of focal degradation is the high concentration of hydrolytic enzymes in highly purified lysosomes from kidney and liver (Straus, 1956; Sawant *et al.*, 1964c). Although a moderate degree of concentration of enzymes is explicable by van Lancker's hypothesis of passive accumulation, the high concentrations actually found are similar to the concentrations of hydrolytic enzymes in zymogen granules. Recent observations by Locke and Collins (1965) suggest a way by which Golgi vesicles may bring lysosomal enzymes to sequestered regions of the cytoplasm. In the fat body of a butterfly larva studied by Locke and Collins, small regions of the endoplasmic reticulum, mitochondria, or other cytoplasmic areas, were surrounded by numerous Golgi vesicles. The vesicles fused, thus isolating these areas from the rest of the cytoplasm. Subsequently, the inner of the two "isolation membranes" was resorbed, and processes of coalescence and lysis within the "isolation bodies" were observed. The process described by Locke and Collins also seems to explain why "autophagic vacuoles" are surrounded sometimes by one and sometimes by two membranes.

V. Lysosomes and Ingestion Processes (Phagosomes and Phago-Lysosomes)

For reasons discussed by de Duve (1959b), lysosomal enzymes are probably concerned with the degradation of metabolites, rather than with transfer or synthetic reactions. This idea is confirmed by the relationship between lysosomes and the reticulo-endothelial system. As will be discussed in the following paragraphs, the material broken down, in many cases, had been taken up from the environment by pinocytosis or phagocytosis. Even in the case of physiological breakdown of cell components, a type of intracellular pinocytosis or phagocytosis might be involved ("autophagy", de Duve, 1963a, b).

Thus the problem of the lysosomes is closely related to fields of inquiry which have been known to cytologists and physiologists for many years under the terms of pinocytosis, phagocytosis, and "*Speicherung*". Many investigations dating back over 50 years showed that macrophages and endothelial cells in spleen, lymph nodes, bone marrow, thymus, liver, connective tissue and some other organs could take up such diverse materials as bacteria, colloidal carbon, metal oxides, dyes, and proteins, and segregate them in intracellular granules or vacuoles. Granules in the epithelial cells of the kidney and intestine were found to take part in the "*Speicherung*" of ingested material (Metchnikoff, 1892; Aschoff, 1924; Möllendorff, 1920, 1925; Gérard and Cordier, 1934; Kedrowski, 1933a, b). These properties were related to the defence of the organism against harmful agents.

Whereas phagocytosis, i.e., the ingestion of solid particles, was studied mainly in mammalian macrophages and leucocytes, pinocytosis, i.e., the ingestion of fluid droplets, was first observed in tissue culture cells by Lewis (1931, 1937) and in amoebae by Mast and Doyle (1934). During the last 10 years better understanding of pinocytosis is mainly due to pioneering investigations on amoebae by Holter (1960, 1961) and by Chapman-Andresen. Chapman-Andresen (1962), in her recent studies on pinocytosis in amoebae has also briefly reviewed ingestion processes in other types of cells. The submicroscopic aspects underlying pinocytosis and phagocytosis are essentially similar. Therefore, a general term, "phagosome" (Straus, 1958, 1959) may be used to designate granules in which material ingested by pinocytosis or phagocytosis is segregated.

A. MICROSCOPIC AND SUBMICROSCOPIC ASPECTS OF INGESTION PROCESSES

During the last few years, a better understanding of ingestion processes has been obtained, mainly by electron microscopy. Application of fluorescence microscopy and cytochemistry have also led to useful new information. On the basis of these observations, the following stages of ingestion processes could be distinguished: (a) adsorption of macromolecular or colloidal material on to the cell surface and formation of pinocytic or phagocytic vesicles (phagosomes) containing this material, (b) fusion of phagosomes with lysosomes, (c) intracellular transport or transport across cells by small phagosomes.

1. Adsorption Phase and Formation of Micropinocytic Vesicles

Observations by light microscopy of cells in tissue culture or of amoebae show the appearance of funnels or invaginations of the cell surface, or the engulfment of droplets of the surrounding medium by thin extensions of the cell surface. By electron microscopy, pinocytic vesicles of submicroscopic size (Clark, 1957, 1959; Miller, 1960; Farquhar and Palade, 1961) or larger vacuoles of microscopic size (Ericsson, 1964) could be distinguished close to the cell surface. It is now generally recognized that macromolecular material, prior to its ingestion, is adsorbed onto a layer of the cell surface. Thus, Brandt and Pappas (1960, 1962) and Nachmias and Marshall (1962) demonstrated by electron microscopy the adsorption of colloidal particles of ThO_2 and ferritin onto the surface layer of amoeba, and Brandt (1958) and Chapman-Andresen (1962) described adsorption of fluorescent-labelled proteins onto this layer. Jansco (1955) demonstrated adsorption of colloidal gold onto the surface of Kupffer cells of liver. The adsorption of proteins onto the cell surface of mammalian cells was reported by Ryser, Aub and Caulfield (1962), Easton, Goldberg and Green (1962a), Cormack and Ambrose (1962), Straus (1962a, 1963) and Aronow, Danon, Shahar and Aronson (1964). It would be of considerable interest to determine the chemical nature of the surface layer of absorbing cells. Mucopolysaccharides are often considered to be characteristic constituents of this layer (Bairati and Lehmann, 1953; Brandt, 1962; Bennett, 1963, Rambourg, Neutra and Leblond, 1966). O'Neill (1964) succeeded in isolating the surface coat of *Amoeba proteus*, after shrinking the cells by concentrated sucrose solution. The surface preparation contained 35% lipid, 26% protein, and 16% polysaccharide. No sialic acid, which is present in the surface membranes of mammalian cells, was found (O'Neill, 1964).

Following the adsorption of material onto the cell surface, small pinocytic vesicles develop which transport the adsorbed materials into the cells. This process may be related to "membrane vesiculation" as suggested by Bennett (1956). The vesicles often coalesce into larger vacuoles in which the ingested material accumulates. In some cases, phagocytic vacuoles (phagosomes) of relatively large size seem to develop directly at the cell surface (Straus, 1962a, 1964a; Ericsson, 1964). Interesting details of structures forming at the cell surface during thc uptake of ThO_2 by the endothelial cells of the liver may be seen in electron micrographs of Törö, Rusza and Röhlich (1962), during the uptake of blood proteins by oocytes in electron micrographs by Roth and Porter (1964) and during the ingestion of ferritin in spinal ganglion cells of toads in electron micrographs of Rosenbluth and Wissig (1964). In recent work, when glutaraldehyde instead of OsO_4 was used for the fixation of tissues for electron microscopy, microtubules were detected in many cells. According to Sandborn, Koen, McNabb and Moore (1964), these structures may serve in the transport of fluids or of suspended solids, and micropinocytic vesicles may represent dilatations of this microtubular system. It is interesting to note that microtubules were also seen in lysosomes and phagosomes. Journey (1964) demonstrated their presence by electron microscopy in lysosomes and phagocytic vacuoles of peritoneal macrophages, and Boler and Arhelger (1966) found them in lysosomes of rabbit kidney cells but not in the kidneys of rats and humans. (A diagram showing the relationships of the various structures is given in Fig. 2.)

2. Fusion of Phagosomes with Lysosomes

The first observations on the fusion of pinocytized fluid droplets with structures, probably lysosomes, were made by Robineaux and Frederic (1955) on leucocytes, and by Rose (1957) on HeLa cells in tissue culture, using phase contrast cinematography. Hirsch (1962) who identified the specific granules of leucocytes as lysosomes (Cohn and Hirsch, 1960a), also observed by phase contrast cinematography the fusion of phagocytic vacuoles with lysosomes after ingestion of bacteria into these cells. The relationship between lysosomes and phagosomes in the tubule cells of the kidney and in Kupffer cells of the liver was studied by the writer (Straus, 1963, 1964a, b), and the fusion of both types of granules was reported. This could be demonstrated by staining phagosomes and lysosomes in contrasting colours in the same cells, after injection of plant peroxidase (Section IX. c.). It should be noted that the size of phagosomes and lysosomes which fuse may vary widely from sub-microscopic to microscopic dimensions. Thus, small or large phagosomes may fuse with small or large lysosomes. Recently, the fusion of phagosomes with lysosomes in the tubular cells of the kidney was also demonstrated by electron microscopy by Ericsson (1964) after *intravenous* injection of homologous haemoglobin into male rats. Figures 9 and 10 are electron micrographs made by Dr. J. E. L. Ericsson (Ericsson, 1964, figures 101 and 102) showing kidney cells of animals injected intravenously with homologous haemoglobin. In Fig. 9, a lysosome (acid phosphatase reaction product) lies close to a phagosome, and in Fig. 10, fusion of a lysosome with a phagosome seems to occur. Miller and Palade (1964) reported the simultaneous appearance of *intraperitoneally* injected haemoglobin together with the acid phosphatase reaction product in granules of kidney cells in a study by electron microscopy. They did not detect fusion of large phagosomes with pre-existing lysosomes. It may be suggested that in the experiments of Miller and Palade, haemoglobin after *intraperitoneal* injection into female mice was reabsorbed gradually over an extended period of time. Under these conditions, mainly *small* phagosomes (micropinocytic vesicles) probably fused with pre-existing lysosomes, as was also suggested by Miller and Palade. The fusion of *large* phagosomes (phagocytic vacuoles) with lysosomes probably occurred infrequently or not at all. After *intravenous* injection of proteins, on the other hand, the initial load of the injected protein is much higher than after

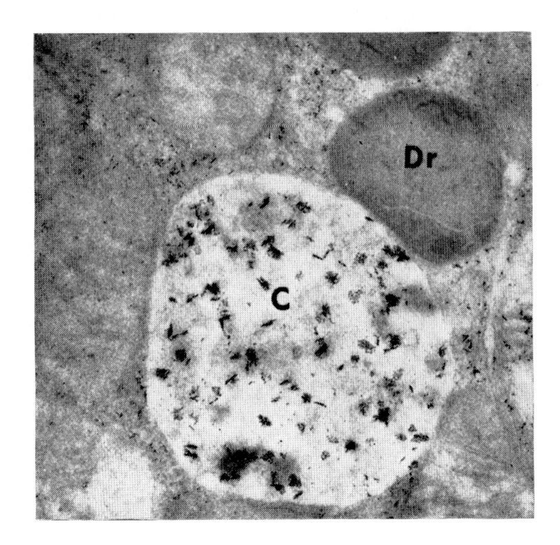

FIG. 9. Two hr after injection of homologous haemoglobin. The lysosome (cytosome) (C), incubated in the Gomori medium for acid phosphatase appears light after post-fixation in 2% osmium tetroxide buffered with s-collidine (Ericsson, 1964). The phagosome (Dr) containing reabsorbed haemoglobin appears dark. ×27,000. (Reprinted with permission from *Acta path. Microbiol. Scand.*; cf. Ericsson, 1964.)

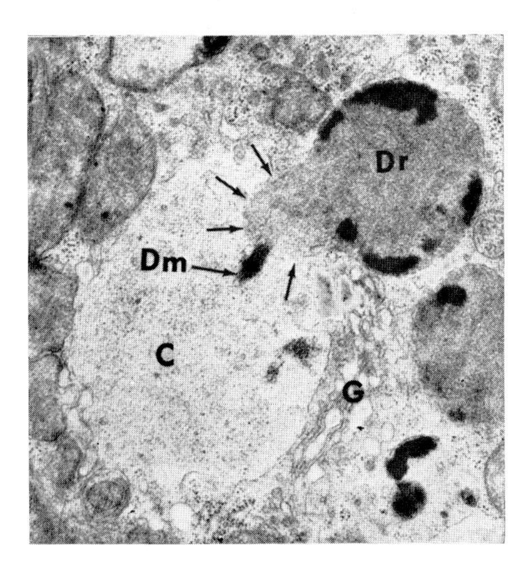

FIG. 10. Four hr after injection of homologous haemoglobin. The lysosome (C) appears light and the phagosome (Dr) appears dark. Some of the dense material (Dm) from the droplet seems to have been delivered to the lysosome. ×14,700. (Reprinted from *Acta path. Microbiol. Scand.*; cf. Ericsson, 1964.)

intraperitoneal injection, and many large phagosomes fuse with pre-existing lysosomes at about the same time (20 to 40 minutes after intravenous injection of peroxidase (Straus, 1963, 1964a, b) thus facilitating detection of fusion by electron microscopy (Ericsson, 1964). In addition, the glomerulus of male rats may be more permeable to proteins than the glomerulus of female mice. It should also be noted that the molecular weight of peroxidase (44,000) is lower than that of haemoglobin (68,000), and that therefore more peroxidase than haemoglobin may pass the glomerulus, and arrives at the absorbing surfaces of the epithelial cells. Fusion of phagosomes containing ingested bacteria with lysosomes was also observed by electron microscopy in macrophages and in leucocytes (North and Mackaness, 1963a, b; Lockwood and Allison, 1964a, b; Zucker-Franklin and Hirsch, 1964; Aronow *et al.*, 1964; Horn, Spicer and Wetzel, 1964). Gordon, Miller and Bensch (1965) showed that newly formed phagosomes, labelled with gold-DNA-protein complexes, fused with late phago-lysosomes, pre-labelled with colloidal iron, in fibroblasts in culture. In studying the fusion of cytoplasmic granules, it has to be considered that this probably is a rapid process (see Hirsch, 1962) and therefore may be difficult to detect in fixed tissue. Section VIII. A. will discuss the fusion of lysosomes and phagosomes in pathologically altered tissue to form enlarged phagolysosomes.

Food vacuoles of protozoa may be considered to be types of phagolysosomes. Rosenbaum and Rolon (1960a) and Müller, Röhlich, Tóth and Törö (1963) investigated the appearance of lysosomal enzymes in food vacuoles by cytochemical methods. Although the transfer of the enzymes into the vacuoles has not yet been studied in detail, a close analogy with the fusion of lysosomes and phagosomes in mammalian cells may be expected.

An important new aspect of the fusion of lysosomes with phagosomes was revealed by the work of Wollman *et al.* (1964) on thyroid cells. Whereas the phago-lysosomes mentioned in the preceding paragraphs functioned in the segregation and degradation of *exogenous* material, the phago-lysosomes in the thyroid cells seem to function in the degradation of an *endogenous* protein containing the precursor of a hormone. As was shown by Wollman *et al.* (1964) and by Nadler, Sarkar and Leblond (1962), colloid material containing thyroglobulin is ingested from the lumen into the thyroid cells by pinocytosis and accumulates in apically located droplets. These may be considered to be types of phagosomes. Wollman and co-workers induced the new formation of colloid droplets by injection of thyroid stimulating hormone (TSH) into hypophysectomized animals, and they identified lysosomes by the cytochemical reaction for acid phosphatase and esterase. Characteristic changes in the location of granules suggested a fusion of lysosomes with phagosomes. The investigators discussed the possibility that thyroxine is split from thyroglobulin within the phago-lysosomes, and is then released. The relationship between colloid droplets and lysosomes in the thyroid was investigated with the electron microscope by Wetzel, Spicer and Wollman (1965). Since lysosomes are numerous in the epithelial cells of endocrine glands, the liberation of active portions of hormones, after partial degradation in phago-lysosomes, may take place in other endocrine organs. This possibility was also discussed by Jung and Nüsslein (1962) when they observed a high concentration of proteolytic enzymes in endocrine tissues.

3. Intracellular Transport and Transport across Cells by small Phagosomes

Several investigators have observed that proteins and colloidal particles can be transported by micropinocytic vesicles across endothelial cells of blood vessels (Alksne, 1959; Farquhar and Palade, 1961, 1962; Staubesand, 1963; Fernando and Movat, 1964; Palade and Bruns, 1964; Casley-Smith, 1965, see also Majno, 1965). Moore and Ruska (1957) proposed the term "cytopempsis" for transport across cells. Transport of fat (Palay and Karlin 1959) and of proteins (Clark, 1959) in small vesicles across the intestinal epithelial cells was observed by electron microscopy. The transport of injected plant peroxidase across the epithelial cells of the liver, probably by small pinocytic vesicles, was suggested from cytochemical observations (Straus, 1962a, 1963, 1964a). In the tubule cells of the kidney, small pinocytic vesicles containing injected peroxidase or haemoglobin probably originate at the base of the brush border and fuse into large apical phagosomes (Miller, 1960; Straus, 1961, 1962a, Ericsson, 1964). In addition small phagosomes may originate from the large apical phagosomes themselves (Straus, 1964b). This latter process may be similar to the formation of small vesicles from food vacuoles as observed in protozoa by Roth (1960) and by other investigators. It was suggested that the injected peroxidase was transported by these small phagosomes either to pre-existing lysosomes, or that it was transported across the cells to the infoldings of the basal cell membranes and was brought back to the peritubular capillaries or spaces. (Straus, 1961, 1962a, 1964b). Menefee, Mueller, Miller, Myers and Bell (1964b) made observations that suggested transport of injected globin by small pinocytic vesicles across the tubular epithelial cells of the kidney.

It is of interest that Marchesi and Barrnett (1963, 1964) demonstrated nucleoside phosphatase activity in micropinocytic vesicles of some but not all endothelial cells. These investigators discussed the possible relationship of this enzyme to transport processes. Pappas and Kaye (1964) reported ATPase activity on the lateral membranes and in intercellular spaces, but not in the micropinocytic vesicles, of the corneal endothelium. Adenosine triphosphatase was also present in the apical infolding of the epithelium in the ciliary processes of the eye (Pappas and Kaye, 1964).

B. UPTAKE OF PROTEINS

Although it is now accepted by most investigators that ingested proteins are segregated in phagosomes and phago-lysosomes, the mitochondria were thought for several years to be sites of protein uptake (Zollinger, 1950; Oliver, McDowell and Lee, 1954; Rhodin, 1954; Straus and Oliver, 1955; Haurowitz, Reller and Walter, 1955; Gansler and Rouiller, 1956). This opinion is still encountered in recent publications (Fisher, 1964; Menefee, Mueller, Miller, Myers and Bell, 1964). However, the hypothesis that mitochondria take up proteins was questioned by Gitlin, Landing and Whipple (1951), Straus (1954, 1956, 1957a, b, 1959), Ingraham (1955), and Jansco (1955). The sequestration of mitochondria together with phagolysosomes in "composite bodies" in some cases (Section VI. and VIII.) may explain, in part, the earlier confusion.

The problem of protein uptake, specially in kidney cells and amoebae, was clarified to a large extent by electron microscopy. As was mentioned in Section V. A. 1, proteins are adsorbed onto the surface layers of pinocytizing cells, and small vesicles which seem to pinch off from infoldings of the surface membranes carry the proteins to apical vacuoles. The electron micrographs by Miller (1960) and Ericsson (1964) showing absorption of haemoglobin in the tubule cells of the kidney, and those by Roth and Porter (1964) showing uptake of proteins by oocytes, illustrate in a striking way the submicroscopic aspects of this process. Small and large phagosomes probably are related to the smooth-surfaced endoplasmic reticulum seen in electron micrographs. It is not known whether or not ingested proteins can also penetrate to the rough-surfaced endoplasmic reticulum. If antigens can penetrate into cell nuclei as reported by Coons (1956) and Wellensiek and Coons (1964), the passage of proteins through the rough-surfaced endoplasmic reticulum is possible. As was pointed out by Holtzer and Holtzer (1960) and by Straus (1964d), nuclei may adsorb proteins during fixation or isolation, and hence the apparent localization of injected proteins (antigens) in cell nuclei may be an artifact.

In the following paragraphs, the uptake of proteins by various tissues is reviewed briefly.

1. Cells of the Reticulo-Endothelial System

Kedrowski (1933b) in his investigations on the "segregation apparatus" in cells of the reticulo-endothelial system also demonstrated the uptake of haemoglobin in granules of macrophages and histiocytes of the connective tissue. Sabin (1939), Kruse and McMaster (1949), and Latta, Gitlin and Janeway (1951) noted uptake of dye-protein complexes in granules of the reticulo-endothelial cells throughout the organism. Schiller, Schayer and Hess (1953), and Mancini *et al.* (1961) using fluorescent-labeled proteins observed their deposition in the cells of the reticulo-endothelial system. Jansco and Jansco-Gabor (1952, 1954) made interesting observations on the uptake of proteins by macrophages and histiocytes and the effect of histamine on it. Straus (1964a, Fig. 8 and 1964d, Fig. 10) demonstrated the uptake of injected plant peroxidase by "phago-lysosomes" of macrophages in liver (Kupffer cells) and spleen. In Figs. 5–7 of the present article the uptake of peroxidase by phago-lysosomes in macrophages of thymus, lung, and prostate (connective tissue) is illustrated. In this context, it may be of interest to mention the probable existence of phagosomes (peroxidase uptake) in mast cells (Fig. 11; see legend to Fig. 11 for further comment, and Section IX. C. for more details on the peroxidase technique).

2. *Kidney Cells*

The ability of kidney cells of amphibians to segregate proteins in intracellular granules was investigated by Lambert (1936). Randerath (1947) discussed the uptake of abnormal proteins (Bence-Jones) by human kidney. The early literature on "hyaline droplets" in relation to protein uptake was reviewed by Rather (1952) who himself investigated the re-absorption of haemoglobin and its effects on droplet formation (Rather, 1948). Smetana (1947) and Kruse and McMaster (1949) observed deeply-stained droplets in kidney cells after injection of dye-protein complexes. Oliver *et al.* (1954) investigated the appearance of protein absorption droplets by staining with iron-haematoxylin and the Gram method. Straus (1957b, 1959, 1961, 1962a, 1964a, b) studied the uptake of plant peroxidase in "phagosomes" of the tubule cells and their differentiation from mitochondria. Niemi and Pearse (1960) demonstrated the appearance of injected egg white conjugated with a fluorescent dye in kidney droplets and the different localization of protein absorption droplets and mitochondria. Miller (1960) and Miller and Palade (1964) investigated the reabsorption of heterologous haemoglobin, and Ericsson (1964) studied the uptake of injected homologous haemoglobin and the development of phagosomes and phago-lysosomes.

The formation of protein absorption droplets in the cells of the glomerulus was investigated by electron microscopy after injection of ferritin by Farquhar and Palade (1960, 1961, 1962), and by the fluorescent antibody technique by Kurtz and Feldman (1962). Aggregates of injected globin were observed in the glomerulus during passage through the basement membrane, by Menefee and co-workers (1964a).

The uptake of fluorescent-labelled proteins by kidney slices was reported by Holtzer and Holtzer (1960), and the uptake of plant peroxidase by Miller *et al.* (1965).

3. *Liver Parenchymal Cells*

In contrast to the Kupffer cells, most investigators have found that liver parenchymal cells do not take up proteins in significant amounts. Mayersbach (1957), however, concluded that fluorescent-labelled proteins entered the liver cells. Straus (1962a, 1963, 1964a) observed that only small amounts of injected plant peroxidase entered the normal liver cells, especially in the peri-portal areas, and appeared in phago-lysosomes close to the bile capillaries (see also Section VIII. A.).

4. Intestinal Epithelial Cells of New-born Animals

As was shown by Brambell (1958), antibodies from the milk of the mother are absorbed through the mucosa of the intestine of new-born animals, and this process ceases after 10–20 days. Clark (1959) and Payne and Marsh (1962) fed fluorescent-labelled proteins to new-born rats, mice and pigs and demonstrated their localization in large vacuoles of the intestinal epithelial cells. Figures 12 and 13 show that plant peroxidase, fed to new-born rats appears in apical phagosomes and, later, in large phago-lysosomes.

5. Nerve Cells

Rosenbluth and Wissig (1964) observed by electron microscopy that nerve cells of the toad take up ferritin by pinocytosis. It will be interesting to see whether the capacity for pinocytosis of proteins is a general property of nerve cells.

6. Erythroblasts

Bessis (1963) showed that erythroblasts take up ferritin by pinocytosis from the surrounding reticulo-endothelial cells of the bone marrow.

7. Embryonic Cells

The uptake of proteins by embryonic cells was reviewed by Schechtman (1956) and by Brambell (1958). Telfer (1961) studied the uptake of blood proteins by oocytes of saturniid moths by staining with fluorescent-labelled antibodies. Roth and Porter (1964) investigated by electron microscopy the uptake of protein by oocytes of a mosquito. In these species, proteins formed outside the egg are absorbed into the egg from the blood; in other species, yolk protein is synthesized in the egg itself. Dalq (1963), on the basis of his own work and that of Pasteels and Mulnard, presented an interesting discussion on the formation of yolk bodies in relation to phagosomes and lysosomes.

8. Tumour Cells In vivo

Easty (1964) tested the uptake of two fluorescent-labelled proteins by 12 different tumours and did not observe a significant uptake by most of them. Easty reported uptake mainly by the cells of the reticulo-endothelial system and by the tubule cells of the kidney. However, Raimondi (1964) observed by electron microscopy that radio-iodinated serum albumin, injected into two patients with brain tumour (astrocytoma), appeared in pinocytic vesicles of the neoplastic and endothelial cells.

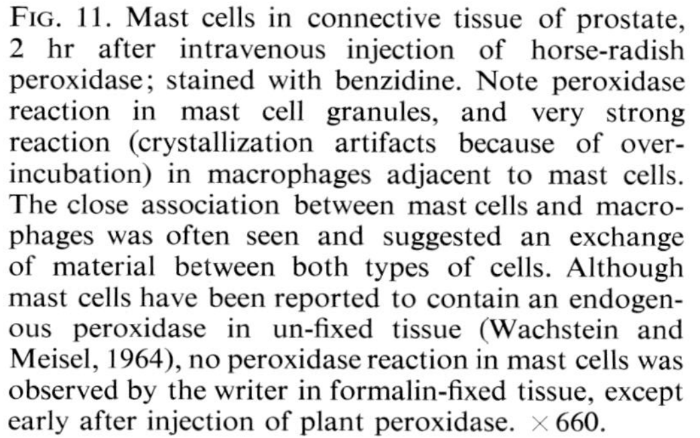

FIG. 11. Mast cells in connective tissue of prostate, 2 hr after intravenous injection of horse-radish peroxidase; stained with benzidine. Note peroxidase reaction in mast cell granules, and very strong reaction (crystallization artifacts because of over-incubation) in macrophages adjacent to mast cells. The close association between mast cells and macrophages was often seen and suggested an exchange of material between both types of cells. Although mast cells have been reported to contain an endogenous peroxidase in un-fixed tissue (Wachstein and Meisel, 1964), no peroxidase reaction in mast cells was observed by the writer in formalin-fixed tissue, except early after injection of plant peroxidase. ×660.

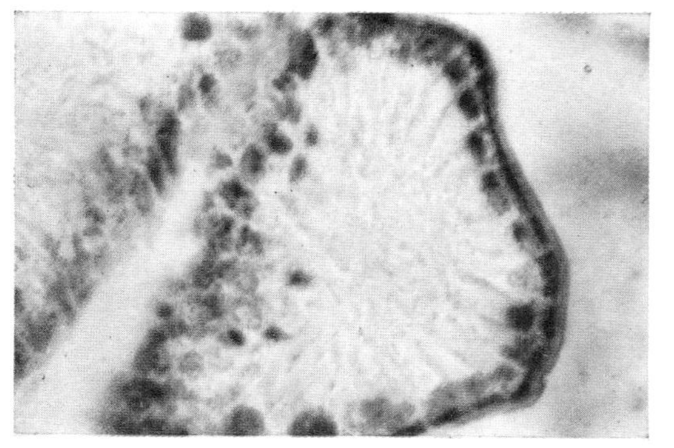

FIG. 12. Phagosomes in intestinal epithelial cells of newborn rat, 2 hr after feeding plant peroxidase by mouth; staining with benzidine. Note newly-forming phagosomes at the base of brush border, and larger phagosomes in the interior of cells. ×660.

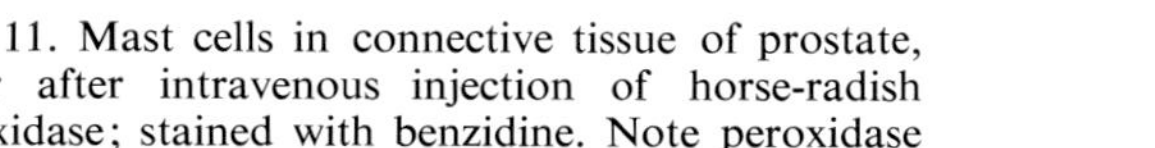

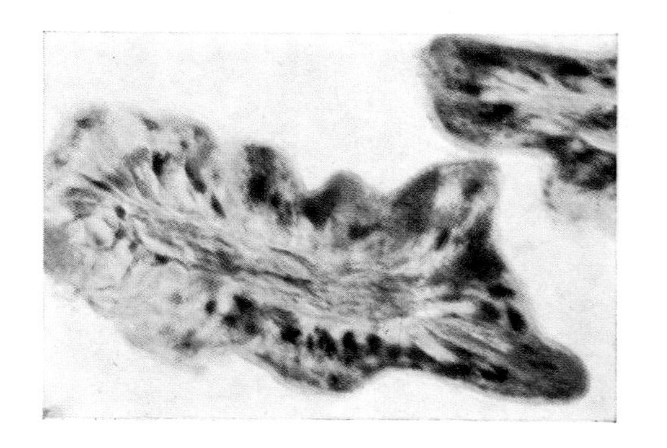

FIG. 13. Phago-lysosomes in intestinal epithelial cells of newborn rat, 3 hr after feeding horse-radish peroxidase by mouth; double staining for acid phosphatase and peroxidase. Note blue and red reaction products for both enzymes in the same granules. ×660.

9. Amoebae and Lower Animals

The uptake of proteins by amoebae has been discussed in the excellent reviews by Holter (1960, 1961) and by Chapman-Andresen (1962). The uptake of egg albumin in food vacuoles of protozoa and the relationship to lysosomes was investigated by Müller *et al.* (1963). The uptake of horse-radish peroxidase in the kidney phagosomes of fishes was investigated by Zeidenberg and Janoff (1964), in the gland cells of snails by Rosenbaum and Ditzion (1963), and in planarians by Rosenbaum and Rolon (1960b) and by Osborne and Miller (1962).

10. Cells in Tissue Culture and In vitro

It is known from the work of Lewis (1931, 1937) that tissue culture cells are very active in pinocytosis. Ryser, Caulfield and Aub (1962) and Easton *et al.* (1962a) observed the uptake of ferritin and ferritin-antibody complexes by ascites tumour cells in tissue culture. Ryser *et al.* (1962a) and Ryser (1963) elaborated a quantitative method to measure uptake of I^{131}-labelled albumin. Robineaux and Pinet (1960), Cormack, Easty and Ambrose (1961), Cormack and Ambrose (1962), and Easty, Yarnell and Andrews (1964) investigated the uptake of fluorescent-labeled proteins by kidney cells and tumour cells in tissue culture. Maeir (1961) studied the uptake of horse-radish peroxidase by peritoneal monocytes in culture and Chapman-Andresen (1957) demonstrated uptake of fluorescent-labelled protein by leucocytes *in vitro*.

C. UPTAKE OF LIPIDS

Many investigations in recent years have been concerned with the process by which lipids are taken up by various cells. Although several observations seem to indicate that fat particles are taken up by pinocytosis or phagocytosis, no experiments have been reported involving the lysosomes in the degradation of the ingested fat. The question as to whether or not lipolytic enzymes are present in phago-lysosomes is mentioned briefly in Section V. K.

The technical difficulties in the study of fat uptake should be mentioned. The identification of ingested fatty substances, by staining with Sudan, by radioactive labels, or in electron micrographs, is often difficult. It is not possible to establish in electron micrographs, for example, whether fatty material in droplets or small vesicles represent triglycerides, monoglycerides, or fatty acids. Artifacts of adsorption or redistribution *in vitro* may occur in cell fractionation experiments. An important difficulty arises from the rapid metabolic changes which fat seems to undergo during, or soon after ingestion. Contradictory results reported in the literature may also be due to differences in the physical state and in the composition of various lipid preparations administered to animals. Various preparations of triglycerides, fatty acids, chylomicra, lipoproteins, and cholesterol, used by different investigators, may have differed in the degree of dispersion (particle size) or in their hydrophilic properties, and this may have influenced their uptake.

1. Uptake of Fat by Reticulo-Endothelial Cells

Uptake of emulsified cholesterol, cedar oil, and olive oil by phagocytic cells of the connective tissue was reported by Kedrowski (1933a) in his studies on the "segregation apparatus". Whereas colloidal carbon, silver, or trypan blue, after simultaneous injection, often appeared in the same granules of macrophages and fibroblasts, the ingested fat was localized in separate granules. More recently, the uptake of fat by cells of the reticulo-endothelial system was studied by Waddell, Geyer, Clarke and Stare (1954), French and Morris (1960), and by Day (1964). Elsbach (1964, 1965a, b) concluded that leucocytes and macrophages took up free fatty acid only, and that triglycerides were hydrolysed and re-esterified *at the surface* of macrophages prior to being ingested. Casley-Smith and Day (1966) demonstrated by electrou microscopy and radioactive assay that lipids and lipoproteins were taken up by macrophages *in vitro* into small and large vesicles. They concluded that the degradation of the fatty material took place within the vesicles and not extracellularly.

The possible relationship of fat uptake to the development of atherosclerotic changes has been the subject of many investigations. It was observed by electron microscopy that macrophages filled with fat accumulated in atherosclerotic lesions (Balis, Houst and More, 1964; Suzuki, Greenberg, Adams and O'Neal, 1964). Increased phagocytic activity and fat uptake by endothelial cells of atherosclerotic aortae was reported by Parker (1960), Friedman, Byers and St. George (1962), and Hess and Staubli (1963). The latter authors noticed greatly increased acid phosphatase activity in endothelial cells during fat accumulation. Gonzales (1963) studied atherosclerotic lesions histochemically and reported the presence of lipid-laden cells in the intima with pronounced cytochemical reactions for acid phosphatase and esterase. Although lipids can be synthesized in the aorta (Zilversmit, McCandless, Jordan, Henly and Ackerman, 1961), the opinion is often expressed in the literature that the excess of lipid and cholesterol in atherosclerotic lesions of the aorta is derived from the blood. If this is correct, lysosomes and phagosomes of endothelial cells and macrophages may play a role in atherosclerotic changes.

2. Uptake of Fat by the Intestine

Palay and Karlin (1959) and Palay and Revel (1964) reported the presence of micropinocytic vesicles containing absorbed fat in the intestinal epithelial cells. However, Lacy and Taylor (1962) and Rostgaard and Barrnett (1965) demonstrated in electron micrographs that fat appeared in micellar form in intestinal epithelial cells during absorption. As suggested by Ladman,

Padykula and Strauss (1963) and by Strauss (1963), pinocytosis may play only a minor role in the absorption of fat by the intestinal cells. Barka (1964) did not observe significant changes in the lysosomes of intestinal epithelial cells during fat absorption.

3. Uptake of Fat by Liver Cells

Ashworth, Stembridge and Sanders (1960) and Ashworth (1963) detected relatively little lipid in micropinocytic vesicles of liver cells, after feeding to rats. Jordan (1964) and Trotter (1965) reported increased pinocytosis of fat after hepatectomy. Rodbell, Scow and Chernick (1964) studied the uptake of radioactive-labelled triglyceride emulsions by perfused liver, and concluded that fat was removed from the blood and taken up by liver cells without de-esterification. McBride and Korn (1964) and Robinson (1964) reported that triglycerides in chylomicra were taken up by liver cells rapidly *in vivo*. However, Green and Webb (1964) concluded from experiments with isolated liver cells that fat was adsorbed and broken down at the surface of liver cells before being ingested. Information on the uptake of lipids in liver and in other cells may be found in the publications of a recent symposium on lipid transport (Meng, 1964).

D. UPTAKE OF NUCLEIC ACIDS AND VIRUSES

The uptake of foreign nucleic acid by mammalian cells has been investigated by several authors in studies on whether the genetic properties of cells can be altered by such means. Such an effect would be similar to the influence of viruses on the genetic apparatus of the cell.

1. Uptake of Nucleic Acids

Bensch and King (1961) observed by autoradiography that heterologous DNA, complexed with protein, but not DNA alone, was taken up by strain L fibroblasts in tissue culture. Borenfreund and Bendich (1961) reported uptake of heterologous, tritium-labelled DNA by HeLa cells in tissue culture, and the appearance of radioactivity in the nucleus. Chorazy, Bendich, Borenfreund, Ittensohn and Hutchison (1963) investigated by autoradiography the uptake of tritium-labelled chromosomes, isolated from mouse leukemia cells, by macrophages, HeLa cells, and fibroblasts. Although the radioactivity during the first few hours was mainly in the cytoplasm, some activity appeared in the nucleus after 16 hr. Other authors also noted radioactivity in the nucleus after treatment with homologous or heterologous DNA. However, it was not certain whether whole DNA molecules, or degradation products had entered the nucleus. Bensch, Gordon and Miller (1964) and Gordon, Miller and Bensch (1965) studied the ingestion of DNA protein coacervates by strain L fibroblasts, using electron microscopy combined with staining for acid phosphatase and esterase. The ingested DNA-protein complex, labelled by colloidal gold, could be recognized within the phagocytic vacuoles. The fusion of phagosomes with small Golgi vesicles or with dense bodies, and the digestion of the DNA-protein complex within the phago-lysosomes, were demonstrated by these investigators in a striking series of electron micrographs (see also Fig. 8 of this paper).

2. Uptake of Viruses

The process by which virus particles are taken up by mammalian cells has been clarified to a certain extent by Dales and Choppin (1962), Dales (1962, 1963), and by Epstein, Hummeler and Berkaloff (1964). These investigators showed by electron microscopy and autoradiography that influenza, vaccinia, herpes, and adeno-virus were adsorbed onto the surface of HeLa cells and, after ingestion by membrane invagination, appeared within phagocytic vesicles and vacuoles. The virus particles, within the vacuoles lost an outer coat derived from the surface membranes of the cells of origin. The stripped central portion of the virus left the phagocytic vacuoles by an unknown process and appeared free in the cytoplasm. The passage of the virus core out of the vacuoles distinguished the virus from other macomolecular or colloidal particles. Colloidal gold particles, for example, used by Epstein *et al.* (1964) as a comparison, remained within the vesicles. The possibility was discussed by Epstein and co-workers and by Dales and Kajioka (1964) that the outer coat of the virus particles was digested within the vacuoles by lysosomal enzymes.

E. UPTAKE OF OTHER MACROMOLECULAR AND COLLOIDAL PARTICLES

Over the last 50 years the segregation of particles of colloidal carbon, metals, bacteria, or other foreign substances in granules of the cells of the reticulo-endothelial system or of absorbing epithelial cells has been studied by many investigators. The rate of clearance from the blood of such particles has often been determined as a measure of uptake by the reticulo-endothelial system. The work of Kedrowski (1933a, b) on macrophages of the connective tissue, and of Gérard and Cordier (1934) on "athrocytosis" of colloidal particles by the epithelial cells of frog kidneys, has already been mentioned.

The relationship between cell granules in which foreign material is sequestered, and lysosomes has been recognized only recently. By analogy with the fusion of lysosomes with phagocytic vacuoles containing ingested bacteria (Zucker-Franklin and Hirsch, 1964) and phagosomes containing ingested proteins (Straus, 1963, 1964a, b; Ericsson, 1964), it may be suspected that newly formed phagosomes containing other particles also fuse with pre-existing lysosomes (Section V. A. 2.). In recent studies by electron microscopy a relationship was suggested between lysosomes (or better, acid phosphatase-positive dense granules) and the uptake of colloidal ThO_2 by frog kidney cells (Trump, 1961), the uptake of colloidal HgS in carcinoma cells *in vitro* (Takahashi and Mottet, 1962), and the uptake of dextran or iron-dextran by liver cells (Daems, 1962; Kent *et al.*, 1963a, b).

By analogy with the role of small phagosomes in protein transport (Section V. A. 3.), it may be expected that small phagosomes containing ingested colloidal particles either transport the particles to pre-existing lysosomes, or, independent of lysosomes, transport the particles across and out of the cells. In electron micrographs of certain endothelial cells, many small phagosomes but not many lysosomes can be seen. Electron micrographs of such cells show ingested colloidal thorotrast, gold, or iron oxide within many micropinocytic vesicles (Palade, 1961; Kaye and Pappas, 1962; Jennings, Marchesi and Florey, 1962; Pappas and Tennyson, 1962; Palade and Bruns 1964). Data on the transport of ingested materials across endothelial cells may be found in the excellent review by Majno (1965). Two studies by electron microscopy on the transport of colloidal particles across epithelial cells may be mentioned: transport of polystyrene particles across the intestinal epithelial cells (Sanders and Ashworth, 1961), and transport of colloidal HgS and ThO_2 across the liver parenchymal cells (Hampton, 1958).

F. UPTAKE OF VITAL DYES

Many early investigations were concerned with the concentration ("*Speicherung*") of vital stains in intracellular granules or vacuoles, after administration to living animals. Studies on the uptake of neutral red have given rise to many controversies, especially concerning the relationship of neutral red granules to the Golgi apparatus and to mitochondria. Pioneering work in this field was performed by Möllendorff (1918, 1920, 1925), Evans and Scott (1921), and Chlopin (1927, 1928). Recently, vital staining was discussed in an excellent review by Stockinger (1964).

The close relation between the uptake of vital dyes and the uptake of proteins and colloidal materials may be seen from the fact that all seem to be concentrated in granules of the same type of cells: the cells of the reticulo-endothelial system (macrophages and endothelial cells) and the epithelial cells of the kidney and intestine.

A distinction has to be made between the uptake of acid dyes of low degree of dispersion such as trypan blue or lithium carmin, and the uptake of highly dispersed basic dyes, such as neutral red or toluidine blue. After injection into animals, the former probably are bound to serum proteins and are then taken up by pinocytosis as protein-dye complexes, and they accumulate in phagosomes and phago-lysosomes (Kojima and Imai, 1962; Schmidt, 1962). The mechanism of ingestion of highly dispersed basic dyes, such as neutral red or acridine orange, is less well understood. As was pointed out by Mitchison (1950), Rustad (1959), and Chapman-Andresen (1962), basic dyes (such as toluidine blue, basic fuchsin, neutral red and acridine orange) are adsorbed and oriented at the surface of amoebae, and they induce pinocytosis. Andresen (1945a, b) demonstrated the formation *de novo* of neutral red vacuoles during vital staining of amoebae. Quertier and Brachet (1959) studied in amoebae the formation of metachromatic vacuoles containing ingested toluidine blue and discussed the relation to lysosomes and phagosomes. In some cases, the adsorption of the positive dyes onto the negatively charged surface coat of cells may induce the formation of phagosomes, as during pinocytosis of proteins. In other cases, however, basic dyes were observed to stain the cytoplasm in a diffuse manner before they were segregated in vacuoles or granules (Wittekind, 1958; Schmidt, 1962; Tanaka, 1962; Stockinger, 1964; Robbins, Marcus and Gonatas, 1964). It is not certain whether the diffuse appearance of the ingested dye is due to its localization in sub-microscopic, pinocytic vesicles, or because it dissolves in the soluble phase of the cytoplasm. Tanaka (1962) thought that the entry of basic dyes into the cytoplasm was different from pinocytosis. The subsequent segregation of basic dyes in granules or vacuoles ("segresomes", Tanaka, 1962), however, may occur by a kind of intracellular pinocytosis (see also Robbins *et al.*, 1964). In this case, the dyes might first accumulate in newly formed "segresomes" and the latter then fuse with pre-existing lysosomes. The relationship of the granules that contain dye to lysosomes was shown clearly in the investigations by Schmidt (1962), Koenig (1963b), Stockinger (1964), and Robbins *et al.* (1964). Earlier, Weiss (1955) and Trump (1961) had investigated the appearance of injected trypan blue and neutral red in bodies corresponding to cytosomes in electron micrographs of pancreas and kidney cells. Byrne (1964a, b) studied neutral red granules of mouse pancreas by electron microscopy and demonstrated the reaction product of acid phosphatase in the granules. It may also be mentioned in this context that Allison amd Malucci (1964a) and Allison and Young (1964) reported concentration of hydrocarbon carcinogens and fluorescent drugs in the lysosomes of monkey kidney cells in tissue culture.

It should be mentioned that basic dyes (neutral red, for example) also stain isolated kidney lysosomes post-vitally (Straus, 1954). The question has to be raised, therefore, whether such a passive staining, as distinct from active uptake, may also occur *in vivo* after cell injury.

The similarities between dye-inclusion bodies and phago-lysosomes seem to extend also to the pathological enlargement of the bodies (see Section VIII.). Enlarged bodies containing injected dyes may correspond to the "*Krinom Komplex*" described by Chlopin (1927) (see Schmidt, 1962, and Stockinger, 1964, for more details). Perhaps, the toxicity of certain dyes causes cell injury. These injured cytoplasmic areas may be sequestered together with lysosomes and dye-containing phagosomes, and may be extruded, in some cases.

G. FACTORS INFLUENCING PERMEABILITY, UPTAKE, AND TRANSPORT

In the following paragraphs, certain factors are discussed which influence the uptake of macromolecular or colloidal materials by pinocytosis or phagocytosis. Since most of these factors are little understood and since the literature could only be reviewed incompletely on this subject, only some of them will be mentioned.

1. Iso-electric Point of Proteins and Surface Charge of Colloidal Particles

The studies of Chapman-Andresen (1962) showed that pinocytosis of proteins in amoeba is strongly affected by the iso-electric point of the protein. The proteins are probably bound to a negatively charged surface coat of the cells before being ingested. Non-charged compounds did not induce pinocytosis in amoeba.

In the uptake of non-protein particles by mammalian cells, the surface charge of these particles is probably important for their binding to serum proteins or to the surface layers of the cells.

2. Opsonins

It is known that certain serum factors, opsonins, accelerate the uptake of antigens by the cells of the reticulo-endothelial system in immunized animals. These factors are related to antibodies and complement (Nelson and Lebrun, 1956; Mabry, Wallace, Dodd and Wright, 1956; Robineaux and Pinet, 1960). In addition, non-specific opsonins (serum factors) exist which combine with inert particles (colloidal carbon, for example) before these particles are taken up by the cells of the reticulo-endothelial system (Jenkin and Rowley, 1961; Vaughan, 1965). Since proteins are probably the main pinocytosis-inducing compounds (Chapman-Andresen, 1962), it is possible that non-protein particles of macromolecular, colloidal, or microscopic size require surface films of adsorbed proteins from the serum or medium before they can be ingested (Murray, 1963). However, Sbarra and Karnovsky (1959) observed that the uptake of polystyrene particles by phagocytic cells, in contrast to the uptake of starch granules, did not seem to require serum factors, and they suggested that the lipophilic surface of polystyrene particles may be the cause for this.

3. Hormones

According to Nicol, Vernon-Roberts and Quantock (1965) oestrogens increase the phagocytic activity of the reticulo-endothelial system. It would be important to clarify whether these, and other effects of hormones on uptake processes reported in the literature are specific or unspecific. Barrnett and Ball (1960) observed that administration of insulin increased the formation of microvilli and pinocytic vesicles in adipose tissue. Easty *et al.* (1964), however, did not detect a significant influence of insulin on pinocytosis of fluorescent-labelled proteins in tissue culture cells.

4. Specificity of Uptake

The question of recognition of "self" and "non-self" in relation to immunity will be mentioned in Section VII. It is of considerable theoretical interest to determine whether homologous proteins are taken up in a different manner from heterologous proteins. The initial clearance rates of injected proteins from the blood are often used as a measure of uptake by the reticulo-endothelial system. Although the initial clearance rates of injected heterologous proteins seem to be somewhat faster than those of the corresponding homologous proteins (Weigle, 1960; McFarlane, 1964), the differences do not seem to be great enough to be significant. The kidney phagosomes may also be considered as a test system for this problem. It is known that injected foreign proteins (egg white, horse-radish peroxidase) are concentrated in the phagosomes. Endogenous proteins are also taken up by these granules as may be seen from the observation that the size and number of the "protein droplets" increased in male rats in old age and in nucleoside nephrosis in relation to proteinuria (Fisher, 1964). Ericsson (1964) demonstrated uptake of homologous haemoglobin in kidney phagosomes. Thus no significant difference seems to exist between the uptake of homologous and heterologous proteins by these granules.[1]

A certain unspecificity of pinocytic uptake may also be deduced from experiments by Chapman-Andresen and Holter (1955) and by Holter (1960). Whereas amoebae did not take up radioactive glucose alone, they did ingest glucose when pinocytosis was induced by protein. The "piggy-back" phagocytosis described by Sbarra, Shirley and Bardawil (1962) seems to be a related process.

In other cases, however, a certain selectivity of uptake between different proteins has been observed. Telfer (1961) reported that the uptake by oocytes of one female blood protein was 20 times greater than that of other blood proteins. He suggested that a differential adsorption (affinity) of the proteins to the surface layers of the cells might be related to this process. Roth and Porter (1964) suggested that the "bristle coat" which they observed in electron micrographs of certain micropinocytic vesicles might be related to selectivity of protein uptake. Perhaps, specificity is also involved in the process described by Brambell (1958) and by Clark (1959) in which it was shown that the uptake and transport of proteins through the intestinal epithelial cells cease for unknown reasons 1–2 wks after birth, and that the closure could be influenced to a certain extent by cortisone. According to Holter (1961) selectivity may operate, not during the entry of the proteins

[1] see Note Added in Proof, p. 319.

into the pinocytic vacuole, but during their exit from the vacuole into the cytoplasm.

5. Permeability Barriers; Molecular Weight (Particle Size); Load

Macromolecular or colloidal matter injected into the blood stream will be taken up by pinocytosis or phagocytosis preferentially by those cells which have direct access to the blood stream (for example endothelial cells and Kupffer cells). The transport of macromolecular or colloidal particles across endothelial cells takes place by means of micropinocytic vesicles (Sections V. A. 3., V. E.) or, less frequently, by passage through intercellular spaces (Majno, 1965). Permeability barriers might exist at these two sites. Majno and Palade (1961) and Majno, Palade and Schoefl (1961) demonstrated in interesting studies on the cremaster muscle that histamine and serotonine increased the passage of colloidal particles through the intercellular spaces of endothelial cells. The basement membranes might constitute another barrier. Proteins with molecular weights above 70,000, for example, are retained for the most part in the glomerulus (Lambert, 1936), and the basement membrane seems to be the barrier (Farquhar and Palade, 1961). Investigations by electron microscopy showed morphological alterations of the basement membranes of the glomerulus during pathological changes of permeability (Trump and Benditt, 1962). On the other hand, Menefee *et al.* (1964a) observed by electron microscopy that relatively large aggregates of injected globin crossed the basement membranes of the cells in the glomerulus. Red and white blood cells are known to pass through the basement membranes without causing injury. Thus the exact way in which the basement membrane acts as filter or barrier does not seem to be completely understood as yet.

The passage or retention of a protein in the glomerulus will affect, of course, its clearance rate from the blood and excretion in the urine, its reabsorption by the tubule cells and degradation in the phago-lysosomes there, and therefore will influence indirectly the amounts taken up by the cells of other tissues. In general, the extent of uptake of macromolecular materials by the cells close to the blood stream will determine whether low or high concentrations of them will reach the phagocytic cells of the connective tissue or the epithelial cells, or whether they will not arrive there at all. In this context, it would be important to know more about the effect of the concentration (or load) on the uptake. By using horse-radish peroxidase as a marker protein, the increase of uptake with dose by the kidney was measured colorimetrically (Straus, 1962b). After injection of relatively low amounts of peroxidase, the protein appeared in the phagosomes of the kidney (Straus, 1964a). However, much smaller doses were sufficient for uptake by the phagosomes in the endothelial cells and Kupffer cells of the liver (Straus, 1962b, Table II). Nossal, Ada and Austin (1964) injected a very low dose of ^{131}I-labelled antigen and detected its uptake by macrophages of lymph nodes by autoradiography. If it is correct that the adsorption of proteins onto the surface layers of the cells initiates pinocytosis, this process may be started by very low concentrations.

Different types of phagocytic cells seem to take up particles of different sizes. Whereas macrophages take up particles of macromolecular, colloidal, and microscopic dimensions up to several micra in diameter, fibroblasts and endothelial cells seem mainly to take up particles of macromolecular and colloidal size.

6. Competition (Interference)

If macromolecular or colloidal material is injected into animals and, subsequently, another substance of a more or less similar type is injected, an interesting phenomenon of competition (interference) can be observed in many cases. A few examples from the recent literature may be cited. Thorbecke, Maurer and Benacerraf (1959) reported that the clearance of proteins from the blood was slower after a previous injection of colloidal carbon. Wagner and Iio (1964) noted that after blockage of the reticulo-endothelial system with large doses of serum albumin or gelatin, the clearance rate from the blood of similar particles was lower than the clearance of dissimilar particles. Brambell, Halliday and Morris (1958) observed that the intestinal absorption of antibodies by suckling animals was interfered with by the simultaneous administration of serum proteins. The release phenomenon described by Sicot, Afifi, Benhamou and Fauvert (1963) seems to be related to interference. After intravenous injection of colloidal radioactive gold (^{198}Au) into the ear vein of rabbits, the plasma level of the injected metal dropped to low levels, probably after uptake by the reticulo-endothelial cells. If stable Au, or colloidal carbon together with gelatin, or gelatin alone, were injected subsequently, the plasma levels of ^{198}Au increased for a short time (20 min). If, however, the injection of stable Au was made 24 hr after the injection of radioactive gold, no release occurred. Rapid binding of Au to the cell surface and release from it, and stable inclusion of Au in phagosomes at a later time, might explain these different effects.

The process of interference may be better understood by cytological observations. After intraperitoneal injection of haemoglobin, colloidal carbon, colloidal silver, and vital stains in different combinations and concentrations, Kedrowski (1933a, b) observed that two of the injected materials often appeared in the same granules of fibroblasts or macrophages of the connective tissue but that in many cases effects of interference were evident. Rustad (1959) reported that the molecular orientation of toluidine blue at the surface membrane of amoeba was interfered with by the previous adsorption of a basic protein. Straus (1962b, 1964a) observed by biochemical and cytochemical methods that the uptake of injected plant peroxidase was strongly interfered with, in the phago-lysosomes of the kidney tubule cells and Kupffer cells, if the granules contained previously administered egg white. The clearance of peroxidase from the blood and excretion in the urine were also altered by previous treatment with egg white.

Interference may also play a role in ingestion processes in normal animals. The cell surface, or the phagosomes or phago-lysosomes may contain endogenous materials which may interfere with the uptake of other endogenous material. Theoretically, interference may involve three cytoplasmic processes (a) the adsorption of materials onto the cell surface and the formation of phagosomes there, (b) the fusion of phagosomes with lysosomes, (c) digestion within the phago-lysosomes or the formation of new lysosomes.

H. METABOLISM OF UPTAKE PROCESSES

Pioneering investigations on the biochemistry of phagocytosis have been carried out by Karnovsky and his co-workers. Karnovsky (1962) has presented a lucid review of his own work and the work of others on this problem. Therefore, only a few points will be mentioned here.

On the basis of inhibition studies, Karnovsky and co-workers concluded that the energy for phagocytosis in polymorphonuclear leucocytes was derived mainly from glycolysis. Sbarra and Karnovsky (1959) observed a 7-fold increase in the conversion of the carbon-1 atom of glucose to CO_2 during phagocytosis in leucocytes. Cagan and Karnovsky (1964) and Evans and Karnovsky (1961) reported that the increase of respiration during phagocytosis was related to the activation of a CN-insensitive NADH-oxidase (1.6.1.–), located in the cytoplasm (supernatant fluid) of the leucocytes, and to the activation of a NADPH-linked lactate dehydrogenase (1.1.1.27). This latter enzyme would regenerate NADP, which is rate-limiting for the hexose monophosphate shunt in leucocytes. The activation of both enzymes might be induced, according to Evans and Karnovsky, by the depression of the pH by lactate formation during glycolysis. Karnovsky and Wallach (1961) observed that the incorporation of [^{32}P]orthophosphate into phosphatidic acid, inositolphosphatide, and phosphatidylserine was increased several times during phagocytosis of starch granules by leucocytes. From these results Karnovsky and Wallach discussed the possible involvement of phosphatides in the turnover of membranes during phagocytosis and secretion. Oren, Farnham, Saito, Milofsky and Karnovsky (1963) compared the metabolism of polymorphonuclear leucocytes, peritoneal macrophages, and alveolar macrophages during phagocytosis. Whereas the first two cell types depended only on glycolysis for energy supply during phagocytosis, alveolar macrophages depended to a large extent on oxidative phosphorylation.

Rossi and Zatti (1964b, 1966) confirmed several of the major findings of Cagan and Karnovsky on leucocytes. They suggested, however, that the stimulation of oxygen uptake during phagocytosis was not dependent on a

NADPH-linked lactic dehydrogenase, and was related to the increased activity of a NADPH-oxidase released from the leucocyte granules during phagocytosis. Roberts and Quastel (1964) attributed the increased NADPH oxidation during phagocytosis to the action of peroxidase in the leucocyte granules.

Schumaker (1958) and de Terra and Rustad (1959) studied the influence of low temperature and metabolic inhibitors on channel formation and protein uptake in amoeba, and concluded that it depended, in part, on aerobic metabolism. Miller, Hale, and Alexander (1965) observed that the uptake of horse-radish peroxidase by kidney slices was abolished by cold, anoxia, 2,4-dinitrophenol, and fluoroacetate, and was inhibited when the kidney slices were incubated in a salt-free medium.

I. SECRETION AS REVERSED PINOCYTOSIS

The opinion has been expressed repeatedly in the literature that secretion is similar to pinocytosis acting in reverse. Although several aspects of pinocytosis are not understood sufficiently to make a valid comparison, a few aspects of the two processes may be compared.

Observations by electron microscopy and biochemical experiments have shown that secretory enzymes are synthesized in the endoplasmic reticulum; they are assembled into secretory granules in the Golgi zone, transported to the cell surface and excreted there (Caro and Palade, 1964). In pinocytosis, on the other hand, small vesicles form at the cell surface and transport ingested material into the cells. Ingested material often appears in the Golgi region of the cells. Thus, the last stages of secretion resemble pinocytosis acting in reverse. It is not known, however, whether or not ingested materials reach the rough-surfaced endoplasmic reticulum. Only in such a case, would it be possible to consider the combined rough-surfaced and smooth-surfaced endoplasmic reticulum as a system for transport of material into and out of cells. If a teleological speculation might be permitted, it does not seem probable that cells would allow foreign materials to penetrate into the rough-surfaced endoplasmic reticulum, except, perhaps, for antibody formation (Section VII.). Rather, it seems that an important function of the phagolysosomes is to prevent foreign materials from penetrating further into the cytoplasm. On the other hand, useful *endogenous* materials might not be restricted in such a way, and macromolecules such as hormones might enter the cells by the same channels as those used in secretion but in the opposite direction.

It was mentioned in Section IV. that the young lysosomes might be a type of secretory granule. It is of interest, therefore, that Woodin and Wienecke (1964a, b) observed the extrusion of lysosomes from leucocytes under certain conditions and compared the process to secretion (Woodin,

1963) (Section VIII. c.). The similarities between the metabolism of uptake and secretion have been discussed by Karnovsky (1962).

J. DEGRADATION OF INGESTED MACROMOLECULAR MATERIAL, ESPECIALLY OF PROTEINS

1. Intracellular vs *Extracellular Digestion*

Whereas the bulk of ingested food is digested outside the cells, in the cavities of the stomach and intestine, the discovery of the lysosomes has revealed the existence of an *intra*cellular digestive system. In extracellular digestion, hydrolytic enzymes, produced in gland cells of the stomach, pancreas, and intestine, and accumulated in secretory granules, are exported out of the cells into ducts and cavities. In the case of the lysosomes, hydrolytic enzymes (with maximal activity in acid pH) are used in the same cells in which they are produced. Thus, lysosomes (phago-lysosomes) may be considered to be a type of digestive vacuoles similar to the food vacuoles of protozoa. It is not known what are the natural substrates for the lysosomal enzymes in normal animals. The types of enzymes present may give some clues. They point to the degradation of proteins, nucleic acids, and mucopolysaccharides.

It should be noted that the uptake of proteins (or other material) by the phagosomes, and the degradation of these proteins in phago-lysosomes are two different processes. Factors which affect the uptake (Section V. G.) should therefore be distinguished from those which affect the degradation of ingested material within the phago-lysosomes (for example pH, or the concentration of hydrolytic enzymes).

2. Catabolism of Native vs *Denatured Proteins*

It is known that denatured proteins are cleared from the blood or subsequently catabolized more rapidly than native proteins (Gordon, 1957; Freeman, Gordon and Humphrey, 1958). According to Okunuki (1961), degradation of most proteins begins only after denaturation. Antigens are cleared from the blood or degraded more quickly after immunization (Sorkin and Boyden, 1959; Patterson, Suszko and Pruzansky, 1962). As was mentioned in the preceding paragraph, these changes might be due either to increased uptake or to increased degradation, or to both (see also Section VII. B.).

3. Catabolism of Serum Proteins

It is not known where serum proteins are catabolized. Possible sites of breakdown are the kidneys, especially in pathological states (Sellers, Katz and Rosenfeld, 1961; Solomon, Waldman, Fahay and McFarlane, 1964), the liver (Cohen, Freeman and McFarlane, 1961), and the intestine (Wetterfors *et al.*, 1960; Franks, Mosser and Anstadt, 1963). Jansco and Jansco-Gabor (1954) suggested that plasma proteins might be broken down in the granules of macrophages of the connective tissue. Katz, Rosenfeld and Sellers (1961), however, thought that there is no evidence for the participation of the reticulo-endothelial system and of cathepsin in the breakdown of serum proteins.

4. Catabolism of Injected Plant Peroxidase and of Other Foreign Proteins in Phago-Lysosomes

The gradual disappearance of injected peroxidase in the phago-lysosomes could be estimated by double staining of the same granules for peroxidase and acid phosphatase (Straus, 1962b). The peroxidase reaction disappeared in the phago-lysosomes of kidney tubule cells and in macrophages of spleen, thymus, lymph nodes, and connective tissue after 2–4 days. Miller and Palade (1964) estimated from observations by electron microscopy that the disposal of injected haemoglobin in the kidney lysosomes required 2 days approximately. The pH within the phago-lysosomes may not be low enough for maximal catheptic activity, and this may explain the relatively slow rate of degradation of ingested proteins.

The degradation of radioactive-labelled antigens by leucocytes *in vitro* was studied by Cochrane, Weigle, and Dixon (1959). Riddle and Barnhart (1964) investigated the dissolution of fibrin clots by leucocytes in "skin windows". In this procedure, a microscope cover glass containing a sterile fibrin clot is placed over a very small, denuded area of the skin, and the infiltrating blood cells are removed with the cover glass and examined. The leucocyte granules involved in these catabolic processes probably were lysosomes. Mego and McQueen (1965) studied the degradation of injected ^{131}I-albumin and ^{74}As-arsono-azoalbumin by particulate fractions of mouse liver. The release of radioactivity during incubation at 37° had a pH optimum of about 5, was stimulated by cysteine and inhibited by iodoacetamide and was attributed to the cathepsins of the lysosomes.

5. Partial Degradation of Proteins into Biologically Active Fragments

As could be seen from the experiments with injected plant peroxidase, native proteins are degraded in the phago-lysosomes at a relatively slow rate. This might facilitate the *partial* degradation within the phago-lysosomes of ingested proteins into biologically active fragments. The splitting of thyroxine from thyroglobulin in the phago-lysosomes of the thyroid as suggested by Wollman *et al.* (1964) (Section V. A. 2.), and the partial degradation of antigens into antibody-inducing fragments (Section VII.) might be such functions of the phago-lysosomes.

6. Degradation of Ingested Bacteria and Other Material in Phago-Lysosomes

The degradation of bacterial proteins, lipids, and nucleic acids in leucocytes and macrophages was investigated by Cohn (1963a, b, 1964), using radioactive labels and antigenic properties. Thorpe and Marcus (1964) observed that rabbit mononuclear phagocytes of immune animals had significantly increased intracellular destructive capacity towards *Pasteurella tularensis* as compared to normal animals. The breakdown of ingested erythrocytes, probably by lysosomal enzymes, in phagocytic vacuoles of macrophages was studied by Essner (1960).

The autolytic breakdown of cell constituents in areas of focal degradation will be discussed in Section VI.

7. Non-Lysosomal Enzymes involved in Protein Catabolism

In addition to the catheptic breakdown of ingested proteins in the phago-lysosomes, other proteolytic enzymes seem to function in protein catabolism at other cytoplasmic sites. Protein degradation at a neutral pH which was energy-dependent and stimulated by ATP or CoA was reported by Simpson (1953), Steinberg and Vaugham (1956) and Penn (1960, 1961). Differences between acid proteinases in lysosomes and neutral proteinases at other cytoplasmic sites were reported by Marks and Lajtha (1963) and by Nagel and Willig (1964a, b) on the basis of experiments with brain and kidney fractions.

K. ACCUMULATION OF POORLY DIGESTIBLE MATERIAL IN PHAGO-LYSOSOMES (RESIDUAL BODIES)

Material, segregated within phago-lysosomes, will accumulate there if no corresponding lytic enzymes are present to digest it. This explains the relationship between "residual bodies" and storage ("*Speicherung*", "athrocytosis"). It may be expected that the cells attempt to get rid of residual bodies by extrusion.

Lysosomes seem to lack lipolytic enzymes, except the slowly acting esterase. The occurrence of myelin-like structures in electron micrographs of many lysosomes may be related to the lack of lipolytic enzymes. Nilsson (1962c) and Fuxe and Nilsson (1963) made interesting observations by electron microscopy and cytochemistry on the role of lysosomes in lipid decomposition in the uterine epithelium of spayed mice. They suggested that acid phosphatase might participate in the degradation of phosphatides and esterase in the degradation of glycerides. It would be of interest to establish with certainty whether lipase is located in the lysosomes of macrophages and leucocytes as suggested by Cohn and Wiener (1963a) and Elsbach and Rizak (1963).

According to Gedigk and Fischer (1958, 1959), products of autoxidation or polymerization of unsaturated fatty acids accumulate in the ceroid and lipofuscin granules, and these granules give positive histochemical reactions for acid phosphatase and esterase. Essner and Novikoff (1960) and Novikoff (1961) concluded from observations by electron microscopy that lipofuscin granules in liver were related to lysosomes. A similar conclusion was reached by Miyawaki (1965) in the case of cells of mouse mammary glands. Siebert *et al.* (1962) isolated lipofuscin granules from heart muscle and reported that they had a chemical composition of 50% protein, 20% lipid, and 30% melanin-like residue (after HCl digestion). Lipofuscin granules isolated from human cardiac muscle by Björkerud (1964) were composed of 50% lipid and 30% protein. Hendley and Strehler (1965) isolated lipofuscin granules from human heart, human liver, and beef heart. They found cathepsin, but not acid phosphatase and esterase to be concentrated in the lipofuscin granule fractions. Loss of enzymes during the isolation procedure was not excluded by these investigators. The small fraction of phago-lysosomes, observed by Straus (1959, Fig. 19D) to float after high-speed centrifugation on the surface of liver homogenates, obtained from rats injected with horse-radish peroxidase, may have corresponded to lipofuscin granules. In addition to peroxidase-positive granules, this layer sometimes contained higher concentrations of acid phosphatase than the homogenate (W. Straus, unpublished observation).

Baudhuin, Hers and Loeb (1964) observed by electron microscopy, that glycogen accumulated in large vacuoles of liver cells in certain types of glycogen disease. They suggested that this process might be related to the lack of glucosidase, normally present in lysosomes of liver.

VI. Autolysis and Autophagy

A. Physiological Autolysis

Involvement of lysosomes in the breakdown of tissues and cells has been observed in certain cases of physiological regression. These processes are of special interest since they seem to be under hormonal control. In these studies the question arose whether the breakdown of tissue was caused by lysosomes in regressing epithelial cells or in macrophages, or in both.

1. Regression of the Mullerian Duct

These embryonic organs regress in the male between the ninth and thirteenth day of development. Brachet, Decroly-Briers and Hoyez (1958) were the first to show that the total and free activities of 3 lysosomal enzymes (acid phosphatase, cathepsin and ribonuclease) increased in chick embryos during the regression of the Mullerian duct. This was confirmed by Scheib (1963). Scheib (1964), in cytochemical studies, observed an increase in the number and size of the lysosomes in the regressing epithelial cells. In addition, there was an increase in the number of macrophages in the connective tissue. Scheib suggested that the increase in size of the lysosomes was related to the release of lytic enzymes *in vivo*.

2. Regression of the Tail in Xenopus *Larvae*

This process, which is under the control of the thyroid hormone, was investigated by Weber and co-workers. Weber (1963) reported that the amount of cathepsin, a lysosomal enzyme, markedly increased in the regressing tail. Salzmann and Weber (1963) demonstrated cytochemically that most of the lysosomes in the regressing tail were in macrophages. A marked increase of several lysosomal enzymes during the regression of *Xenopus* tails was also observed by Eekhout (1964). Recent electron microscope studies by Weber (1964) indicated that changes in muscle fibres and mitochondria of the regressing tissue occurred independently of the lysosomes. Thus, the lysosomes are not the primary cause of autolysis during the regression process in *Xenopus* tail. However, the lysosomes of the macrophages in the regressing tail function in cleaning up injured or dead cells.

3. Muscular Dystrophy

Invading macrophages and leucocytes seem to be the main source for the increase of lysosomal enzymes in dystrophic muscle (Tappel *et al.*, 1963). The lysosomes of macrophages participate in the breakdown of injured or dying cells. There is no indication so far that lysosomes in muscle cells are involved in the regression of the tissue.

4. Post-Partum Involution of the Uterus and Regressive Changes in the Ovaries and the Endometrium during the Oestrous Cycle

Post-partum involution of the uterus was investigated by Lobel and Deane (1962) using acid phosphatase as a cytochemical marker for lysosomes. Although little enzyme activity was observed in the uterus of virgin or pregnant rats, marked activity developed immediately post-partum in placental detachment sites, epithelial cells, in macrophages, and stromal cells of all uterine layers. Lobel and Dean suggested that the lysosomes function in the resorption and reorganization of the tissue in post-partum involution. Banon, Brandes and Frost (1964) investigated the changes in lysosomes in the ovaries and endometrium during the oestrous cycle. In the corpora lutea undergoing involution, the lysosomal activity was found to be concentrated in surrounding macrophages rather than in the lutein cells themselves. However, during folicular atresia and also in the endometrium during the oestrous cycle, the lysosomes of the regressing cells themselves increased in size and seemed to take part in the lytic process.

5. Holocrine Secretion in Sebaceous Glands

Brandes, Bertini and Smith (1965) made an interesting study of this process by histochemical electron microscopy. With the maturation of the cells, the number of lysosomes increased greatly. They seemed to be derived from small vesicles of the Golgi apparatus. During the disintegration of the mature cells, the lysosomes released their enzymes into the general cytoplasm.

6. *Bone Resorption*

Electron micrographs by Scott and Pease (1956), Gonzales and Karnovsky (1961) and Hancox and Boothroyd (1961) indicated that osteoclasts were very active in phagocytosis (ruffled border) and contained numerous phagocytic vacuoles. Schajowicz and Cabrini (1958) observed that acid phosphatase activity was marked in osteoclasts and chondrioclasts. Handelman, Morse, and Irving (1964) found acid phosphatase activity concentrated in a junctional band between osteoclasts and bone, suggesting that hydrolytic enzymes may be active at the surface of the osteoclasts. Vaes (1965) reported that lysosomal enzymes were released into the external medium during resorption of bone cells in tissue culture in the presence of parathyroid hormone. Woods and Nichols (1965a, b) found that particulate fractions isolated from rat bone were rich in collagenase activity. All these observations seem to indicate that the lysosomes of osteoclasts take part in bone resorption and that the degradation of bone material may take place either in phagolysosomes within the cells, or by lysosomal enzymes secreted at the cell surface, or at both sites.

7. *Autolysis* In vitro

The autolytic effect of purified lysosomes on isolated nuclei, mitochondria and microsomes was investigated by Sawant, Desai and Tappel (1964b), and the release of lysosomal enzymes during autolysis *in vitro* was investigated by Van Lancker and Holtzer (1959a, 1963).

B. AUTOPHAGY

Enlargement of lysosomes has been observed in many types of cells under various physiological and pathological conditions. The formation and function of these enlarged bodies is not yet completely understood. In electron micrographs, areas of the cytoplasm containing lysosomes (cytosomes), mitochondria, endoplasmic reticulum, or other cell components have been seen to be sequestered from the rest of the cytoplasm by a surrounding membrane (Ashford and Porter, 1962; Novikoff and Essner, 1962; Moe and Behnke, 1962; Hruban *et al.*, 1963). Acid phosphatase activity has often been detected in these areas. Similar bodies containing residual mitochondrial structures had been described previously in kidney cells by Rhodin (1954), Clark (1957), Novikoff (1959), and Miller (1960). These enlarged bodies were called "cytolysomes" by Novikoff and Essner (1962), "areas of focal degradation" by Hruban *et al.* (1963), and "autophagic vacuoles" by de Duve (1963b). Novikoff and Essner (1962) and Napolitano (1963) discussed whether these bodies have a function in the regulation of normal cell metabolism or only in the autolysis of abnormal cell components. The presence of such enlarged bodies in normal liver cells (Daems, 1962; Novikoff and Shin, 1964), in the regressing Mullerian duct (Scheib, 1964), in the foetal rat duodenum during certain stages of differentiation (Behnke, 1963), and in brown adipose tissue during rapid mobilization of lipid (Napolitano, 1963), indicate that "autophagic vacuoles" may develop in normal cells during marked alterations of metabolism. An interesting example of such a process was demonstrated by Elliot and Bak (1964) in *Tetrahymena*. During ageing of these cells, a proportion of the mitochondria was broken down in autophagic vacuoles showing acid phosphatase activity. The electron micrograph from the paper of Miller and Palade, reproduced in Fig. 14, shows such enlarged bodies in kidney cells. The reaction product for acid phosphatase, remnants of mitochondria, and increased density, probably indicating ingested haemoglobin (see Section VIII. A.), may be recognized in the bodies. In other cases in which enlargement of lysosomes has been reported, drastic treatments which might be expected to induce pathological changes were used. The formation of many of these bodies, therefore, may be related to pathological alterations (see Section VIII.).

VII. Relationship of Lysosomes and Phagosomes to Immune Processes

A. Antibody Formation

It is known from the work of Sabin (1939), Coons, Leduc and Kaplan (1951) and others, that foreign proteins are taken up by the cells of the reticulo-endothelial system by pinocytosis. Although the cells of the reticulo-endothelial system themselves are not thought by most investigators to produce antibodies, an important role of macrophages in antibody formation was suggested by Fishman. In Fishman's experiments, antibody formation against T_2 bacteriophage occurred in cultures of lymph node cells to which cell-free filtrates from homogenized macrophages had been added. The macrophages had been incubated previously with T_2 bacteriophage (Fishman, 1961). The antibody-inducing material from the macrophages was ribonuclease-sensitive (Fishman and Adler, 1963). The investigators suggested that the antigen taken up by the macrophages was partially degraded there, and that complexes of antigen fragments and ribonucleic acid were transferred to antibody-producing cells. A similar theory has been put forward by Campbell and Garvey (1961, 1963),

If partial degradation of antigens precedes antibody formation, this may well take place in the phago-lysosomes of the reticulo-endothelial cells. It is of interest in this context that Lapresle and Durieux (1957) found that antibodies against serum albumin react with split products of serum albumin degraded by spleen extracts. After partial degradation, the antigen fragments, according to this hypothesis, leave the phagolysosomes and reach the endoplasmic reticulum or the nucleus. It is of interest, therefore, that according to Coons (1956) and Wellensiek and Coons (1964), antigens often penetrate

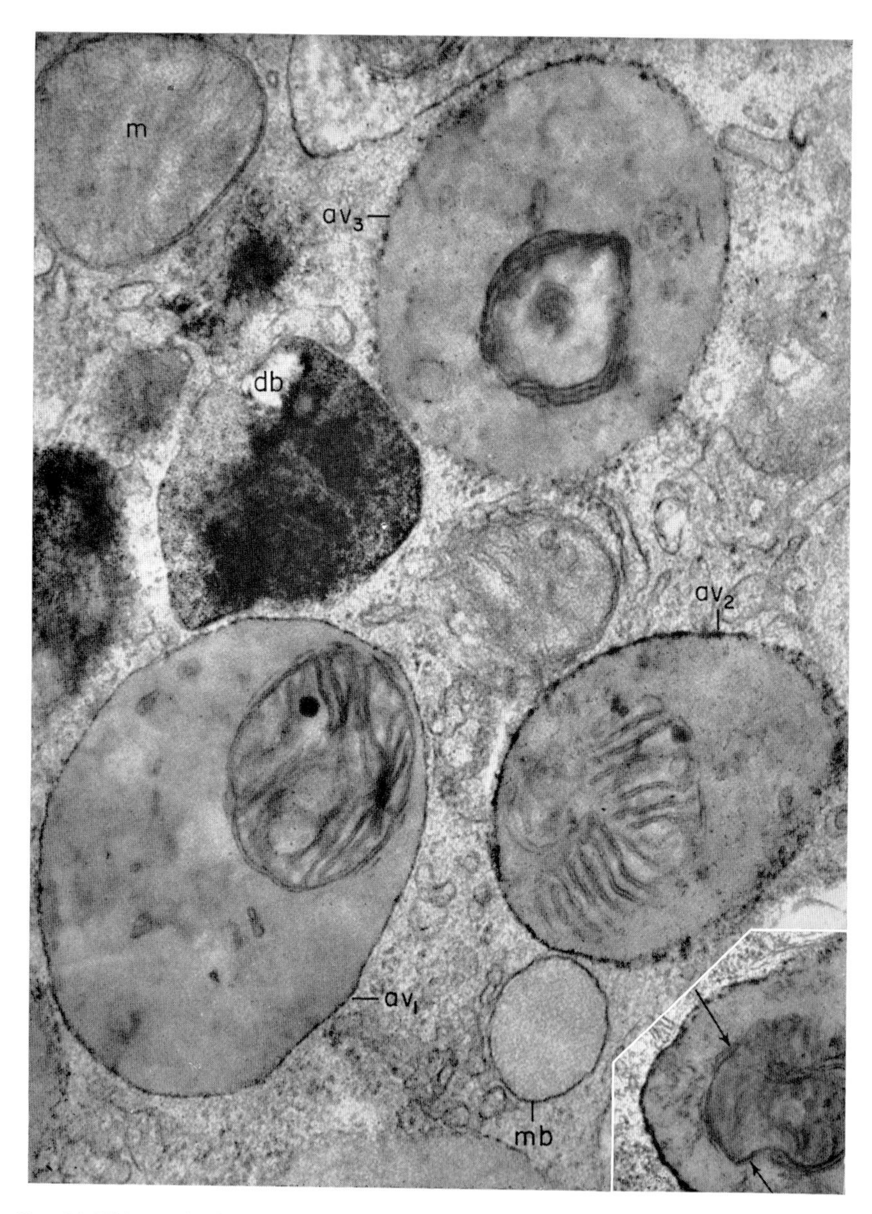

FIG. 14. Kidney cell of female mouse injected with haemoglobin intraperitoneally two hours previously. Note the reaction product for acid phosphatase around the bodies and remnants of mitochondria in the interior. The increased density suggests segregation of injected haemoglobin. Magnification × 45,000; inset × 30,000. (Electron micrograph from the paper of Miller and Palade (1964), reproduced with the permission of *J. Cell Biol.*)

into the nuclei of cells. Fishman's theory is also supported by Nossal, Ada and Austin (1964) who used a sensitive autoradiographic method to detect ^{131}I-labelled antigen in the cells of lymph nodes during antibody formation. They observed that radioactivity was present only in the macrophages but not in the surrounding plasma cells. Lymphocytes, now generally considered to be precursors of antibody-producing cells, normally contain few lysosomes. However, the number of lysosomes in lymphocytes increased greatly after sensitization with dead tubercle bacilli or exposure to haemagglutinin (Hirschhorn *et al.*, 1965; Allison and Malucci, 1964b; Diengdoh and Turk, 1965).

The transfer of antigenic material from macrophages to plasma cells might take place via cytoplasmic bridges. Schoenberg, Mumaw, Moore and Weissberger (1964) demonstrated that such bridges exist, especially in immunized animals. They occur between phagocytic and antibody-producing cells in lymph nodes and spleen. Deane (1964) also noted by electron microscopy that there was a close association between plasma cells and macrophages in the lamina propria of the gut, suggesting exchange of material between both types of cells.

The question of the persistence of antigens or antigen fragments is often raised in relation to the theories of antibody formation. Garvey and Campbell (1957) concluded that small amounts of S^{35} which persisted in the liver for several years, belonged to antigenic fragments bound to ribonucleic acid, of the original S^{35} labelled protein (see also Campbell and Garvey, 1963). Speirs (1963) found radioactivity derived from tritiated tetanus toxin in lymphocytes and macrophages as long as 9 months after injection into immunized mice. If there is persistence of such antigenic fragments, this may be at cell-sites far from lysosomes, or the fragments may be resistant to lytic enzymes.

B. UPTAKE OF ANTIGEN-ANTIBODY COMPLEXES

Sabesin (1963) showed by electron microscopy that eosinophils were especially active in the uptake of antigen (ferritin)-antibody complexes. This calls to mind the well-known eosinophilia of allergic states. The uptake of I^{131}-labelled antigen-antibody complexes by spleen cells *in vitro* was investigated by Patterson *et al.* (1962). These authors made the interesting observation that the larger complexes formed from equivalent amounts of antigen and antibody were pinocytized more readily than the smaller complexes formed in antigen excess. Tahaba, Kinuwaki and Hayashi (1964) studied the uptake of soluble complexes of bovine-serum-albumin and rabbit-anti-bovine-serum-albumin by tissue culture cells using phase contrast microscopy. They observed cell damage and greatly increased pinocytosis.

In some of the above cases the antigen-antibody complexes probably were digested in the phago-lysosomes of the phagocytic cells, and then did not cause further injury. However, antigen-antibody complexes may also be deposited at tissue sites where they are inaccessible to lysosomal enzymes. In such cases they may cause various kinds of cellular injury. McCluskey and Benacerraf (1959) produced acute glomerulo-nephritis, necrotizing arteritis, and acute endocarditis by injection of soluble antigen-antibody complexes (ovalbumin and bovine serum albumin) into mice, and they demonstrated the deposition of the complexes at the sites of tissue damage. Andres, Seegal, Hsu, Rothenberg and Chapeau (1963) produced glomerular nephritis in rabbits by repeated injection of serum albumin. With the aid of ferritin-labelled antibody the investigators showed that ferritin aggregates, probably representing antigen-antibody complexes, were trapped between the basement membrane and the epithelial cells of the glomerulus. The ratio of antigen to antibody (particle size of the complex) was an important factor for the development of glomerulo-nephritis. Dixon (1963) discussed the possible involvement of antigen-antibody complexes in the causation of glomerulo-nephritis, rheumatoid arthritis, lupus erythematosus, and polyarteritis nodosa. He pointed out that weak antigens. which produce soluble antigen-antibody complexes, are most likely to cause chronic injury. The essential factor, according to Dixon, is not the antigenic specificity, but the formation of aggregates which may form insoluble deposits at inaccessible sites (close to the basement membranes, for example). The possible involvement of humoral factors was also discussed by Dixon.

The so-called Arthus reaction and reversed Arthus reaction, associated with focal vascular necrosis, is related to the deposition of antigen-antibody complexes in the vessel walls and the uptake of the complexes by phagocytic cells. Daems and Oort (1962) studied this process by electron microscopy and noticed degranulation of leucocytes and the formation of phagocytic vacuoles containing the immune complex. Movat, Urinhara, MacMorine and Burke (1964) and Urinhara and Movat (1964) demonstrated by electron microscopy that the Arthus reaction was initiated by the precipitation of antigen-antibody complexes in the lumen and walls of blood vessels. They showed that during the uptake of antigen-antibody complexes by leucocytes and the subsequent fusion of lysosomes with phagosomes, a permeability-increasing factor was released.

Those antigens which are good antibody producers may give rise to anaphylactic shock when the antigen combines with antibody. This may be associated with the release of histamine, probably from mast cells (Cochrane, 1963a, b). Movat, Mustard, Taichman and Urinhara (1965) noted that antigen-antibody complexes were phagocytized by blood platelets, and that during their subsequent aggregation histamine, serotonin, and lysosomal enzymes were released. Fennell and Santamaria (1962) reported that anaphylactic shock in rats was associated with liver cell necrosis in the mid-zonal and peri-portal areas and with alterations in the size and location of the lysosomes. Blackwell (1965) also observed that after injection of antigen into sensitized rats, the liver contained large eosinophilic globules, positive for acid phosphatase and esterase, and related them to the optically clear vacuoles of the same cells. Treadwell (1965) found a 6–7 fold increase of acid phosphatase activity in blood serum soon after induction of anaphylactic shock and a concomitant decrease of 3 lysosomal enzymes in liver homogenates.

C. ANTIGEN AND ANTIBODY REACTIONS AT THE CELL SURFACE

As mentioned in Section V. A. 1., foreign proteins (antigens) may become adsorbed onto the cell surface during the early phase of pinocytosis. Greater adsorption onto membranes occurs in injured cells (Holtzer and Holtzer, 1960; Straus, 1964d). Easton, Goldberg and Green (1962b) noted that antibody and complement reacted with antigen at the cell surface causing injury to the cell membrane and leakage of material ("immune cytolysis"). The fragility of lysosomes was increased when ascites tumour cells in tissue culture were exposed to immune serum and complement (Bitensky, 1963). Dumonde, Roodyn, and Prose (1965) observed that immune ferritin-labelled antibodies from rabbits were bound at the surface of isolated lysosomes and mitochondria from rat liver together with large amounts of complement. No release of acid phosphatase from lysosomes and no change of oxidative phosphorylation in mitochondria was noted during the complement fixation reaction. Boyden and Sorkin (1960) and Boyden, Sorkin and Spärk (1960) reported that immune cells were coated at their surface by a cytophilic, non-precipitating antibody which cause increased uptake of antigen. These diverse observations show that our understanding of the processes taking place during immune reactions at the cell surface are far from clear.

D. RECOGNITION OF "SELF"

The distinction of "self" and "non-self" is one of the least understood properties of the immune process. Boyden (1962) pointed out that phagocytic cells, too, have the ability of self recognition. They do not take up cells of the same organism, except if the cells are injured. Boyden suggested that humoral factors might be involved and might be similar to those involved in chemotaxis. Perkins and Leonard (1962) observed that, in the absence of type-specific antisera, erythrocytes from more distantly related animals were phagocytized preferentially by mouse peritoneal mononuclear cells *in vitro*, and that this selectivity could be modified by the addition of specific erythrocyte antibodies. Nelson and Buras (1963) also noted that homologous and heterologous erythrocytes were phagocytized at different rates, and that spleen cells preferentially took up homologous, and liver cells heterologous erythrocytes.

The most interesting observations on this problem were made by Ada, Nossal and Austin (1964), who injected [^{131}I]-labelled antigenic, minimally antigenic, and non-antigenic (homologous) proteins and studied their distribution in the polipleal lymph nodes by autoradiography. They observed that the macrophages in the medulla took up antigenic and non-antigenic proteins to an equal extent but that the macrophages in the lymphoid follicles only took up antigenic proteins (with the exception of autologous gamma-globulin). They concluded that this selectivity was due to opsonins (see Section V. G. 2. above) rather than to inherent properties of the phagocytic cells themselves.

VIII. Pathological Changes Related to Lysosomes and Phagosomes

A. Pathological Enlargement of Lysosomes and Phagosomes

The possibility has already been discussed (Section VI.) that enlarged lysosomes ("areas of focal degradation", "autophagic vacuoles") may develop during sudden drastic changes in metabolism under relatively physiological conditions. However, many of the reports of enlargement of lysosomes were after treatment which must have caused cell injury. Hruban *et al.* (1963) observed areas of focal degradation in kidney, liver, and pancreas after potassium deficiency, after partial occlusion of the renal vein, or after injection of toxic agents. Novikoff (1959) reported enlargement of lysosomes in kidney cells after ligation of the ureter. Enlargement of lysosomes in nerve cells after axon section has also been observed (Barron and Sklar, 1961). The question has to be raised in each case, therefore, whether injury to cell components preceded the development of autophagic vacuoles and was the cause of it, or whether autophagic vacuoles developed without cell injury and functioned in the autolytic degradation or turnover of *normal* cell components.

Another point does not seem to have been considered sufficiently in the interpretation of areas of focal degradation. This is the possibility that materials ingested from the blood stream may have participated in the enlargement of bodies resulting from the fusion of phagosomes with lysosomes. Lysosomes and phagosomes are enlarged after fusion. A further enlargement of the bodies occurred in kidney and liver cells after injection of relatively high doses, or toxic preparations of horse-radish peroxidase (Straus, 1961, 1963, 1964a, b and Figs. 15 and 16 of the present article). The enlarged bodies in kidney cells often seemed to include cell fragments, mitochondria, or nuclei, in addition to lysosomes and phagosomes ("composite bodies", Straus, 1961). In a similar way, uptake of injected material

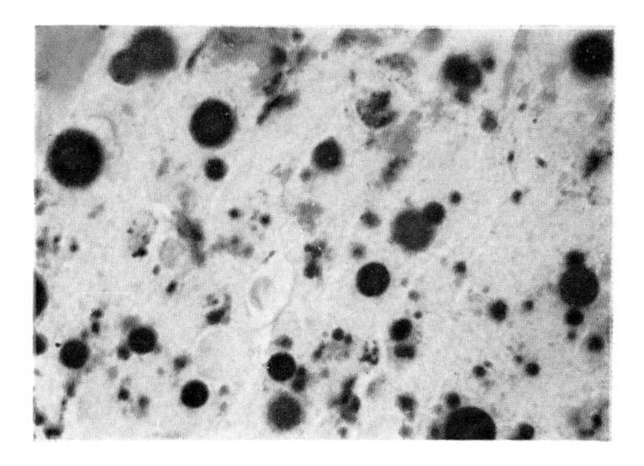

FIG. 15. Pathologically enlarged phagosomes in liver cells, 3 hr after intravenous injection of a high dose of plant peroxidase; staining with benzidine. The same large granules also showed acid phosphatase activity (not demonstrated here). ×660.

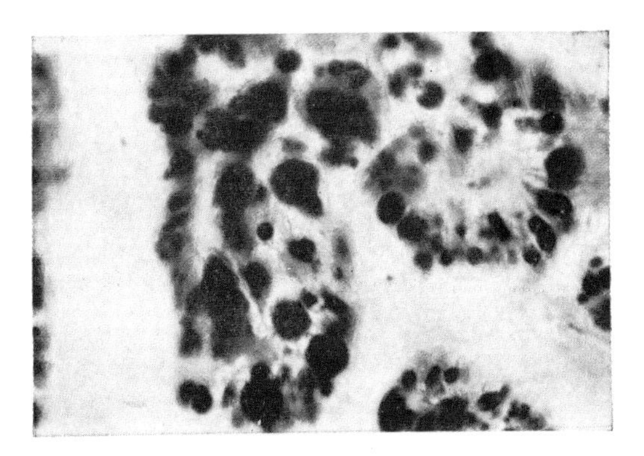

FIG. 16. Pathologically enlarged lysosomes (phagolysosomes) in kidney cells, 48 hr after intravenous injection of plant peroxidase; stained for acid phosphatase. The animal had pyelo-nephritis. ×660.

in phagosomes and its transport to areas of focal degradation may have occurred in the experiments of Ashford and Porter (1962) with glucagon, in the experiments of Hruban *et al.* (1963) with toxic substances, and in the experiments of Novikoff and Essner (1962) with the neutral detergent Triton WR-1339. Even when cell injury was induced by treatments mentioned on the preceding page not involving injection of foreign compounds (K^+ deficiency, partial occlusion of the renal vein, ligation of the ureter, axon section), the treatments may have caused the entry of increased amounts of serum proteins, and these, segregated in phagosomes, may have been transported to the areas of focal degradation.

Enlarged phago-lysosomes probably are related to the so-called "watery vacuoles" of liver cells. According to Trowell (1946), increase of sinusoidal pressure and anoxia are main factors in the development of these vacuoles. Doniach and Weinbren (1952), Aterman (1958), Nairn, Chadwick and McEntegart (1958), Anderson, Cohen and Barka (1961) and Schlicht (1963), demonstrated the presence of injected substances or of serum proteins in such vacuoles of liver cells. In recent years, several investigators have discussed the formation of enlarged vacuoles in relation to lysosomes and ingestion processes. Vacuoles of this type in liver cells have been described after injection of dextran and iron-dextran (Daems, 1962; Kent *et al.* (1963a, b), mercury sulphide (Oudea, 1963), and sucrose (Wattiaux, Wattiaux-DeKoninck, Rutgeerts and Tulkens, 1964; Brewer and Heath, 1963). Similar large vacuoles have been observed in kidney cells after the injection of sucrose (Janigan and Santamaria, 1961; Trump and Janigan, 1962; Maunsbach, Madden and Latta, 1962). It is possible that in all these cases, cell injury was associated with the increased uptake of material from the blood and with the fusion of numerous lysosomes with phagosomes.

How are enlarged phago-lysosomes, which appear round and relatively empty in electron micrographs, related to the irregularly-shaped areas of focal degradation, which appear relatively dense in electron micrographs? This may depend on the size and number of lysosomes and phagosomes that fuse and on whether other cell components (such as mitochondria, endoplasmic reticulum or microbodies) are sequestered together with phagosomes and lysosomes. The following hypothesis was suggested for the formation of enlarged phago-lysosomes (Straus, 1964b). During the pathological fusion of several lysosomes with several phagosomes some of the hydrolytic enzymes may be released into the cytoplasm *in vivo*. The cells might react against this irritation by sequestering these areas from the rest of the cytoplasm and surrounding them with a membrane. Although acid phosphatase activity can often be demonstrated in these areas by histochemical electron microscopy, no distinct cytosomes (lysosomes) can be seen in electron micrographs of many of these structures. This may indicate that the hydrolytic enzymes have been released from the lysosomes within these areas.

Although phagosomes may participate in many cases in the formation of areas of focal degradation, it is possible, of course, that other autophagic vacuoles may develop independently of phagosomes.

B. RELEASE OF LYSOSOMAL ENZYMES *IN VIVO*

In his earlier work on lysosomes, de Duve suggested that lysosomes might be a frequent cause of cell injury and necrosis ("suicide bags"). However, this opinion has not yet been confirmed (de Duve 1963a) and a general release of lysosomal enzymes into the cytoplasm *in vivo* does not occur frequently.

It is not difficult to detect the release of lysosomal enzymes from isolated cell fractions, and several agents causing release have been mentioned in Section II. C. It is difficult, however, to determine whether lysosomal enzymes have been released *in vivo*. The membranes of pathologically enlarged lysosomes are probably very fragile so that enzymes are released during homogenization or fractionation of the tissue (see Sections II. A., II. C.). Another source of error arises if pathological tissues contain areas of necrosis surrounded by viable tissue. The lysosomal enzymes from the necrotic areas only may be released *in vivo* (van Lancker and Holtzer, 1959a) and may then appear in the supernatant fluid after fractionation of the tissue. Rigorous correlation of cytological (or cytochemical) and biochemical observations is therefore essential for the study of lysosomes in pathological tissues.

The occurrence of a limited release of lysosomal enzymes in sequestered areas of the cytoplasm *in vivo* was suggested in Section VIII. A. (see the discussion on "composite bodies"). A similar effect in shock and in virus infection will be mentioned below (Section VIII. D.). General release of enzymes from the lysosomes into the cytoplasm is known to occur in leucocytes after treatment with streptolysin (Hirsch *et al.*, 1963).

C. EXTRUSION OF LYSOSOMES AND OF LYSOSOMAL ENZYMES

Interesting observations on the extrusion of lysosomes and of their enzymes in leucocytes have been made by Woodin and Wienecke (1964a, b). They showed that after treatment of leucocytes with leucocidin or with vitamin A, the contents of the granules were extruded into the medium, and their membranes fused with the surface membranes of the cells. Calcium and adenosine triphosphate seemed to play a role in the process of extrusion which showed certain analogies to a secretion process.

The extrusion of enlarged phago-lysosomes from the convoluted tubule cells of the kidney into the lumen was observed after high doses of peroxidase (Straus, 1964b). This process may be similar to "potocytosis" observed by Rouiller and Modjtabai (1958) after repeated injection of ovalbumin or sucrose, resulting in extrusion of cell fragments together with hyaline droplets. Novikoff (1959) reported extrusion of enlarged lysosomes from kidney cells after ligation of the ureter. The extrusion of cell material from kidney cells occurring *in vivo* has to be distinguished from the extrusion that occurs *post mortem* (Hanssen, 1960; Parker, Swann and Sinclair, 1962; Longley and Burstone, 1963).

D. POSSIBLE ROLE OF LYSOSOMES IN INFLAMMATION, NECROSIS, SHOCK AND VIRUS INFECTION

Biochemical studies on lysosomal enzymes during liver injury by carbon tetrachloride and other poisons were reported by Slater and Greenbaum (1965) and Dianzani (1963). The relatively uncharacteristic changes in lysosomal enzymes were not considered by these authors to be primary causes of cell injury.

Nagel and Willig (1964a) concluded that lysosomal enzymes did not play an essential part in ischaemic necrosis of rat kidney. These investigators observed a rapid inactivation of lysosomal cathepsins during the development of necrosis, whereas aminopeptidases and neutral proteases of the microsomal fraction retained most of their activity 48 hr after treatment. Trump, Goldblatt and Stowell (1962), made careful observations by electron microscopy of early cytoplasmic alterations in parenchymal cells of mouse liver during necrosis *in vitro*, and concluded that the release of lysosomal enzymes was a secondary process, damage to mitochondria and endoplasmic reticulum occurring earlier. Van Lancker and Holtzer (1959a, 1963), by studying the release of lysosomal enzymes during autolysis *in vitro*, also concluded that the lysosomes were not the primary cause of necrosis.

A role of lysosomes of leucocytes in inflammation is indicated by the work of Janoff and Zweifach (1964). They observed that cationic proteins isolated from lysosomes of leucocytes caused the adhesion of leucocytes along the endothelial surface of capillaries and venules and the subsequent emigration of the leucocytes into the perivascular tissue. Janoff, Schaefer, Scherer, and Bean (1965) reported that a lysosomal extract from leucocytes caused disruption of mast cells. The increase of vascular permeability by lysosomes of leucocytes may be mediated, according to these investigators, by the release of vasotropic agents from mast cells.

According to Halpern (1964), tissue damage in haemorrhagic shock may be due to the release of cathepsin, since the effect was counteracted by a polypeptide possessing anti-protease activity. Bitensky, Chayen, Cunningham and Fine (1963) proposed that in haemorrhagic shock the lysosomes of the reticulo-endothelial cells were disrupted and were no longer able to detoxify endotoxin. Janoff, Weissmann, Zweifach and Thomas (1962) earlier suggested that endotoxin caused release of lysosomal cathepsin.

A possible injurious effect of released lysosomal enzymes during virus infection was discussed by Allison and Sandelin (1963), who observed the release of these enzymes in mouse liver cells and monkey kidney cells infected with hepatitis and vaccinia virus, respectively. Release of lysosomal enzymes was also reported in virus hepatitis by Pagliaro, Giglio, Moli, Catania and Citarrella (1964) by taking needle biopsies from human liver.

IX. Appendix: Methods for the Study of Lysosomes and Phagosomes

A. Isolation of Lysosomes

The theoretical problems connected with the isolation of lysosomes and of other cytoplasmic particles by differential centrifuging or by density gradient centrifuging have been discussed by Schneider and Hogeboom (1951), Allfrey (1959), de Duve (1964), and by others. In the following paragraphs, only a few practical considerations on the isolation of lysosomes and some examples from the literature will be given.

In view of their heterogeneity, it will be difficult or even impossible to isolate most of the lysosomes from a complex tissue as a homogeneous population of granules. Only a proportion of them which fall into a similar range of size and density, may be separated as a single, highly purified fraction. The lysosomes which fall outside this range have either to be discarded, or to be collected separately. This difficulty does not exist for mitochondria or nuclei which can be isolated as a relatively homogeneous class of particles.

The difficulties in preparing highly purified lysosomes may be illustrated by a few examples from the literature. The "light mitochondrial" fraction, prepared by de Duve and co-workers (1955) contained 30–40% of the lysosomal enzymes of the total homogenate but was contaminated by an excess of mitochondria. The lysosomes of this fraction were probably derived mainly from the parenchymal cells, whereas the larger lysosomes from the Kupffer cells sedimented with the heavy mitochondrial or nuclear fraction. Sawant *et al.* (1964c), by using a combination of differential centrifuging and of discontinuous gradient density centrifuging, succeeded in obtaining relatively pure lysosomes from liver (with 6 to 10% microsomal contamination) in relatively good yields. Kidney lysosomes were separated from "nuclear" and "mitochondrial" fractions by differential centrifuging in good purity but poor yield (Straus, 1954, 1956). Fractions containing "large" lysosomes (1–5 μ diameter), "intermediate-sized" lysosomes (0·5–1·5 μ diameter), and "small" lysosomes (0·1–0·5 μ diameter) were collected separately from the same homogenate. Figure 1 shows one of the fractions. The purification was facilitated by the property of the granules to form compact, dark brown layers at the bottom of the centrifuge tubes. Smaller lysosomes, present in the microsomal fraction of kidney homogenates could not be purified sufficiently. Very large lysosomes (phago-lysosomes) formed after injection of egg white were purified by differential centrifuging combined with filtration through a column of non-absorbent cotton. The contaminating nuclei were retained by the cotton, whereas the "droplets" passed into the filtrate (Straus, 1954). Kidney lysosomes of high purity, but probably low yield, were also isolated by Maunsbach (1964a) using differential centrifuging. Shibko and Tappel (1965) have recently obtained highly purified preparations of large kidney lysosomes. These investigators also made use of the fact that the granules form dark brown sediments at the bottom of the centrifuge tube during differential centrifuging. The purified fractions contained 1–2% of all the lysosomes of the homogenate. Since the lysosomes are larger and more numerous in the kidney than in the liver, the purification of kidney lysosomes presents much less difficulty than that of liver lysosomes. Lysosomes isolated from blood cells, pancreas (van Lancker and Holtzer, 1959b), thymus, brain, and intestine (see Section III.) probably were contaminated by other types of granules.

The degree of purification of isolated lysosomes can be estimated by comparing the specific activities of lysosomal enzymes in the isolated fractions with those of the original homogenates. The increase in specific activity depends not only on the degree of purification but also on the number of lysosomes in the tissue. Of course, only those enzyme activities which are predominantly localized in the lysosomes can be used as a measure of purity. For example, the specific activities of hydrolytic enzymes in "pure" liver lysosomes were found to be 60 times higher than in the original homogenate (Sawant *et al.*, 1964c) and in "pure" kidney cortex lysosomes 10–15 times higher than in the original homogenate (Straus, 1956). From these figures, it may be estimated that there are five times more lysosomes in the kidney cortex than in the liver, if the average concentration of hydrolytic enzymes in the lysosomes of both tissues is approximately the same. The contamination of isolated lysosomes by other cytoplasmic granules can be estimated by using characteristic markers for the contaminating particles: DNA for nuclei and cytochrome oxidase (1.9.3.1.) for mitochondria, for example.

The purification of "large" and "intermediate-sized" kidney lysosomes may be followed by microscopic examination. The large kidney lysosomes can be distinguished in sucrose suspension from mitochondria by their greater refractility and greater resistance towards swelling (Straus, 1954, 1956). Isolated fractions of small lysosomes ($0{\cdot}5\,\mu$ or smaller) require electron microscopy for morphological evaluation of their purity. Electron micrographs of liver lysosomes, reproduced from the paper by de Duve (1963a) and of kidney lysosomes, reproduced from the paper by Shibko, Pangborn and Tappel (1965), are shown in Figs. 17 and 18.

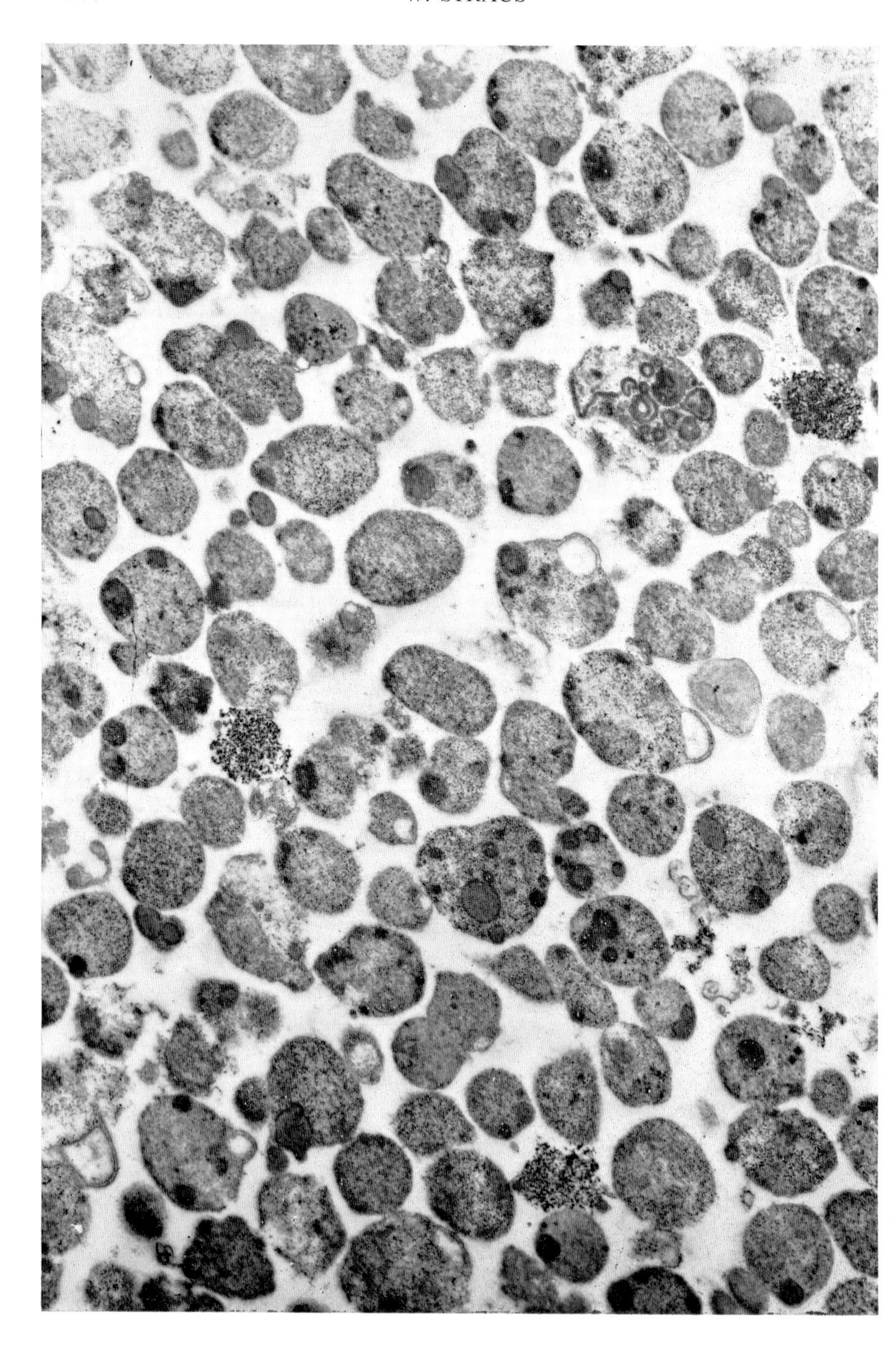

B. CYTOCHEMICAL REACTIONS FOR LYSOSOMAL ENZYMES

Cytochemical staining reactions for acid phosphatase and, to a lesser extent, for esterase, have greatly facilitated the study of the lysosomes (see Novikoff, 1963). Many applications of these cytochemical methods may be found in the articles cited in Section III. Since the technical details for the staining procedures for acid phosphatase are discussed in books on histochemical methods (Pearse, 1960; Burstone, 1962; Barka and Anderson, 1963), only a few comments on the principles of the methods will be made here.

The factors should be pointed out which have made these cytochemical methods so valuable for the study of the lysosomes. There are adequate histochemical methods for relatively few enzymes, among them for phosphatases and esterases. Until a few years ago, the resolution of staining reactions for most of these enzymes was only sufficient for localization at the level of the tissue but not in individual cells. The methods for phosphatases and esterases have now been improved to such an extent that the cellular sites of these enzymes may be visualized not only under the light microscope but also at the dimension of the electron microscope. The second favorable factor for enzyme cytochemistry of lysosomes is the fact that a considerable portion of lysosomal enzymes survives the conventional fixation of the tissue with formaldehyde (approximately 50% in the case of acid phosphatase, Holt and Hicks, 1961a). Thus the cytochemical reactions for lysosomal enzymes may be performed with well-fixed tissues. It may be mentioned, for comparison, that significant amounts of mitochondrial enzymes usually do not survive formaldehyde fixation and that cytochemistry for mitochondrial enzymes, therefore, is only possible with poorly fixed, disintegrating mitochondria.

The procedure for acid phosphatase described by Gomori (1952) is based on the principle that inorganic phosphate, split by acid phosphatase from β-glycerophosphate, or from another phosphate ester, is trapped as lead phosphate at the site of acid phosphatase activity, and lead phosphate is transformed into lead sulphide by ammonium sulphide. Lysosomes are stained dark brown by this reaction. Since the reaction product (lead sulphide) is electron dense, the reaction can be applied to studies at the electron microscope level (Holt and Hicks, 1961b; Miller and Palade, 1964). In the azo-dye procedures (Burstone, 1958; Barka and Anderson, 1962), phosphate esters of naphthol-AS compounds are used as substrates. Enzymic hydrolysis of the phosphate group of the ester by acid phosphatase releases the naphthol-AS compound which couples with a diazotized aromatic amine resulting in a deeply-stained reaction product at the site of acid phosphatase activity.

Staining for esterase also permits precise localization of lysosomes in

FIG. 17. Electron micrograph of isolated liver lysosome, ×21,000. (Reproduced from the paper of de Duve, 1963a, with the permission of the Ciba Foundation.)

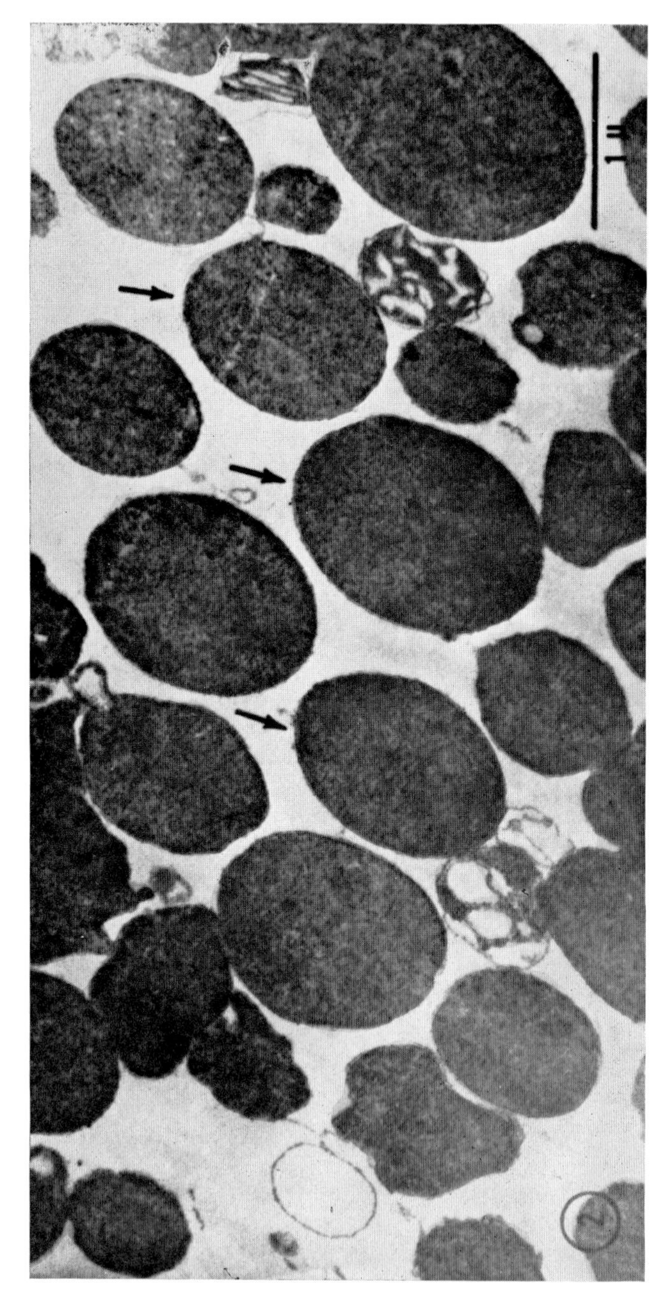

FIG. 18. Electron micrograph of isolated kidney lysosomes. × 22,000. (Reproduced from the paper by Shibko, Pangborn and Tappel (1965), with the permission of the *J. Cell Biol.*)

many tissues although the exact nature of the enzyme and its lysosomal localization is still controversial (see Section II. A. 1., and Holt, 1963). Methods for esterase using indoxyl acetate as substrate were developed by Holt (1956). These and the azo-dye procedures for esterase were used by Shnitka and Seligman (1961) who identified esterase-positive granules (lysosomes) in kidney cells.

Great progress was made recently in the cytochemical procedure for β-glucuronidase using naphthol-AS-BI-glucuronide as substrate and hexazonium pararosanilin as coupler (Hayashi, Nakajima and Fishman, 1964; Hayashi, 1964). Hayashi (1964) reported that the granules which were stained for β-glucuronidase in different tissues had a similar intracellular location as those stained for acid phosphatase. This confirmed the results of biochemical experiments showing that both enzymes were present in lysosomes. Pugh and Walker (1961) had previously used naphthol-AS derivatives for the demonstration of β-glucuronidase in granules of chondriocytes of hyaline cartilage.

Aronson, Hempelman and Okada (1958) and Vorbrodt (1961) reported a cytochemical method for acid deoxyribonuclease, another lysosomal enzyme. In this procedure, highly polymerized DNA and acid phosphatase were added to an incubation mixture similar to the one used for acid phosphatase by Gomori. According to the principle envisaged by Aronson *et al.* (1958), nucleotides split from DNA by deoxyribonuclease are further degraded by acid phosphatase, and the inorganic phosphate is transformed into lead sulfide at the site of DNase activity. The same principle was applied by Atwal, Enright and Frye (1964) to the cytochemical detection of ribonuclease, and ribonuclease-positive granules were demonstrated by these authors in leucocytes.

Naphthol-AS-BI-sulfate-K was introduced recently by Woohsmann and Hartrodt (1964) as substrate for the cytochemical detection of aryl-sulphatases. Goldfischer (1965) demonstrated aryl sulphatase activity in lysosomes of various tissues by using *p*-nitrocatechol sulphate as substrate in conjunction with lead nitrate in a Gomori-type reaction. Hopsu, Arstila and Glenner (1965) detected aryl sulfatase activity in lysosomes at the electron microscope level by using sulphate esters of 8-hydroxy-quinoline or *p*-nitrophenol as substrates and barium salts as precipitating agents.

Hayashi (1965) was able to demonstrate *N*-acetyl-β-glucosaminidase (3.2.1.29) in lysosomes of kidney and liver employing naphthol AS-BI *N*-acetyl-β-glucosaminide as substrate.

C. MARKERS FOR PINOCYTIC AND PHAGOCYTIC VACUOLES

Since foreign materials (proteins, colloidal particles of heavy metals, carbon, or dyes) ingested by pinocytosis or phagocytosis are concentrated in phagosomes, such materials may be used as markers. Many examples may be found in the articles reviewed in Sections V. B., V. D., V. E., and V. F. The most useful markers are those ingested materials which can be detected by light or by electron microscopy in very low concentrations. By labelling proteins or other materials with fluorescent or radioactive groups, it is often possible to determine the intracellular location of these substances by fluorescence microscopy or autoradiography. Phagosomes, marked by fluorescent-labelled proteins may be seen with great clarity (see for example, illustrations in Holtzer and Holtzer, 1960; Cormack and Ambrose, 1962; and Payne and Marsh, 1962). Antigens may be detected in isolated phagosomes by reaction with antisera (Straus and Oliver, 1955). or, *in situ*, by the fluorescent antibody procedure (Coons, 1956). Injected ferritin is a useful marker for phagosomes since it can be identified in electron micrographs by the characteristic shape of the iron micelles (see, for example, Farquhar and Palade, 1962). Ingested haemoglobin, segregated in phagosomes, may be recognized in electron micrographs by its slightly increased electron density (Miller, 1960; Miller and Palade, 1964; Ericsson, 1964). Colloidal particles of HgS, ThO_2, Au, or carbon, segregated in phagocytic vesicles or vacuoles, may also be readily identified in electron micrographs. Iron compounds such as iron-dextran and iron saccharide may be stained in phagosomes by the Prussian blue reaction (Kent *et al.*, 1963a). Wattiaux, Wibo and Baudhuin (1963) observed that rat liver lysosomes accumulated injected Triton WR-1339. The altered granules showed increase in size and decrease in density which facilitated their separation from mitochondria and microsomes by density gradient centrifuging.

The use of an enzyme (ribonuclease) as a marker protein for the study of pinocytosis was introduced by Brachet (1956). An injected enzyme is a good marker for phagosomes if no similar enzyme is present in the cells before treatment, if it can be visualized by a sensitive cytochemical staining reaction, and if it is not toxic to animals. Plant peroxidase was found to be favorable in this respect, and cytochemical procedures for the use of this marker were reported (Straus, 1964c, d). Staining for injected peroxidase with benzidine can be combined with staining for acid phosphatase, or another lysosomal enzyme, by an azo-dye procedure in the same tissue section. Since the staining for the two enzymes can be made to result in contrasting colors, the reaction products for the exogenous protein (peroxidase) and for the endogenous lysomal enzyme can be recognized in the same granules, after fusion of phagosomes with lysosomes. In the experiments illustrated in Figs. 3, 11, 12 and 15 staining for injected plant peroxidase with benzidine alone was applied and Figs. 5, 6, 7 and 13 show successive staining for acid phosphatase and for peroxidase in the same tissue sections. Graham and Karnovsky (1966) observed that 3,3′-diaminobenzidine resulted in a very electron dense reaction product with peroxidase after post-fixation with OsO_4. They used this substrate for the detection of injected horse-radish peroxidase by electron microscopy. A further advantage of the use of peroxidase as a protein marker is the possibility of correlating cytochemical observations with colorimetric assays for the injected protein in isolated fractions (Straus, 1957b, 1962b). As mentioned in Section V. A. 3., an endogenous enzyme, nucleoside phosphatase, may also be used in some cases as a marker for studying phagosomes under the electron microscope.

X. Concluding Comments

In looking back on lysosome research over the last ten years, it must be recognized that great progress has been made. With the aid of cytochemistry and electron microscopy, lysosomes have been identified in many types of epithelial and reticulo-endothelial cells. The heterogeneity of the granules is better understood, and the definition of lysosomes has been broadened to include sub-types of different age and function. It has been observed that pre-existing lysosomes fuse with newly-formed phagosomes, and that this might be a general process by which phagocytizing or absorbing cells incorporate such diverse materials as proteins, vital dyes, various colloidal particles, or bacteria. These results have established the relationship of the lysosomes to the well-known function of the reticulo-endothelial system in protecting the organism against foreign material. The discovery of the fusion of newly-formed phagosomes with pre-existing lysosomes has settled the old controversy as to whether ingested material is segregated in newly-formed or in pre-formed cytoplasmic granules. It has also been found that lysosomes, in addition to breaking down ingested foreign material, take part in the autolytic degradation of components of the cells themselves. This may reflect a role in the renewal of cell constituents or in the alteration of metabolic pathways. Finally, lysosomes have been shown to take part in physiological regression, for example, in the resorption of embyronic organs.

Other properties of the lysosomes are less well understood. Our ignorance applies especially to the origin of lysosomes and of their enzymes. The relationship of the lysosomes to cell injury requires further study, especially the question as to whether cell injury is initiated by the release of lysosomal enzymes *in vivo*. The possible role of lysosomes in the early stages of antibody formation has not yet been studied systematically. It also is not known whether lysosomes take part in the degradation of serum proteins.

The function and properties of the phagosomes also raises new questions. The relationship between small and large phagosomes, pre-existing lysosomes, and the fusion of phagosomes and lysosomes in the same cells, (i.e., the relationship between uptake, intracellular digestion, and transport into and out of the cells) is not fully understood. The process of fusion between lysosomes and phagosomes has not been sufficiently studied. Although it has been observed that uptake of injected proteins occurs mainly in the cells of the reticulo-endothelial system, in the epithelial cells of the kidney, and in the intestine of new-born animals, electron micrographs show micropinocytic vesicles and vacuoles in many other cells. This raises the question as to whether *endogenous* macromolecules are absorbed by this pathway, and if so, what is the significance of this process.

Thus there is no lack of problems waiting for those interested in the study of lysosomes and phagosomes.

Acknowledgement

I should like to thank Dr. J. Brachet, Brussels, and Dr. G. E. Palade, New York, for making valuable suggestions for the improvement of the manuscript, and to thank Dr. Bensch, Dr. de Duve, Dr. Ericsson, Dr. Palade, and Dr. Tappel for allowing the reproduction of figures from their papers.

References

Ada, G. L., Nossal, G. J. V. and Austin, C. M. (1964). *Aust. J. exp. Biol. med. Sci.* **42**, 331.
Alksne, J. F. (1959). *Q. Jl exp. Physiol.* **44**, 51.
Allen, J. M. and Gockerman, J. (1964). *Ann. N.Y. Acad. Sci.* **121**, 616.
Allen, J. M. and Slater, J. J. (1958). *Anat. Rec.* 731.
Allfrey, V. (1959). *In* "The Cell" (J. Brachet and A. E. Mirsky, eds.) Vol. 1, p. 193, Academic Press, New York and London.
Allison, A. C. and Malucci, L. (1964a). *Nature, Lond.* **203**, 1024.
Allison, A. C. and Malucci, L. (1964b). *Lancet* **7374,** 1371.
Allison, A. C. and Sandelin, K. (1963). *J. exp. Med.* **117**, 879.
Allison, A. C. and Young, M. R. (1964). *Life Sci.* **3**, 1407.
Anderson, P. J. and Song, S. K. (1962). *J. Neuropath. exp. Neurol.* **21**, 263.
Anderson, P. J., Cohen, S. and Barka, T. (1961). *Archs Path.* **71**, 89.
Andres, A. G., Seegal, B. C., Hsu, K. C., Rothenberg, M. S. and Chapeau, M. L. (1963). *J. exp. Med.* **117**, 691.
Andresen, N. (1945a). *C.r. Trav. Lab. Carlsberg* **25**, 147.
Andresen, N. (1945b). *C.r. Trav. Lab. Carlsberg* **25**, 169.
Archer, G. T. and Hirsch, J. G. (1963). *J. exp. Med.* **118**, 277.
Aronow, R., Danon, D., Shahar, A. and Aronson, M. (1964). *J. exp. Med.* **120**, 943.
Aronson, J., Hempelman, L. H. and Okada, S. (1958). *J. Histochem. Cytochem.* **6**, 255.
Aronson, N. N. and Davidson, E. A. (1965). *J. biol. Chem.* **240**, 3222.
Aschoff, L. (1924). *Ergebn. inn. Med. Kinderheilk.* **26**, 1.
Ashford, T. P. and Porter, K. R. (1962). *J. Cell Biol.* **12**, 198.
Ashworth, C. T. (1963). *Expl mol. Pathol.* (suppl.) **1**, 83.
Ashworth, C. T., Stembridge, V. A. and Sanders, E. (1960). *Am. J. Physiol.* **198**, 1326.
Aterman, K. (1958). *Lab. Invest.* **7**, 577.
Atwal, O. S., Enright, J. B. and Frye, F. L. (1964). *Proc. Soc. exp. Biol. Med.* **115**, 744.
Bainton, D. F. and Farquhar, M. G. (1966). *J. Cell Biol.* **28**, 277.
Bairati, A. and Lehmann, F. E. (1953). *Expl Cell Res.* **5**, 220.

Balis, J. U., Houst, M. D. and More, R. H. (1964). *Expl mol. Path.* **3**, 511.
Banon, P., Brandes, D. and Frost, J. K. (1964). *Acta Cytol.* **8**, 416.
Barka, T. (1961a). *J. Histochem. Cytochem.* **9**, 542.
Barka, T. (1961b). *J. Histochem. Cytochem.* **9**, 564
Barka, T. (1964). *J. Histochem. Cytochem.* **12**, 229.
Barka, T. and Anderson, P. (1962). *J. Histochem. Cytochem.* **10**, 741.
Barka, T. and Anderson, P. J. (1963). "Histochemistry" Hoeber, New York.
Barka, T., Schaffner, F. and Popper, H. (1961). *Lab. Invest.* **10**, 590.
Barrnett, R. J. and Ball, E. G. (1960). *J. biophys. biochem. Cytol.* **8**, 83.
Barron, K. D. and Sklar, S. (1961). *Neurology, Minneap.* **11**, 866.
Battifora, H., Eisenstein, R., Laing, G. H. and McCreary, P. (1966). *Am. J. Path.* **48**, 421.
Baudhuin, P., Hers, H. G. and Loeb, H. (1964). *Lab. Invest.* **13**, 1139.
Beaufay, H. and de Duve, C. (1959). *Biochem. J.* **73**, 604.
Becker, N. H., Goldfischer, S., Shin, W. Y. and Novikoff, A. B. (1960). *J. biophys. biochem. Cytol.* **8**, 649.
Behnke, O. (1963). *J. Cell Biol.* **18**, 251.
Behnke, O. and Moe, H. (1964). *J. Cell Biol.* **22**, 633.
Bennett, H. S. (1956). *J. biophys. biochem. Cytol.* (suppl.) **2**, 99.
Bennett, H. S. (1963). *J. Histochem. Cytochem.* **11**, 14.
Bensch, K. and King, W. D. (1961). *Science, N.Y.* **133**, 381.
Bensch, K., Gordon, G. and Miller, L. (1964). *J. Cell Biol.*. **21**, 105.
Berthet, J. (1965). *Arch. Biol.* **76**, 367.
Bertini, F. and Brandes, D. (1964). *Fed. Proc.* **23**, 332.
Bessis, M. C. (1963). *Harvey Lect.* **58**, 125.
Birns, M. (1960). *Expl Cell Res.* **20**, 202.
Bitensky, L. (1963). "Ciba Foundation Symposium on Lysosomes", p. 362, Churchill, London.
Bitensky, L., Chayen, J., Cunningham, G. J. and Fine, J. (1963). *Nature, Lond.* **199**, 493.
Björkerud, S. (1964). *Expl mol. Path.* **3**, 369.
Blackwell, J. B. (1965). *J. Path. Bact.* **90**, 259.
Boler, R. K. and Arhelger, R. B. (1966). *Lab. Invest.* **15**, 302.
Borenfreund, E. and Bendich, A. (1961). *J. biophys. biochem. Cytol.* **9**, 81.
Bowers, W. E. (1964). "Fourth Annual Meeting of the American Society of Cell Biology", Cleveland.
Boyden, S. V. (1962). *J. theoret. Biol.* **3**, 123.
Boyden, S. V. and Sorkin, E. (1960). *Immunology* **3**, 272.
Boyden, S. V., Sorkin, E. and Spärk, J. V. (1960). Proceedings of a Symposium "Mechanism of Antibody Formation", p. 237, Publishing House, Czechoslovakian Academy of Science, Prague.
Brachet, J. (1956). *Expl Cell Res.* **10**, 255.
Brachet, J., Decroly-Briers, M. and Hoyez, J. (1958). *Bull. Soc. Chim. biol.* **40**, 2039.
Brambell, F. W. R. (1958). *Biol. Rev.* **33**, 488.
Brambell, F. W. R., Halliday, R. and Morris, F. G. (1958). *Proc. R. Soc.* **B149**, 1.
Brandes, D. (1963). *Lab. Invest.* **12**, 290.
Brandes, D. (1965). *J. Ultrastr. Res.* **12**, 63.
Brandes, D., Groth, D. P. and Gyorkey, F. (1962). *Expl Cell Res.* **28**, 61.
Brandes, D., Buetow, D. E., Bertini, F. and Malkoff, D. B. (1964). *Expl mol. Path.* **3**, 583.

Brandes, D., Bertinin, F. and Smith, E. W. (1965). *Expl mol. Path.* **4**, 245.
Brandt, P. W. (1958). *Expl Cell Res.* **15**, 300.
Brandt, P. W. (1962). *Circulation* **26**, 1075.
Brandt, P. W. and Pappas, G. D. (1960). *J. biophys. biochem. Cytol.* **8**, 675.
Brandt, P. W. and Pappas, G. D. (1962). *J. Cell Biol.* **15**, 55.
Brewer, D. B. and Heath, D. (1963). *Nature, Lond.* **198**, 1015.
Burstone, M. S. (1958). *J. natn. Cancer Inst.* **21**, 523.
Burstone, M. S. (1962). "Enzyme Histochemistry", Academic Press, New York and London.
Byrne, J. M. (1964a). *Q. Jl microsc. Sci.* **105**, 219.
Byrne, J. M. (1964b). *Q. Jl microsc. Sci.* **105**, 343.
Cagan, R. H. and Karnovsky, M. L. (1964). *Nature, Lond.* **204**, 255.
Campbell, D. H. and Garvey, J. S. (1961). *Lab. Invest.* **10**, 1126.
Campbell, D. H. and Garvey, J. S. (1963). *Adv. Immunol.* **3**, 261.
Carasso, N., Favard, P. and Goldfischer, S. (1964). *J. Microsc.* **3**, 297.
Caro, L. G. and Palade, G. E. (1964). *J. Cell Biol.* **20**, 473.
Casley-Smith, J. R. (1965). *Brit. J. exp. Path.* **46**, 35.
Casley-Smith, J. R. and Day, A. J. (1966). *Q. Jl exp. Physiol.* **51**, 1.
Chapman-Andresen, C. (1957). *Expl Cell Res.* **12**, 397.
Chapman-Andresen, C. (1962). *Compt. Rend. Trav. Lab., Carlsberg* **33**, 73.
Chapman-Andresen, C. and Holter, H. (1955). *Expl Cell Res.* (suppl.) **3**, 52.
Chlopin, N. (1927). *Arch. exp. Zellforsch.* **4**, 462.
Chlopin, N. (1928). *Arch. exp. Zellforsch.* **6**, 324.
Chorazy, M., Bendich, A., Borenfreund, E., Ittensohn, O. L. and Hutchison, D. J. (1963). *J. Cell Biol.* **19**, 71.
Chouinard, L. A. (1964). *Can. J. exp. Zool.* **42**, 103.
Clark, S. L., Jr. (1957). *J. biophys. biochem. Cytol.* **3**, 349.
Clark, S. L., Jr. (1959). *J. biophys. biochem. Cytol.* **5**, 41.
Cochrane, C. G. (1963a). *J. exp. Med.* **118**, 489.
Cochrane, C. G. (1963b). *J. exp. Med.* **118**, 503.
Cochrane, C. G., Weigle, W. O. and Dixon, F. J. (1959). *J. exp. Med.* **110**, 481.
Cohen, S., Freeman, T. and McFarlane, A. S. (1961). *Clin. Sci.* **20**, 161.
Cohn, Z. A. (1963a). *J. exp. Med.* **117**, 27.
Cohn, Z. A. (1963b). *J. exp. Med.* **117**, 43.
Cohn, Z. A. (1964). *J. exp. Med.* **120**, 869.
Cohn, Z. A. and Hirsch, J. G. (1960a). *J. exp. Med.* **112**, 983.
Cohn, Z. A. and Hirsch, J. G. (1960b). *J. exp. Med.* **112**, 1015.
Cohn, Z. A. and Benson, B. (1965a). *J. exp. Med.* **121**, 153.
Cohn, Z. A. and Benson, B. (1965b). *J. exp. Med.* **121**, 835.
Cohn, Z. A. and Wiener, E. (1963a). *J. exp. Med.* **118**, 991.
Cohn, Z. A. and Wiener, E. (1963b). *J. exp. Med.* **118**, 1009.
Confer, D. B. and Stenger, R. J. (1964). *Am. J. Path.* **45**, 533.
Coons, A. H. (1956). *Int. Rev. Cytol.* **5**, 1.
Coons, A. H., Leduc, E. H. and Kaplan, M. H. (1951). *J. exp. Med.* **93**, 173.
Cormack, D. H. and Ambrose, E. J. (1962). *J. R. Microsc. Soc.* **81**, 11.
Cormack, D. H., Easty, G. C. and Ambrose, E. J. (1961). *Nature, Lond.* **190**, 1207.
Daems, W. T. (1962). "Mouse Liver Lysosomes and Storage" Thesis, University of Leiden, Holland. "Luctor et Emergo", Leiden.
Daems, W. T. and Oort, J. (1962). *Expl Cell Res.* **28**, 11.
Dales, S. (1962). *J. Cell Biol* **13**, 303.

Dales, S. (1963). *J. Cell Biol.* **18**, 51.
Dales, S. and Choppin, P. W. (1962). *Virology* **18**, 489.
Dales, S. and Kajioka, R. (1964). *Virology* **24**, 278.
Dalq, A. M. (1963). "Ciba Foundation Symposium on Lysosomes", p. 226. Churchill, London.
Dannenberg, A. M., Jr., Burstone, M. S., Walter, P. C. and Kinsley, J. W. (1963). *J. Cell Biol.* **17**, 465.
Davies, J. (1954). *Am. J. Anat.* **94**, 45.
Day, A. J. (1964). *J. Atherosclerosis Res.* **4**, 117.
Deane, H. W. (1964). *Anat. Rec.* **149**, 453.
de Duve, C. (1959a). *In* "Subcellular Particles" (T. Hayashi, ed.), p. 128. Ronald Press, New York.
de Duve, C. (1959b). *Expl Cell Res.* (suppl.) **7**, 169.
de Duve, C. (1963a). "Ciba Foundation Symposium on Lysosomes", p. 1. Churchill, London.
de Duve, C. (1963b). *Sci. Am.* **208**, 64.
de Duve, C. (1964). *J. theoret. Biol.* **6**, 33.
de Duve, C. and Wattiaux, R. (1966). *Ann. Rev. Physiol.* **28**, 435.
de Duve, C., Pressman, B. C., Gianetto, R., Wattiaux, R. and Appelmans, F. (1955). *Biochem. J.* **60**, 604.
de Duve, C., Wattiaux, R. and Wibo, R. (1962). *Biochem. Pharmacol.* **9**, 97.
de Reuck, A. V. S. and Cameron, P. (1963). "Ciba Foundation Symposium on Lysosomes", Churchill, London.
Desai. I. D., Sawant, P. L. and Tappel, A. L. (1964). *Biochim. biophys, Acta* **86**, 277.
de Terra, N. and Rustad, R. C. (1959). *Expl Cell Res.* **17**, 191.
Dianzani, M. U. (1963). "Ciba Foundation Symposium on Lysosomes", p. 335, Churchill, London.
Diengdoh, J. V. and Turk, J. L. (1965). *Nature, Lond.* **207**, 1405.
Dingle, J. T. (1963). "Ciba Foundation Symposium on Lysosomes", p. 384. Churchill, London.
Dingle, J. T., Sharman, I. M. and Moore, T. (1966). *Biochem. J.* **98**, 476.
Dixon, F. J. (1963). *Harvey Lect.* **58**, 21.
Doniach, J. and Weinbren, K. (1952). *Brit. J. exp. Path.* **33**, 499.
Dumonde, D. C., Roodyn, D. B. and Prose, P. H. (1965). *Immunology* **9**, 177.
Easton, J. M., Goldberg, B. and Green, H. (1962a). *J. Cell Biol.* **12**, 437.
Easton, J. M., Goldberg, B. and Green, H. (1962b). *J. exp. Med.* **115**, 275.
Easty, G. C. (1964). *Brit. J. Cancer* **18**, 368.
Easty, G. C., Yarnell, M. M. and Andrews, R. D. (1964). *Brit. J. Cancer* **18**, 354.
Eekhout, Y. (1964). *Arch. internat. Physiol. Biochim.* **72**, 316.
Elliot, A. M. and Bak, I. J. (1964). *J. Cell Biol.* **20**, 113.
Elsbach, P. (1964). *Biochim. biophys. Acta* **84**, 8.
Elsbach, P. (1965a). *Biochim. biophys. Acta* **98**, 402.
Elsbach, P. (1956b). *Biochim. biophys. Acta* **98**, 420.
Elsbach, P. and Rizak, M. A. (1963). *Am. J. Physiol.* **205**, 1154.
Epstein, M. A., Hummeler, K. and Berkaloff, A. (1964). *J. exp. Med.* **119**, 291.
Ericsson, J. L. E. (1964). *Acta path. microbiol. Scand.* (suppl.) **168**, 1.
Ericsson, J. L. E. and Trump, B. F. (1964). *Lab. Invest.* **13**, 1427.
Essner, E. (1960). *J. biophys. biochem. Cytol.* **7**, 329.
Essner, E. and Novikoff, A. B. (1960). *J. Ultrastr. Res.* **3**, 374.
Evans, W. H. and Karnovsky, M. L. (1961). *J. biol. Chem.* **236**, PC30–PC32.
Evans, H. M. and Scott, K. (1921). *Contr. Embryol.* **10**, 1.

Farquhar, M. G. and Palade, G. E. (1960). *J. biophys. biochem. Cytol.* **7**, 297.
Farquhar, M. G. and Palade, G. E. (1961). *J. exp. Med.* **114**, 699.
Farquhar, M. G. and Palade, G. E. (1962). *J. Cell Biol.* **13**, 55.
Fedorko, M. E. and Morse, S. I. (1965). *J. exp. Med.* **121**, 39.
Fennell, R. H. and Santamaria, A. (1962). *Am. J. Path.* **41**, 521.
Fernando, N. V. P. and Movat, H. Z. (1964). *Expl mol. Path.* **3**, 87.
Fisher, E. R. (1964). *Expl mol. Path.* **3**, 304.
Fishman, M. (1961). *J. exp. Med.* **114**, 837.
Fishman, M. and Adler, F. L. (1963). *J. exp. Med.* **117**, 595.
Flaxman, A. and Mulnard, J. (1961). *Archs Biol.* **72**, 573.
Frankland, M. and Wynn, C. H. (1962). *Biochem. J.* **84**, 20P.
Franks, J. J., Mosser, E. L. and Anstadt, G. L. (1963). *J. gen. Physiol.* **46**, 415.
Freeman, T., Gordon, A. H. and Humphrey, J. H. (1958). *Brit. J. exp. Path.* **39**, 459.
French, J. E. and Morris, B. (1960). *J. Path. Bact.* **79**, 11.
Friedman, M., Byers, S. O. and St. George, S. (1962). *J. Clin. Invest.* **41**, 828.
Fuxe, K. and Nilsson, O. (1963). *Expl Cell Res.* **32**, 109.
Gansler, H. and Rouiller, C. (1956). *Schweiz. Z. allg. Path. Bakt.* **19**, 217.
Garvey, J. S. and Campbell, D. H. (1957). *J. exp. Med.* **105**, 361.
Gedigk, P. and Fischer, R. (1958). *Virchows Arch. path. Anat. Physiol.* **331**, 341.
Gedigk, P. and Fischer, R. (1959). *Virchows Arch. path. Anat. Physiol.* **332**, 431.
Gérard, P. and Cordier, R. (1934). *Biol. Rev.* **9**, 110.
Gianetto, R. and de Duve, C. (1955). *Biochem. J.* **59**, 433.
Gitlin, D., Landing, B. H. and Whipple, A. (1951). *Proc. Soc. exp. Biol. Med.* **78**, 631.
Gitzelmann, R., Davidson, E. A. and Osinshak, J. (1964). *Biochim. biophys. Acta* **85**, 69.
Goldfischer, S. (1965). *J. Histochem. Cytochem.* **13**, 520.
Gomori, G. (1941). *Archs Pathol.* **32**, 189.
Gomori, G. (1952). "Microscopic Histochemistry" The University of Chicago Press.
Gonzales, I. E. (1963). *Ann. Histochim.* **8**, 335.
Gonzales, F. and Karnovsky, M. J. (1961). *J. biophys. biochem. Cytol.* **9**, 299.
Gordon, A. H. (1957). *Biochem. J.* **66**, 255.
Gordon, G. B., Miller, L. R. and Bensch, K. G. (1965). *J. Cell Biol.* **25**, 41.
Graham, R. C. and Karnovsky, M. J. (1966). *J. Histochem. Cytochem.* **14**, 291.
Green, M. H. and Verney, E. L. (1956). *J. Histochem. Cytochem.* **4** 106.
Green, C. and Webb, J. A. (1964). *Biochim. biophys. Acta* **84**, 404.
Halpern, B. N. (1964). *Proc. Soc. exp. Biol. Med.* **115**, 273.
Hampton, J. C. (1958). *Acta Anat.* **32**, 262.
Hancox, N. M. and Boothroyd, B. (1961). *J. biophys. biochem. Cytol.* **11**, 651.
Handelman, C. S., Morse, A. and Irving, J. T. (1964). *Am. J. Anat.* **115**, 363.
Hanssen, O. E. (1960). *Acta path. microbiol. Scand.* **49**, 297.
Haurowitz, F., Reller, H. H. and Walter, H. (1955). *J. Immunol.* **75**, 417.
Harrington, J. F. and Altschul, A. M. (1963) *Fed Proc* **22**, 475 (abstract).
Hayashi, M. (1964). *J. Histochem. Cytochem.* **12**, 659.
Hayashi, M. (1965). *J Histochem. Cytochem.* **13**, 355.
Hayashi, M., Nakajima, Y. and Fishman, W. H. (1964). *J. Histochem. Cytochem.* **12**, 293.
Hendley, D. D. and Strehler, B. L. (1965). *Biochim. biophys. Acta* **99**, 406.
Hess, R. and Pearse, A. G. E. (1958). *Brit. J. exp. Path.* **39**, 292.
Hess, R. and Stäubli, W. (1963). *Am. J. Path.* **43**, 301.

Hirsch, J. G. (1962). *J. exp. Med.* **116**, 827.
Hirsch, J. G., Bernheimer, A. W. and Weissmann, G. (1963). *J. exp. Med.* **118**, 223.
Hirschhorn, R., Kaplan, J. M., Goldberg, A. F., Hirschhorn, K. and Weissmann, G. (1965). *Science, N.Y.* **147**, 55.
Holt, S. J. (1954). *Proc. R. Soc.* **B142**, 160.
Holt, S. J. (1956). *J. Histochem. Cytochem.* **4**, 541.
Holt, S. J. (1963). "Ciba Foundation Symposium of Lysosomes", p. 114. Churchill, London.
Holt, S. J. and Hicks, R. M. (1961a). *J. biophys. biochem. Cytol.* **11**, 31.
Holt, S. J. and Hicks, R. M. (1961b). *J. biophys. biochem. Cytol.* **11**, 47.
Holt, S. J. and Withers, R. F. J. (1952). *Nature, Lond.* **170**, 1012.
Holter, H. (1960). *Int. Rev. Cytol.* **8**, 481.
Holter, H. (1961). *Proc. Fifth Int. Congr. Biochem., Moscow*, Vol. 2, p. 248. Pergamon Press, New York.
Holtzer, H. and Holtzer, S. (1960). *C.r. Trav. Lab. Carlsberg* **31**, 373.
Hopsu, V K., Arstila, A. and Glenner, G. G. (1965). *Ann. Med. Exp. Fenn.* **43**, 114.
Horn, R. G., Spicer, S. S. and Wetzel, B. K. (1964). *Am. J. Path.* **45**, 327.
Hosoya, T. (1963). *J. Biochem. Japan* **53**, 86.
Hruban, Z., Spargo, B., Swift, H., Wissler, R. W. and Kleinfeld, R. G. (1963). *Am. J. Path.* **42**, 657.
Hsu, L. and Tappel, A. L. (1964). *J. Cell Biol.* **23**, 233.
Ingraham, J. S. (1955). *J. infect. Dis.* **96**, 129.
Jacques, P., Ennis, R. S. and de Duve, C. (1964). "Fourth Annual Meeting American Society of Cell Biology", Cleveland.
Janigan, D. T. and Santamaria, A. (1961). *Am. J. Path.* **39**, 175.
Janoff, A. and Zweifach, B. W. (1964). *Science, N.Y.* **144**, 1456.
Janoff, A., Weissmann, G., Zweifach, B. W. and Thomas, L. (1962). *J. exp. Med.* **116**, 451.
Janoff, A., Schaefer, S., Scherer, J. and Bean, M. A. (1965). *J. exp. Med.* **122**, 841.
Jansco, M. (1955). *Acta med. Hung.* **7**, 173.
Jansco, M. and Jansco-Gabor, A. (1952). *Experientia* **8**, 465.
Jansco, M. and Jansco-Gabor, A. (1954). *Experientia* **10**, 256.
Jenkin, C. R. and Rowley, D. (1961). *J. exp. Med.* **114**, 363.
Jennings, M. A., Marchesi, V. T. and Florey, H. (1962). *Proc. R. Soc.* **B156**, 14.
Jordan, S. W. (1964). *Expl mol. Path.* **3**, 183.
Journey, L. J. (1964). *Cancer Res.* **24**, 1391.
Jung, G. and Nüsslein, G. (1962). *Endokrinologie* **42**, 137.
Kanczak, N. M., Krall, J. I., Hayes, E. R. and Elliot, W. B. (1965). *J. Cell Biol.* **24**, 259.
Karnovsky, M. L. (1962). *Physiol. Rev.* **42**, 143.
Karnovsky, M. L. and Wallach, D. F. H. (1961). *J. biol. Chem.* **236**, 1895.
Katz, J., Rosenfeld, S. and Sellers, A. L. (1961). *Am. J. Physiol.* **200**, 1301.
Kaye, G. I. and Pappas, G. D. (1962). *J. Cell Biol.* **12**, 457.
Kedrowski, B. (1933a). *Z. Zellforsch. mikrosk. Anat.* **17**, 547.
Kedrowski, B. (1933b). *Z. Zellforsch. mikrosk. Anat.* **17**, 587.
Keiser, H., Weissmann, G. and Bernheimer, A. W. (1964). *J. Cell Biol.* **22**, 101.
Kent, G., Volini, F. I., Orfei, E., Minick, O. T. and de la Huerga, J. (1963a). *Lab. Invest.* **12**, 1094.
Kent, G., Minick, O. T., Volini, F. I., Orfei E. and de la Huerga, J. (1963b). *Lab. Invest.* **12**, 1102.

Klamer, B. and Fennell, R. A. (1963). *Expl Cell Res.* **29**, 166.
Koenig, H. (1962). *Nature, Lond.* **195**, 782.
Koenig, H. (1963a). *J. Histochem. Cytochem.* **11**, 556.
Koenig, H. (1963b). *J. Cell Biol.* **19**, 87A.
Koenig, H. and Jibril, A. (1962). *Biochim. biophys. Acta* **65**, 543.
Koenig, H., Gaines, D., McDonald, T., Gray, R. and Scott, J. (1964). *J. Neurochem.* **11**, 729.
Kojima, M. and Imai, Y. (1962). *Tokuhu J. exp. Med.* **76**, 161.
Kruse, H. and McMaster, P. D. (1949). *J. exp. Med.* **90**, 425.
Kuff, E. L. and Schneider, W. C. (1954). *J. biol. Chem.* **206**, 677.
Kurtz, S. M. and Feldman, J. D. (1962). *Lab. Invest.* **11**, 167.
Lacy, D. and Taylor, A. B. (1962). *Am. J. Anat.* **110**, 155.
Ladman, A. J., Padykula, H. A. and Strauss, E. W. (1963). *Am. J. Anat.* **112**, 389.
Lagunoff, D. (1964). *C.r. Trav. Lab. Carlsberg* **34**, 433.
Lambert, P. P. (1936). *Archs Biol.* **47**, 125.
Lapresle, C. and Durieux, J. (1957). *Ann. Inst. Pasteur* **92**, 62.
Latta, H., Gitlin, D. and Janeway, C. A. (1951). *Arch. Path.* **51**, 260.
Leake, E. S. and Myrvik, Q. N. (1964). *Brit. J. exp. Path.* **45**, 384.
Leduc, E. H. and Wilson, J. W. (1959). *J. Histochem. Cytochem.* **7**, 8.
Lejeune, N., Thinès-Sempoux, D. G. and Hers, H. G. (1963). *Biochem. J.* **86**, 16.
Lewis, W. H. (1931). *Johns Hopkins Hosp. Bull.* **49**, 17.
Lewis, W. H. (1937). *Am. J. Cancer* **29**, 666.
Lindner, E. (1957). *Z. Zellforsch. mikrosk. Anat.* **45**, 702.
Lobel, B. L. and Deane, H. W. (1962). *Endocrinology* **70**, 567.
Locke, M. and Collins, J. V. (1965). *J. Cell Biol.* **26**, 857.
Lockwood, W. R. and Allison, F. (1964a). *Brit. J. Exp. Path.* **44**, 593.
Lockwood, W. R. and Allison, F. (1964b). *Brit. J. Exp. Path.* **45**, 294.
Longley, J. B. and Burstone, M. S. (1963). *Am. J. Path.* **42**, 643.
Lucy, J. A., Dingle, J. T. and Fell, H. B. (1961). *Biochem. J.* **79**, 500.
Lutzner, M. A. (1964). *Fed. Proc.* **23**, 441.
Mabry, D. S., Wallace, J. H., Dodd, M. C. and Wright, C. S. (1956). *J. Immunol.* **76**, 62.
Maeir, D. M. (1961). *Expl Cell Res.* **23**, 200.
Majno, G. (1965). "Ultrastructure of the Vascular Membrane", *In* "Handbook of Physiology", Vol. 3, Sect. 2., Baltimore, Williams and Wilkins, p. 2293
Majno, G. and Palade, G. E. (1961). *J. biophys. biochem. Cytol.* **11**, 571.
Majno, G., Palade, G. E. and Schoefl, G. I. (1961). *J. biophys. biochem. Cytol.* **11**, 607.
Mancini, R. E., Villar, O., Gomez, C., Delaccha, J. M., Davidson, O. W. and Castro, A. (1961). *J. Histochem. Cytochem.* **9**, 356.
Marchesi, V. T. and Barrnett, R. J. (1963). *J. Cell Biol.* **17**, 547.
Marchesi, V. T. and Barrnett, R. J. (1964). *J. Ultrastr. Res.* **10**, 103.
Marcus, A. J., Zucker-Franklin, D., Safier, L. B. and Ullman, H. L. (1966). *J. clin. Invest.* **45**, 14.
Marks, N. and Lajtha, A. (1963). *Biochem. J.* **89**, 438.
Mast, S. O. and Doyle, W. L. (1934). *Protoplasma* **20**, 555.
Matile, P., Balz, J. P. and Semadeni, E. (1965). *Z. Naturforsch.* **20b**, 693.
Maunsbach, A. B. (1964a). *Nature, Lond.* **202**, 1131.
Maunsbach, A. B. (1966). *J. Ultrastr. Res.* **14**, 167.

Maunsbach, A. B., Madden, S. C. and Latta, H. (1962). *Lab. Invest.* **11**, 421.
Mayersbach, H. (1957). *Z. Zellforsch. Mikrosk. Anat.* **45**, 483.
McBride, O. W. and Korn, E. D. (1964). *J. Lipid Res.* **5**, 459.
McCluskey, R. T. and Benacerraf, B. (1959). *Am. J. Path.* **35**, 275.
McFarlane, A. S. (1964). *In* "Metabolism of Plasma Proteins" (H. N. Munro, and J. B. Allison, eds.) p. 297. Academic Press, New York and London.
Mego, J. L. and McQueen, J. D. (1965). *Biochim. biophys. Acta* **100**, 136.
Menefee, M. G., Mueller, C. B., Bell, A. L. and Myers, J. K. (1964a). *J. exp. Med.* **120**, 1129.
Menefee, M. G., Mueller, B., Miller, T. B., Myers, J. K. and Bell, A. L. (1964b). *J. exp. Med.* **120**, 1139.
Meng, H. C. (editor) (1964). "Lipid Transport", Charles Thomas, Springfield, Illinois.
Metchnikoff, E. (1892). "Pathologie Comparée de l'Inflammation", Masson, Paris.
Miller, A. T., Jr., Hale, D. M. and Alexander, K. D. (1965). *J. Cell Biol.* **27**, 305.
Miller, F. (1960). *J. biophys. biochem. Cytol.* **8**, 689.
Miller, F. (1962). *Proc. Fifth Intern. Congr. Electron Microsc.*, Philadelphia, Vol. 2, Q-2.
Miller, F. and Palade, G. E. (1964). *J. Cell Biol.* **23**, 519.
Mitchison, J. M. (1950). *Nature, Lond.* **166**, 313.
Miyawaki, H. (1965). *J. natn. Cancer Inst.* **34**, 601.
Moe, H. and Behnke, O. (1962). *J. Cell Biol.* **13**, 168.
Möllendorff, W. v. (1918). *Arch. Mikrosk. Anat.* **90**, 463.
Möllendorff, W. v. (1920). *Ergebn. Physiol.* **18**, 141.
Möllendorff, W. v. (1925). *Z. Zellforsch. Mikrosk. Anat.* **2**, 127.
Moore, B. W. and Angeletti, P. U. (1961). *Ann. N.Y. Acad. Sci.* **94**, 659.
Moore, D. H. and Ruska, H. (1957). *J. biophys. biochem. Cytol.* **3**, 457.
Morrison, A. B. and Panner, B. J. (1964). *Am. J. Path.* **45**, 295.
Movat, H. Z., Urinhara, T., MacMorine, D. L. and Burke, J. S. (1964). *Life Sci.* **3**, 1025.
Movat, H. Z., Mustard, J. F., Taichman, N. S. and Urinhara, T. (1965). *Proc. Soc. exp. Biol. Med.* **120**, 232.
Müller, M. (1962). *Acta biol. hung.* **13**, 283.
Müller, M. and Törö, I. (1962). *J. Protozool.* **9**, 98.
Müller, M., Tóth, J. and Törö, I. (1962). *Acta biol. Hung.* **13**, 105.
Müller, M., Röhlich, P., Tóth, J. and Törö, I. (1963a). "Ciba Foundation Symposium on Lysosomes", p. 201. Churchill, London.
Müller, M., Törö, I., Polgár, M. and Druga, A. (1963b). *Acta biol. Hung.* **14**, 209.
Mulnard, J. (1961). *Archs Biol.* **72**, 523.
Munro, T. R., Daniel, M. R. and Dingle, J. T. (1964). *Expl Cell Res.* **35**, 515.
Murray, M. I. (1963). *J. exp. Med.* **117**, 139.
Nachmias, V. T. and Marshall, J. M., Jr. (1962). IUB/IUBS "Symposium on Biological Structure and Function", Vol. 2, p. 605.
Nadler, N. J., Sarkar, S. K. and Leblond, C. P. (1962). *Endocrinology* **71**, 120.
Nagel, W. and Willig, F. (1964a). *Nature, Lond.* **201**, 617.
Nagel, W. and Willig, F. (1964b). *Naturwissenschaften* **51**, 115.
Nairn, R. C., Chadwick, C. S. and McEntegart, M. G. (1958). *J. Path. Bact.* **76**, 143.
Napolitano, L. (1963). *J. Cell Biol.* **18**, 478.
Neil, M. W. and Horner, M. W. (1964a). *Biochem. J.* **92**, 217.

Neil, M. W. and Horner, M. W. (1964b). *Biochem. J.* **93**, 220.
Nelson, E. L. and Buras, N. S. (1963). *J. Immunol.* **90**, 412.
Nelson, R. A. and Lebrun, J. (1956). *J. Hyg., Camb.* **54**, 8.
Nicol, T., Vernon-Roberts, D. C. and Quantock, D. C. (1965). *J. Endocrinol.* **33**, 365.
Niemi, M. and Pearse, A. G. E. (1960). *J. biophys. biochem. Cytol.* **8**, 279.
Nilsson, O. (1962a). *J. Ultrastr. Res.* **6**, 413.
Nilsson, O. (1962b). *J. Ultrastr. Res.* **6**, 422.
Nilsson, O. (1962c). *Expl Cell Res.* **26**, 334.
North, R. J. and Mackaness, G. B. (1963a). *Brit. J. Exp Path.* **44**, 601.
North, R. J. and Mackaness, G. B. (1963b). *Brit. J. Exp. Path.* **44**, 608.
Nossal, G. J. V., Ada, G. L. and Austin, C. M. (1964). *Aust. J. exp. Biol. med. Sci.* **42**, 311.
Novikoff, A. B. (1959). *J. biophys. biochem. Cytol.* **6**, 136.
Novikoff, A. B. (1960). *In* "Developing Cell Systems and their Control" (D. Rudnick, ed.), p. 167. The Ronald Press, New York.
Novikoff, A. B. (1961). *In* "The Cell" (J. Brachet and A. E. Mirsky, eds.), Vol. 2, p. 423. Academic Press, New York and London.
Novikoff, A. B. (1963). "Ciba Foundation Symposium on Lysosomes", p. 36. Churchill, London.
Novikoff, A. B. and Essner, E. (1962). *J. Cell Biol.* **15**, 140.
Novikoff, A. B. and Shin, W. Y. (1964). *J. Microsc.* **3**, 187.
Novikoff, A. B., Beaufay, H. and de Duve, C. (1956). *J. biophys. biochem. Cytol.* (suppl.) **2**, 179.
Ogawa, K., Mizuno, N. and Okamoto, M. (1961). *J. Histochem. Cytochem.* **9**, 202.
Okunuki, K. (1961). *Adv. Enzymol.* **23**, 29.
Oliver, J., McDowell, M C. and Lee, Y. C. (1954). *J. exp. Med.* **99**, 589.
Olszewska, M. J. and Gabara, B. (1964). Protoplasma **59**, 163.
O'Neill, C. H. (1964). *Expl Cell Res.* **35**, 477.
Oren, R., Farnham, A. E., Saito, K., Milofsky, E. and Karnovsky, M. L. (1963). *J. Cell Biol.* **17**, 487.
Osborne, P. J. and Miller, A. T., Jr. (1962). *Biol. Bull.* **123**, 589.
Osinshak, J. (1964). *J. Cell Biol.* **21**, 35.
Oudea, P. R. (1963). *Lab. Invest.* **12**, 386.
Pagliaro, L., Giglio, F., Moli, S. L., Catania, A. and Citarrella, P. (1964). *J. Lab. clin. Med.* **63**, 977.
Paigen, K. and Griffiths, S. K. (1959). *J. biol. Chem.* **234**, 299.
Palade, G. E. and Bruns, R. R. (1964). *In* "Small Blood Vessel Involvement in Diabetes Mellitus" (Marvin D. Siperstein *et al.*, eds.), p. 39, Am. Inst. of Biological Sciences, Washington, D.C.
Palade, G. E. (1961). *Circulation* **24**, 368.
Palay, S. L. and Karlin, L. J. (1959). *J. biophys. biochem. Cytol.* **5**, 373.
Palay, S. L. and Revel, J. P. (1964). *In* "Lipid Transport" (H. C. Meng, ed.), p. 33, Charles Thomas, Springfield, Illinois.
Pappas, G. D. and Kaye, G. I. (1964). *Eleventh Int. Congr. Cell Biol.*, Providence, Rhode Island.
Pappas, G. D. and Tennyson, V. M. (1962). *J. Cell Biol.* **15**, 227.
Parker, F. (1960). *Am. J. Path.* **36**, 19.
Parker, J. W., Wakasa, H. and Lukes, R. J. (1965). Lab. Invest. **14**, 1736.
Parker, M. V., Swann, H. G. and Sinclair, J. G. (1962). *Tex. Rep. Biol. Med.* **20**, 425.

Patterson, R., Suszko, J. M. and Pruzansky, J. J. (1962). *J. Immunol.* **89**, 471.
Payne, L. C. and Marsh, C. L. (1962). *J. Nutr.* **76**, 151.
Pearse, A. G. E. (1960). "Histochemistry", Little, Brown, and Co., Boston.
Penn, N. W. (1960). *Biochim. biophys. Acta* **37**, 55.
Penn, N. W. (1961). *Biochim. biophys. Acta* **53**, 490.
Perkins, E. H. and Leonard, M. R. (1962). *J. Immunol.* **90**, 228.
Pugh, D. and Walker, P. G. (1961). *J. Histochem. Cytochem.* **9**, 105.
Quertier, J. and Brachet, J. (1959). *Archs Biol.* **70**, 153.
Rahman, Y. E. (1962a). *J. Cell Biol.* **13**, 253.
Rahman, Y. E. (1962b). *Proc. Soc. exp. Biol. Med.* **109**, 378.
Rahman, Y. E. (1964). *Biochim. biophys. Acta* **90**, 440.
Raimondi, A. J. (1964). *Archs Neurol.* **11**, 174.
Rambourg, A., Neutra, M. and Leblond, C. P. (1966). *Anat. Rec.* **154**, 41.
Randerath, E. (1947). *Virchows Arch. path. Anat. Physiol.* **314**, 388.
Rather, L. J. (1948). *J. exp. Med.* **87**, 163.
Rather, L. J. (1952). *Medicine* **31**, 357.
Reid, E. and Nodes, J. T. (1963). *Nature, Lond.* **199**, 176.
Rhodin, J. (1954). "Correlation of Ultrastructural Organization and Function in Normal and Experimentally Changed Proximal Convoluted Tubule Cells of Mouse Kidney." Karolinska Institutet, Stockholm, Aktiebolaget Godvil.
Riddle, J. M. and Barnhart, M. L. (1964). *Am. J. Path.* **45**, 805.
Robbins, E., Marcus, P. I. and Gonatas, N. K. (1964). *J. Cell Biol.* **21**, 49.
Roberts, J. and Quastel, J. H. (1964). *Nature, Lond.* **202**, 85.
Robineaux, R. and Frederic, J. (1955). *C.r. Soc. Biol.* **149**, 486.
Robineaux, R. and Pinet, J. (1960). "Ciba Foundation Symposium on Cellular Aspects of Immunity", p. 5. Churchill, London.
Robinson, D. S. (1964). *In* "Lipid Transport" (H. C. Meng, ed.), p. 194. Charles C. Thomas, Springfield, Illinois.
Rodbell, M., Scow, R. O. and Chernick, S. C. (1964). *J. biol. Chem.* **239**, 385.
Rose, G. G. (1957). *J. biophys. biochem. Cytol.* **3**, 697.
Rosenbaum, R. M. and Ditzion, B. (1963). *Biol. Bull.* **124**, 211.
Rosenbaum, R. M. and Rolon, C. I. (1960a). *Biol. Bull.* **118**, 315.
Rosenbaum, R. M. and Rolon, C. I. (1960b). *Anat. Rec.* **137**, 389.
Rosenbaum, R. M. and Rolon, C. I. (1962). *Histochemie* **3**, 1.
Rosenbaum, R. M. and Wittner, M. (1962). *Arch. Protistenk.* **106**, 223.
Rosenbluth, J. and Wissig, S. L. (1964). *J. Cell Biol.* **23**, 307.
Rossi, F. and Zatti, M. (1964a). *Experientia* **20**, 21.
Rossi, F. and Zatti, M. (1964b). *Brit. J. exp. Path.* **45**, 548.
Rossi, F. and Zatti, M. (1966). *Biochim. biophys. Acta* **113**, 395.
Rostgaard, J. and Barrnett, R. J. (1965). *Anat. Rec.* **152**, 325.
Roth, L. E. (1960). *J. Protozool.* **7**, 176.
Roth, T. F. and Porter, K. R. (1964). *J. Cell Biol.* **20**, 313.
Rouiller, C. and Modjtabai, A. (1958). *Ann. Anat. Path.* **3**, 223.
Roy, A. B. (1958). *Biochem. J.* **68**, 519.
Rustad, R. C. (1959). *Nature, Lond.* **183**, 1058.
Ruyter, J. H. C. (1964). *Histochemie* **3**, 521.
Ryser, H. (1963). *Lab. Invest.* **12**, 1009.
Ryser, H., Aub, J. C. and Caulfield, J. B. (1962a). *J. Cell Biol.* **17**, 437.
Ryser, H., Caulfield, J. B. and Aub, J. C. (1962b). *J. Cell Biol.* **14**, 255.
Sabesin, S. M. (1963). *Proc. Soc. exp. Biol. Med.* **112**, 667.
Sabin, F. R. (1939). *J. exp. Med.* **70**, 67.

Salzmann, R. and Weber, R. (1963). *Experientia* **19**, 352.
Sandborn, E., Koen, P. F., McNabb, J. D. and Moore, G. (1964). *J. Ultrastr. Res.* **11**, 123.
Sanders, E. and Ashworth, C. T. (1961). *Expl Cell Res.* **22**, 137.
Sawant, P. L., Desai, I. D. and Tappel, A. L. (1964a). *Archs Biochem. Biophys.* **105**, 247.
Sawant, P. L., Desai, I. D. and Tappel, A. L. (1964b). *Biochim. biophys. Acta* **85**, 93.
Sawant, P. L., Shibko, S., Kumta, U. S. and Tappel, A. L. (1964c). *Biochim. biophys. Acta* **85**, 82.
Sbarra, A. J. and Karnovsky, M. L. (1959). *J. biol. Chem.* **234**, 1355.
Sbarra, A. J., Shirley, W. and Bardawil, W. A. (1962). *Nature, Lond.* **194**, 255.
Schajowicz, F. and Cabrini, R. L. (1958). *Science, N.Y.* **127**, 1447.
Schaub, M. C. (1964). *Helv. physiol. pharmac. Acta* **22**, 271.
Schechtman, A. M. (1956). *Int. Rev. Cytol.* **5**, 303.
Scheib, D. (1963). "Ciba Foundation Symposium on Lysosomes", p. 264. Churchill, London.
Scheib, D. (1964). *Ann. Histochem.* **9**, 99.
Schiller, A. A., Schayer, R. W. and Hess, E. L. (1953). *J. gen. Physiol.* **36**, 489.
Schlicht, I. (1963). *Virchows Arch. path. Anat. Physiol.* **336**, 342.
Schmidt, W. (1962). *Z. Zellforsch.* **58**, 573.
Schneeberger, E. E. and Morrison, A. B. (1965). *Lab. Invest.* **14**, 674.
Schneider, W. C. and Hogeboom, G. H. (1951). *Cancer Res.* **11**, 1.
Schneider, W. C. and Kuff, E. L. (1954). *Am. J. Anat.* **94**, 209.
Schoenberg, M. D.. Mumaw, V. R., Moore, R. D. and Weissberger, A. S. (1964). *Science, N.Y.* **143**, 964.
Schuel, H. and Anderson, N. G. (1964). *J. Cell Biol.* **21**, 309.
Schulz, H. (1958). *Beitr. Path. Anat.* **119**, 71.
Schumaker, V. N. (1958). *Expl Cell Res.* **15**, 314.
Scott, B. L. and Pease, D. C. (1956). *Anat. Rec.* **126**, 465.
Seaman, G. R. (1961). *J. biophys. biochem. Cytol.* **9**, 243.
Sedgwick, B. and Hübscher, G. (1965). *Biochim. biophys. Acta* **106**, 63.
Sellers, A. L. Katz, J. and Rosenfeld, S. (1961). *Nature, Lond.* **192**, 562.
Sellinger, O. Z., Beaufay, H., Jacques, P., Doyen, A. and de Duve, C. (1960). *Biochem. J.* **74**, 450.
Sheldon, H., Zetterquist, H. and Brandes, D. (1955). *Expl Cell Res.* **9**, 592.
Shibko, S. and Tappel, A. L. (1963). *Biochim. biophys. Acta* **73**, 76.
Shibko, S. and Tappel, A. L. (1964). *Archs Biochem. Biophys.* **106**, 259.
Shibko, S. and Tappel, A. L. (1965). *Biochem. J.* **95**, 731.
Shibko, S. Caldwell, K. A., Sawant, P. L. and Tappel, A. L. (1963). *J. cell comp. Physiol.* **61**, 85.
Shibko, S., Pangborn, J. and Tappel, A. L. (1965). *J. Cell Biol.* **25**, 479.
Shnitka, T. K. and Seligman, A. M. (1961). *J. Histochem. Cytochem.* **9**, 504.
Sicot, C., Afifi, F., Benhamou, J. P. and Fauvert, R. (1963). *Revue fr. Étud. clin. biol.* **8**, 786.
Siebert, G. (1963). "Ciba Foundation Symposium on Lysosomes", p. 306. Churchill, London.
Siebert, G., Diezel, P. B., Jahr, K., Krug, E., Schmitt, A., Grünberger, E. and Bottke, I. (1962). *Histochemie* **3**, 17.

Simpson, M. V. (1953). *J. biol. Chem.* **201**, 143.
Slater, T. F. and Greenbaum, A. L. (1965). *Biochem. J.* **96**, 484.
Smetana, H. (1947). *Am. J. Path.* **23**, 255.
Smith, R. E. (1963). "Third Annual Meeting American Society of Cell Biology", New York.
Smith, A. D. and Winkler, H. (1966). *J. Physiol.* **183**, 179.
Solomon, A., Waldman, T. A., Fahay, J. L. and McFarlane, A. S. (1964). *J. Clin. Invest.* **43**, 103.
Sorkin, E. and Boyden, S. V. (1959). *J. Immunol.* **82**, 332.
Speirs, R. S. (1963). *Science, N.Y.* **140**, 71.
Staubesand, J. (1963). *Z. Zellforsch.* **58**, 915.
Steinberg, D. and Vaugham, M. (1956). *Archs Biochem. Biophys.* **65**, 93.
Stockinger, L. (1964). *Protoplasmologia* **2**, D1, 1.
Straus, W. (1954). *J. biol. Chem.* **207**, 745.
Straus, W. (1956). *J. biophys. biochem. Cytol.* **2**, 513.
Straus, W. (1957a). *J. biophys. biochem. Cytol.* **3**, 933.
Straus, W. (1957b). *J. biophys. biochem. Cytol.* **3**, 1037.
Straus, W. (1958). *J. biophys. biochem. Cytol.* **4**, 541.
Straus, W. (1959). *J. biophys. biochem. Cytol.* **5**, 193.
Straus, W. (1961). *Expl Cell Res.* **22**, 282.
Straus, W. (1962a). *Expl Cell Res.* **27**, 80.
Straus, W. (1962b). *J. Cell Biol.* **12**, 231.
Straus, W. (1963). "Ciba Foundation Symposium on Lysosomes", p. 151. Churchill, London.
Straus, W. (1964a). *J. Cell Biol.* **20**, 497.
Straus, W. (1964b). *J. Cell Biol.* **21**, 295.
Straus, W. (1964c). *J. Histochem. Cytochem.* **12**, 462.
Straus, W. (1964d). *J. Histochem. Cytochem.* **12**, 470.
Straus, W. and Oliver, J. (1955). *J. exp. Med.* **102**, 1.
Strauss, E. W. (1963). *J. Cell Biol.* **17**, 597.
Suzuki, M., Greenberg, S. D., Adams, J. G. and O'Neal, R. M. (1964). *Expl mol. Path.* **3**, 455.
Swingle, K. F. and Cole, L. J. (1964). *J. Histochem. Cytochem.* **12**, 442.
Tahaba, Y., Kinuwaki, Y. and Hayashi, H. (1964). *Proc. Soc. exp. Biol. Med.* **115**, 906.
Takahashi, N. and Mottet, N. K. (1962). *Lab. Invest.* **11**, 743.
Tanaka, H. (1962). *Tokuhu J. exp. Med.* **76**, 144.
Tappel, A. L. and Ragab, H. (1964). *Fed. Proc.* **23**, 488.
Tappel, A. L., Sawant, P. L. and Shibko, S. (1963). "Ciba Foundation Symposium on Lysosomes", p. 78. Churchill, London.
Telfer, W. H. (1961). *J. biophys. biochem. Cytol.* **9**, 747.
Tessenow, W. (1964). *Naturwissenshaften* **51**, 417.
Thorbecke, G. J., Maurer, P. H. and Benacerraf, B. (1959). *Brit. J. exp. Path.* **41**, 190.
Thorbecke, G. J., Old, L. J., Benacerraf, B. and Clarke, D. A. (1961). *J. Histochem. Cytochem.* **9**, 392.
Thorpe, B D. and Marcus, S. (1964). *J. Immunol.* **92**, 657.
Torack, R. M. and Barrnett, R. J. (1962). *Expl Neurol.* **6**, 224.
Törö, I., Rusza, P. and Röhlich, P. (1962). *Expl Cell Res.* **26**, 101.
Treadwell, P. E. (1965). *J. Immunol.* **94**, 692.
Trotter, N. L. (1965). *J. Cell Biol.* **25**, 41.

Trowell, O. A. (1946). *J. Physiol.* **105**, 268.
Trump, B. F. (1961). *J. Ultrastr. Res.* **5**, 291.
Trump, B. F. and Benditt, E. P. (1962). *Lab. Invest.* **11**, 753.
Trump, B. F., Goldblatt, P. J. and Stowell, R. E. (1962). *Lab. Invest.* **11**, 986.
Trump, B. F. and Janigan, D. T. (1962). *Lab. Invest.* **11**, 395.
Underhay, E., Holt, S. J., Beaufay, H. and de Duve, C. (1956). *J. biophys. biochem. Cytol.* **2**, 635.
Urinhara, T. and Movat, H. Z. (1964). *Lab. Invest.* **13**, 1057.
Vaes, G. (1965). *Expl Cell Res.* **39**, 470.
Vaes, G. and Jacques, P. (1965). *Biochem. J.* **97**, 380; 389.
Van Lancker, J. L. (1964). *Fed. Proc.* **23**, 1050.
Van Lancker, J. L. and Holtzer, R. L. (1959a). *Am. J. Path.* **35**, 563.
Van Lancker, J. L. and Holtzer, R. L. (1959b). *J. biol. Chem.* **234**, 2359.
Van Lancker, J. L. and Holtzer, R. L. (1963). *Lab. Invest.* **12**, 102.
Vaughan, R. B. (1965). *Brit. J. exp. Path.* **46**, 71.
Viala, R. and Gianetto, R. (1955). *Can. J. Biochem. Physiol.* **33**, 839.
Vorbrodt, A. (1961). *J. Histochem. Cytochem.* **9**, 647.
Wachstein, M. and Meisel, E. (1964). *J. Histochem. Cytochem.* **12**, 538.
Wachstein, M., Meisel, E. and Ortiz, J. (1962). *Lab. Invest.* **11**, 1243.
Wächtler, K. and Pearse, A. G. E. (1966). *Z. Zellforsch. Mikrosk. Anat.* **69**, 326.
Waddell, W. R., Geyer, R. P., Clarke, E. and Stare, F. J. (1954). *Am. J. Physiol.* **177**, 90.
Wagner, H. N. and Iio, M. (1964). *J. Clin. Invest.* **43**, 1525.
Walek-Czernecka, A. (1962). *Acta Soc. Bot. Pol.* **31**, 539.
Walek-Czernecka, A. (1963). *Acta Soc. Bot. Pol.* **32**, 405.
Walek-Czernecka, A. (1965). *Acta Soc. Bot. Pol.* **34**, 573.
Walkinshaw, C. H. and van Lancker, J. L. (1964). *Lab. Invest.* **13**, 513.
Wattiaux, R. (1962). *Arch. int. Physiol.* **70**, 765.
Wattiaux, R. and de Duve, C. (1956). *Biochem. J.* **63**, 606.
Wattiaux, R. and de Duve, C. (1966) *Ann. Rev. Physiol.* **28**, 435.
Wattiaux, R., Wibo, M. and Baudhuin, P. (1963). Ciba Found. Symp., "Lysosomes", p. 176. Churchill, London.
Wattiaux, R., Wattiaux-DeKoninck, S., Rutgeerts, M. J. and Tulkens, P. (1964). *Nature, Lond.* **203**, 757.
Weber, R. (1963). "Ciba Foundation Symposium on Lysosomes", p. 282. Churchill, London.
Weber, R. (1964). *J. Cell Biol.* **22**, 481.
Weigle, O. (1960). "Mechanism of Antibody Formation", p. 53. Publishing House Czechoslovak Academy of Sciences.
Weiss, J. M. (1955). *J. exp. Med.* **101**, 213.
Weiss, L. P. and Fawcett, D. W. (1953). *J. Histochem. Cytochem.* **1**, 47.
Weissmann, G. (1965). *New England J. Med.* **273**, 1084, 1143.
Weissmann, G. and Dingle, J. (1961). *Expl Cell Res.* **25**, 207.
Weissmann, G. and Thomas, L. (1962). *J. exp. Med.* **116**, 433.
Weissmann, G. and Thomas, L. (1963). *J. clin. Invest.* **42**, 661.
Weissmann, G., Keiser, H. and Bernheimer, A. W. (1963). *J. exp. Med.* **118**, 205.
Wellensiek, H. J. and Coons, A. H. (1964). *J. exp. Med.* **119**, 685.
Wetterfors, J., Gullberg, R., Liljedahl, S. O., Plantin, L. O., Birke, G. and Olhagen, B. (1960). *Acta med. scand.* **168**, 347.
Wetzel, B. K., Spicer, S. S. and Wollman, S. H. (1965). *J. Cell Biol.* **25**, 593.
White, J. G., Krivit, W. and Vernier, R. (1964). *Fed. Proc.* **23**, 238.

Wilgram, G. F. and Kennedy, E. P. (1963). *J. biol. Chem.* **238**, 2615.
Wittekind, D. (1958). *Z. Zellforsch.* **49**, 58.
Woessner, J. F., Jn. (1965a). *In* "International Review of Connective Tissue Research" (D. A. Hall, ed.) Vol. III, p. 201. Academic Press, New York and London.
Woessner, J. F., Jn. (1965b). *Biochem. J.* **97**, 855.
Wollman, S. H., Spicer, S. S. and Burstone, M. S. (1964). *J. Cell Biol.* **21**, 191.
Woodin, A. M. (1963). *In* "The Structure and Function of the Membranes and Surfaces of Cells". Biochem. Soc. Symp. no. 22, p. 126. Cambridge Univ. Press.
Woodin, A. M. and Wienecke, A. A. (1964a). "Ciba Foundation Symposium on Cellular Injury", p. 30. Churchill, London.
Woodin, A. M. and Wienecke, A. A. (1964b). *Biochem. J.* **90**, 498.
Woods, J. F. and Nichols, G., Jn. (1965a). *Nature, Lond.* **208**, 1325.
Woods, J. E. and Nichols, G. (1965b). *J. Cell Biol.* **26**, 747.
Woohsmann, H. and Hartrodt, W. (1964). *Histochemie* **4**, 336.
Zilversmit, D. B., McCandless, E. L., Jordan, P. H., Henly, W. S. and Ackerman, R. F. (1961). *Circulation* **23**, 370.
Zeidenberg, P. and Janoff, A. (1964). *Comp. Biochem. Physiol.* **12**, 429.
Zeya, H. J. and Spitznagel, J. K. (1963). *Science, N.Y.* **142**, 1085.
Zollinger, H. U. (1950). *Schweiz. Z. allg. Path. Bakt.* **13**, 146.
Zucker-Franklin, D. and Hirsch, J. G. (1964). *J. exp. Med.* **120**, 569.

Note Added in Proof

Maunsbach [*J. Ultrastr. Res.* **15,** 197-241 (1966)]. presented the first definite proof for this hypothesis by microperfusing single proximal tubules of the kidney with I^{125}-labelled homologous serum albumin and by demonstrating the uptake of the protein in apical phagosomes of the proximal convoluted tubules by electron microscopic autoradiography.

Chapter 6

MEMBRANE SYSTEMS

E. REID

Biochemistry Department, University of Surrey, London, England

I. INTRODUCTION

A. NATURE OF MEMBRANE SYSTEMS

1. Scope of Present Survey

The term "membrane" in the present context connotes the classical triple-layer structure made up of lipid sandwiched between two protein layers, which is manifest in electron micrographs as two electron-opaque lines in close apposition and which limits the passage of solutes. On morphological grounds, the survey excludes cell walls (for which there are examples of penetrability even by substances of high molecular weight); reviews that deal with cell walls include those of Wolfe (1964) for bacteria and of Whaley, Mollenhauer and Leech (1960) for plant cells, It should be noted that some authors regrettably use the term "cell membrane" as a synonym not for the plasma membrane but for cell wall; the term is best avoided altogether.

The term "systems" has now been interpreted rather arbitrarily as including the plasma membrane (cell membrane, plasmalemma), the cytomembranes (here treated as distinct from the plasma membrane; see Section I. B.), and the nuclear membrane, but as excluding the membrane of mitochondria, lysosomes and other organelles (see Fig. 3). The fact that the lysosomal membrane appears to arise from the plasma membrane illustrates the arbitrary nature of the distinction now made.

There is, of course, no nuclear membrane in bacteria. Cytomembranes, in contrast with free ribosomes, are lacking or deficient in most bacteria, but are present to at least some extent in some algae, fungi and other protista, as well as in the cells of higher organisms. However, they are lacking in mammalian red blood cells, and are scanty in leucocytes, in some plantc ells (Whaley *et al.*, 1960) and in the proximal tubule cells of kidney. The plasma membrane may well be the basic structure from which, with the evolution of large and metabolically complex cells, cytomembranes and the nuclear membrane have developed. The plasma membrane itself may become differentiated; for example, the myelin of nerve tissue represents regularly layers of Schwann-cell plasma membranes.

To lay a foundation for the rest of the chapter, the nature of membranes and of membrane systems will be touched on. Consideration will then be given to cytochemical work and (in more detail) to biochemical work on enzymes, with emphasis on liver—a tissue which lends itself to biochemical experimentation (partly because it is fairly homogeneous in cell type) and which has furnished a core of knowledge to which observations on other tissues can be related. Questions of terminology will be considered in later sections. Locke (1964) may be consulted for fuller treatments, particularly of morphological aspects. Recently, Ryter and Jacob (1966) have discussed bacterial membranes.

2. General Morphological Background

(*a*) *Membranes.* The "unit membrane" as described by Robertson (1959) represents a universal structure which, however, has been said to vary in dimensions from 4·5 to 11 mμ (Stoeckenius, 1964). Typical thicknesses for the plasma membrane are 7·5—9 mμ overall and 2 mμ for each of the two protein layers. Tashiro (1957) quotes values of the order of 20 mμ (for overall thickness), but Palade and Siekevitz (1956) and other authors reckon the thickness to be about 5 mμ; cytomembranes and especially the nuclear and mitochondrial membranes are reckoned to be thinner than the plasma membrane. From work with model systems it is conceivable that membranes vary in respect to the nature of the lipid and of the outer (protein) layers; the latter may differ from each other, and there may even be partial replacement of protein by other macromolecules such as polysaccharides (Stoeckenius, 1964). From measurements of sterol incorporation with a sterol-requiring insect, Clayton (1965) concluded that there may well be "a repeating unit of membrane structure which is common to all of the subcellular membrane systems of a particular tissue or cell type, but is different in different tissues."

The idea that there is homogeneity in a given membrane with respect to appearance is now back in the melting-pot. There may be a "mosaic" structure with local specializations. There are thought to be actual "pores", which in excitable membranes are reckoned to have a diameter of the order of 1 $m\mu$ (Mullins, 1960). Moreover, Sjöstrand (1963) has performed high resolution electron microscopy on mitochondrial membranes and smooth-surfaced cytomembranes in kidney cells (Fig. 1), and has interpreted the pictures. particularly those with permanganate staining, as showing globular components, possibly lipid micelles, separated by stained septa, possibly protein. From observations such as the latter, and from biochemical findings (e.g. that the protein and lipid are intimately linked), Green and Perdue (1966) have argued—with force rather than lucidity—that a membrane is "a fused continuum of repeating particles" each made up of an invariant base-piece and of a "detachable sector"; the former, on exposure to a bile salt or acetone, can form a "vesicle". The term "vesicle" may here have a different connotation to that conveyed by the use of the term elsewhere in this chapter.

(*b*) *Membrane systems.* The following structures will be considered (cf. Figs. 2 and 3):

Plasma membrane. This structure is distinguished here from cytomembranes (cf. Novikoff, 1964), although a sharp distinction is difficult to make (Emmelot, Bos, Benedetti and Rumke, 1964b). The erythrocyte membrane, as discussed by Wolfe (1964) *inter alia*, will not be considered.

Cytomembranes. This term is used here in preference to e.r., an abbreviation for either the "endoplasmic reticulum" (the equivalent in skeletal muscle being the "sarcoplasmic reticulum") or the "ergastoplasm" [although the latter term refers to basophilic material and hardly includes "smooth-surfaced" cytomembranes; see Haguenau (1958)]. Cytomembranes are subclassified in Section I.B.1.

Nuclear envelope. This term will not be restricted here merely to the outer of the two membranes surrounding the nucleus.

The outer surface of the plasma-membrane is usually closely apposed to that of adjoining cells, and actual structural links (desmosomes) are reckoned to exist. In many cell types there are convolutions of the membrane, forming tubular ramifications (micro-villi) which comprise, in the case of hepatic cells, the bile canaliculi, and in the case of kidney tubular-cells, the "brush-border" structures. These "convolutions" are thicker than the non-convoluted portion of the plasma membrane, and may have hair-like protrusions; the lumen may contain hollow rod-like elements.

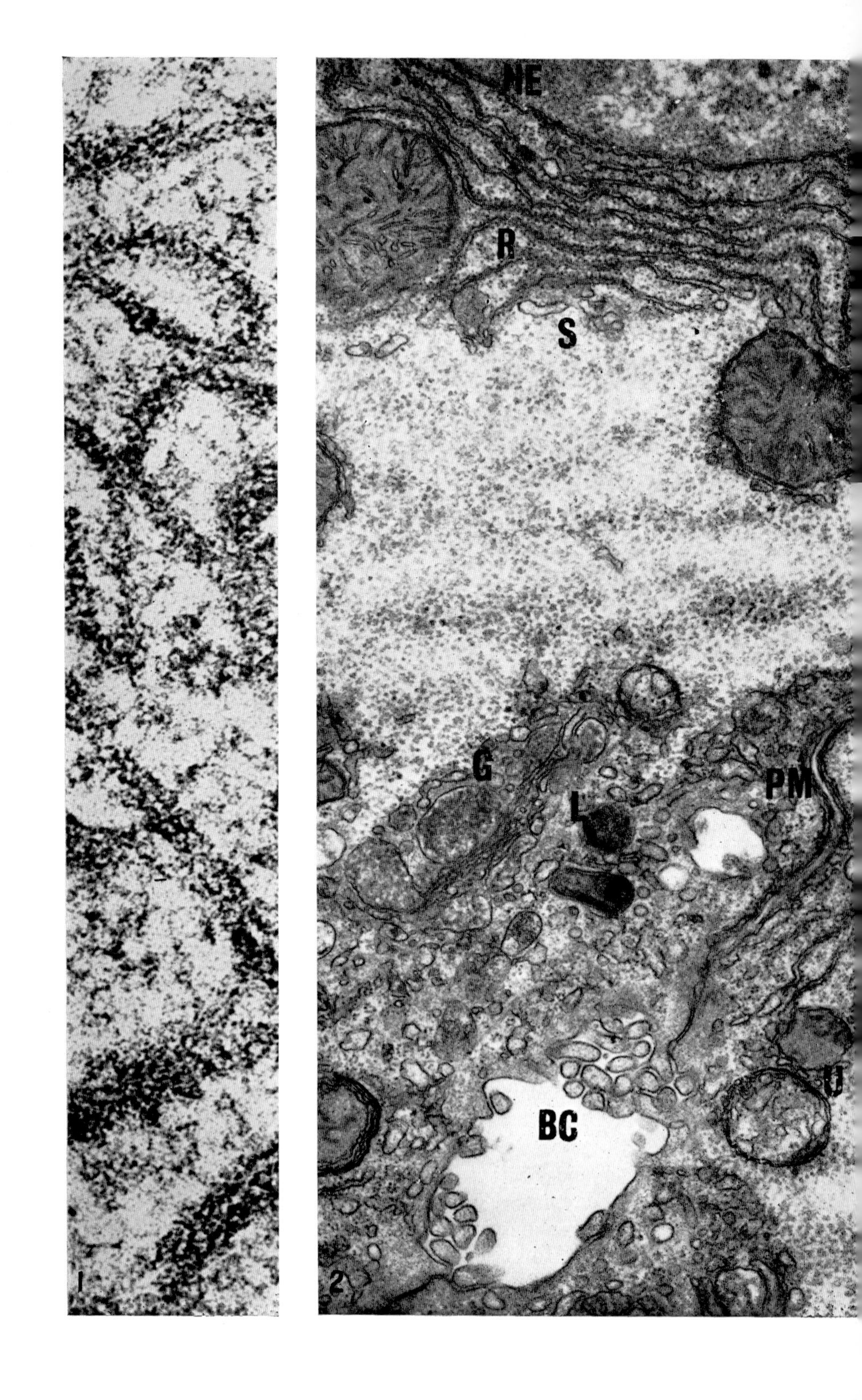
NE
R
S
G
L
PM
U
BC
1
2

The cytomembranes, which will be further considered in Section I.B. [for historical aspects see Haguenau (1958)], are paired structures, and in sections of tissues such as liver they appear as parallel lines (often grouped in stacks) or as flattened rings. The central space between each pair of membranes is termed a cisterna. Each element (sac) can be visualized as resembling a rubber hot-water bottle; the appearance of a particular unit thus depends on the plane in which it is cut. In qualifications of this simile, it should be pointed out that the sacs are typically interconnected with one another to form sponge-like reticulum, ramifications of which connect with the nuclear envelope and probably also with pores in the plasma membrane. Evidently, then, the perinuclear space enclosed by the double membranes is in direct communication with the exterior of the cell, by way of the cisternae of the cytomembranes, so that there is a line of transport which has no direct link with the cell sap (Watson, 1960). Cytomembranes are often seen to surround the mitochondrial membrane (Fig. 2), but no actual connections appear to exist.

In the nuclear envelope there appear to be pores which directly connect the nuclear sap to the cell sap (Fig. 2), and which might well serve to allow passage of macromolecules [Watson (1960); see also Haguenau (1958) and Whaley *et al.* (1960)]. Biochemical evidence largely supports the view that there is little restriction on interchange of molecules between nucleus and cytoplasm; observations such as the high concentration of Na^+ ions in nuclei are explicable in terms of binding to nuclear constituents (Siebert and Humphrey, 1965).

B. CYTOMEMBRANES

1. *Classification*

On the basis of their appearance (Figs. 2 and 3), depending particularly on whether or not the membrane surface is studded with ribosomes, cytomembranes may be classified as follows (see, *inter alia*, Haguenau, 1958):—

Rough-surfaced (*granular*, *α*-) *cytomembranes*. These are commonly found in regular stacks, and in profile each pair typically has a thickness of 50 mμ where the cisternal separation is narrowest.

Smooth-surfaced (*agranular*, *β*- *and* *γ*-) *cytomembranes*. These form irregular ramifying agglomerations located largely in the outer areas of the cytoplasm and in profile, have a narrowest thickness ranging from 40 to 100 mμ; the prefix β- has been used to denote cytomembranes which seem to represent

◄

FIG. 1. (left). Smooth-surfaced cytomembranes in proximal convoluted tubule of mouse kidney, showing globular pattern in both oblique and tangential sections. From Sjöstrand (1963), courtesy of author. × 400,000.

FIG. 2 (right). Portion of a mouse liver cell. BC = bile canaliculus, G = Golgi region, L = lyosome, NE = nuclear envelope, PM = plasma membrane, R = rough-surfaced cytomembranes, S = smooth-surfaced cytomembranes, U = uricosome (micro-body). Courtesy of M. S. C. Birbeck. × 32,000.

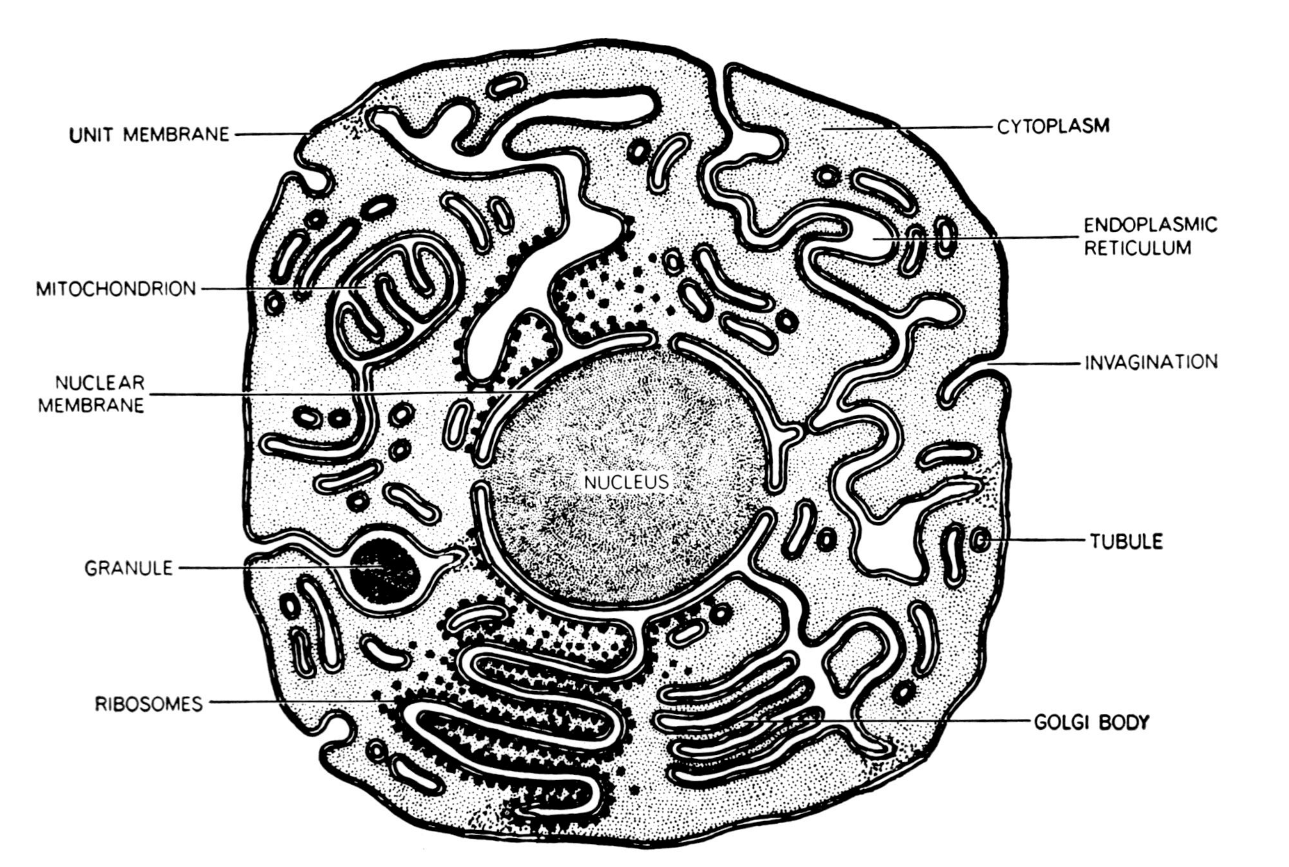

FIG. 3. Idealized cell. Note that "pores" are shown in the nuclear envelope, that smooth-surfaced and rough-surfaced cytomembranes are shown in contiguity, and that the intracisternal spaces of the cytomembranes (endoplasmic reticulum)—within one of which a secretion granule is depicted—are represented as communicating both with the interior of the nuclear envelope and with the cell exterior. Courtesy of J. D. Robertson.

invaginations of the plasma membrane. *Golgi cytomembranes* comprise a cluster ("complex") of membranes designated γ-cytomembranes lying close to (but not linked with) the plasma membrane, and are particularly evident in some plant cells (Whaley *et al.*, 1960).

The above sub-division is somewhat arbitrary, particularly since tissue sections sometimes show rough- and smooth-surfaced strips in contiguity on the same membrane; the different types of cytomembrane may in fact be local differentiations within a common system, and be interconvertible in the day-to-day functioning of the cell. There may be tissue differences, to which scant regard can be paid here; thus in certain secretory organs such as pancreas, zymogen granules may be present in the cisternal spaces within smooth-surfaced cytomembranes (Palade and Siekevitz, 1958a), and in liver these cytomembranes are concerned in glycogen storage. In at least some plant cells there appear to be few rough-surfaced cytomembranes (Whaley *et al.*, 1960). Steroid-producing and certain other vertebrate organs likewise contain cells in which there is a predominance of smooth-surfaced cytomembranes—an "organelle" the functions of which have been reviewed by Jones and Fawcett (1966).

In the living cell, as distinct from the dead cells seen in tissue sections, structures such as mitochondria and the plasma membrane may change from one moment to another in shape and position. It is difficult to obtain evidence for a dynamic state of the cytomembranes [see however, III. A.3 (*a*) and IV. B. 1]. From work *in vitro* there are indications that they may have osmotic properties (Palade and Siekevitz, 1956; Wallach and Kamat. 1964), but some of this work (e.g. Packer and Rahman, 1962; Tedeschi, James and Anthony, 1963) is rather inconclusive, and it has been claimed that there is no hindrance to access of sucrose (Share and Hansrote, 1960). Certainly it would be unwise to assume a static, structural role which would justify the term "cytoskeleton", which is sometimes used; probably "the membranes . . . are in constant flux and motion" (Siekevitz, 1965).

2. *Experimental Approaches*

(*a*) *Cytochemistry*. Ideally a cytochemical assay should be available for each enzyme investigated which, with whole tissue, furnishes quantitative unambiguous information about its intracellular location. This ideal is a long way from fulfilment; nevertheless for some enzymes there are qualitative but valuable cytochemical observations which are probably unblemished by artifacts such as can arise from diverse causes—e.g. partial destruction during fixation, or diffusion of the reaction product. Such observations, as briefly surveyed in Section II, have come mainly from the recent application of electron microscopy to this field. Little attention will be paid to histochemical observations, since their value lies largely in ascertaining which type of cell in a heterogeneous tissue contains the enzyme question; nevertheless the light microscope can give cytochemical help unless the enzyme in question is associated with cytomembranes.

Some workers (see Haguenau, 1958) have subjected suspensions of intact cells to ultracentrifugation and obtained distinct layers, one of which represents cytomembranes as judged by staining properties and electron-microscopic appearance. This early approach, semi-biochemical in nature, may warrant rescue from its recent neglect.

(*b*) *Biochemistry*. The only strictly biochemical approach at present available for the study of intracellular membranes entails deliberate breakdown of the plasma membrane. The method of choice is gentle homogenization in an isotonic or hypertonic medium, so that there is minimal damage to organelles such as mitochondria. Techniques for homogenization and for differential centrifugation will be surveyed in Section III. Here it may be pointed out that, as discussed by J. Chauveau and colleagues in 1955 [see Haguenau (1958)], Palade and Siekevitz (1956) and Tashiro (1957), there appears to be fragmentation of the cytomembranes, the double membranes becoming transected and "pinched together" at intervals so as to furnish "vesicles". These vesicles are the main constituent (on a weight basis) of the "microsomal fraction" isolated from the homogenate by high-speed centrifugation (Figs. 7a and 8a).

A tissue made up of diverse types of cells is hardly likely to give clear-cut biochemical results. Even with liver, which consists substantially of cells, of one type (parenchymal, hepatic), non-parenchymal cells could conceivably be the source of certain enzymes—although there is no well authenticated example. Moreover, there may be diversity of enzyme pattern within each lobule (cf. Fig. 4). With other tissues, cytological interpretation of biochemical results needs great caution, although for tissues such as kidney and pancreas the difficulties are not insurmountable.

The pioneer work, such as that by Claude (1946) suffered from the handicap that the morphology of microsomal fractions—i.e. of the pellet obtained after removal of mitochondria—could be studied only by light microscopy, which is a poor tool for this purpose. Even in 1952, the possibility that "microsomes" are really fragmented mitochondria was serious mooted (Green, 1952). Only with the advent of electron microscopy did it become clear that

FIG. 4. Enzyme localizations in liver lobule as shown by light microscopy. From Reid *et al.* (1964), courtesy of Walter de Gruyter & Co. Top left (a) 5′-Nucleotidase with UMP as substrate: there is some staining in nuclei (an artifact probably due to an insufficiently high concentration of Pb^{2+} ions) as well as in plasma membranes and sinusoids; the cytoplasm is unstained. × 540. Bottom (b). 5′-Nucleotidase: note that the staining is predominantly in centrolobular plasma membranes and sinusoids, and in blood-vessel walls (central vein at top right, portal vein at bottom left). × 270. Top right (c). Glucose-6-phosphatase: note that there is staining throughout the cytoplasm but not in nuclei. × 270.

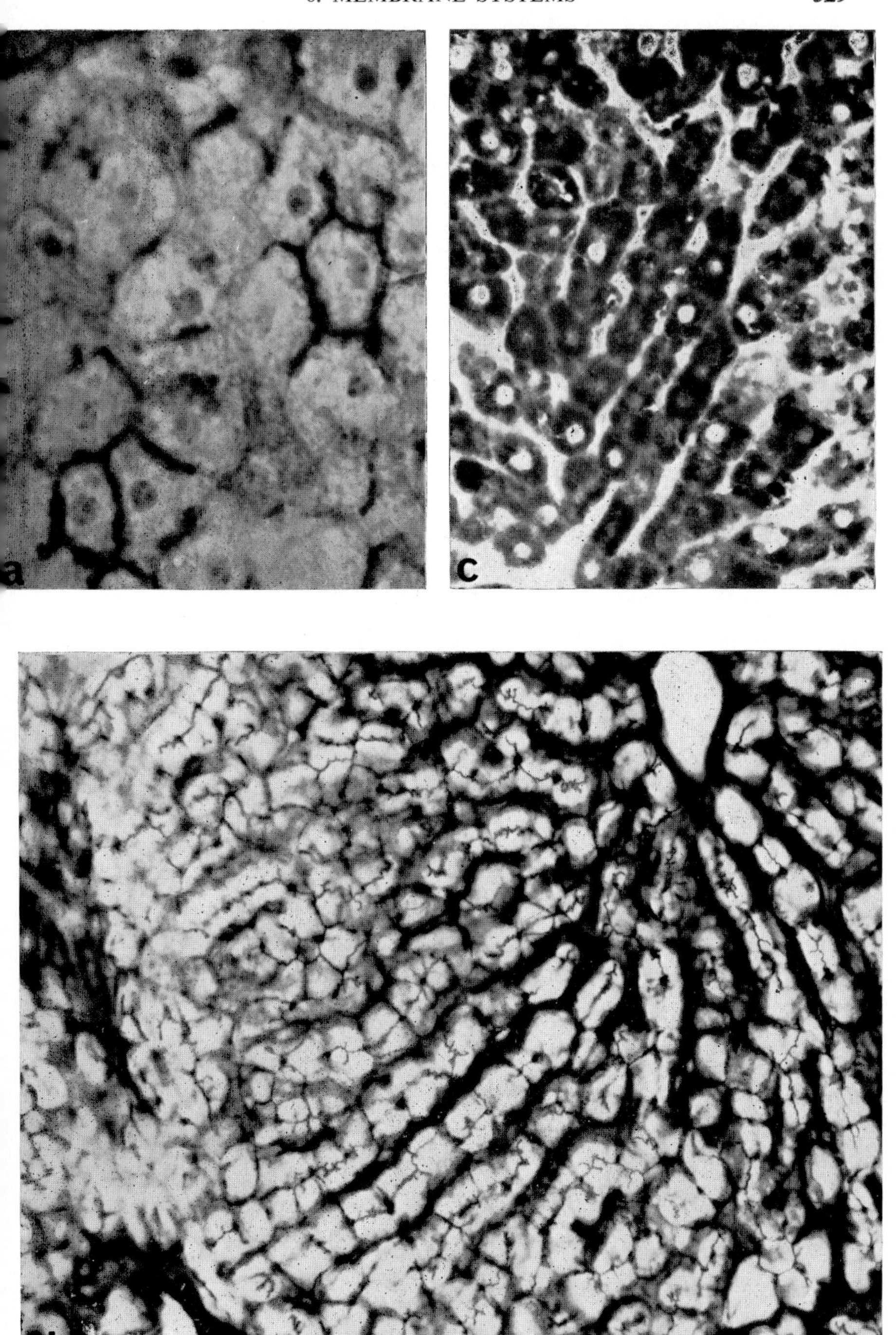
a
c
b

microsomal fractions from liver and certain other tissues consist mainly of fragmented cytomembranes. For some tissues this conclusion may not hold, and indeed the lack of electron-microscopic data in some biochemical studies (even with liver) seriously detracts from their value. There has even been neglect to verify in each case the classical observation that material designated "microsomal" (by the criterion of centrifugal behaviour) is characteristically rich in RNA and phospholipid.

(*c*) *Nomenclature for sub-cellular fractions.* In pleading for the term "microsomal fraction" rather than "microsomes", de Duve (1964) has pointed out: "Possibly the most troublesome word in our cytological vocabulary is represented by the term 'microsomes'. Originally proposed as synonym of 'small granules', it had a purely operational significance and served simply as an inclusive denomination for all subcellular entities requiring a relatively high centrifugal force for complete sedimentation. The term . . . . was used by many biochemists to designate what was believed to be a distinct group of intracellular components . . ." But in view of the heterogeneity disclosed by the electron microscope, ". . . the term 'microsomes' has become extremely hard to define since it has lost part of its purely operational significance, while gaining nothing in cytological precision." Occasional use of the term is perhaps permissible if it is clearly intended as an abbreviation for "microsomal fraction".

The term "ribosome", which has superseded the term "Palade particle", can justifiably be used both in cytological and in operational contexts. However, use of the classical term "RNP particle" is sometimes still warranted, for example when the material has been exposed to detergents that might have damaged the ribosomes (Campbell, Colper and Hicks, 1964).

As discussed in Section III., there is no uniformity in the centrifugal conditions used by different workers to sediment the microsomal fraction. Where different microsomal "cuts" were taken, the later cuts have been variously termed "light-microsomal", "ultra-microsomal" and "post-microsomal". The policy in the present chapter (cf. Table III.) is to avoid the term "ultra-microsomal", to use the term "light-microsomal" where "heavy" and "light" fractions each rich in phospholipid have been prepared, and to apply the term "post-microsomal" to a final cut known or inferred to be poor in phospholipid. The terms as used in this article are in some instances different from those used by the authors concerned.

II. Cytochemical Observations

A. Methodology

The general procedure in cytochemical work is to incubate a tissue sample, already lightly fixed, in the presence of the substrate and of a "product-capturing" agent (usually Pb^{2+} ions, if phosphate liberation is being studied); finally, after further fixation, sections are examined. (No consideration will be given here to autoradiography, of which increasing use is being made by electron microscopists.) One point which warrants mention is that tissue sections (cut frozen) are now preferred (Goldfischer, Essner and Novikoff, 1964) to the tissue blocks used for incubation in early cytochemical work with the electron microscope (Persijn, Daems, de Man and Meijer, 1961). Another approach, which is complementary to this rather than an alternative, is to incubate suspensions of subcellular fractions for the enzyme assay, re-centrifuge them and then section the pellet (see Section IVA. 1).

There has been much debate about the merits of different fixatives (Sabatani, Miller and Barrnett, 1964), and it is doubtful whether a perfect fixative of universal applicability will ever be discovered. Many authors have neglected to check biochemically the degree of enzyme destruction caused by the fixative, or indeed, by the incubation medium itself. For example, there can be significant impairment of nucleotide dephosphorylation through the use of OsO_4 (Persijn *et al.*, 1961) or glutaraldehyde (Goldfischer *et al.*, 1964). Destruction may be less with formaldehyde-Ca^{2+} or hydroxyadipaldehyde, and indeed the latter fixative is thought to be advantageous if glucose-6-phosphatase (3.1.3.9) or mitochondrial ATPase (3.6.1.3.) is to be demonstrated (Goldfischer *et al.*, 1964; Sabatani *et al.*, 1964; cf. Tice and Barrnett 1962). However, results of fair reliability can often be obtained by judicious use of glutaraldehyde, as in Figs. 5 and 6 below; for glucose-6-phosphatase the pre-fixation should be very brief (Ericsson, 1966).

One serious source of unreliability is possible diffusion of the reaction product. Such diffusion could explain an observation shown in Fig. 4—that nuclear staining for 5′-nucleotidase (3.1.3.5.) is sometimes demonstrable (and also for glucose-6-phosphatase), for example if the concentration of Pb^{2+} ions in the incubation medium is rather low (Reid, El-Aaser, Turner and Siebert, 1964) or if the incubation time is prolonged (Goldfischer *et al.*, 1964).

Much patient work is necessary to eliminate the diverse sources of artefacts. Even the choice of buffer may be important; β, β-dimethylglutarate is a particularly versatile buffer for which, unlike *tris*, no inhibitory actions have so far been reported. With over-incubation there may be cellular damage which, with intestine, could be a cause of failure to detect invertase (3.2.1.26) in the brush border (Hübscher, West and Brindley, 1965).

B. ENZYMES ASSOCIATED WITH MEMBRANES

As is evident from Table I and Figs, 5 and 6, only dephosphorylating enzymes have so far been shown to have a membranous location, at least with electron as distinct from light microscopy. [For hepatic cytochemistry with

TABLE I. *Enzymes shown cytochemically, usually by electron microscopy, to be associated with membranes*

Enzyme	Tissue	Cell type	Cell structures showing activity[a]	Other sites in the tissue	Reference and remarks
Alkaline phosphatase (3.1.3.1)	Liver intestine and other tissues	parenchymal epithelial	Cytomembranes—mainly Golgi; *pl*—especially villi, e.g. brush-border		Persijn *et al.* (1961)—activity low in liver; Goldfischer *et al.* 1964); cf. Hübscher *et al.* (1965)
Glucose-6-phosphatase (3.1.3.9)	Liver kidney	parenchymal tubular	Cytomembranes, *rs* and *ss* (including *ne*)—especially in periportal cells (liver); not in Golgi.		Tice and Barrnett (1962); Goldfischer *et al.* (1964); El-Aaser (1965); Ericsson (1966)
Nucleoside diphosphatases[b] acting on ADP or CDP (3.6.1.)	Liver	parenchymal	Only *bc*		Goldfischer *et al.* (1964); El-Aaser (1965)
Nucleoside diphosphatases[b] acting on GDP, IDP, or UDP (3.6.1.6)	Liver and other tissues	parenchymal	Cytomembranes, at least *rs* including *ne* (liver)—but usually only Golgi in tissues other than liver[c]; *pl*—especially *bc* (liver)		Goldfischer *et al.* (1964); El-Aaser (1965)

Nucleoside triphosphatase(s) acting on ATP or UTP[b, d] (3.6.1.4)	Liver kidney and other tissues	parenchymal tubular	Only *bc* or brush-border villi[c]; especially in periportal cells (liver)	Blood-vessel walls, biliary epithelium	Persijn *et al.* (1961); Novikoff and Goldfischer. (1961); Goldfischer *et al.* (1964); El-Aaser (1965)
5′-Nucleotidase (nucleoside-5′-monophosphatase) (3.1.3.5)	Liver and other tissues	parenchymal	Only *pl*, mainly (in liver) *bc*; especially in centrolobular cells	Walls of central veins and sinusoids; connective tissue	Goldfischer *et al.* (1964); El-Aaser (1965)—results for AMP and UMP identical
Thiamine pyrophosphatase (3.6.1.)	Liver and other tissues	parenchymal	Cytomembranes, usually only Golgi in tissues other than liver; also *pl*, at villi (intestine)		Goldfischer *et al.* (1964)—Golgi activity lowered by glutaraldehyde

[a] Abbreviations used:—*bc*, bile canaliculi (micro-villi); *ne*, nuclear envelope (membrane); *pl*, plasma membrane; *rs*, rough-surfaced elements; *ss*, smooth-surfaced elements.

[b] Assay performed in presence of Mg^{2+} ions

[c] In striated muscle, the site of dephosphorylation of ATP, of IDP and (at pH 5) of CMP appears to be "in the region of the sarcoplasmic reticulum that lies adjacent to the intermediary vesicle" (Goldfischer *et al.*, 1964).

[d] With formaldehyde-Ca^{2+} fixation, some ATPase activity has also been demonstrated in rough-surfaced cytomembranes (Wachstein and Fernandez, 1964; not confirmed in unpublished work by A. A. El-Aaser); activity in mitochondria is sometimes also demonstrable (Goldfischer *et al.*, 1964).

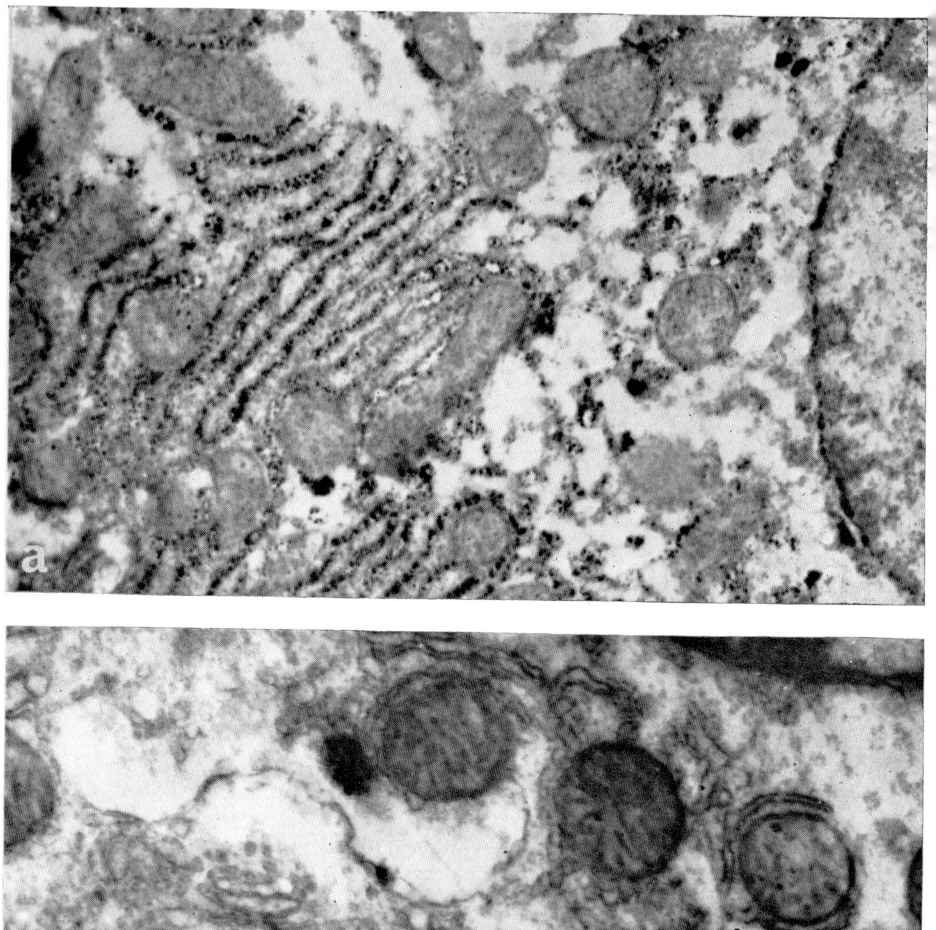
a

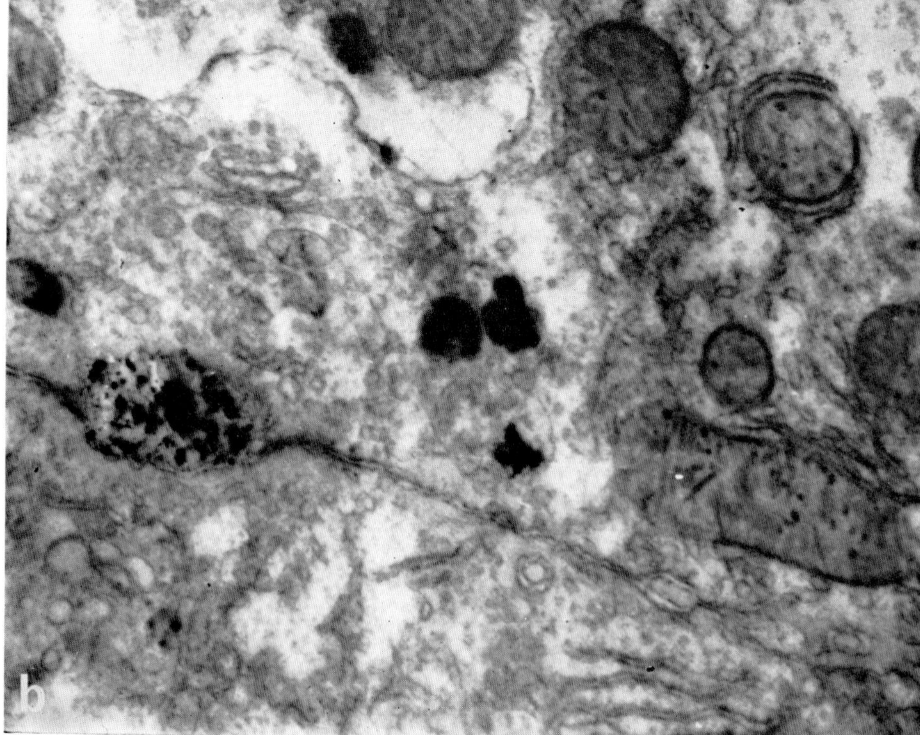
b

the latter, see Wachstein (1963). Note that the references in Table I are not intended to be comprehensive]. So far, there has been no demonstration of a membrane-linked esterase. It will be noted that the difference shown by light microscopy between glucose-6-phosphatase and 5′-nucleotidase—the former having a diffuse cytoplasmic distribution and the latter being apparently in the plasma membrane (Fig. 4)—is supported and clarified by electron-microscope findings. These are shown in Figs. 5 and 6 together with findings for nucleoside di- and tri-phosphatases. (It may be pointed out here that the term "nucleoside phosphatase", as used by same authors, is ambiguous and deplorable.) The illustrations support the conclusions of Goldfischer *et al.* (1964) and other authors, that glucose-6-phosphatase is intimately associated with cytomembranes, whereas "UTPase" (or 'ATPase') and "ADPase" (not illustrated) are mainly in the plasma membrane especially where it folds to form micro-villi, and that the enzyme which splits GDP, IDP or UDP (but not ADP or CDP) is located in *both* cytomembranes and the plasma membrane, Enzymic activities demonstrable in cytomembranes are likewise evident in the nuclear envelope. Staining in the non-convoluted regions of the plasma membrane, with substrates such as UTP or UMP, may not be obvious if a short incubation time is employed.

From an examination of some 60 cell types in Novikoff's laboratory, the generalization has been made that in no cells can the cytomembranes dephosphorylate nucleoside mono- or tri-phosphates or ADP or CDP (Goldfischer *et al.*, 1964). Novikoff (1964) also believes that plasma membrane contains a 5′-nucleotidase which is activated by Co^{2+} as distinct from Mg^{2+} ions. Thiamine pyrophosphatase is unusual in being located in the Golgi cytomembranes; staining sometimes observed in other structures (such as the nucleus) may be an artefact (Goldfischer *et al.*, 1964).

The electron microscopic results for cytomembrane-located enzymes point to a location on the side of the membrane facing into the cisterna (see Figs. 5a and 6a), but do not conclusively establish that the enzymes indeed have a membranous as distinct from a cisternal location (Tice and Barrnett, 1962; Goldfischer *et al.*, 1964; Ericsson, 1966).

This cursory survey, in which scant justice has been done to individual papers as distinct from reviews, serves a background to the biochemical observations which will now be presented (Section III), and which later (Section IV.A.) will be dove-tailed with the cytochemical observations.

◄———

FIG. 5. Enzyme localizations in hepatic cells as shown by electron microscopy (El-Aaser, 1965), courtesy of A. A. El-Aaser and M. S. C. Birbeck. × 73,000.
Top (a). Glucose-6-phosphatase: note staining in cytomembranes and nuclear envelope.
Bottom (b). 5′-Nucleotidase with UMP as substrate: note staining in bile canaliculus, in plasma membrane (faint), and also (due to the non-specific acid phosphatase) in lysosomes.

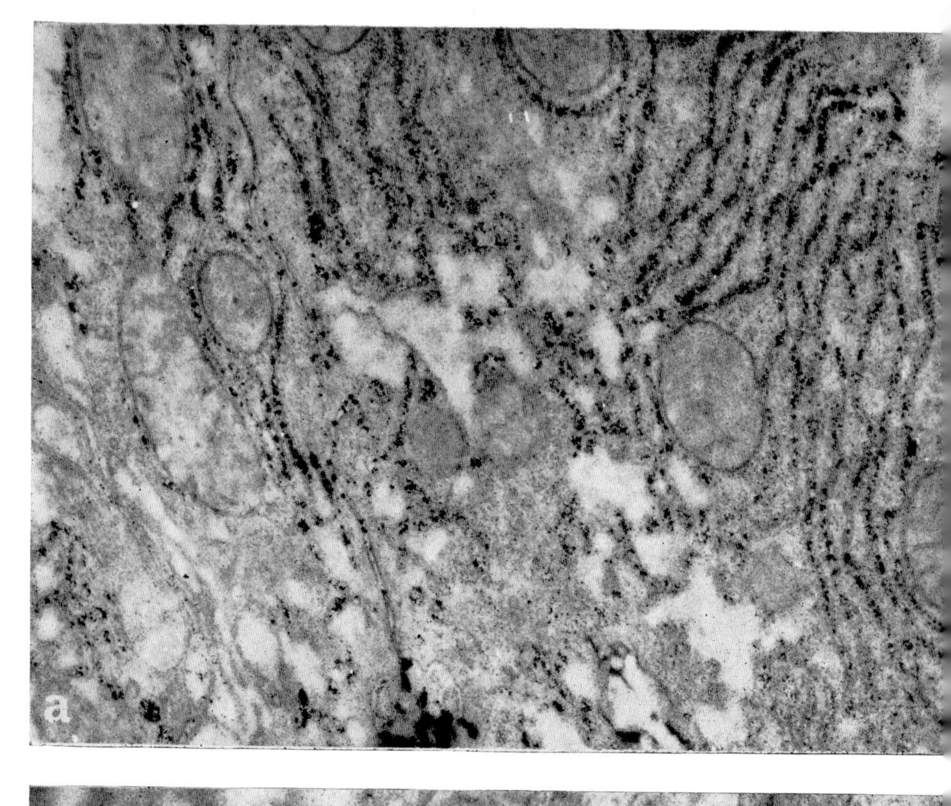
a

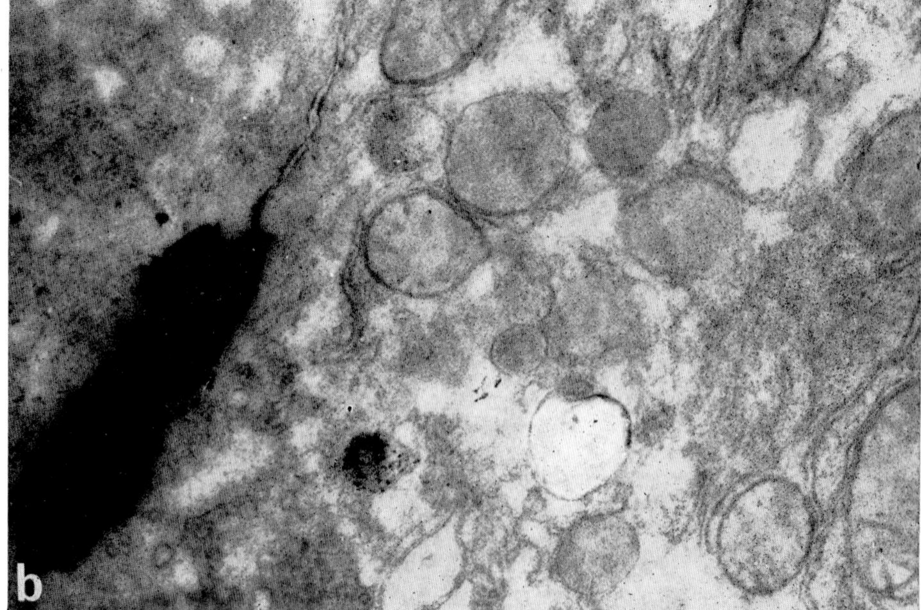
b

III. Biochemical Observations

A. Isolation and Characteristics of Hepatic Microsomal Fractions

1. Methods of Isolation

(*a*) *Preparation of homogenates.* Mild conditions such as are necessary to preserve the integrity of nuclei, mitochondria and lysosomes are desirable and effective for cytomembranes also. The formation of vesicles from cytomembranes by a pinching-off process is in artifact that must be accepted, and at least this process appears to be reproducible if use is made of a Potter-Elvejhem homogenizer and of a sucrose medium. Surprisingly, however, this homogenizer was found by Leadbeater and Davies (1964) to be inferior to a Waring blendor with respect to the yield and the stability *in vitro* of microsomal demethylases [see also III.c.1(c) below].

For subtle studies such as are considered in III. B. below, 0·88 M sucrose may be preferable to 0·25 M, although its effectiveness in giving flattened rather than swollen vesicles (cf. Palade and Siekevitz, 1956) may be poor (see Fig. 7a compared with Fig. 8a). However, for routine preparation of microsomal fractions, 0·25 M sucrose solution is more convenient, a trace of EDTA being permissible (de Duve, Pressman, Gianetto, Wattiaux and Appelmans, 1955). Rupture of a high proportion of the cells is not too hard to achieve, particularly if the first "nuclear" pellet is rehomogenized (de Duve *et al.*, 1955) and the *g* values adopted for the centrifugations need be only about half of those for isolation in 0·88 M sucrose. Some workers compromise by using 0·4 M sucrose for example. Unfortunately the "nuclear fraction" will, whichever strength of sucrose is used, be contaminated with cytomembrane fragments, as shown by assays for glucose-6-phosphatase; only by the use of unorthodox media such as organic solvents (the tissue being first freeze-dried) or 2·2 M sucrose can virtually uncontaminated nuclei be obtained, at the expense of the integrity of the nuclear envelope [Reid *et al.* (1964), *inter alia*].

With a raffinose-dextran-heparin medium which Birbeck and Reid (1956) found to be advantageous for isolation of mitochondria, contamination of the mitochondria with cytomembrane fragments is low, but the latter are themselves not readily sedimentable and, moreover, are rounded in shape just as with 0·25 M sucrose medium. Media lacking sucrose but containing salt(s), as used by A. Claude and other pioneers of differential centrifugation (cited by Moulé, Rouiller and Chauveau, 1960), may result in less sharp separations

◄

Fig. 6. See legend to Fig. 5.

Top (a). Nucleoside diphosphatase with UDP as substrate: note staining in cytomembranes and also in bile canaliculus. Bottom (b). Nucleoside triphosphatase with UTP as substrate: note staining in bile canaliculus and adjoining plasma membrane, with little cytomembrane staining.

than sucrose media although for microsomal fractions the difference may be small (Takanami, 1959). Nevertheless, trial of such media may be helpful where it is suspected that the presence of an enzyme in the microsomal fraction is an artefact due to redistribution (de Duve, Wattiaux and Baudhuin, 1962; Schneider, 1963; see also Section III. B. 4). Recently, particularly in studies of amino-acid incorporation, increasing use has been made of media containing both sucrose and a trace of neutral buffer, often supplemented with KCl and $MgCl_2$ (*inter alia*, Campbell *et al.*, 1964; cf. Tashiro and Ogura, 1951; and Dallner, 1963); with such a medium in place of a medium containing sucrose only there is a difference in the yield of the 'post-microsomal'" sediment, if not of the microsomal sediment (Hird, McLean and Munro, 1964). The $MgCl_2$ serves to minimize damage to ribosomes and polysomes. With use of phosphate instead of bicarbonate or *tris* buffer, there is some risk of inadvertent elution of cytomembrane constituents (Reid, 1961a).

(*b*) *Centrifugation conditions.* Some comments are warranted about the choice and description of conditions of centrifugation. Firstly, the choice of g-min values [namely the product of the relative centrifugal force (g) value and the time of centrifugation, ideally with allowance for the acceleration and deceleration periods] for differential centrifugation inevitably entails some compromise. This depends on whether the aim is an "analytical fractionation" (so that a balance-sheet can be drawn up for the distribution of the constituents analysed) or a "preparative fractionation" with interest centred on a particular fraction (de Duve, 1964). Secondly, runs at nominally similar g-min values may differ in effectiveness according to the geometry of the rotor (de Duve, Berthet and Beaufay, 1959). Fortunately, however, angle (rather than swing-out) rotors of comparable geometry have been used in much of the published work entailing high-speed centrifugation without a density gradient. Thirdly, some authors regard the value g as referring to the bottom of the tube and others as referring to the centre (sometimes with disregard of any incompleteness of filling of the centrifuge tube); such ambiguity is lessened if the actual rotors, speeds and times employed are specified. Nevertheless, values for g-min are valuable as a rough yardstick for comparing results obtained in different laboratories at least for the non-gradient procedures now to be considered. [Gradient centrifugation represents a powerful tool (see de Duve *et al.*, 1959, and de Duve, 1964), but only certain applications will be considered here.]

Light microscopy, as used in the pioneer work, is almost valueless in connection with cytomembrane studies. The classical paper of de Duve *et al.* (1955) illustrates how careful application of purely biochemical techniques, particularly enzymic, may be rewarding in choosing conditions for differential centrifugation (see also de Duve, 1964). In connection with the desirability of obtaining a cytomembrane fraction little contaminated with lysosomes, enzymic assays (particularly for acid phosphatase (3.1.3.2.)) are in fact the only reliable guide, since lysosomes are too few to be enumerated by electron microscopy. For examination of actual cytomembrane elements electron-microscopy is, of course, a valuable tool, unfortunately neglected in some studies where its use would have been of notable benefit.

When use is made of the conventional 0·25 M sucrose medium, with no gradient, removal of mitochondria together with the bulk of the lysosomes can be effected by centrifugation at a value in the range of 100,000–170,000 *g*-min, as used by most authors [de Duve *et al.* (1955) and Ernster, Siekevitz and Palade (1962), *inter alia*]. With a higher value—e.g. 600,000 *g*-min, as deliberately used by Campbell *et al.* (1964) with 0·3 M sucrose medium—there is likely to be complete sedimentation of all lysosomes, but some loss of microsomal material must be accepted. When 0·88 M sucrose medium is used, a value of the order of 500,000 *g*-min is suitable.

The nature of the "fluffy layer" which overlies the mitochondrial pellet, at least with 0·25 M sucrose medium, was a matter of controversy until it became evident that it consists largely of damaged mitochondria. In the method of Palade and Siekevitz (1956) it is left behind with the mitochondria, but some authors transfer it to the next centrifugation stage or give no indication of the procedure followed.

For sedimentation of the microsomal fraction with 0·25 M sucrose, a value of the order of 6,000,000 *g*-min has been used by many authors (e.g. Ernster *et al.*, 1962); a 60-min run at 40,000 revs/min in a No. 40 angle rotor (Spinco) is customary, the nominal *g* being 105,000 (centre of tube) or 145,000 (bottom of tube). However, a value of only 3,000,000 *g*-min—as used in the study by de Duve *et al.* (1955)—is ample as judged by assays for glucose-6-phosphatase, and in at least some contexts use of a much lower *g*-min value (as employed by the pioneer workers in the field) may be sufficient or even preferable [see Reid (1961a), and Section III. B. 1(*b*)]. Moulé *et al.* (1960) shrewdly remark:—"The centrifugal forces used to carry out the sedimentation of microsomes have increased as the centrifuges have become more powerful so that, with 0·88 M sucrose as suspending medium, the earlier centrifugation of 2 hours at 40,000*g* . . . has been changed to 130,000*g* . . . and 145,000*g* for varying times . . . [Usually] no rational explanation has ever been given to justify the experimental conditions." One practice common when only low-speed centrifuges were available—the sedimentation of microsomal material by acidification and low-speed centrifugation—is now used only occasionally (e.g. Görlich and Heise, 1962, 1963) and is to be deprecated. Thus with citrate at pH 5·4 there can be enzyme solubilization (Gosselin, Podber-Wagner and Waltregny, 1962); on the other hand, a study of deCDP-choline synthesis indicated that there was enzyme adsorption on to microsomal material at pH 5·2 (Schneider, 1963).

The microsomal fraction is normally obtained as a gelatinous reddish pellet, underlying which there may be a pellet of particulate glycogen if the animals were not fasted. There is no excuse for the vagueness in some papers as to whether the animals were fasted, or for apparent surprise that the "microsomal" pellet showed a lower layer which consisted of glycogen (cf. Trivus and Spirtes, 1964). Most authors do take the sensible precaution of fasting the animals overnight, although this failed to depress the level of total glycogen in one strain of rat (E. Reid and G. Siebert, unpublished experiments). The microsomal material is best resuspended mechanically with a loosely fitting pestle, inserted directly into the centrifuge tube after addition of the suspending medium.

2. *Characteristics of the microsomal fraction* (*liver*)

(*a*) *Morphology*. Electron-microscope examination of sections of microsomal material, after fixation (usually with osmium tetroxide) and resedimentation, discloses mainly "vesicles" representing cytomembrane fragments—these are mostly "rough-surfaced" (i.e. ribosome-studded)—together with free ribosomes (Fig. 4a). From cytochemical examination (see Fig. 10), glucose-6-phosphatase appears to be localized in the membrane fragments, in accordance with the cytochemical observations on intact cells. Besides the vesicles—which are rounded, at least if a medium containing 0·25 M sucrose is employed, there is a varying admixture of miscellaneous elements, including so-called amorphous material.

(*b*) *Constituents found by chemical analysis*. In view of the morphological heterogeneity of microsomal fractions and of differences between laboratories in the conditions of isolation, it would be pointless to tabulate values for protein or RNA. Typically, about $\frac{1}{5}$ of the liver protein and up to $\frac{2}{5}$ of the RNA is recovered in the microsomal fraction, the "microsomal" RNA being mainly in the ribosomes. The ratios of RNA and of phospholipid to protein (or to lipid-free solids) are relatively high for the microsomal fraction (for examples of actual values, see Table III). The microsomal fraction is likewise rich in free and esterified cholesterol (for values see Reid, 1961b).

Microsomal fractions are low in cytochrome *c*, but contain cytochrome b_5 (cytochrome *m*) (Palade and Siekevitz, 1956; Ernster *et al*., 1962) and a CO-binding pigment (citations in Orrenius, Ericsson and Ernster, 1965). They contain coenzymes Q_9 and Q_{10} (Leonhaüser *et al*., 1962), and NADP (see Reid, 1961b) and inosine (Siekevitz, cited by Ernster and Jones, 1962). Immunological studies on microsomal material (Westrop and Green, 1960; see also Reid, 1961b and Jungblut, 1963) have not so far been very illuminating, except to demonstrate the existence of proteins characteristic of this material. Recently discovered constituents include cytochrome P-450 ("reticulochrome"; Omura, Sato, Cooper, Rosenthal and Estabrook, 1965) and *o*-tyrosine (Feuer, Golberg and Gibson, 1965).

(*c*) *Enzymes*. Table II lists enzymic activities reported to be present in microsomal fractions, not necessarily in a high amount relative to the

TABLE II. *Enzymic activities of liver microsomal fractions*

Enzyme or enzyme system	% in microsomal fraction[a]	% activity / % glucose 6-phosphatase	References and remarks [including references to evidence for location of enzyme in membranous elements].[b]
Oxidoreductases			
L-Amino-acid oxidase (1.4.3.2)			Struck and Sizer, (1960)—microsome fraction prepared by pH 5·5 pptn. from blender homogenate
Aryl-4-hydroxylase (1.14.1.1)	~100%		Booth and Boyland, (1957)
Ascorbate-forming system (1.1.1.19)			Kanfer *et al.*, (1959)
Azo-dye reductase (1.6.6.7)			Emanoil-Ravicovitch and Herisson-Cavet (1963) [*iidem*]
Catalase (1.11.1.6)	14%		Higashi and Peters (1963a)—% value with enzymic assay lower than with immunochemical assay [Peters (1962)];
Cystine reductase (1.6.4.1)			Myers and Worthen (1961)
Desmosterol reductase (1.3.1.)	35%		Avignon and Steinberg (1961)
Diaphorases—NAD_2, $NADH_2$ (1.6.99)			See Reid (1961a), de Duve *et al.* (1962), Ernster *et al.* (1962)
Glucuronolactone reductase (1.1.1.20)			Suzuki *et al.* (1960)
β-Hydroxy-β-methylglutaryl-CoA reductase (1.1.1.34)			Bucher *et al.* (1960)—trace of 'condensing enzyme' also present [Siperstein and Fagan (1964)]
Lipid-peroxidation system (ADP-activated; coupled to $NADPH_2$ oxidase)			Hochstein and Ernster (1964) [*iidem* speculate that location is in cytomembranes surrounding mitochondria]
Malate dehydrogenase (1.1.1.37)	low		Shull (1959)—cf. de Duve *et al.* (1962)
$NADH_2$-cytochrome b_5 reductase (1.6.2.2)			Strittmatter and Velick (1956)
$NADH_2$-cytochrome *c* reductase (1.6.99.3)	62%	0·83	de Duve *et al.* (1955), *inter alia* [Palade and Siekevitz, 1956; Chauveau *et al.*, 1962; Ernster *et al.*, 1962; Dallner, 1963]

TABLE II—*continued.*

Enzyme or enzyme system	% in microsomal fraction[a]	% activity / % glucose 6-phosphatase	References and remarks [including references to evidence for location of enzyme in membranous elements].[b]
$NADPH_2$-cytochrome *c* reductase	64%		Phillips and Langdon (1962)—but much lower %-value reported in earlier studies (e.g. de Duve *et al.*, 1955) [Ernster *et al.*, 1962]
$NADPH_2$ peroxidase ($NADPH_2$ oxidase) (1.11.1.2)			Gilette *et al.*, (1957)
Stearate-dehydrogenating system (→ oleate)			Marsh and James (1962)
Steroid reductase (→ 5α isomer)			Tomkins (1959)
Sulphite oxidase (1·8·3·1)	75%		McLeod *et al.*, (1961)—but diffuse distribution found for rat liver by Baxter *et al.* (1958).
Transferases			
3-Acylglycerophosphorylcholine acyltransferase (from acyl-CoA) (2.3.1.)			Lands and Merkl (1963)
deCDPcholine:1,2-diglyceride cholinephosphotransferase (2.7.8)	50%		Schneider (1963) [*idem*]
CDPcholine:1,2-diglyceride cholinephosphotransferase (2.7.8.2)	97%	1·0	Wilgram and Kennedy (1963); for earlier work on incorporation of choline, myoinositol and serine see Hübscher (1962)
Diglyceride acyltransferase (2.3.1.20)	84%	0·9	Wilgram and Kennedy (1963)
Glycerolphosphate acyltransferase (2.3.1.15)			Kornberg and Pricer (1954); Stein *et al.* (1957)
Phenylphosphate: cytidine phosphotransferase			Brawerman and Chargaff (1955)—data do not exclude lysosomal location
Phosphatidylinositol kinase (2.7.1.)			R. H. Michell and J. N. Hawthorne (1965)
Pyrophosphate:glucose phosphotransferase[c]			Nordlie and Arion (1964); Stetten (1964)

Ribonuclease ("alkaline") (2.7.7.)	low		Morais and de Lamirande (1965), and earlier studies (see de Duve *et al.*, 1962); apparently ribosomal (Roth, 1960) [but membranes *do* contain a ribonuclease, and also a phosphodiesterase—de Lamirande *et al.*, 1966]
Transmethylases, to phosphatides or other acceptors (2.1.1.)			Bremer and Greenberg (1961), Cooksey and Greenberg (1961)
UDPglucose:glycogen glucosyltransferase (2.4.1.11)			Leloir and Goldemberg (1960); Luck (1961) [*iidem*—associated with glycogen granules, *not* membranes]
UDP glucuronyltransferase (2.4.1.17)	high		Dutton and Storey (1956); Strominger *et al.* (1957)
Hydrolases			
Alkaline phosphatase (3.1.3.1)	42%		Allard *et al.* (1957)—%-value lower if Mg^{2+} ions in assay
Amylase (3.2.1.1)	52%	1·15	Brosemer and Rutter (1961)—shows latency as for lysosomal enzymes
Arginase (3.5.3.1)	41%		Rosenthal *et al.* (1956)—contaminant?
Arylsulphatase C (3.1.6.1)	62%		Dodgson *et al.* (1955) [Dodgson *et al.* (1957)]
ATPase (ATP pyrophosphohydrolase; Mg^{2+} ions in assay) (3.6.1.4)	low		Novikoff *et al.* (1953)—liver contains at least one other ATPase [See Nucleoside triphosphatase entry, also Börnig (1964)]
N-Deacylase ("esterase") (3.5.1.)			Krisch (1963)—cf. Remmer (1962)
Esterases: Acetylcholinesterase ("true cholinesterase") (3.1.1.7)	46%		Goutier and Goutier (1955)
	62%	0·85	Underhay *et al.* (1956)
Esterases: Arylesterase (3.1.1.2)	58%		McCann (1957) [Carruthers and Baumler (1962); Chauveau *et al.* (1962)—cf. Takanami (1959)]
	67%	0·9	Underhay *et al.* (1956)
	85%	1·0	Carruthers and Baumler (1962)
Esterases: Benzoylcholinesterase (3.1.1.9)	46%		Goutier and Goutier-Pirotte (1955)
Esterases: Carboxylesterase (3.1.1.1)			Omachi *et al.* (1948); Heller and Bargoni (1950)
Esterases: Cholesterol esterase (3.1.1.13)	112%		Schotz *et al.* (1954)
Esterases: Vitamin A esterase (3.1.1.12)			Gauguly and Devel (1953); Shibko and Tappel (1964)

TABLE II—*continued.*

Enzyme or enzyme system	% in microsomal fraction[a]	% activity / % glucose 6-phosphatase	References and remarks [including references to evidence for location of enzyme in membranous elements].[b]
Glucose-6-phosphatase (3.1.3.9)[c]	74%	(1·0)	de Duve *et al.* (1955) [Chauveau *et al.* (1962), *inter alia*]
β-Glucuronidase (3.2.1.31)	40%		Walker (1952) } probably lysosomal and microsomal enzymes different
	37%	0·5	de Duve *et al.* (1955) } probably lysosomal and microsomal enzymes different
Inorganic pyrophosphatase (3.6.1.1)[c]	80%	1·0	Stetten (1964); Nordlie and Arion (1964)—cf. Ernster and Jones (1962) and El-Aaser *et al.* (1966b), probably different from mitochondrial enzyme
Lipoprotein lipase (3.1.1.)	40%		Shibko and Tappel (1964)
Lysophospholipase (lysolecithinase) (3.1.1.5)	61%		Shibko and Tappel (1964)
NAD nucleosidase (3.2.2.5)	93%		Jacobson and Kaplan (1957)
NAD pyrophosphatase (3.6.1.9)	63%		Jacobson and Kaplan (1957)
NADP pyrophosphatase (3.6.1.)	52%		Jacobson and Kaplan (1957)
Nicotinamide deamidase (3.5.1.)			Petrack *et al.* (1963) [*iidem*]
Nucleoside diphosphatases: ADPase (≡CDPase ?) (3.6.1.)			Ernster and Jones (1962); Novikoff and Heus (1963) [Ernster and Jones (1962); Börnig (1964)]
Nucleoside diphosphatases: GDP/IDP/UDPase (3.6.1.6)	>50%		Ernster and Jones (1962) [*iidem*]
	71%-IDP		Novikoff and Heus (1963) [*iidem*]
5′-Nucleotidase (nucleoside-5′-monophosphatase) (3.1.3.5)	35–40%		Novikoff *et al.* (1953); de Lamirande *et al.* (1958)
	47%	1·0	El-Aaser (1965) [*idem*, and Ku and Wang (1963)]
Nucleoside triphosphatase (3.6.1.4)	23%-ITP		Novikoff and Heus (1963)
Phosphodiesterase I (3.1.4.1)	29%		Razzell (1961) [de Lamirande *et al.* (1966)]
Thiamine pyrophosphatase (3.6.1.)	~50%		Novikoff and Heus (1963)—cf. Kiessling and Tilander (1960)

Uronolactonase (3.1.1.19)	94%		Winkelman and Lehringer (1958)—see also Yamada *et al.* (1959), and de Duve *et al.* (1962).
Other enzymes or enzyme systems			
Acyl-CoA synthetase (6.2.1.3)			Creasey (1962)—apparently different from enzyme in nuclear fraction
Cholyl-CoA synthetase (6.2.1.7)			Bremer (1955); Elliott (1956)
Demethylase (azo-dye)			Conney *et al.* (1957); see also Leadbeater and Davies (1964)
Epoxysteroid lyase			Breuer *et al.* (1963)
Ether(aromatic)-cleaving system			Axelrod (1956)
Fumarase (fumarate hydratase) (4.2.1.2)	28%	0·4	de Duve *et al.* (1955, 1962)—activity possibly due to contamination
Glutamine synthetase (6.3.1.2)	47%		Wu (1963) [*idem*]
Prothrombin-forming system			Goswami and Munro (1962) [*iidem*]

[a] %-values are given only where there were reasonably quantitative results and the recovery of homogenate activity in the fractions was satisfactory. Where low %-values have been included, the homogenate may have contained at least two enzymes one of which was truly associated with the microsomal fraction. On the other hand some fairly high values in the literature have been excluded because, for example, the enzyme is really lysosomal. (Note that most of the results are with rat liver.)

[b] Evidence for location of enzyme in membranes [ref. given as *idem* or *iidem* if exactly as for ref. just given] was obtained by methods indicated later in the Chapter. For other surveys, not centred on the microsomal fraction, see Reid (1961b), de Duve *et al.* (1962), Rouiller (1964) and Roodyn (1965). For various microsomal "mixed function oxidases" see Mason *et al.* (1965).

[c] It has been argued that three supposedly distinct enzymic activities may be due to a single enzyme (concerned with glucose-6-phosphate metabolism) (Nordlie and Arion, 1964; Stetten, 1964); Feuer *et al.* (1965) and El-Aaser *et al.* (1966b) have obtained supporting evidence.

original homogenate. For enzymes which have been sought in microsomal fractions and found *absent*, see Reid (1961b), de Duve *at al.* (1962), or Roodyn (1965). The evidence for a membranous rather than a ribosomal location, as indicated thus [ ] in the table, is based on methods reviewed in Section III. B. Some implications of the enzymic pattern found in microsomal fractions are dealt with in Section IV. A. 2 in connection with cytomembrane function.

Many of the tabulated studies of microsomal fraction enzymes are open to criticism, for example because of failure to make use of glucose-6-phosphatase as a "marker" enzyme or even to express quantitatively the results of the enzyme actually studied. On the other hand, even where close parallelism with glucose-6-phosphatase *has* been established, as in the case of arylesterase (3.1.1.2) and 5′-nucleotidase, it cannot be assumed that the enzyme is located in the same cytomembrane elements as glucose-6-phosphatase; indeed, some such enzymes (e.g. 5′-nucleotidase) are now known to be associated with plasma-membrane fragments, the presence of which in microsomal fractions has long been suspected by many investigators. Relevant cytochemical evidence, already surveyed in Section II, will be re-considered in Section III. E. in connection with the biochemical evidence. In qualification of the statement that arylesterase and 5′-nucleotidase sediment in parallel with glucose-6 phosphatase during differential centrifugation, it may be pointed out that the lowering of sedimentability observed when the usual 0·25 M sucrose medium is supplemented with glycerol is more striking for arylesterase than for glucose-6-phosphatase (Carruthers, Wernley, Baumler and Lilga, 1960), and that with 0·25 M sucrose a centrifugal "cut" preceding the main microsomal fraction may differ in the ratio glucose-6-phosphatase/5′-nucleotidase (El-Aaser, 1965). It was through shrewd attention to such discrepancies that de Duve *et al.* (1955) built up evidence for the existence of lysosomes; here there is a clear moral for work on microsomal fractions.

Consideration will now be given to the actual membranous elements present in microsomal fractions, the main aim being to ascertain whether elements differing in morphology differ in enzyme content.

B. ISOLATION AND CHARACTERISTICS OF HEPATIC CYTOMEMBRANE FRACTIONS WITH FEW RIBOSOMES

1. Special Conditions of Centrifugation

(*a*) *Approach to the problem.* For the purpose of isolating "intact" cytomembrane fragments with few attached ribosomes (intact in the sense of retaining the appearance found in homogenates), the use of carefully controlled centrifugation alone is the method of choice, since additives such as deoxycholate (see 2.) may damage the membrane if not actually dissolve it (Siekevitz, 1962). Mere centrifugal procedures, without such additives, would hardly be expected to furnish ribosome-free fragments derived mainly from the rough-surfaced cytomembranes; yet in some of the studies mentioned below, the yield of smooth-surfaced as distinct from rough-surfaced fragments was so high in relation to the proportion estimated from electron micrographs, that one wonders whether certain rough-surfaced cytomembranes readily lose their ribosomes *in vitro* (at the outset of homogenization?) The postulate is that among the rough-surfaced cytomembranes it is only a particular population that is highly prone to lose its ribosomes. Thus the postulate is not incompatible with the observation that drastic mechanical treatments which might have been expected to cause detachment of ribosomes —e.g. homogenization with a blendor (Dallner, 1963), or freezing of a microsomal suspension (Fig. 8, a and b; cf. Moulé *et al.*, 1960)—have shown, by electron microscopy, no detectable differences from normal microsomal fractions.

In pioneer experiments in this field (e.g., Petermann, Mizen and Hamilton, 1956) in which membrane fragments ("microsomes") and ribosomes were isolated in separate centrifugal fractions, the use of an alkaline buffered medium could well have led inadvertently to detachment of almost all the ribosomes, through chemical rather than mere mechanical loosening.

Before considering subtle methods of centrifugation, for example with a density gradient, mention should be made of experiments in which in place of a single microsomal centrifugation step, there were successive centrifugations, so as to furnish different microsomal "cuts". (The term "cuts" here implies a multi-step primary fractionation; a term such as "microsomal sub-fractions" or "sub-microsomal fractions" might be taken as signifying recentrifugation of a primary fraction.)

(*b*) *Sedimentation with no gradient* (Differential centrifugation). In one study (Barbieri and di Marco, 1963), a "woolly" layer overlying the packed pellet was found to be particularly rich in phospholipid, but was not examined morphologically. The study of Trivus and Spirtes (1964) will be disregarded, since the lower layer termed the "heavy microsomal fraction" apparently consisted of glycogen. In a few studies, of which recent examples are shown in Table III, successive centrifugations have been made, and each pellet examined by electron microscopy. It appears that, at least with 0·88 M sucrose medium, smooth-surfaced cytomembranes may be the predominant constituent of "light" as distinct from "heavy" microsomes. Since the "light" fraction obtained in 0·88 M sucrose medium is not rich in free ribosomes (Fig. 7b), the fact that the RNA concentration is no lower than in the "heavy" fraction is compatible with the possibility that some RNA is present in actual membranes [see Section III. B. 3.(*a*)]. It is, of course, not surprising that the two fractions are equally rich in phospholipid. It should be emphasized that with 0·25 M sucrose in place of 0·88 M, or with saline, successive

TABLE III. *Characteristics of successive rat-liver microsomal "cuts" obtained as pellets after centrifuging in sucrose media of uniform density*[a]

Parameter studied	Microsomal "cut"			References and Remarks
	"Heavy microsomes" *g*–min × 10^{-6}: 1–5 in 0·25 M sucrose and 2–10 in 0·88 M sucrose	"Light microsomes" *g*–min × 10^{-6}: 6–12 in 0·25 M sucrose and 11–25 in 0·88 M sucrose	"Post-microsomes" *g*–min × 10^{-6}: >12 in 0·25 M sucrose and >25 in 0·88 M sucrose	
Electron-microscopic appearance (*especially of membranous fragments*)				
	"Wide variety of subcellular constituents"	Smooth-surfaced fragments plus ribosomes		Decken and Campbell (1961)
	Fragments, some rough-surfaced	Small fragments, plus amorphous material; few ribosomes		M. S. C. Birbeck and E. Reid (unpublished)
	Smooth- and rough-surfaced fragments (in each cut)	Various fragments, ribosomes, plus ferritin (5 mμ granules)		Kuff and Dalton (1957)—4 cuts
	Fragments, some rough-surfaced	Small fragments, usually few ribosomes	Amorphous material, plus ferritin	Moulé *et al.* (1960)—0·88 M
Chemical constituents				
Solids, mg/g liver	30	14		Reid (1956)
% Phospholipid-P	0·85	0·64		

% RNA-P	0·61	0·76		
N × 6·25, mg/g liver	15, 6·5	4·4, 4·3		Kuff and Dalton (1957)—4 cuts
% RNA-P	0·9, 1·15	1·7, 1·4		
N × 6·25, mg/g liver	14, 8, 5	5	1·0, 0·8	Moulé *et al.* (1960)—0·88 M (several cuts)
% Phospholipid-P	1·7, 1·4, 1·55	1·7	0·27, 0·17	
% RNA-P	0·95, 1·15, 1·25	1·25	0·63, 0·3	
% Phospholipid-P		2·9	3·2	Hird *et al.* (1964)
% RNA-P		1·2	2·0	
Serum albumin (newly synthesized), specific radioactivity of purified protein	++[b]	++		Peters (1962)—0·88 M
Prothombin, units/g protein	3·7	2·8		Goswami and Munro (1962)
"X" (factor stimulating amino-acid incorporation; apparently not RNA)	Contains system for incorporation		Contains "X"	Mizrahi (1965)—previously studied in M. B. Hoagland's laboratory
Enzymic activities[c]				
Amino-acid activating enzymes	±		++	Hird *et al.* (1964)
Arginase	++, ++			Rosenthal *et al.* (1956)—2 cuts
Ascorbate-forming systems (from D-glucuronolactone or L-gulonolactone)	++, +, ±	0		Kar *et al.* (1962) and Ghosh *et al.* (1963)—4 cuts; meagre enzymic data

TABLE III—*continued*

Parameter studied	Microsomal "cut"			References and Remarks
	"Heavy microsomes" g–min × 10^{-6}: 1–5 in 0·25 M sucrose and 2–10 in 0·88 M sucrose	"Light microsomes" g–min × 10^{-6}: 6–12 in 0·25 M sucrose and 11–25 in 0·88 M sucrose	"Post-microsomes" g–min × 10^{-6}: >12 in 0·25 M sucrose and >25 in 0·88 M sucrose	
Glucose-6-phosphatase	++ (little activity in other 3 cuts)			Emanoil-Ravicovitch and Herisson-Cavet (1963)—procedure of Moulé *et al.* (1960)
Azo-dye reductase	+, ++	+		
Cholesterol-synthesizing system (especially lathosterol → cholesterol)	++	+		Franz *et al.* (1959)—sup. also necessary
Enzymes oxidizing or hydrolysing drugs, e.g. procaine hydrolysis	+	++		Remmer (1962) [cf. Fouts (1961)]
Esterase (arylesterase)	+	++		Takanami (1959)—saline medium
Esterase (arylesterase)	++	++		Chauveau *et al.* (1962)—0·88 M
Glucose-6-phosphatase	++	++		
$NADH_2$-cytochrome *c* reductase	++	++		
Esterase (arylesterase)	+, ++	+		Carruthers and Baumler (1962)—0·88 M, 3 cuts
Esterase (arylesterase)	+	++		Shibko and Tappel (1964)
Lipoprotein lipase	++	++		

Lysolecithinase	++	++		
Glucose-6-phosphatase	+	++		El-Aaser (1965) —variable
5′-Nucleotidase	+	++		
Lecithin-synthesizing system	++	+		Schneider (1963)—swing-out head
Prothrombin-forming system (endogenous precursor)	++	+		Goswami and Munro (1962)
Squalene-synthesizing system	+, ++	±		Gosselin *et al.* (1962)—3 cuts, similar in RNA/protein ratios
Glucose-6-phosphatase	++, +	0		

[a] This table is not intended to be comprehensive; thus, the data of Palade and Siekevitz (1956) are not included. "M" refers to sucrose molarity. Most experiments were with 0·25 M sucrose.

[b] Albumin in each of several "cuts" had the same specific activity, although more subtle methods of centrifugation with high-density media gave fractions which showed activity differences (Table IV).

[c] ++ good activity, + some activity, ± very little activity, 0 no activity.

centrifugal cuts may be increasingly rich in RNA (at least relative to phospholipid), but that there is no good morphological evidence for a predominance of smooth-surfaced fragments in the lighter fractions. It may also be pointed out when light elements rich in RNA have been sedimented, prolonged centrifugation of the supernatant gives a small pellet which is poor in cytomembrane fragments as shown by electron microscopy (Fig. 8f); it may be low in RNA and phospholipid (Table III). This material truly warrants the term "post-microsomal". This fraction may have abundant ribosomes in the case of pancreas, which is rich in free ribosomes (Palade and Siekevitz, 1958b).

A judicious choice of centrifugal "cuts" in the classical, non-gradient, procedure may, then, be helpful in ascertaining whether different groups of membrane fragments (perhaps differing in morphology) differ in their enzyme content. There may indeed be instances of such enzymic heterogeneity (Table III), but unfortunately in few studies has assessment been made of the morphology of the fractions, or even of biochemical parameters such as phospholipid and glucose-6-phosphatase which would facilitate the interpretation of the results and their comparison with results obtained in other laboratories. In one particularly careful study (Moulé *et al.*, 1960), close parallelism was actually found in the distribution of three different enzymes (Table III); but on applying the procedure of Moulé and co-workers, Carruthers and Baumler (1962) observed that an intermediate "cut"—presumably representing a mixture of rough- and smooth-surfaced fragments—was particularly rich in esterase.

(*c*) *Flotation with no gradient.* Chauveau, Moulé, Rouiller and Schneebeli (1962) ingeniously exploited the relatively low density of phospholipid-containing structures and succeeded in isolating membrane fragments almost free of ribosomes. The starting material was a "light" microsomal fraction already with a low content of ribosomes and rough-surfaced membrane fragments (see above). Centrifugation of this fraction for 20–40 hr in a concentrated sucrose medium gave, besides a pellet, a surface pellicle which was particularly firm and manageable if the sucrose density were 1·25 rather than 1·21. More than half of the nitrogen of the original fraction was recovered in this pellicle, which appeared in the electron microscope to consist of smooth-surfaced cytomembranes (Fig. 7b). The chemical and enzymic composition of such material is given in Table IV.

FIG. 7. Microsomal material isolated from liver in 0·88 M sucrose medium (Moulé, 1964; see also Moulé *et al.*, 1960, and Chauveau *et al.*, 1962). Courtesy of Y. Moulé. Top (a). Whole microsomal fraction. R=rough-surfaced vesicle, S=smooth-surfaced vesicle. ×75,000 approx. Bottom left (b). Fraction obtained at 145,000 *g* (3 hr) after removal of a first microsomal "cut" at 40,000 *g* (2 hr): note predominance of smooth-surfaced vesicles. × 75,000 approx. Bottom right (c). Pellicle (floating) sub-fraction obtained by re-centrifugation, in sucrose medium of high density, of the second microsomal "cut" depicted in b): note that only smooth-surfaced elements are present. × 150,000 approx.

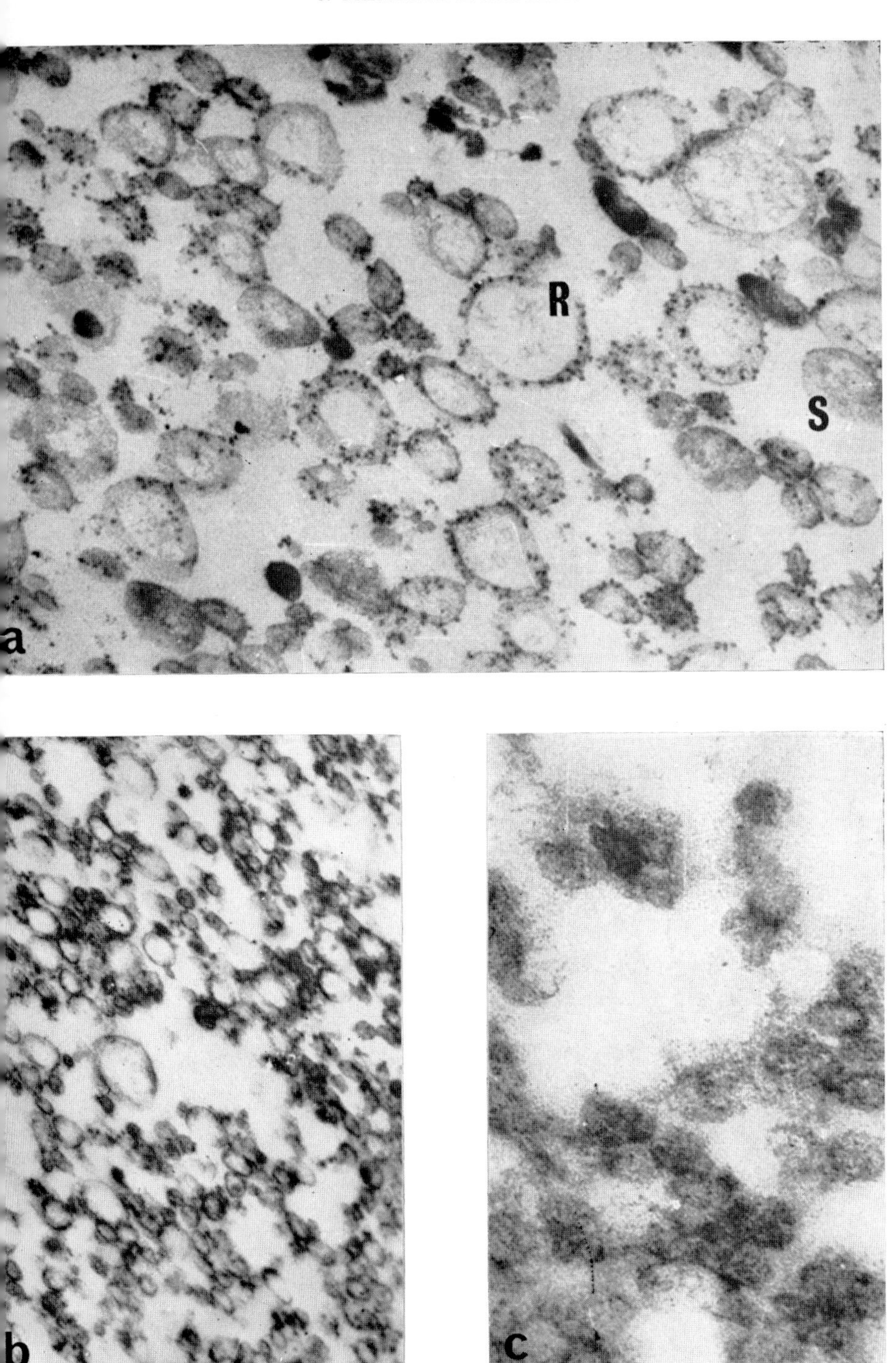
R
S
a
b
c

Table IV. *Characteristics of membrane fractions isolated in high-density media and with low content of ribosomes*[a]

Outline of procedures, particularly final centrifugation	Fraction(s) obtained and morphological appraisal	Inferred density of the material[b]	Yield of protein, mg/g tissue	Phospholipid-P, mg/g protein	RNA-P, mg/g protein	Enzymic and other parameters, especially comparisons between the membrane fractions[c]	References and remarks
0·88 M **S** mic.-sup. diluted to 0·175 M; 2 layers, 7–10 hr—time critical if angle rotor used—essentially method of Rothschild (1961)	*ss* (free from *rs*), [*rs* (free from *ss*) and ribosomes]	1·17 [>1·17]	9 [7]	12 [13]	4·7 [1·9]	*ss* 5 times richer than *rs* in $NADPH_2$ oxidase and 3–5 times richer in drug-oxidizing or -dealkylating enzymes	Fouts (1961)—rabbit liver, analytical values are from Remmer and Merker (1963)
0·88 M **S** "light mic."; then **S** (uniform), 20–40 hr—flotation of *ss*	*ss* (almost ribosome-free) [ribosomes]	⩽1·14 [⩽1·25]	5	27	4·5	*ss* almost as rich as *rs* in glucose-6-phosphatase, $NADH_2$-cyt. *c* reductase, and arylesterase, also azo-dye reductase.	Chauveau *et al.* (1962), also Emanoil-Ravicovitch and Herrisson-Cavet (1963)
0·25 M **S** mic.; then **S** (only 0·25 M) + deoxycholate (only 0·024%)—*ss* as pellet overlying ribosomes	*ss* (almost ribosome-free, but partly derived from *rs*)		5	50	9	*ss* rich in various mic. enzymes, but procedure precludes comparison with *rs* or even with mic	Ernster *et al.* (1962)
0·88 or 0·25 M **S** mic.; then **S**, 3 layers,	*ss* (almost ribosome-free)	>1·17	6–8	12·5	1·5	*ss* the locus of albumin; *rs* the locus	Peters (1962), Higashi and

10–16 hr—flotation of *ss*—essentially method of Rothschild (1961)	[*rs* (rich in ribosomes)]	[>1·17]				of catalase, but mainly in a low-density fraction as if *rs* were non-uniform[e].	Peters (1963*a*)
Procedure of Fouts (1961)	*ss* [and *rs*]		15·5	24·5	0·95	Lecithin synthesis from de-CDPcholine: *ss*: +, *rs*: ++	Schneider (1963)
Procedure of Peters (1962)	*ss* [and *rs*]		7	22·5	1·37	Lecithin synthesis from de-CDPcholine: *ss*: ±, *rs*: ++	Schneider (1963)
Deoxycholate extract of mic. dialysed and freeze-dried; then gradient of NaBr etc, —flotation of *ss*	*ss* (presumed) [also a lipid-poor fraction]			9	<0·1	*ss* eventually, but not initially, the main locus of labelled albumin (*in vivo* precursor)	Jungblut(1963) —the salt medium caused some solubilization
0·3 M S mic.; then layer over 1·5 M S with 10^{-2} M Mg^{2+}, 75 min —re-centrifuge interface of layer of *ss*	*ss* (few ribosomes) [*rs* and ribosomes]	<1·20 [≥1·20]	6·6	22	5	*ss* the locus of 'DT diaphorase', but (compared with *rs*) almost free of glucose-6-phosphatase, low in nucleoside di- and triphosphatases and especially in $NAD(P)H_2$-oxidizing enzymes; *ss* similar to *rs* in CO-binding pigment, $NADPH_2$-cyt. *c* reductase and oxidative-demethylating activity [cf.	Dallner *et al.* (1963); S. Orrenius, G. Dallner & L. Ernster, cited by Orrenius, *et al.* (1965) —cf. IV.B.1 in text

TABLE IV—*continued*

Outline of procedures, particularly final centrifugation	Fraction(s) obtained and morphological appraisal	Inferred density of the material[b]	Yield of protein, mg/g tissue	Phospho-lipid-P, mg/g protein	RNA-P, mg/g protein	Enzymic and other parameters, especially comparisons between the membrane fractions[c]	References and remarks
						Fouts (1961), above]	
Essentially procedure of Dallner *et al.* (1963)	*ss* (with some ribosomes)				4·5	*ss* (showed only low amino-acid incorporation *in vitro*)	Campbell *et al.* (1964)
0·25 M S + 0·013 M phosphate; crude nuclear pellet lysed in phosphate; then KBr (uniform)	*pl* and nuclear-membrane fragments (no electron-microscopy)	<1·22	20			No hexosamine detectable; rich in H-2 (homograft) antigens	Herzenberg and Herzenberg (1961) "Duall" homogenizer, mouse liver
mM $NaHCO_3$[e] low-speed pellet; then S, 5 layers, 75 min in swing-out rotor—flotation of *pl*—method based on Neville (1960)	*pl* (some *ss*-type vesicles present, but thought to represent true association of *ss* with *pl*, rather than contamination)	1·17	0·4	20	<2 (inferred)	*pl* relatively rich[f] in sialic acid, hexosamine, ATPases, 5′-nucleotidase[g] and NAD pyrophosphatase; rather low activities of glucose-6-phosphatase, $NADH_2$-cyt. *c* reductase, NAD nucleosidase ADPase and ribonuclease(s)	Emmelot *et al.* (1964*a*), Emmelot *et al.* (1964*b*)—homogenization has to be gentle

Ehrlich ascites cells: gas-cavitation, 0·25 M S; *ss* prepared from mic, and lysed in *m*M tris: then gradient, preferably Ficoll (dialysed), 12–16 hr, swing-out rotor	*pl* (apparently no study of morphology), cf. *ss* before lysis	1·07 in Ficoll; cf. *ss* before lysis: 1·13 in S, 1·15 in NaBr, 1·03 in Ficoll	(*ss*, 18)	(*ss*, 1·3)	Parameters used as markers: antigens concerned in agglutination, and ATPase (Na^+-, K^+-, Mg^{2+}- activated, ouabain-sensitive)	Wallach and Kamat (1964)—apparently liver not tried. See also Wallach and Ullrey (1964)
Ehrlich ascites cells: essentially as above, but Mg^{2+} present (*m*M in the Ficoll centrifugation step); pH important (8·6)	*pl*[h] (apparently no study of morphology), 1·05	1·05[h]			ATPase largely associated with the antigens, as was part of the $NADH_2$ diaphorase (contaminating cytomembranes?)	Kamat and Wallach (1965)—no citation of Dallner (1963) although relevant
0·25 M S mic-sup. (from 5 g liver); isopycnic centrifugation (6 hr) in zonal centrifuge with S gradient (pH 7·4 *tris*, 5 × 10^{-3} M)	*pl* (vesicles, +some *ss*; semi-separated from *ss* in later work) *ss* (vesicles) [*rs*]	1·14–1·16[i] 1·15–1·17 [1·18–1·19]		1·6	5′-Nucleotidase as marker for *pl* (cytochemical as well as biochemical), though it partly ran with 'soluble' material, as did UDPase; some ATPase but not ADPase associated with *pl*	El-Aaser *et al.* (1966b) *ss* and *rs* not aggregated even if Mg^{2+} added [cf. Kamat and Wallach (1965)] but *ss* and *rs* not then separated
0·08 M S nuclear fraction filtered, then centrifuged in low-speed zonal rotor	*pl*′ (vesicles; contaminated with *rs* and *ss*)	1·10–1·16 (shown by re-centrifugation)			*pl* (5′nucleotidase as marker) contained ATPase and were virtually free of	El-Aaser *et al.* (1966a)-little change if Mg^{2+} pres-

TABLE IV—*continued*

Outline of procedures, particularly final centrifugation	Fraction(s) obtained and morphological appraisal	Inferred density of the material[b]	Yield of protein, mg/g tissue	Phospholipid-P, mg/g protein	RNA-P, mg/g protein	Enzymic and other parameters, especially comparisons, between the membrane fractions[c]	References and remarks
with S gradient (+*tris*)-5′-nucleotidase ran in low-density region (*pl*′), in high-density region (*pl*), and with nuclei	*pl* (with some sheets—bile canaliculi?)	1·16–1·18		6	0·4	glucose-6-phosphatase	ent; 5–10% of 5′-nucleotidase in *pl*
Liver, kidney and intestinal mucosa—procedures of Neville (1960) and Emmelot *et al.* (1964)	*pl* (microvilli, desmosomal regions, etc. identifiable)					*pl* enriched in 5′-nucleotidase, L-leucyl-β-naphthylamidase and (not liver) alkaline phosphatase	Coleman and Finean (1965)—guinea-pig
Epididymis: S+NaCl (latter essential); gradient, 1 hr in swing-out rotor	Golgi fragments (some Golgi bodies apparently intact),	<1·15	5·5	29	10	Alkaline phosphatase present	Schneider and Kuff (1954)—no success with intestine
Intestinal mucosa (epithelial cells) —0·3 M S (+EDTA); filtration, then 2nd of 2 low-speed sediments	*pl* (microvillus sheets, + some nuclei, mitochondria, etc.)					Enrichment in aminopeptidase and invertase; alkaline phosphatase and glucose-6-phos-	Porteous and Clark (1965)—rabbit; "sheets" and nuclei not

taken (*no* high-density medium)					phatase also present	separable, though no physical links seen
Intestinal mucosa—0·3 M S nuclear fraction: washed, then blended in 0·03 M S (pellet taken; *no* high-density medium)	*pl* (from microvilli—intact brush-border structures seen *before* blending; fate of nuclear membrane uncertain)				*ss* (brush border) the site of invertase and alkaline ribonuclease, and one site of alkaline phosphatase and other hydrolases	Hübscher *et al.* (1965)–some data are for guinea-pig, some for cat (methods unsuitable for rat)
Intestinal mucosa—0·3 M S mic.-sup.: essentially procedure of Dallner *et al.* (1963) with use of Cs^+, Mg^{2+}; 3 fractions (I, II and *rs*)	?*pl* (II; long strands, + large *ss* vesicles) *ss* + some *rs* (I) [*rs*]	<1·15 ~1·15	4 15	4 10	II low in glucose-6-phosphatase and acyl-CoA synthetase, and high in phosphatidate phosphatase; I intermediate between *rs* and II	
Leucocytes: 0·34 M S homogenate freed from nuclei and centrifuged over S gradient, giving layer of 'light membranes' ('heavy membranes' also studied)	*pl* (vesicles, some ruptured; nuclear membrane and Golgi as well as *pl*?)	<1·15	15	<0·05	*pl* contained ATPase, ADPase, 5′-nucleotidase (feeble), $NADH_2$-cyt. *c* reductase	Woodin and Wieneke (1966)—rabbit

[Footnotes to Table iv]

[a] Fractions prepared from rat liver unless otherwise stated; for non-hepatic tissues see also Table VI (in which the emphasis is on "primary" microsomal fractions; but all brain and muscle work has been arbitrarily assigned to Table VI). The entries are in roughly chronological order. Abbreviations: mic., microsomal fraction; mic.-sup., supernatant from sedimentation of mitochondria; *pl*, plasma-membrane fragments; *rs*, rough-surfaced cytomembrane fragments, *ss*, smooth-surfaced fragments (typically vesicles; from cytomembranes?); S, sucrose solution.
[b] Density values refer to 20°:—1·18 corresponds to 1·45 M S (48% w/v) and 1·28 to 2·2 M (73%)—for full table see de Duve *et al.* (1959); other fractions obtained simultaneously are indicated in square brackets [ ].
[c] Abbreviations such as *ss* are here used in the non-committal sense of fractions containing the elements in question.
[d] From the original note by Rothschild (1961), it appears that some *ss* had density 1·17 and other *ss* (with lower RNA concentration) <1·17, and that *rs* had density 1·16 and ribosomes 1·17.
[e] Initial treatment with 2·8 M pH 6 citrate was sometimes employed, particularly in work with hepatomas.
[f] The term "relatively" here implies a comparison with mic, with respect to concentrations but side-by-side comparisons are not usually given in the paper cited.
[g] Ku and Wang (1963) and El-Aaser (1965) have also found 5′-nucleotidase in *pl* fractions from liver.
[h] Yield of *pl* almost quantitative as judged by antigen assays; previous density values probably "spuriously high" (nonspecific interactions).
[i] Density values if Mg^{2+} present: *pl*, 1·13 (minor peak) and 1·16–1·17; *ss* and *rs*, 1·16–1·20.

(*d*) *Centrifugation with a gradient.* As has been discussed by de Duve (1964) (see also Wallach and Kamat, 1964), density-gradient centrifugation is a powerful tool if the conditions are judiciously chosen. Even a simple layering procedure may be effective. Examples are given in Table IV. Thus, a technique previously tried by Dallner, Orrenius and Bergstrand (1963) was applied by Campbell *et al.* (1964) to a homogenate in 0·3 M sucrose which had been freed from mitochondria and heavier particles and was then layered over 1·5 M sucrose solution. Centrifugation in an angle rotor for 11×10^6 *g*-min in the presence of 10 mM Mg^{2+} ions gave, in the final upper layer, "smooth vesicles" together with some free ribosomes (see Table IV). As will be considered below, a few workers have already used such procedures while studying the enzymology of microsomal fractions.

Increasing application is likely to be made of the ingenious "zonal" rotors designed by Anderson (see Anderson, Price, Fisher, Canning and Burger, 1964) which, with comparatively short centrifugation times, enable particles to be banded isopycnically or on the basis of differences in sedimentation rate. In experiments designed to separate plasma membrane fragments from cytomembrane fragments, this aim was partly realized (Table IV), but the most notable result was that discrete fractions were obtained corresponding to smooth-surfaced cytomembranes, rough-surfaced cytomembranes and ribosomes (El-Aaser, Reid, Klucics, Alexander, Lett and Smith, 1966b).

2. *Re-centrifugation of Microsomal Material after Special Treatment*

(*a*) *Deoxycholate treatment.* Dallner (1963) and Ernster *et al.* (1962), *inter alia*, have tried deoxycholate concentrations lower than those customarily used to solubilize membrane fragments [see III. B. 4.(*a*)]. [Not only the concentration but also the ratio of deoxycholate to protein is important, as emphasized by Higashi and Peters (1963a) and earlier authors]. When a microsomal suspension in 0·25 M sucrose containing 0·024% deoxycholate was centrifuged (13×10^6 *g*-min), Ernster *et al.* observed a loose reddish sediment ("M fraction") overlying the ribosome-containing pellet. Table IV gives data for this fraction. Its yield was low, and it was not entirely free of ribosomes; possibly the smooth-surfaced cytomembranes had dissolved preferentially on exposure to deoxycholate, leaving behind some of the rough-surfaced membranes. Supernatants thus obtained after deoxycholate treatment may themselves yield a pellet when exhaustively re-centrifuged after dilution with sucrose medium free of deoxycholate (Ernster *et al.*, 1962), or after addition of extra deoxycholate (Moulé, Bouvet and Chauveau, 1963). The latter workers (see also Moulé *et al.*, 1960) were able to obtain such a deoxycholate-treated pellet from a smooth-membrane fraction as described in Section III. B. 1.(*b*) and (*c*), and considered it to represent particles derived from the smooth membranes and distinct from ribosomes. These particles were apparently smaller and (or just lighter) than ribosomes, and were lower in RNA content and richer in phospholipid than ribosomes; they could be obtained even in the presence of Mg^{2+} ions [see also III. B. 3.(*a*)]. Particles rich in phospholipid have likewise been reported by Pollak and Shorey (1964) who obtained them by treating rough-surfaced microsomal material (Dallner, 1963) isolated from embryonic chick liver with deoxycholate and also with ribonuclease. Their assumption that the particles, given the unfortunate term "reticulosomes", have physiological significance is not well founded.

(*b*) *Iso-octane treatment.* Use of iso-octane as described by Hawtrey and Schirren (1962) was found by Campbell *et al.* (1964) to be less effective than gradient centrifugation [see III. B. 1.(*d*)] in giving membranes free from ribosomes. In brief, a microsomal fraction was treated with iso-octane, and the emulsion centrifuged to give an interface layer containing membrane fragments. However, with a "light" microsomal fraction prepared by the method of Moulé *et al.* (1960) as the starting material, Hallinan and Munro (1964, 1965) obtained a centripetally migrating fraction which was thought to represent smooth-surfaced membrane fragments of good morphological purity.

(*c*) *Other treatments.* Dallner (1963) tried re-centrifugation in the presence of Cs^{+} and Mg^{2+} ions, which were believed to show selectivity in their binding to vesicles. The smooth-membrane fractions thus obtained—one binding Mg^{2+} but not Cs^{+}, and the other binding neither cation—showed some difference in antigen and esterase patterns (Lundkvist and Perlman, 1965).

When microsomal fractions are treated with certain buffers at an alkaline pH, there may be differential effects on the membranes, although such effects are not seen with *tris* (Palade and Siekevitz, 1956). Membrane rather than ribosomal constituents are perhaps preferentially extracted by bicarbonate buffer, as judged from assays for glucose-6-phosphatase (Hultin, 1957), or by Mg^{2+}-containing phosphate or glycine buffer [Reid, 1961a; see III. B. 4.(*c*) and Fig. 8]. The converse may occur on extraction with pyrophosphate buffer (Sachs, 1958; Goswami, Barr and Munro, 1962). The latter treatment,

FIG. 8. Microsomal material isolated from liver in 0·25 M sucrose medium at 20,000 *g* (90 min) (Reid, 1961a). × 50,000 approx. Top left (a). Freshly isolated microsomal fraction. Top right (b). Microsomal fraction after freezing and thawing.
Centre (c, d). Residue from microsomal fraction after extraction with Mg^{2+}-containing alkaline buffers (pH 7·4 phosphate, then pH 9·0 glycine—(c) and (d) respectively): note persistence of the ribosomes.
Bottom left (e). Residue from further extraction of the residue depicted in (d) with EDTA-containing pH 9·0 glycine buffer: note disappearance of the ribosomes.
Bottom right (f). Material sedimented at 145,000 *g* (1 hr) from 20,000 *g* supernatant i.e. after removal of material shown in a): note paucity of ribosomes.

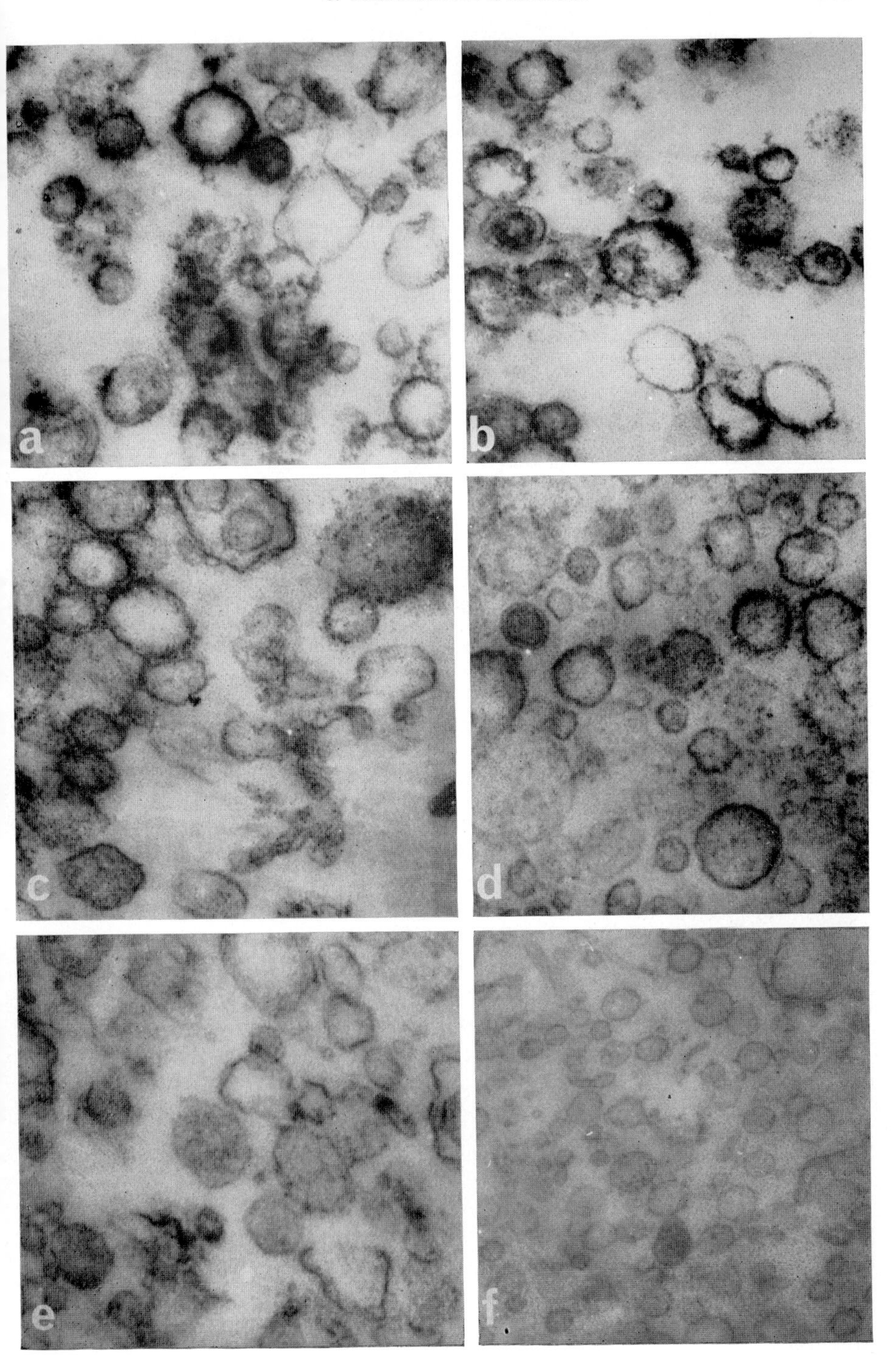
a
b
c
d
e
f

as for treatment with EDTA (Palade and Siekevitz, 1958; Siperstein and Fagan, 1964) leaves mainly membrane fragments with a few attenuated, ribosomes; such fragments, like those in preparations mentioned under (*a*) and (*b*) above, would presumably correspond to rough-surfaced as well as smooth-surfaced cytomembranes. The lipoprotein "ghost" fractions obtained by Westrop and Green (1960) have not been adequately characterized, particularly with respect to morphology.

3. Constituents of Cytomembranes

(*a*) *RNA*. The view that the RNA of microsomal fractions resides almost solely in ribosomes (*inter alia*, Palade and Siekevitz, 1956) was soon challenged [see Reid and Stevens (1958) and Reid (1961a), and citations therein], and is now less tenable—although Campbell *et al.* (1964) consider that contamination with ribosomes could account for the RNA found in a membrane fraction by Hawtrey and Schirren (1962). Moulé *et al.* (1960) found that a "light" microsomal fraction low in ribosomes contained as much RNA as a "heavy" fraction (Table III). Goswami, Barr and Munro (1962), Ernster *et al.* (1962), Dallner (1963) and Petrovic, Becarevic and Petrovic (1965) have likewise found RNA in fractions containing few ribosomes, although in some such fractions, such as that prepared by Schneider (1963), the RNA content was low (Table IV). The "light" microsomal fraction (0·88 M sucrose) of Moulé *et al.* resembled the "heavy" fraction in that half, and only half, of the RNA could be recovered in the "ribosomal" pellet obtained with deoxycholate—the implication being that smooth-surfaced membranes may contain RNA which is sedimentable in the presence of deoxycholate. (They mention that the pellet from either fraction consisted of dense granular elements.)

A "light" microsomal fraction prepared in 0·25 M sucrose and with a low content of ribosomes, as shown in Fig. 8f, was subjected to an extraction procedure which, when applied to a heavy fraction, removed one-third of the RNA and left ribosomes apparently intact (Reid, 1961a; Fig. 8). With the "light" fraction, two-thirds of the RNA was recovered in the extract (E. Reid, unpublished results). Moreover, the light fraction differs from the heavy fraction, and resembles the supernatant fraction, in its changes in the level of RNA and in the labelling of RNA after injection of [^{14}C]-orotate in response to a change in hormonal status (Reid, 1956, 1961a).

When fractions that are rich in RNA supposedly derived from membranes are prepared from fasting rats given [^{14}C]-orotate, or [^{32}P]-orthophosphate the RNA labelling differs from that found for ribosome-rich fractions. For example, the labelling tends to be relatively high soon after orotate administration (Reid, 1961a, and Petrovic *et al.*, 1965; see also Fig. 2 in Bouvet and Moulé, 1964), an observation compatible with the possibility that the RNA is of "messenger" type, as suggested by Pitot (1964). A similar result was obtained with fed rats given [^{14}C]-adenine, but the converse was found with fasted rats (Hallinan and Munro, 1964). In the latter connection it is of interest that in the experiments of Petrovic *et al.* (1965) with labelled orotate, the turnover of membrane RNA was particularly high if the rats were fed rather than fasted. The RNA of membrane-rich fractions was found by Chauveau *et al.* (1962). to differ from the RNA of ribosomal or supernatant (cell sap) fractions in having a higher content of guanine and a lower content of uracil; but no such difference was encountered by Reid (1961a).

Cytomembranes do, then, appear to contain RNA, values for which are given in Table IV; but little morphological change other than aggregation is evident after treatment *in vitro* with ribonuclease (Palade and Siekvitz, 1956; Tashiro and Ogura, 1957). However, little attack on this RNA by nucleases was found by Pitot (1964), who suggested that the RNA was a stabilized form of "messenger RNA". It may be of interest that in many ways the properties of RNA in the cytomembranes resemble those of membrane-bound mitochondrial RNA (see Chapter 3).

(*b*) *Various chemical constituents.* Table IV gives phospholipid/protein ratios for membrane fractions isolated in apparently "intact" condition so that their morphology could be examined. It is generally agreed that microsomal phospholipid is located in membranes rather than ribosomes, but the published values differ widely; perhaps there may have been inadvertent elution of protein from membrane fractions in some studies. The phospholipid/protein ratio was 0·08 for the "reticulosomes" of Pollak and Shorey (1964) [not tabulated; see III. B. .2*a*].

From experiments in which the non-ribosomal material of microsomal fractions was solubilized by use of agents such as deoxycholate [see III. B. 4.*b*], it can be inferred that certain specific constituents are present in the membranes. These constituents include cytochrome *b* (cytochrome *m*) (Palade and Siekevitz, 1956; Ernster *et al.*, 1962)—which, however, is only loosely linked to the membranes—coenzyme Q (Leonhäuser *et al.*, 1962), and a haem-protein termed Fe_x (Nebert and Mason, 1964). There is evidence that one protein fraction of the membrane proteins is particularly rich in aspartic and glutamic acids (Cohn and Simson, 1963).

(*c*) *Enzymes.* The biochemical evidence that microsomal enzymes are generally located in the membranes rather than in the ribosomes comes mainly from experiments in which the membranes were solubilized. There is the complication that some enzymes become unstable when, through use of an agent such as deoxycholate, the membranes are solubilized or are stripped of ribosomes, as in the "M fraction" of Ernster *et al.* (1962) (see also Siekevitz, 1962). Such instability is to be distinguished from that encountered for certain microsomal enzymes, e.g. demethylases (Leadbeater and Davies, 1964), when "intact" microsomal fractions are stored or incubated.

In the list of enzymes in liver microsomal fractions (Table II), attention is drawn to those enzymes for which there is evidence for location in the membranes. This evidence is based rather insecurely in some instances on results of analysis of different microsomal "cuts" (Table III). However the evidence is particularly good for glucose-6-phosphatase, $NADH_2$-cytochrome *c* reductase and azo-dye reductase, which have been assayed in a membrane fraction obtained without use of deoxycholate by the procedure of Chauveau *et al.* (1962).

The warning should be given that many experiments using deoxycholate without Mg^{2+} ions have been unaccompanied by electron-microscopy or even by RNA analyses, and do not show unequivocally that the constituent in question is membrane-linked. In comparison with treatment in the presence of Mg^{2+} ions, treatment in the presence of EDTA with a low concentration of deoxycholate was strikingly effective in solubilizing RNA, glucose-6-phosphatase, and 5′-nucleotidase (El-Aaser, 1965).

There is some evidence (e.g. Tashiro, 1958; Roth, 1960) that the relatively low "latent" ribonuclease activity of microsomal fractions [Morais and de Lamirande (1965), *inter alia*] is associated with ribosomes rather than with membranes, although Takanami (1959) found a light-microsomal fraction to be particularly active. Recent evidence points to a second ribonuclease and to a phosphodiesterase, apparently membrane-located (de Lamirande, Boileau and Morais, 1966). The energy-dependent incorporation of amino acids into microsomal protein [*inter alia*, Campbell *et al.* (1964) and Hallinan and Munro (1965)], and into serum albumin (Decken and Campbell, 1961), is attributable to ribosomes rather than to membranes. Membrane-rich fractions obtained by a density-gradient procedure were able to incorporate phenylalanine into protein, particularly if polyuridylate were added; but it was thought that contaminating ribosomes could be responsible for the activity (Campbell *et al.*, 1964). Hendler (1965) argues that membrane-bound ribosomes are the sites of protein synthesis, both components being indispensable. At least membranes may have some auxiliary role (Campbell *et al.*, 1964; Hendler, 1965); indeed Peters (1962) has shown that newly-formed albumin passes first to the rough-surfaced and thence to the smooth-surfaced membranes, and is secreted without ever being found in the supernatant fraction to a significant extent [also see Jungblut (1963) and, for pancreatic amylase synthesis, Redman, Siekevitz and Palade (1966)]. Pitot (1964) suggests that proteins destined to be secreted are actually synthesized on membranes, with participation of membrane-RNA as distinct from ribosomal RNA.

As has already been indicated [see III. B. 1.*b*], there may be differences between the enzyme patterns of different groups of membrane fragments, and in particular between smooth-surfaced and rough-surfaced elements. This may occur particularly in experimental situations (see IV.B.1 below). Pointers to such differences have come particularly from separations in high-density media as summarized in Table IV; but further study is needed before definite conclusions can be reached, even concerning glucose-6-phosphatase. From experiments with a zonal centrifuge (El-Aaser *et al.*, 1966b and unpublished experiments) it appears that rough-surfaced and smooth-surfaced cytomembrane elements are qualitatively, although not necessarily quantitatively, similar in enzymology, but that within the smooth-surfaced fraction there may be subtle differences in location, for example between glucose-6-phosphatase and nucleoside diphosphatase assayed with UDP as substrate. One difficulty is that the actual membrane fragments arise not only from rough- and smooth-surfaced cytomembranes (including Golgi cytomembranes) but also from the plasma membrane. That certain "microsomal" enzymes such as 5′-nucleotidase are largely attributable to plasma-membrane fragments seems certain from the cytochemical results surveyed in Section II, from the admittedly scanty biochemical evidence concerning plasma-membrane fractions (Table IV, and Section III. C. 1.), and from cytochemical studies of microsomal fractions in which 5′-nucleotidase has been shown to be in elements distinct from the vesicles containing glucose-6-phosphatase (see Section IV. A. 1.). For a different reason the results of Higashi and Peters (1963a) on the distribution of catalase (1.11.1.6) (Table IV) are inconclusive; here the observed activity may well have been in distinct organelles (microbodies, or uricosomes—see Fig. 2) that tend to sediment with cytomembrane fragments (de Duve *et al.*, 1962; See Appendix Chapter 3).

4. *Solubilization and Activation of Microsomal Enzymes*

(*a*) *Approach to the problem.* Attempts to solubilize enzymes in the microsomal-fraction have sometimes been prompted by a suspicion that the enzyme in question was present because of a re-distribution artifact, but more often by the hope that the enzyme could be purified—a hope so far seldom realized. Consideration may first be given to enzymes which are solubilized only with fairly drastic treatment, but which differ in their ease of solubilization. Such differences hardly establish that the enzymes are differently located among the diverse elements that are disclosed by morphological studies. As Tashiro (1957) and other authors have recognized, enzymes in membranes are presumably bound to lipid and resistant to solubilization, unless treatments are applied to remove the lipid or to break the bonds supposedly concerned in the linkage of lipid to protein (presumably hydrogen bonds and Van der Waals forces); the strength of the linkage is hardly likely to be the same for all enzymes. Siekevitz (1962) has pointed out some difficulties in the performance and interpretation of such experiments: "What are the criteria for solubility? When are enzymes soluble? Certainly not when they resist centrifugation in a specified gravitational field, for perhaps a higher field will cause them to sediment. I think that we can only say that a protein is soluble when it is completely surrounded by water molecules. And in too many cases when proteins have been termed 'soluble' there is really no good reason to think this might be so. . . . On the other hand, when we extract and really purify a protein, and measure certain parameters, like pH optima, reaction rates, equilibrium constants, substrate binding constants, we have no idea whether this same protein does have these characteristics when it is embedded within the cell."

Apart from the question of solubility, Siekevitz (1962) and Ernster *et al.* (1962) have discussed the question of the activity of membrane-bound enzymes which may or may not be lipoproteins. The activity of such enzymes is profoundly influenced by their membranous location. Solubilization of a nucleoside diphosphatase was apparently preceded by activation (Ernster and Jones, 1962). It has been suggested that the bound form of glucose-6-phosphatase differs from the free form in substrate affinity. For enzymes having non-polar substrates, concentration of the latter within membrane micelles may have an important effect on activity (Tashiro, 1957).

(*b*) *Deoxycholate*. Biphasic effects on enzyme activity of increasing concentrations of deoxycholate are shown in Table V. [No regard has been paid in Table V to differences between laboratories in the deoxycholate:protein ratio employed (cf. Moulé *et al.*, 1960).] An irreversible fall in activity after treatment with deoxycholate in high concentration would of course be a stumbling-block to enzyme purification if this concentration were required to solubilize the enzyme. The question of reversibility has seldom been checked, but certainly in one instance, $NADH_2$-cytochrome *c* reductase, there may be restoration of activity on lowering the deoxycholate concentration (Ernster *et al.*, 1962; Siekevitz, 1962). Beaufay and de Duve (1954) concluded that solubilization of glucose-6-phosphatase by deoxycholate renders the enzyme more susceptible to thermal denaturation, presumably because of instability of the lipid-free protein. The enzyme becomes more susceptible to destruction by deoxycholate itself when the microsomes are first exposed to bicarbonate buffer (Hultin, 1957).

In view of the cytochemical evidence that ATPase and 5′-nucleotidase are really located elsewhere than in cytomembranes, it is striking that the activity of the former enzyme is depressed even with the low concentration of deoxycholate that activates glucose-6-phosphatase. Also the activity of 5′-nucleotidase is only enhanced with a high concentration of deoxycholate which results in a reduction of glucose-6-phosphatase activity (Table V). [It is puzzling that Ernster and Jones (1962) found negligible 5′-nucleotidase activity in the microsomal fraction for various authors have found high activity even without treatment with deoxycholate (Table II).]

Although deoxycholate may cause instability, or at least inhibition (possibly reversible), even at concentrations that do not solubilize enzymes [Ernster *et al.* (1962)], it is at present unsurpassed as a tool for solubilizing microsomal enzymes. The minimum concentration and amount required depend on the enzyme studied [see III. B. 4.*e*]. Deoxycholate does not

TABLE V. *Effect of deoxycholate on enzymic activities of microsomal fractions.*

Enzyme	Effect with low concentration (<0·1%)	Effect with high concentration (>0·2%)	References and Remarks
Glucose-6-phosphatase	Increase	Decrease	Beaufay and de Duve (1954), Ernster and Jones (1962); Dallner (1963), *inter alia*
$NADH_2$-cyt. *c* reductase	Sometimes increase	Increase or decrease (variable)	Packer and Rahman (1962); Ernster *et al.* (1962)—latter also studied "aging" effects, etc.
$NADPH_2$-cyt. *c* reductase	None	None	
5-Nucleotidase (Nucleoside-5′-monophosphatase)	None	Increase	El-Aaser (1965); cf. Ernster and Jones (1962)
Nucleoside diphosphatase: ADP	Decrease	Decrease	Ernster and Jones (1962)—Lubrol W gave similar effects
Nucleoside diphosphatase: CDP	Increase	Slight decrease	
Nucleoside diphosphatase: GDP, IDP or UDP	Increase	Plateau (increase)	
Nucleoside triphosphatases: ATP	Decrease (inactivation)	Marked decrease	Ernster and Jones (1962)—Lubrol W gave similar effects; see also (*inter alia*) Muscatello *et al.* (1961); Dallner (1963); Jarnefelt (1964)
Nucleoside triphosphatases: CTP or ITP	None	Decrease	
Nucleoside triphosphatases: GTP or UTP	Increase	Decrease	

solubilize the glycogen particles which are the site of attachment of UDP-glucose:glycogen glucosyltransferase (2.4.1.11) (Table II).

(*c*) *Various surface-active agents.* Agents other than deoxycholate are sometimes effective in solubilizing microsomal enzymes without loss of activity. Thus, certain cationic and non-ionic detergents rendered arylsulphatase *c* (3.1.6.1) soluble (and also enhanced its activity), although it became insoluble if the detergent was removed (Dodgson, Rose, Spencer and Thomas, 1957). On the other hand, cationic detergents failed to solubilize arylesterase and indeed destroyed it, whereas anionic or non-ionic detergents were effective (Carruthers and Baumler, 1962). In the latter study, arylesterase and glucose-6-phosphatase differed in ease of solubilization. Ribosomes are readily "dissolved" by some agents other than deoxycholate—e.g. Triton-X-100 (but see Hunter and Korner, 1966)— but survive exposure to the anionic detergent Lubrol W (Cohn and Simson, 1963). In general, the selective fractionation of proteins solubilized by detergents, as tried by Hultin (1957), is not readily achieved.

(*d*) *Various other treatments.* Authors who have tried mere freezing and thawing, as in a study of albumin (Peters, 1962), or drastic homogenization in a blender or sonication, have seldom met with success. However, the nucleoside diphosphatase active towards GDP, UDP and IDP can be activated and partly solubilized merely by blending (Ernster and Jones, 1962). With buffers such as those mentioned in Section III. B. 2.(*c*) above, extraction of enzymes may be inefficient or at least unselective (Palade and Sickevitz, 1956, *inter alia*). This was found when the procedure of Reid (1961a) was applied by El-Aaser (1965), to a study of glucose-6-phosphatase and 5′-nucleotidase. Use of *n*-butanol gave no solubilization of arylsulphatase *c* (Dodgson *et al.*, 1957) or $NADPH_2$-cytochrome *c* reductase (Kamin, Masters, Gibson and Williams, 1965). and caused inactivation of glucose-6-phosphatase (Beaufay and de Duve, 1954). Preliminary experiments by A. A. El-Aaser have suggested that some solubilization and activation of glucose-6-phosphatase occurs merely through exposure to hypertonic sucrose solution.

Little success has been achieved by the use of enzymes, as applied particularly to studies on glucose-6-phosphatase. Lecithinase causes inactivation, and chymotrypsin may give activation, as may trypsin (Görlich and Heise, 1962), but not solubilization (Beaufay and de Duve, 1954). The supposed solubilization by trypsin used in conjunction with deoxycholate (Heise and Görlich, 1963) or with digitonin (Görlich and Heise, 1963) seems far from efficient—this also applies to the use of digitonin in conjunction with deoxycholate (Görlich and Heise, 1962). No success has been achieved with pancreatic ribonuclease, for example in attempts to solubilize arylesterase (Carruthers and Baumler, 1962) and glutamine synthetase (6.3.1.2) (Wu, 1963). Ribonuclease however has been used mainly in a different context in order to verify that RNA is not essential for enzymic activity [Palade and Siekevitz (1956), *inter alia*]. Lipase treatment may be effective for electron-transport proteins (Strittmatter, 1965; Kamin *et al.*, 1965; cf. Green and Perdue, 1966).

(*e*) *Difference between enzymes in ease of solubilization.* The above discussion of techniques has dealt mainly with tightly bound enzymes such as glucose-6-phosphatase, $NADH_2$-cytochrome *c* reductase and nucleoside triphosphatases which, like microsomal albumin (Peters, 1962), are not readily solubilized, although some may be associated with plasma-membrane rather than cytomembrane fragments. At the other extreme, there are enzymes which are so readily solubilized that their presence in the microsomal fraction, and sometimes in the nuclear fraction also, may be an artefact, possibly due to linkage of the enzyme (which may be a basic protein) to nucleic acid. Arginase (3.5.3.1) can be solubilized merely by use of sucrose containing 5 mM divalent cations (Rosenthal, Gottlieb, Gorry and Vars, 1956), fumarate hydratase (4.2.1.2) by ribonuclease treatment (de Duve *et al.*, 1962), and glutamine synthetase merely by use of hypotonic saline (Wu, 1963)—although with brain tissue the proportion of homogenate activity recovered in the microsomal fraction was increased by use of water in place of sucrose medium. The capacity of microsomal fractions to synthesize deCDPcholine may be an artefact, since the activity is low if water rather than sucrose medium is used (Schneider, 1963). de Duve *et al.* (1962) suggest that the presence of uronolactonase (3.1.1.19) in the microsomal fraction may be an adsorption artefact, since no activity was found with a saline medium.

Between these extremes, there are enzymes which may be truly membrane-linked but which are solubilized by use, for example, of deoxycholate in *low* concentration. The nucleoside diphosphatase which attacks GDP, UDP and IDP is one example (Ernster and Jones, 1962); indeed, this enzyme and also 5′-nucleotidase run partly with low-density and soluble material as well as in the plasma-membrane region in the zonal centrifuge even without deoxycholate (El-Aaser *et al.*, 1966b). It is, of course, difficult to know where to draw the line between loose, and possibly fortuitous, association with membrane fragments and true binding. Patient trial of different treatments, including the use of different media for homogenization, may be called for.

C. MEMBRANE-CONTAINING FRACTIONS (VERTEBRATE) OTHER THAN HEPATIC CYTOMEMBRANE FRACTIONS

1. *Plasma-membrane Fractions*

(*a*) *Isolation.* Attempts to isolate plasma-membrane fragments from liver have met with some success (Table IV); but there has been too little discussion of the rationale of the techniques and of possible blind alleys. It seems that if a gentle (mechanical) procedure is used to rupture the cells, plasma-membrane fragments are recoverable largely in the crude "nuclear" fraction obtained by low-speed centrifugation, with preservation of microvillar structure; EDTA may help (Porteous and Clark, 1965). With ultrasonic disintegration a high speed is necessary to sediment the fragments (Herzenberg and Herzenberg, 1961). Moreover, it seems that with conventional sucrose media there is adhesion of mitochondria and of cytomembrane fragments to those plasma-membrane fragments that are recovered in nuclear fractions, but that use of water containing a trace of bicarbonate overcomes this trouble, the nuclear envelope then being broken up (Neville, 1960). Neville mentions that use of young rats gave the best preparations. Perhaps there may be reduced contamination by extracellular fibrous elements too small to be held back on a cloth filter such as is commonly used (cf. Porteous and Clark, 1965).

It is the practice, then, to take a lysed "nuclear" fraction as the starting material for preparing fractions rich in plasma-membrane fragments, the separation of which from other elements is achieved by judicious centrifugation in high-density media, preferably in a swing-out rotor (Table IV). In the fraction obtained by Herzenberg and Herzenberg (1961), fragments of the nuclear membrane were apparently present as judged by the rather inadequate tool of light microscopy. Emmelot, Bos, Benedetti and Rümke (1964b) modified the method of Neville (1960). For example, homogenization was carried out briefly with a Potter–Elvejhem homogenizer, rather than with a Dounce homogenizer. In the fraction they obtained, the plasma-membrane "sheets" were accompanied by some fragments of smooth-surfaced cytomembranes (Fig. 9), which might well have been contiguous with the plasma membrane in the intact cell. That the "sheets" indeed represented plasma-membrane fragments was clear from the electron-microscope finding of attached elements resembling bile canaliculi (Fig. 9). El-Aaser, Fitzsimons, Hinton, Reid, Klucis and Alexander (1966a) likewise obtained "sheets", low in glucose-6-phosphatase (a cytomembrane "marker") but accompanied by unidentified contaminants; plasma-membrane fragments also appeared in a fraction containing nuclei [although, as also found by Porteous and Clark (1965), there seemed to be no actual linkage to the nuclei] and in a vesicular fraction with glucose-6-phosphatase activity. The plasma membrane seems prone to form vesicles *in vitro*, possibly so similar to those from smooth-surfaced cytomembranes that electron-microscopic identification may be risky (cf. Skipski, Barclay, Archibald, Terebus-Kekish, Reichman and Good, 1965).

In place of liver, Wallach and Kamat (1964) used ascites cells as the starting

FIG. 9. Plasma-membrane fraction (Emmelot, Bos, Benedetti and Rümke, 1964b): note that the plasma membranes of adjoining liver cells adhere together in places (by way of desmosomes and terminal bars), and that there are structures resembling the bile spaces; the vesicles attached to or adjoining the plasma membrane may be cytomembrane fragments. Courtesy of the authors and of Elsevier Publishing Co. × 24,000.

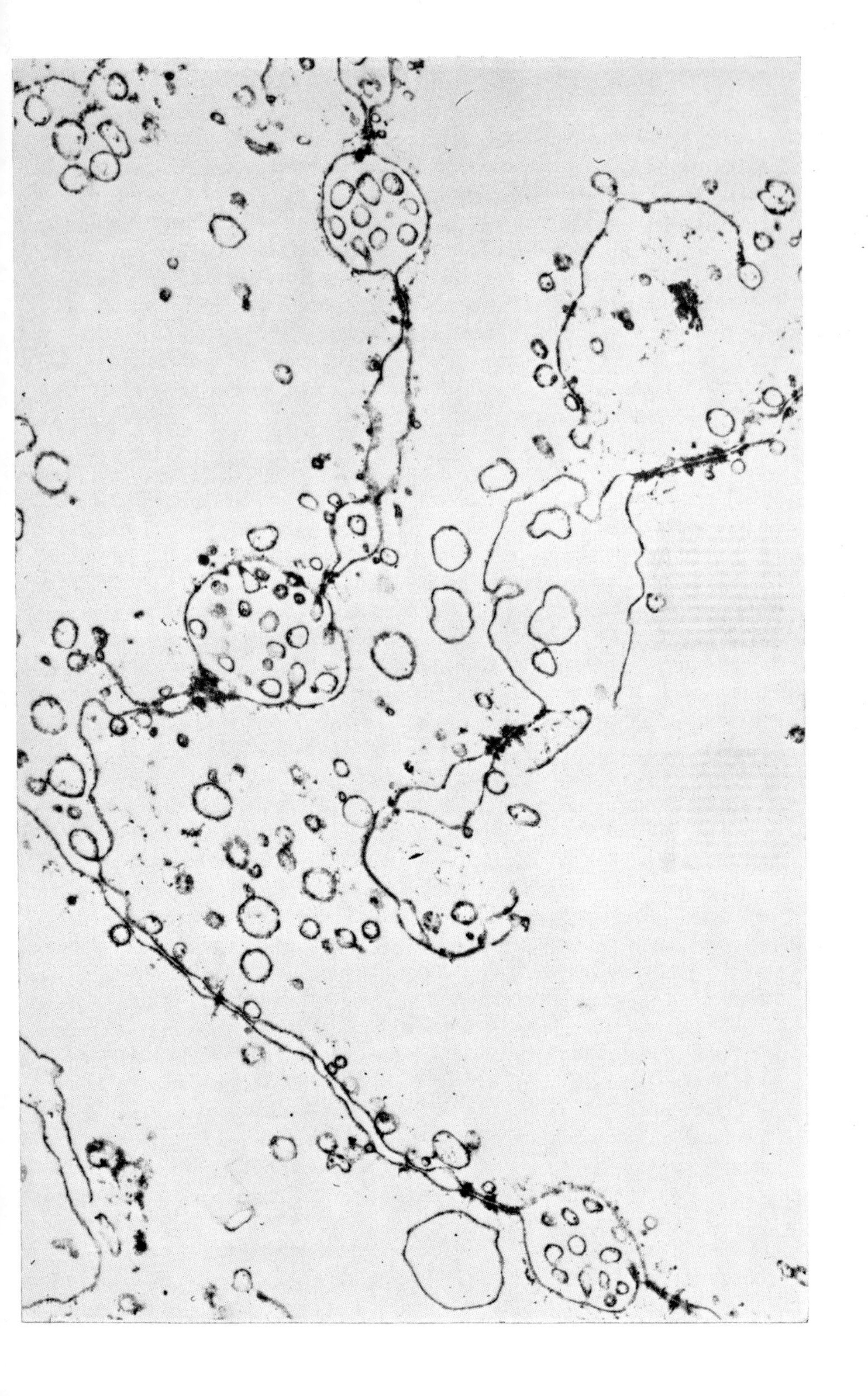

material, with the sucrose polymer Ficoll for the final gradient centrifugation. As they carefully explain, their approach entailed (1) disruption of the plasma membrane by a gas-cavitation procedure, into fragments sufficiently small to sediment with the microsomal rather than the nuclear material, (2) re-centrifugation of the microsomal pellet in a two-layer sucrose medium so that the smooth-surfaced vesicular elements floated, (3) "lysis" of the vesicular elements in a hypotonic medium, and (4) density-gradient centrifugation in Ficoll. The separation of the plasma-membrane elements from the other elements was based on differences in osmotic behaviour, although it is not clear from the paper why "lysed" vesicles should still, as is implied, act as semi-permeable sacs. No electron-microscope description of the final plasma-membrane fraction was given; the identification was based on biochemical criteria mentioned in (*b*) below.

As is indicated in Table IV, Kamat and Wallach (1965) later modified their method: they claimed that with a suitable pH (8·6) and with divalent cations (Mg^{2+}) present there was shrinkage and aggregation of cytomembrane fragments which would otherwise have sedimented with plasma membrane fragments. No such aggregation has been found with rat-liver fractions run in a zonal centrifuge (Table IV), although there were density changes. With microsomal rather than crude-nuclear starting material it is in fact difficult to separate plasma-membrane fragments from cytomembrane fragments in the case of liver; nevertheless a vesicle-containing fraction low in glucose-6-phosphatase and rich in 5′-nucleotidase has now been obtained [unpublished experiments; cf. El-Aaser *et al.* (1966b) and Table IV].

(*b*) *Biochemical characteristics.* In the pioneer study of Neville (1960), no biochemical parameters were investigated. Herzenberg and Herzenberg (1961) found the expected high level of certain homograft antigens in their membrane fractions. With heterologous antibody preparations, Wallach and Kamat (1964) and Emmelot, Bendetti and Rümke (1964a) likewise found characteristic cell-surface antigens in their fractions, together with a Mg^{2+}-activated ATPase dependent on Na^+ and K^+ (and sensitive to ouabain); indeed, these parameters were the criteria used by Wallach and Kamat in developing their isolation procedure. These authors, unlike Emmelot and co-workers, were unable to detect hexosamine in their fractions. It is unfortunate that in the thorough survey made by Emmelot and co-workers of the biochemical characteristics of their fractions (Table IV) side-by-side assays on other cell fractions for sialic acid and for some of the enzymes studied were not reported. Emmelot and co-workers conclude, presumably taking into account past work in their laboratory, that plasma-membrane fractions are rich in certain enzymes such as 5′-nucleotidase, and contain small but significant amounts of other enzymes which are indicative of the presence of cytomembrane elements in their fractions, not necessarily as a "contaminant". Their results for ribonucleases, which showed definite activity particularly at an alkaline pH, are hard to assess, since the results of assays based on RNA breakdown are given in arbitrary values although they are valid for comparative purposes. RNA itself seems to be virtually absent (Table IV). Observations by Peters (1962) preclude the possibility that the plasma membrane plays a role in the synthesis of protein, or at least of serum albumin. Plasma-membrane fractions typically have a high cholesterol/phospholipid ratio (Coleman and Finean, 1965). In a comparison with whole liver, Skipski *et al.* (1965) found a relatively high content of free cholesterol (other lipids were also determined); their value for plasma-membrane phospholipid was 11mg P/g protein.

Emmelot *et al.* (1964b) subjected their plasma-membrane fractions to deoxycholate treatment, which enhanced the activities of certain enzymes such as phosphodiesterase (3.1.4.1) (measured at pH 8·9) and arylesterase, and to extraction with saline, whereby the "phospholipo-glycoprotein core" was stripped of antigenic proteins which are positively charged at a neutral pH but not of enzymes such as 5′-nucleotidase and ATPases. The activity of the latter, which—unlike that of 5′-nucleotidase—is altered after partial removal of the sialic acid by neuraminidase, may be governed by a protein-phospholipid orientation (Emmelot and Bos, 1965; cf. Green and Perdue, 1966). Emmelot and Bos (1966) also consider that glycolytic enzymes associated with plasma-membrane fragments from normal liver are not true components of the plasma membrane, in disagreement with a view based largely on erythrocyte experiments (Green, Murer, Hultin, Richardson, Salmon, Brierly and Baum, 1965).

(*c*) *Indications for future work.* The "rule-of-thumb" approach prevalent in past work on the isolation of plasma-membrane fractions is hardly excusable in view of the ease with "marker" enzymes such as 5′-nucleotidase can be assayed. (The latter may, however, be partly extracellular in location.) Ideally it should not be necessary to use, at least in the initial stages of isolation, special conditions that preclude the simultaneous isolation of the conventional sub-cellular fractions, for example, exposure of the various organelles to an osmotically unfavourable environment.

Exploitation of slight differences in density seems to be the best hope; the density values, of which estimates are given in Table IV, vary with different media. With conventional procedures for homogenization and differential centrifugation using 0·25 M sucrose as an isolation medium, followed by the "mitochondrial supernatant" being centrifuged in high-density media, the different types of vesicles in the microsomal material may prove to be resolvable without recourse to a hypotonic medium at any stage. This may also hold for the crude nuclear fraction, although there is the difficulty that the surviving plasma-membrane "sheets" seem to be rather similar in density to intact mitochondria and erythrocytes.

Apparent disagreements between laboratories over density values for plasma-membrane fragments (Table IV) could hinge not only on the medium, but also on the tissue and the conditions of homogenization. These conditions —although not (for a given Potter-type homogenizer) the number of strokes *per se*—may govern the distribution of plasma-membrane fragments between the nuclear and the microsomal fractions. Thus, with intestinal mucosa the use of a blendor tends to convert brush-border elements into vesicles which sediment partly with microsomal material (Hübscher *et al.*, 1965). However, since the vesicles which arise from the plasma membrane seem not to fall into a unique density class (the density being estimated from the position of 5′-nucleotidase activity after isopycnic centrifugation), some of the vesicles conceivably arise from particular regions (unconvoluted?) of the plasma membrane, rather than indiscriminately from any region through over-drastic homogenization. Further cytochemical work, with due regard to isolated fractions, may elucidate the origin of vesicles in terms of biochemically specialized regions in the plasma membrane of a given cell. Hand-in-hand with such work, progress in the separation of different types of vesicles will come from increasing use of zonal rotors, which achieve large-scale yet sharp separations without the need for repeated centrifugations.

2. *Golgi Cytomembrane Fractions*

The γ-cytomembranes in the Golgi region are presumably recovered in the microsomal fraction, and in smooth-membrane fractions prepared therefrom. Epididymis is the only tissue from which a Golgi fraction has been successfully prepared (Schneider and Kuff, 1954). The fraction was rich in alkaline phosphatase (3.1.3.1), as would be expected from cytochemical observations, and in RNA, which is unexpected. The concentration of phospholipid was within the range found for other smooth-membrane fractions (Table IV). These results are best regarded as essentially a useful starting-point for future work. With liver, unfortunately, no unique enzymic "marker" is available.

3. *Non-Hepatic Microsomal Fractions*

(*a*) *Approach to the problem.* Differential centrifugation has been applied successfully to the isolation of mitochondrial and lysosomal fractions from tissues other than liver, but has so far given few definitive results for cytomembranes. The fact that the latter are scanty in some non-hepatic tissues has not been recognized by certain biochemists who have isolated a "microsomal" fraction but have failed to examine it under the electron-microscope. Indeed, a blind eye has often been turned to the serious heterogeneity of cell type found histologically in most non-hepatic tissues. However, as will be evident from the following brief survey, a few biochemists have in fact performed thorough studies with due attention to histological and cytological aspects. In most tissues other than liver and kidney, glucose-6-

phosphatase is unfortunately not present in adequate amount to serve as a cytomembrane "marker" [guinea-pig, but not rat, intestine contains some (de Duve, 1964)]; but phospholipid determinations can be a useful guide. With many non-hepatic tissues, notably muscle, there is the difficulty that use of a conventional homogenizer and of a sucrose medium may fail to give efficient cell breakage; one consequence is that a reliable "balance-sheet" of enzyme activities in the various sub-cellular fractions can hardly be drawn up. For skeletal muscle, the procedure of Kono and Colowick (1961) seems to be particularly reproducible; the product has awaited characterization by electron microscopy until recently (Table VI). Muscle preparations are liable to contain mitochondrial cristae (Siekevitz, 1965).

(*b*) *Isolation and characteristics.* The data given in Table VI should be viewed in the light of the above cautionary remarks, and the original publications should be consulted hopefully for details not given here. One particularly common sin is vagueness about assay conditions, e.g. the presence or absence of Mg^{2+} ions, as in work on alkaline phosphatase. In summary, "microsomal" fractions have been empirically prepared from diverse tissues, and shown to be rich in enzymes such as alkaline phosphatase, "ATPase", acetylcholinesterase (3.1.1.7), and $NADH_2$- or $NADPH_2$-cytochrome *c* reductase; the differences from liver are quantitative rather than qualitative. In some experiments not in Table VI, Morton (1954a) isolated a microsomal fraction from buttermilk which, as judged from electron microscopy and enzyme assays, seemingly represented lipoprotein particles derived from the cytomembranes of the mammary-gland cells concerned in milk secretion.

Few workers have verified that their "microsomal fraction" was rich in phospholipid. Some workers have verified that their fraction was rich in RNA (Table VI), but it would be pointless to cite values in view of the contamination with ribosomes. Microsomal fractions from intestinal mucosa may show few ribosomes but may be rich in RNA, possibly membrane-located (Porteous and Clark, 1965). There is the complication that with tissues such as kidney (Reid, 1956) a fairly high proportion of the RNA is recovered in the supernatant rather than the microsomal fraction. The converse has been found for xanthine oxidase (1.2.3.2), which seems to be microsomal in mammary gland and milk (Morton, 1954a) and possibly in intestinal mucosa (Morton, 1954b), but is recovered in supernatants from liver (Reid, 1961b; de Duve *et al.*, 1962).

Increasing use is being made of electron microscopy, for example in connection with the fractionation of brain (Whittaker, 1963; Wolfe, 1964) and of muscle (Muscatello, Anderson-Cedegren, Azzone and Decken, 1961), but too often the accompanying biochemical examination of the fractions has been inadequate in scope. It is a reasonable hope that membrane fractions well characterized *both* morphologically and biochemically, will soon become available even with intractable tissues such as muscle and that for the latter

TABLE VI. *Fractions from vertebrate tissues other than liver shown to contain or presumed to contain cytomembrane or other membrane fragments.*

Tissue	Special procedures, if any; designation and morphological assessment of fraction[a]	Enzymes and other constituents	References[b] and remarks
Adrenal	mic.	Acetylcholinesterase	Hagen (1955)
	mic.	$NADPH_2$-cyt. *c* reductase	Spiro and Ball (1961)
	mic.	5′-Nucleotidase (high in other fractions also); 3′,5′-AMP phosphodiesterase	Hilf *et al.* (1961)
	mic.	Coenzyme Q_{10}	Leonhäuser *et al.* (1962)
	mic.	Glucose-6-phosphatase (all in mic.)	Breuer *et al.* (1963)
	mic.	Cytochrome P-450	Omura *et al.* (1965)
Brain	mic.	RNA; acetylcholinesterase; ATPase (Na^+, K^+, Mg^{2+}—activated);	Schwartz *et al.* (1962)
	mic.	NAD nucleosidase (95% in mic.)	Jacobson and Kaplan (1957)
	mic.	$NADPH_2$-cyt. *c* reductase; sphingosine-forming system	Giuditta and Strecker (1959)
	mic.—cyto- and plasma-membrane fragments; and ("light" mic.) ribosomes	RNA; phospholipid (21 mg P/g protein); cytochrom b_5 (adsorption artefact?); acetylcholinesterase (in membranes)	Toschi (1959)
	mic.	RNA; acetylcholinesterase; cholinesterase; arylesterase; ATPase (Na^+, K^+, Mg^{2+}-activated)	Aldridge and Johnson (1959); Aldridge (1962)
	mic.	Glutamine synthetase (but partly in supernatant; readily detached from mic.)	Sellinger and Verster (1962)
	"P_2" pellet sub-fractionated and myelin, etc. removed; "P_3" pellet also prepared (mic.); P_2 gave rough-surfaced cyto-	Acetylcholinesterase; ATPase	Whittaker (1963); Whittaker *et al.* (1964); Hosie (1965) —guinea-pig forebrain—main aim was study of

	membrane and synaptosome-membrane fragments		synaptic elements; see also Weinstein *et al.* (1963)
	mic.	Phosphatidylinositol kinase; diphosphoinositide kinase	Colodzin and Kennedy (1964); M. Kai, J. N. Hawthorne, unpublished
	low-speed mic. (centrifuged in 0·25 M S containing 0·05 M *tris*) —fragments of torn membranes and small vesicles	ATPase, solubilized (deoxycholate) and purified—needed phospholipid as well as Na^+, K^+ and Mg^{2+}, and was sensitive to oubain; contaminated with an ATPase which was insensitive to Na^+, K^+ or phospholipid	Tanaka and Strickland (1965)—beef cerebral cortex
Intestine (mucosa)	mic.	Alkaline phosphatase ($+Mg^{2+}$)	Morton (1954b)—calf
	mic.	Alkaline phosphatase(s) (56% in mic. if no Mg^{2+}, 72% if $+Mg^{2+}$)	Allard *et al.* (1957) see also Porteous and Clark (1965) and Hübscher *et al.* (1965) cf. Table IV
	mic.	Acyl CoA synthetase (6.2.1.3)	Senior and Isselbacher (1961)
	mic. (Ultra Turrax homogenate)	β-Glucosidase (3.2.1.21), β-fructofuranosidase (3.2.1.26)	Dahlqvist (1961)—hog
Kidney	mic.	Alkaline phosphatase (no Mg^{2+}) (85% in mic.); glucose-6-phosphatase (75%)	Hers *et al.* (1951)
	mic.	Lactonase (3.1.1.17.) (tri-acetic) (70% in mic.)	Meister (1952)
	mic.	RNA; phospholipid (8·5 mg P/g solids)	Reid (1956)
	mic.	Acylase (*N*-chloracetyltyrosine)	Hanson *et al.* (1959)
	mic.	Alkaline phosphatase (Mg^{2+} in assay)[d], cysteinylglycinase[d] (3.4.3.5.) and glutathionase[c]—each >100% recovery	Binkley (1961)
	mic.	Phosphodiesterase I (60% in mic.)	Razzel (1961)
	mic.	Cystine reductase	Myers and Worthen (1961)

TABLE VI—*continued*

Tissue	Special procedures, if any; designation and morphological assessment of fraction[a]	Enzymes and other constituents	References[b] and remarks
	mic.	Phosphatidate phosphatase (3.1.3.4) (63%)	Coleman and Hübscher (1962)—hog
	mic.—smooth-surfaced vesicles	$NADH_2$ oxidase, ATPase (Na^+, K^+, Mg^{2+}-activated)	Landon and Norris (1963)
Mammary gland	mic.	Alkaline phosphatase (+Mg^{2+})	Morton (1954b)—cow
	mic.	Fatty acid synthesis from acetate	Abraham *et al.* (1963)
Muscle (heart)	mic.	RNA; arylesterase; NAD nucleosidase; $NADH_2$-cyt. *c* reductase	Hulsman (1961)
Muscle (myometrium)	mic.—cytomembrane fragments, rough- and smooth-surfaced	RNA; phospholipid (15 mg P/g solids, if latter = N × 6·25); ATPase (Mg^{2+}-activated)	Wakid (1960)
Muscle (skeletal)	Filtrate from mince centrifuged in salt media at pH 8·2–8·4; 0·12 × 10^6 *g*–min sediment fractionated in LiBr solutions—empty tubular membrane-like structures (partly from mitochondria?)	Protein, mainly collagen-like (of dry wt. of fraction, 65% was protein and 15% lipid, largely phospholipid); ATPase (Mg^{2+}-activated, oubain-insensitive), phosphohexose-isomerase, $NADH_2$ oxidase; negligible activity of glucose-6-phosphate dehydrogenase (1.1.1.49.) and certain other enzymes	Kono and Colowick (1961)
	0·88 M sucrose; sarcotubular—vesicles + granules	RNA; ATPase (Mg^{2+}-activated), probably same as ATPase of Kielley and Meyerhof	Muscatello *et al.* (1961)—frog muscle
	mic.	Acetylcholinesterase	Smith *et al.* (1960)—human muscle
	0·1 M KCl, blender; 2 mic. fractions—sarcoplasmic reticulum fragments	ATPases (discussed in relation to Ca^{2+} transport)	Martonosi and Feretos (1964)

	KCl-K acetate; mic.	Transphosphorylating system, phosphocreatine or phosphoenolpyruvate (ATP formation); ATPase	Molnar and Lorand (1962)—rabbit
	Method like that of Kono and Colowick (1961)—sarcolemma fragments + granules; no collagen fibres seen (removed by hypertonic KCl treatment)	Protein (67% of dry wt.; high in aspartate and glutamate but low in proline and hydroxyproline), lipids (16%; phospholipids and cholesterol in equal amount), polysaccharide (0·9%; "appears to be an integral part of the membrane"), nucleic acids ($\leqslant$1·5%)	Abood *et al.* (1966)—frog muscle; "the overall chemical nature of the frog and rat materials appears to be qualitatively similar"
Pancreas (*fed* guinea-pigs)	0·88 M sucrose; 2 mic. fractions —the second being rich in free ribosomes	Protease and ribonuclease (concentration in "heavy mic." as high as in zymogen granules)	Palade and Siekevitz (1958a)—zymogen-granule fraction also prepared
Pituitary	mic.—a few rough-surfaced vesicles, and free ribosomes	RNA; phospholipid; alkaline phosphatase (no Mg^{2+}) (53% in mic.), "alkaline protease"	LaBella and Brown (1959)
Prostate	EDTA essential in order to get mic. rich in RNA; mic.	RNA; $NADH_2$-cyt. *c* reductase (36% in mic.) and $NADPH_2$-cyt. *c* reductase (33% in mic.)	Harding and Samuels (1961)
Spleen	mic.	ATPase (+Ca^{2+}) (50% in mic.)	Maxwell and Ashwell (1953)
	mic.	$NADH_2$-cyt. *c* reductase (40% in mic.)	Eichel (1957)
Thymus	Phosphate + Mg^{2+}; mic.—probably from lymphocyte cytomembranes	Yield (solids) only 5 mg/g; RNA and lipid (high)	Hess and Lagg (1963)

[a] "mic." has no morphological connotation: it denotes a microsomal fraction (not necessarily given this designation by the authors concerned) analogous in its mode of preparation to that from liver. There is some "overlap" between this Table and Table IV.

[b] The order of listing is chronological, except that results from a given laboratory are tabulated together. For other surveys, not centred on cytomembrane fractions, see Reid (1961b) and de Duve *et al.* (1962). Enzymes are numbered only if not listed in Table II.

[c] Apparently associated with membranous elements.

[d] Apparently associated with deoxycholate-insoluble RNP particles, but the alkaline phosphatase could have been in brush-border fragments from tubular cells.

tissue it will not be necessary to have recourse to the hypotonic conditions used by some workers, which yield "ghosts" that are possibly diverse in cytological origin.

(*c*) *Linkage of enzymes to membranes.* In efforts to establish the localization and fastness of attachment of enzymes, techniques analogous to those employed for liver (Section III. B. and III. C.) have been applied to brain and certain other tissues. Thus, Schwartz, Bachelard and McIlwain (1962) observed parallel behaviour for ATPase, acetylcholinesterase and cholesterol when various centrifugal procedures were applied to brain, and found apparent solubilization of the ATPase when the microsomal fraction was treated with deoxycholate, Lubrol W or digitonin. Further work, particularly on solubilization (Swanson, Bradford and McIlwain, 1964; Bradford, Swanson and Gammack, 1964), led to the conclusion that "structural factors are important in determining the behaviour of the Na^+ ion-activated adenosine triphosphatase"; with agents other than digitonin, "the relative proportions of different lipids solubilized were similar to those present in the original microsomes".

For solubilization of the Mg^{2+}-dependent alkaline phosphatase of mammary gland or intestinal mucosa, recourse must be had to agents such as proteolytic enzymes or *n*-butanol (Morton, 1954b). The β-glucosidase (3.2.1.21) of intestinal microsomal fractions could be solubilized by use of trypsin or deoxycholate, but not by freezing and thawing or by exposure to water (Dahlqvist, 1961). With pancreatic microsomal fractions, in contrast to zymogen-granule fractions, solubilization of amylase, with release of "latent" activity, could not be achieved by use of Triton-X-100, but was effected by exposure to pH 6–7 buffer (Holtzer and van Lancker, 1963).

D. MEMBRANE-CONTAINING FRACTIONS (INVERTEBRATE)

1. *Micro-organisms*

For the study of amino-acid incorporation, Godson, Hunter and Butler (1961) [see also Hendler (1965)] exposed lysed protoplasts of *Bacillus megatherium* to hypotonic buffers and obtained a centrifugal pellet, designated "membrane complex", which was apparently derived from the plasma membrane. Ultrasonic disintegration and re-centrifugation gave a ribosome-containing supernatant, together with a pellet the upper layer of which showed twisted, irregular filaments by electron microscopy and consisted of muco-protein; the lower layer consisted of phospholipid with virtually no RNA, and showed very small membranous vesicles.

Enzyme studies on this "membrane complex" are needed, in view of the evidence given by Neu and Heppel (1964) that various degradative enzymes such as alkaline phosphatase and a 5′-nucleotidase are indeed present on bacterial plasma membranes but are readily detachable. For example, exposure of bacterial cells to *tris* buffer at pH 8 and then to water caused such detachment, leaving viable spheroblasts behind. It is arguable whether such enzymes, as are also found in yeast, can be validly regarded as membrane-located. However, the possibility remains that protoplast membranes are indeed the site *in vivo* of certain enzymes characteristic of membrane fractions [as documented by Wolfe (1964)]. These enzymes include an ATPase and various oxidases together with cytochromes. One reason for caution is that membrane fractions obtained in the early studies did contain measurable amounts of RNA, unlike the "membrane complex" prepared by Godson *et al.* (1961).

Green *et al.* (1965) centrifuged a French-press homogenate of fermenting yeast cells, in a hypotonic medium, and obtained finally a pellet of "membranous nature" with a phospholipid:protein ratio of 0·2. Besides glycolytic enzymes, it contained nucleotide phosphorus in "substantial amounts". Their assumption that the material arose from the plasma membrane is a dubious one, and their conclusion (based also on erythrocyte experiments) that the plasma mambrane truly contains a complete "glycolytic complex" has been questioned (Emmelot and Bos, 1966).

2. Plant Tissues

The scanty literature on membrane-containing (as distinct from ribosomal) fractions from plants was reviewed by Whaley *et al.* (1960) with particular reference to meristematic tissues. Beet-petiole and wheat-root microsomal fractions were studied by Martin and Morton (1956, 1957), who briefly discuss "evidence for the origin of the plant microsomes from a primitive type endoplasmic reticulum." The fractions were rich in certain cytochromes (particularly b_3), in $NADH_2$-cytochrome reductases (assayed with cytochrome *c*, although possibly linked with b_3 *in vivo*), and in xanthine oxidase.

In a careful study of pea-seedling root tips, with media containing 0·4 M or 0·5 M sucrose, Loenig (1961) recovered half of the RNA of the homogenate in a microsomal pellet which could be sharply sub-divided into "heavy" and "light" components. The former consisted mainly of ribosomes, whereas the latter contained vesicles together with some ribosomes and "dense bodies" and had a lower ratio ($<0·8$) of RNA to protein. The proteins of the "light" fractions had more complex electrophoretic patterns after ribonuclease treatment than those of the "heavy" fractions.

A microsomal fraction isolated from a fungus (*Neurospora crassa*) by Shulman and Bonner (1962), showed membranous structures covered with ribosome-like particles in the electron microscope and was found to contain DNA. This was present as a "hybrid" with RNA and was perhaps nuclear in origin and fortuitously associated with the microsomal material. Neither this fraction nor those isolated by Loenig (1961) were examined for enzymes.

3. Invertebrate Animal Tissues

Except in a few studies centred on ribosomes and protein synthesis, such tissues have received little attention with respect to the biochemistry of microsomal fractions. Clayton's study of fractions from various tissues of a roach (*Eurycotis floridan*), prompted by an interest in sterol incorporation into membranes, did include analyses for constituents such as RNA, and points the way to future work (Clayton, 1965).

IV. Correlations and Functional Aspects

A. Normal Tissues

1. Biochemical Findings in Relation to Cytochemical Findings

The membranous locations established cytochemically for phosphatases (Table I) broadly match those established biochemically (Tables II and VI), given that there is now much evidence to support the often expressed warning that fragments of the plasma membrane as well as of cytomembranes may be present in the microsomal fraction. As is evident from cytochemical examination of hepatic microsomal fractions (Fig. 10), 5′-nucleotidase activity is associated with structures distinct from the vesicles showing glucose-6-phosphatase activity, that might have arisen from micro-villi. Similar observations have been made with crude nuclear fractions, except that 5′-nucleotidase activity is found in sheet-like structures (Fig. 10c), the source of which awaits firm identification—there being possible artefacts in cytochemical work on isolated fractions (A. A. El-Aaser and M. S. C. Birbeck, unpublished work). For nucleoside di- and tri-phosphatases it cannot yet be asserted that biochemical findings fully agree with cytochemical findings: in zonal centrifuge runs, ATPase and UTPase do not consistently run in the same positions as AMPase and ADPase, nor does UDPase consistently coincide with glucose-6-phosphatase—whereas the cytochemical literature suggests that ATPase is largely in the plasma membrane and that UDPase is largely in cytomembranes (El-Aaser *et al.*, 1966b).

For most microsomal-fraction enzymes there is no independent information from cytochemical study of the whole tissue, mainly because of technical

FIG. 10. Enzyme localization in sub-cellular fractions incubated with substrate after isolation. Courtesy of A. A. El-Aaser. Top left (a). Glucose-6-phosphatase, microsomal fraction: note staining of the vesicles. × 73,000. Top right (b). 5′-Nucleotidase with UMP as substrate, microsomal fraction: note that staining is not in vesicles as in (a), but in elements conceivably derived from bile canaliculi × 73,000. Bottom (c). 5′-Nucleotidase, crude nuclear fraction: note that staining is in sheets which resemble plasma-membrane fragments as shown in Fig. 9 (although a derivation from fibrous tissue cannot be ruled out). × 23,000.

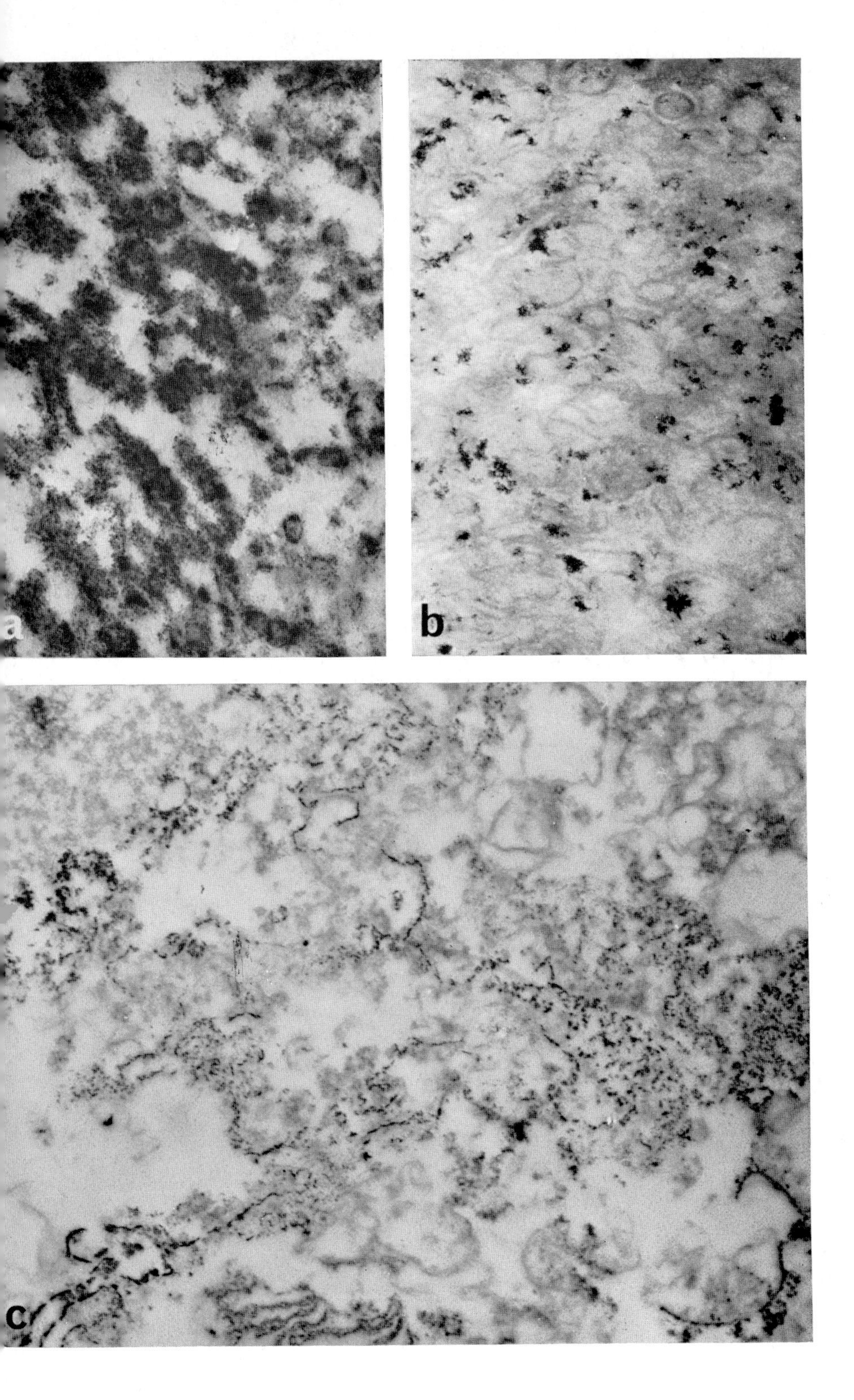
a
b
c

difficulties. Conversely, as mentioned by Novikoff (1964), there may be cytochemical information but "difficulties in isolating the relevant cell structures", although in the case of liver such difficulties are being overcome. For arylesterase there is an unresolved conflict between the biochemical finding of a microsomal location and the cytochemical finding with the light-microscope of activity in organelles, possibly lysosomes, clustered near bile canaliculi (Underhay, Holt, Beaufay and de Duve, 1956). Conceivably there may be leakage from lysosomes in homogenates and subsequent adsorption on to microsomal material (P. C. Barrow and S. J. Holt, unpublished work). There may, however, be several distinct arylesterases (Lundkvist and Perlmann, 1965), certain of which might be destroyed or inhibited by cytochemical procedures. This appears to be the case for mitochondrial ATPase.

Plurality of enzymes is a problem often encountered in efforts to localize the site of a particular activity. For some of the enzymes listed as present in the microsomal fractions (Tables II and VI), the proportion of the total activity in the homogenate attributable to microsomal material is rather low; yet only in a few instances, for example catalase, is there reason to suspect that the microsomal activity may be due to elements other than membrane fragments. As suggested for β-glucoronidase (3.2.1.31) (de Duve *et al.*, 1955), it is likely that there is a subtle difference between the nature of the enzyme in the microsomal fraction and that elsewhere. Microsomal fractions as prepared in some laboratories contain a high proportion of the lysosome population; but for the sake of clarity, reports of the presence in microsomal fractions of enzymes now known to be lysosomal (e.g. acid phosphatase) have been disregarded in the compilation of Tables II and VI.

These tables have been drawn up emphasizing the enzyme rather than the cell structure. Novikoff (1964) has given a useful summary from the viewpoint of locations (Table VII).

It should be noted that this list represents an amalgamation of cytochemical with some biochemical observations, but excludes most of the observations derived purely from biochemical studies. One such omission is $NADH_2$-cytochrome *c* reductase; cytochemical observations by light microscopy on diaphorase activity are compatible with, but hardly reinforce, the biochemical evidence for a cytomembrane location of this enzyme (Chang, 1960; Novikoff, 1960). Apart from any metabolic function that it may have, the Golgi complex may serve "to stock products elaborated elsewhere in the cell" (Rouiller, 1964) or, in connection with this, "to pump in or out the ingredients necessary to segregate or concentrate the final substance" (Haguenau, 1958).

The above list does not touch on the question of possible enzymic differences between smooth-surfaced cytomembranes (other than Golgi) and rough-surfaced cytomembranes. For glucose-6-phosphatase, cytochemistry has disclosed no difference (Tice and Barrnett, 1962) and the biochemical evidence is contradictory [Tables III and IV; see Section III. A. 2.*c*]. Smooth-surfaced

TABLE VII.

Plasma Membrane

Nucleoside phosphatases. Variable in different tissues: nucleoside triphosphatases (Mg^{++}-stimulated; Mg^{++}, Na^{+}, K^{+}-stimulated, ouabain-sensitive); Co^{++}-stimulated nucleoside monophosphatase; Mg^{++}-stimulated nucleoside monophosphatase; etc.

Alkaline phosphatase. Present in only a few cell types.

Endoplasmic Reticulum

Nucleoside diphosphatase. Hydrolyses disphosphates of guanosine, uridine and inosine. Present in liver, kidney, steroid-producing cells, and other cells in which glucuronyl transferase activity has been demonstrated. In these cells the endoplasmic reticulum also shows low levels of thiamine pyrophosphatase.

Glucose-6-phosphatase, esterase, and acid phosphatase are present in the endoplasmic reticulum of some tissues.

Golgi Apparatus

Golgi saccules: Nucleoside diphosphatase (diphosphates of guanosine, uridine, and inosine are hydrolysed more rapidly than those of adenine and cytosine); thiamine pyrophosphatase; acid phosphatase (in some saccules) is being demonstrated in increasing number of cells; alkaline phosphatase is found in intestinal mucosa.

Golgi vesicles: Acid phosphatase is being demonstrated in an increasing number of cells.

---

cytomembranes appear to be concerned particularly "in hydroxylation and elimination of lipid-soluble drugs, and in cholesterol and steroid hormone metabolism" (Jones and Fawcett, 1966).

2. *Functional Aspects of Membrane-bound Enzymes*

(*a*) *Energy supply*. Shull (1959) stressed the need to examine microsomal fractions for dehydrogenases lending to the reduction of NAD; but present-day evidence (Table II) indicates that $NADPH_2$ is more important than $NADH_2$ in cytomembrane enzymology. Various authors have commented on the apparent paradox that cytochrome *c* is lacking in microsomal fractions although $NADPH_2$-cytochrome *c* reductase and particularly $NADH_2$-cytochrome *c* reductase are present. Phillips and Langdon (1962) suggest that the latter enzyme acts *in vivo* merely as a mediator of electron transport, without participation by cytochrome *c*, in certain hydroxylations and reductions that require $NADPH_2$. There is the complication that microsomal fractions contain a diaphorase effective with either $NADH_2$ or $NADPH_2$ but ineffective if cytochrome *c* is used (Ernster *et al.*, 1962; Orrenius. Ericsson and Ernster, 1965).

There is still no reason to believe that cytomembranes make a large contribution to the energy supply of the cell, even by electron transfer through cytochrome b_5 instead of cytochrome *c*. The b_5 pathway may lead to "ion flux" rather than to ATP formation [Siekevitz, 1965; for this and other pathways see also Strittmatter (1965) and Omura *et al.* (1965)]. The re-oxidation of $NADH_2$ is considered to be mediated by substrate shuttles, in particular the "glycerophosphate shuttle" with mitochondrial participation, and that of $NADPH_2$ by mechanisms such as the pentose-phosphate cycle, involving soluble enzymes. Nevertheless, the tendency for mitochondria to be closely surrounded by cytomembranes (Fig. 2) points to some kind of co-operation in energy metabolism; perhaps mitochondria supply cytomembranes with cytochrome *c*, and this is lost by leakage when microsomal fractions are isolated, a situation which would explain the above paradox.

There is no evidence that cytomembranes are concerned in the synthesis as distinct from the storage of glycogen, or in glycogenolysis; morphological findings (Porter and Bruni, 1959) are inconclusive (Jones and Fawcett, 1966). The fact that glucose-6-phosphatase—the determinant of supply of glucose to the blood stream—is located in cytomembranes is of uncertain significance. As yet there is no evidence for the possibility that glucose formed by the enzyme may leave the cell by way of cisternae which lead to the exterior, rather than by way of the extra-membranous cytoplasm or cell sap. The membranous location of the enzyme—a lipoprotein—may offer the possibility of control of enzyme activity through subtle changes in the configuration of the enzyme protein itself or of contiguous macromolecules; such changes, perhaps analogous to those effected by agents such as deoxycholate, might well be induced or suppressed by specific hormones as a means of regulating glucose output (cf. Siekevitz, 1962). This line of thought is essentially a facet of the concept that one role of cellular membranes is to limit and regulate the interaction between enzyme and substrate; for the cytomembranes under discussion, it may be a matter of the membrane serving to accommodate within itself a lipoprotein enzyme, rather than to bar access to a soluble enzyme within the cisternal space. Moreover, within or adjoining the membrane there might be a pH very different from that in the cell sap but particularly appropriate for the action of certain membrane-located enzymes.

(*b*) *Transport.* The preceding speculation that the membranous location of glucose-6-phosphatase may enable glucose to be channelled out of the cell, comes into the realm of transport. A further possibility, likewise unsubstantiated although often mooted, is that the entry of glucose into the cell is mediated by a hexokinase situated in the plasma membrane—a possibility which is by no means disproved by the biochemical evidence that most of the hexokinase activity of the liver cell is *not* membrane-linked. The prevailing view concerning active transport of monosaccharides across membranes is that the process requires the mediation of a "carrier" which is possibly mobile and the nature of which is unknown (Quastel, 1964).

More generally, "It is becoming increasingly evident that cell membranes are equipped with substances that play specific roles in conveying organic molecules from the cell exterior to the cell interior or from one compartment in the cell to another. These substances play roles as important as those of the enzymes in the control of cell growth, metabolism, and function" (Quastel, 1964). The transport of K^+ and Na^+ ions, a process particularly important in nerve function, has received especial attention (Conway, 1960). The picture is still unclear, but the "ATPase" which shows optimal activity in the presence of Mg^{2+}, K^+ and Na^+ seems to be implicated and does not act hydrolytically (Järnefelt, 1964). "Labile phosphate" is involved. "ATP is involved in the process of transport though the involvement is probably an indirect one, which is reflected in the rapid and parallel turnover of phosphate in certain types of phospholipid and of a phosphoprotein fraction. Whether these two phenomena involve a precursor which is common also to a transport carrier or whether either is the actual carrier, is presently a matter of speculation, experimentation, and a challenge to the imagination" (Scholefield, 1964; see also Wolfe, 1964). As Whittam (1964) points out, "Membranes are spatially asymmetrical in selecting Na^+ for outward transport and K^+ for inward transport, and one kind of ion movement does not generally occur without the other. . . . Perhaps the simplest working hypothesis for the mechanism of action of the ATPase is to regard Na^+ and K^+ as enzymatic co-factors which must be in the right location to exert their effects. . . ."

Given that "ATPase" is concerned in active transport and that the cytoplasm differs sharply from the extra-cellular space and the nucleus in ionic composition, it is curious that in parenchymal cells the activity is highest in the micro-villi of bile canaliculi. Usually cytochemists have found poor activity in the smooth portions of the plasma membrane and especially in cytomembranes including the nuclear envelope, although under some conditions, for example with formaldehyde fixation, ATPase activity is demonstrable in cytomembranes (Wachstein and Fernandez, 1964; A. A. El-Aaser disagrees, from unpublished work). Although there is a vast literature on "ATPase", especially on its biochemical aspects, continued and more subtle study is needed. The observed ATP-splitting activity reflects a more complex process or processes, concerned not only with transport but also with the contraction and relaxation of myofibrils, mitochondria, and perhaps membrane systems in general. One "ATPase" seems to be phospholipid-dependent (Tanaka and Strickland, 1965; see also Emmelot and Bos, 1965).

The transport of Ca^{2+} seems to be mechanistically similar to that of Na^+ (Martonosi and Feretos, 1964). With regard to water movement, at present it seems likely that the passage of water molecules across membranes is a passive process, secondary to the movement of ions or of metabolites (Scholefield, 1964). In the transport of phosphate the Golgi elements might play a role by virtue of their high alkaline phosphatase activity, but proof is so far lacking (Goldfischer *et al.*, 1964).

It should be stressed that substances not carried by active transport are either excluded by membranes, a function no less important than active transport, or go through by "passive transport". There is a further but unproved possibility that substances may pass between the cell exterior and the nuclear envelope via cisternal channels (cf. Section II. B.) without ever passing across membranes. It should also be emphasized that "active transport", as briefly surveyed above, can hardly be considered as a self-contained topic. Electron-transport as mentioned under (*b*) may be concerned in active transport, possibly ATP-dependent (cf. Ernster *et al.*, 1962; Ernster and Jones, 1962), and processes such as bile formation, now to be touched on, may entail active transport.

(*c*) *Conjugation processes*; *bile secretion*; *nucleoside mono- and di-phosphatases*. These items are considered together because there are certain correlations between them, although the reason for this is as yet unknown. For different tissues there is a parallelism between the occurrence of nucleoside diphosphatase in membranes and that of UDP glucuronyl transferase (2.4.1.17) in microsomal fractions (Goldfischer *et al.*, 1964). (The authors fail to draw attention to the different techniques for study of the enzymes.) That microsomal enzymes are important in furnishing bile constituents is indicated not only by their ability to convert bilirubin into its glucuronide, but also by their capacity to build the steroid skeleton (Table III; see also Siperstein and Fagan, 1964) and to convert cholic acid into its CoA derivative and thence into glyco- or tauro-cholic acid (Table II).

Another intriguing observation is that the concentration of deoxycholate that activates 5′-nucleotidase (cf. Table IV) is of the same order as the concentration of deoxycholate in rat bile (Ku and Wang, 1963). Since the enzyme is located in bile canaliculi, its activity *in vivo* may thus be under biliary control. This line of thought seems more profitable than the suggestion by de Verdier and Potter (see Novikoff, 1960) that, particularly in tumours, the the enzyme serves to furnish nucleic-acid precursors, in the form of nucleosides, to neighbouring cells, the plasma membrane being impermeable to nucleotides.

Other lines of thought have been developed by Ernster and Jones (1962). "The nucleoside diphosphatase may serve a function in promoting reactions involving the corresponding triphosphates as reactants by shifting the equilibrium toward the formation of the diphosphate. It is conceivable that the nucleoside diphosphatase present in the endoplasmic canals may have such a function in removing GDP formed in the GTP-dependent incorporation of amino acids and/or in removing UDP formed in the varios uridine nucleotide-linked conjugation reactions taking place in the membranes. The

GMP and UMP so formed may then leave the membranes and be transformed into the corresponding triphosphatase [*sic*] by way of the nucleoside monophosphate kinase system." They further speculate, because of the microsomal hydrolysis of IDP and an early observation by Siekevitz that bound inosine is present in microsomal fractions, that inosine nucleotides may have a function in microsomal activities. There is no experimental support for the assumption that the "endoplasmic canals" are the site of nucleoside diphosphatase activity.

(*d*) *Various biochemical processes.* From observations mentioned earlier (e.g. in Table II), and also collated by Ernster *et al.* (1962), Rouiller (1964) and Moulé (1964), it appears that microsomal, possibly cytomembrane, elements are concerned in ascorbic acid synthesis, in drug metabolism, for example, $NADPH_2$-dependent hydroxylations, in the reductive stages of fatty acid and steroid synthesis, and in lipid peroxidation—a process that is perhaps "an important factor in the regulation of cytoplasmic structure and function" (Hochstein and Ernster, 1964). Mason, North and Vanneste (1965) have reviewed, with attention to the role of "Fe_x", the complex topic of microsomal "mixed-function oxidases", concerned in the metabolism of fatty acids, steroids and foreign compounds. The hydrophobic character of the membranes might serve to provide the requisite milieu for reactions concerned with lipids; Jones and Fawcett (1966) have emphasized the role of the smooth-surfaced elements.

These are only some of the more important of the diverse processes catalysed by microsomal elements. The interpretation of the esterase activity, notably the acetylcholinesterase, of microsomal fractions is as yet uncertain. Relaxation phenomena in skeletal muscle appear to depend on a transphorylating system as studied in microsomal fractions by Molnar and Lorand (1962). It should be pointed out that, in general, cytomembrane-linked processes do not require a supply of ATP.

There is already some evidence that cytomembranes, and possibly phospholipids, play a role in protein synthesis distinct from and possibly secondary to that of ribosomes [Godson *et al.*, 1961 (bacteria); Higashi and Peters, 1963a, 1963b (liver); Redman *et al.*, 1966 (pancreas); for other citations see Moulé, 1964, and Hendler, 1965]. The postulate of Barbieri and di Marco (1963) that there is actual synthesis *de novo* of protein in smooth membranes has been criticized by Hallinan and Munro (1965). Pitot (1964) postulates that membranous structures serve to stabilize and store the messenger-RNA templates required for the synthesis of proteins that are to be secreted.

3. Physiological Changes in Morphology and Biochemical Pattern.

(*a*) *Development and renewal.* Our knowledge of the primary origin of cellular membranes has advanced little since, in the context of rough-surfaced cytomembranes, Haguenau concluded in 1958 that there was insufficient evidence to decide whether the rough-surfaced cytomembranes were formed from the nucleus, mitochondria or the plasma membrane.

Essner and Novikoff (1962) interpreted their cytochemical observations on diphosphatases and other enzymes as indicating that normally there is transformation of smooth-surfaced cytomembranes into Golgi cytomembranes and thence into primary lysosomes or secretory vacuoles (see also Goldfischer *et al.*, 1964); enzymes such as acid phosphatases, formed on ribosomes initially, were envisaged as accompanying the membranous elements through these transformations. The essence of this concept is the reasonable supposition that the cytomembranes are normally in a state of flux. More speculatively, Essner and Novikoff (1962) suggest that cytomembranes may originate from the nuclear membrane, In kidney there is evidence that the brush border is the source of the phagosomes that appear in response to foreign protein injections (Straus, 1962). However, our knowledge concerning interconversions of membranous structures has really advanced little since Tashiro (1957) put forward an elaborate hypothesis concerning the "reversible microsome system", The subject of the relationship between cell membranes and lyosomes is also considered in Chapter 5.

There is evidence, summarized by Moulé (1964), that during cell differentiation, for example in the hepatocyte, there is development of cytomembranes and ensuing attachment of ribosomes to the latter. Experiments with chick liver led Pollak and Shorey (1964) to conclude that the protein as distinct from the phospholipid moiety of cytomembranes is already present in early development stages, supposedly in the form of granules which they called "reticulosomes".

In common with various other enzymes, some membrane-linked enzymes are low in foetal liver and show a dramatic rise after birth. Such observations have been made for alkaline phosphatase and glucose-6-phosphatase. The rise is not dramatic in the case of 5′-nucleotidase, nucleoside triphosphatase (El-Aaser, 1965), and enzymes concerned in the oxidative metabolism of drugs (Fouts, 1963). The rise in glucose-6-phosphatase has been found with kidney (in the guinea-pig but not in the rat) and intestinal mucosa as well as with liver (Lea and Walker, 1964). For certain electron-transport enzymes there is "the peculiarity that the enzymes do not seem to be all synthesized at the same time" (Siekevitz, 1965).

In regenerating liver, cytomembrane development follows the cell hypertrophy (Moulé, 1964), but the yield of the smooth-membrane fraction actually drops initially (Delhumeau de Ongay, Moulé and Frayssinet, 1965). Rough-surfaced cytomembranes reappear as clumps surrounding the mitochondria (Haguenau, 1958). There are differences among enzymes in the rate and extent of the changes in concentration. There is often a rise in enzyme activity (Reid, 1962), but the few biochemical observations of membrane-linked enzymes show no rise. Indeed, both glucose-6-phosphatase and

5′-nucleotidase were found by El-Aaser (1965) to decrease after partial hepatectomy (for earlier work see Weber, Singhal, Stamm, Fisher and Mentendiek, 1964); moreover, Fouts (1963) found a fall in drug-metabolizing enzymes. Cytochemically, "ATPase activity is increased in the tortuous canaliculi" (Wachstein, 1963). For 5′-nucleotidase in lactation, see Wang (1962).

In Loenig's (1961) study of root-tip microsomal fractions, the yield of the "light" fraction increased rapidly at the onset of differentiation, as a reflection of cytomembrane formation and perhaps, at a later stage, of plasma-membrane development.

For further information on the wide topic of membranes in development, Locke (1964) may be consulted (see also Burch, 1965).

(*b*) *Fasting and dietary changes.* Fawcett (1955) and other authors have observed that with prolonged fasting there is a loss of hepatic cytomembranes, which may be preceded by aggregation of the smooth-surfaced elements to give "hyaline inclusions", and which can be reversed by re-feeding. Accompanying the active protein synthesis during re-feeding, the region adjoining the nuclear envelope becomes filled with rough-surfaced cytomembranes enmeshing mitochondria [Bernhard and Rouiller, cited by Haguenau (1958)].

Mention has already been made in Section III B. 3 (*a*) of reported changes in the incorporation of precursors into non-ribosomal species of RNA, possibly located in the membrane. There is as yet no evidence for a change in the amount of RNA as distinct from changes in the rate of labelling of this RNA, despite a remark by Hallinan and Munro (1964) that the actual amount is influenced by dietary protein. The pattern of deoxycholate-soluble proteins may change with overnight fasting, but apparently becomes normal again if the fast is prolonged (Hultin, 1957).

Glucose-6-phosphatase is a classical example of an enzyme that is, understandably, sensitive to dietary conditions in respect of the level of activity [Weber *et al.* (1964), *inter alia*]. Although prolonged fasting depresses the content of glucose-6-phosphatase in cytomembranes, as represented by deoxycholate-soluble material, there is a rise in the normally low content of enzyme in the ribosomes, as if transfer of new enzyme to cytomembranes had become arrested (Busch, Weill and Mandel, 1960). However, no change in distribution is detectable cytochemically (Tice and Barrnett, 1962). Prolonged fasting also depresses the oxidative (but not the reductive) metabolism of drugs (Fouts, 1963), whereas 5′-nucleotidase is little affected (Terroine, 1961).

In a biochemical and electron-microscope study of guinea-pig pancreas, Palade and Siekevitz (1958a) found that fasting caused changes such as disappearance of intracisternal zymogen granules and a fall in microsomal-fraction ribonuclease.

With a deficiency of essential fatty acids, the incorporation of phosphate (^{32}P) into the lipids of brain microsomal fractions, with or without supplementation with acetylcholine, shows changes which, according to de Pury and Collins (1962), point to a weakening of lipoprotein membranes.

With cysteine feeding, hepatic microsomal fractions showed decreases in RNA and phospholipid and increases in cytochrome reductases; moreover, tissue sections showed disorganization of the rough-surfaced cytomembranes and an increase in smooth-surfaced cytomembranes (Emmelot, Mizrahi, Naccarato and Benedetti, 1962). Cholesterol feeding causes "feed-back inhibition" of the synthesis of mevalonic acid and hence of cholesterol (Siperstein and Fagan, 1964).

(*c*) *Hormonal influences.* One of the few instances of morphological changes in cytomembranes with a change in hormonal status is an observation by E. G. Ball and collaborators (see Tepperman and Tepperman, 1960), that treatment of adipose-tissue cells with insulin *in vitro* results in the formation of a system of canaliculi and vesicles suggestive of a stimulation of pinocytosis. Certain hormonal treatments may alter the level or the labelling of microsomal RNA, sometimes with a difference in this respect between "heavy" and "light" microsomal fractions (Reid, 1961a); but there is no reason to believe that membrane-located RNA as distinct from ribosomal RNA is affected.

There are scattered reports [cited by Tepperman and Tepperman (1960), Reid (1962), and El-Aaser (1965), *inter alia*] concerning hormonal effects on membrane-linked enzyme activities. For example, hyperthyroid liver may show elevated activities of glucuronyl transferase and of glucose-6-phosphatase but not of 5′-nucleotidase, whereas with adrenalectomy or hypophysectomy there may be a decrease in glucose-6-phosphatase and also in the ATPase of microsomal fractions. After the rise in glucose-6-phosphatase activity caused by thyroxine, the stimulatory action of deoxycholate is still demonstrable (El-Aaser, 1965). Similarly, with the rise accompanying experimental diabetes the stimulatory effect of trypsin (see III B. 4) can still be seen (Heise and Görlich, 1963). The rise with diabetes (or with corticoid treatment) truly seems attributable to synthesis of new enzyme protein (Weber *et al.*, 1964). For certain enzymes of drug metabolism a rise may be induced by anabolic steroids (Gilette, 1963), but with alloxan diabetes the oxidative drug-metabolizing enzymes are depressed (Fouts, 1963).

In adrenal tissue, the proportion of the glucose-6-phosphatase activity recovered in the microsomal fraction falls after ACTH treatment, the nuclear and mitochondrial fractions becoming richer in this enzyme (Breuer, Petershof and Knoppen, 1963). The converse is found for 5′-nucleotidase (Hilf, Breuer and Borman, 1961). In the absence of morphological and cytochemical information however, these observations are difficult to interpret.

As pointed out by Emmelot *et al.* (1964a) (see also Leaf, 1960), there is evidence for activation of Na^+ transport by thyroid-stimulating hormones and neurohypophyseal hormones, this activation being perhaps due to hormone interaction with -SH groups in ATPase. Attempts have been made to explain the actions of insulin and of ACTH in terms of transport changes (Tepperman and Tepperman, 1960). Future work may well establish hormonal effects not mediated by a direct hormone-enzyme interaction: "... do hormones act upon a higher order of the cell, upon the framework in which the enzymes are enmeshed?" (Siekevitz, 1962).

B. PATHOLOGICAL TISSUES

1. Effects of Toxins, Carcinogens and Drugs

To summarize a vast body of ultrastructural, cytochemical and biochemical observations (cf. Rouiller, 1964, and Reid, 1965; see also Novikoff and Essner, 1962, and Jones and Fawcett, 1966), there is no early action of carcinogens on hepatic membrane systems that cannot likewise be produced by non-carcinogenic toxins, and there is virtually no action of the latter that cannot likewise be produced by "physiological" treatment such as fasting. "One may infer that the dedifferentiation, frequently occurring in tumours, is not specific for the cancerous process" (Moulé, 1964). It is, nevertheless, striking that for hepatocarcinogens such as azo dyes or dimethylnitrosamine (but perhaps not ethionine), and hepatotoxins such as carbon tetrachloride, the cytomembranes "seem to be the first site of detectable morphological lesions of the hepatic cell" [Rouiller 1964; see also Haguenau (1958)]. The morphological changes may include detachment of ribosomes from rough-surfaced cytomembranes, and swelling, aggregation or other alterations in the actual cytomembranes, with a randomization of their distribution in the cytoplasm. There may also be damage to the plasma membrane, at least if necrosis sets in; this damage could cause leakage which might account for observations such as decreased activities of esterases (Rouiller, 1964). Such leakage of cytomembranous elements, from mammary-gland secretory cells, has been thought to occur even during normal lactation (Morton, 1954a). It is, however, unlikely that such leakage could be the sole cause of hepatoxin-induced enzyme decreases such as the fall in glucose-6-phosphatase mentioned by Feuer *et al.* (1965).

In no hepatoma is there the normal hepatic complement of cytomembranes, yet certain hepatomas have an almost normal level of glucose-6-phosphatase (Pitot, 1962; Reid, 1965), and even of $NADH_2$-cytochrome *c* reductase (Novikoff, 1960). There may be faster synthesis of cholesterol in slices and loss of "feedback inhibition" (Siperstein and Fagan, 1964); but drug-metabolizing enzymes are commonly depressed (Fouts, 1963). Biochemical results of analyses for ATPase and 5′-nucleotidase in precancerous or cancerous liver are difficult to interpret because these enzymes are not confined to the membranes of parenchymal cells; but the cytochemical evidence, which distinguishes the activity in sites such as connective tissues and blood-vessel walls, indicates that in many hepatomas the actual "parenchymal" cells have rather low activity, at least for nucleoside di- and tri-phosphatase as distinct from 5′-nucleotidase (*inter alia* Chang, 1960; Novikoff, 1960; Essner and Novikoff, 1962; Goldfischer *et al.*, 1964; El-Aaser, 1965). On the other hand, a notably high level of alkaline phosphatase has been found in plasma-membrane fractions of hepatomas by Emmelot, Benedetti and Rümke (1964a), who further observed a very low level of esterase in the hepatoma material, and an abnormal antigen pattern. Many, but not all, hepatomas are deficient in actual bile canaliculi (Essner and Novikoff, 1962; Novikoff and Essner 1962; Goldfischer *et al.*, 1964). Even with only 3 weeks of azo-dye feeding, a depletion is cytochemically demonstrable in the ATPase and 5′-nucleotidase of bile canaliculi, whereas alkaline phosphatase is increased (Novikoff and Essner, 1962). Emmelot and Bos (1965, 1966) later concluded that a Na^+- and K^+-insensitive ATPase in hepatoma plasma-membrane fragments is more sensitive to neuraminidase than is the liver enzyme, and that in hepatomas, unlike liver, the plasma membrane truly contains hexokinase.

At least it would appear that, with the more subtle techniques now becoming available, membrane systems warrant continued study as possible primary sites for the actions of carcinogens and toxins. Cytomembranes have long been thought of as major sites for the "binding" of carcinogens such as azo dyes (*inter alia* Hultin, 1957; Westrop and Green, 1960; Rouiller, 1964). There is likely to be a revival of interest in cytomembranes, although the current emphasis is on possible derangements in ribosomal function. Pitot (1964) puts forward the suggestion that "chemical carcinogens would be expected to randomly alter the membrane of the ER [endoplasmic reticulum], leading to random changes in template stability, resulting in a wide variation of molecular and biologic expressions of neoplasia." Among observations which seen unconnected with the intriguing topic of template stability, one which warrants further investigation is that "Fe_x", possibly a haem-protein, was deficient in a microsomal fraction of a hepatoma (Nebert and Mason, 1964; cf. Mason *et al.*, 1965).

Little work has been done on effects of administered detergents. Injection of Triton WR-1339 caused dilation of Golgi saccules but otherwise had little effect on cytomembrane morphology (Novikoff and Essner, 1962).

The effect of injected phenobarbital on hepatic cytomembrane morphology and on microsomal drug-metabolizing enzymes has been studied in the laboratories of Remmer (cf. Remmer, 1962), of Ernster (Orrenius *et al.*, 1965) and of Fouts (Fouts and Rogers, 1965; cf. Hart and Fouts, 1965), with gratifying agreement on the nature of the derangement. It is the smooth-surfaced cytomembranes that preferentially show proliferation and a rise in enzyme content, the rise being attributable to faster synthesis of the enzyme proteins. This synthesis, and that of the actual membrane proteins, perhaps occurs in contiguous rough-surfaced cytomembranes (Jones and Fawcett, 1966). The effect was equally striking with the drug chlordane, but not with benzpyrene or methylcholanthrene (Fouts and Rogers, 1965; Hart and Rogers, 1965). Since, however, the liver is not the target for the actions of these various agents, the results throw little light on their mode of action. Other microsomal changes observed with phenobarbital or chlordane include rises in total protein, in $NADPH_2$ oxidase, and in CO-binding pigment (Hart and Fouts, 1965). For this pigment, and for $NADPH_2$-cytochrome *c* reductase, the rise after phenobarbital again occurs more in the smooth- than in the rough-surfaced fractions, although the fractions have an equivalent content in untreated animals (Orrenius *et al.*, 1965; see also Jones and Fawcett, 1966). Phenobarbital also causes a marked rise in cytochrome b_5 and glucose-6-phosphatase, but not in NADP nucleosidase or in ATPase—even in the smooth-surfaced fraction; this fraction shows a sharp rise in phospholipid but not in RNA (Remmer and Merker, 1963).

2. Effects of Bile-Duct Ligation

There are striking changes with bile-duct ligation, although Novikoff and Essner (1962) emphasize that these may not be of primary importance in the pathogenesis of jaundice. One such change is a complete loss of 5′-nucleotidase activity from many of the bile canaliculi, together with morphological derangement in the latter. There is also a marked fall in ATPase, the fall being reversed if the bilary obstruction is released (Wachstein, 1963). Cytomembrane enzymes concerned in the oxidative metabolism of drugs are likewise depressed (Fouts, 1963), although there is hypertrophy of smooth-surfaced cytomembranes (see Jones and Fawcett, 1966).

In toxin-induced liver damage, loss of canalicular 5′-nucleotidase activity is a late effect, and is less conspicuous than loss of ATPase (Wachstein, 1963)]. Certain other changes, such as an increase in the alkaline phosphatase activity of sinosoidal surfaces, are at sites remote from the bile canaliculi.

V. Summary

Consideration has been given above to the nature of membrane systems, to methods for their study *in situ*, for their isolation, and to problems of nomenclature. Exception was taken to the terms "cell membrane", "microsomes", and "nucleoside phosphatase". The following systems have been surveyed, with emphasis on hepatic cells: the plasma membrane, cytomembranes (of rough- and smooth-surfaced types, the latter including the Golgi complex), and the nuclear envelope, which on morphological and cytochemical grounds can be regarded as part of the cytomembrane system.

Cytochemical work with the electron microscope, so far confined to phosphatases, has shown that cytomembranes, but not the plasma membrane, contain glucose-6-phosphatase and nucleoside diphosphatase, assayed with GDP, IDP or UDP as substrate; that the Golgi complex contains thiamine pyrophosphatase and sometimes alkaline phosphatase but no glucose-6-phosphatase, and that the plasma membrane, especially where it folds to form micro-villi, contains a nucleoside monophosphatase (5′-nucleotidase) together with a nucleoside triphosphatase, although the latter activity may also be demonstrable in cytomembranes and mitochondria. The question of artifacts has been briefly considered.

Biochemical work has given results compatible with the cytochemical findings, assuming that plasma-membrane fragments are recovered in the microsomal fraction and in the crude nuclear fraction and account for the 5′-nucleotidase activity of these fractions. Processes which, from biochemical study of microsomal fractions, are largely attributable to cytomembranes include steps in lipid synthesis, ascorbate formation, deacetylation of acetylcholine, $NADPH_2$-dependent hydroxylation of aromatic compounds, and glucuronide formation. In general, microsomal fractions have low levels of enzymes requiring NAD or ATP as a reactant; they can catalyse the re-oxidation of $NADH_2$ and $NADPH_2$, but cytomembranes can hardly be regarded as important in energy supply, unless a supposition is made that there is transfer of cytochrome *c* from mitochondria to adjoining cytomembranes *in situ*. Cytomembranes seem to be concerned in glycogen storage, in the late steps of protein synthesis and in transfer of the protein to other sites.

Questions yet to be decided include the location of esterase(s), the metabolic role of nucleoside mono- and di-phosphatases, and the extent to which smooth-surfaced cytomembranes, other than Golgi membranes, and rough-surfaced cytomembranes differ in their enzymology. Progress has been made in the isolation of membrane fragments corresponding to individual types of cytomembrane and to the plasma membrane. There is considerable variability among laboratories in the reported values for phospholipid and RNA in membrane fragments with a low content of ribosomes. There are indications that cytomembranes truly contain some RNA, different from that in the ribosomes; but the plasma membrane is low in RNA.

In work with microsomal fractions from tissues other than liver, there has seldom been conjoint biochemical and electron-microscopic examination; there may be differences from liver with respect to the sub-cellular distribution of some constituents (e.g. RNA and xanthine oxidase), but the findings for most microsomal constituents (e.g. acetylcholinesterase) agree with those for liver.

Some enzymes found in membrane fractions are so readily detached that their presence there may be a re-distribution artifact. Other enzymes, such as glucose-6-phosphatase, appear to be tightly-bound lipoproteins; progressive damage to the membrane by an agent such as deoxycholate may first cause activation, but may be ineffective in achieving solubilization without loss of activity. One function of membrane lipids may be to promote ingress and metabolism of lipid substrates.

Aspects of the selective permeability of membranes have been considered. Nucleoside triphosphatase(s) in membranes seem to be important in the "active transport" of cations. Possibly, however, some substances may travel between the exterior of the cell and the interior of the nuclear envelope by way of the cisternal spaces in the cytomembranes, without ever traversing a membrane.

Altogether it appears that membranes, and in particular cytomembranes, have diverse functions, and may themselves be in a state of flux; there is no evidence that would justify use of the term "cytoskeleton". Mention has been made of the appearance, or disappearance, and behaviour of cytomembranes and cytomembrane enzymes in situations such as cell development, fasting and re-feeding, changes in hormonal status, exposure to toxins and carcinogens, and bile-duct ligation. Damage to cytomembranes might well be an important concomitant property of carcinogenesis.

Acknowledgements

Valuable comments on the manuscript were received from Mr. M. S. C. Birbeck and Dr. Y. Moulé, and also from Dr. A. A. El-Aaser whose help and enthusiasm have been particularly appreciated. These and other workers kindly furnished illustrations as is indicated in the legends. The work in the author's laboratory was supported by grants from the British Empire Cancer Campaign for Research, from the Medical Research Council, and from the Wellcome Trust.

References

Abood, L. G., Kurahasi, K., Brunngraber, E. and Koketsu, K. (1966). *Biochim. biophys. Acta* **112**, 330.

Abraham, S., Matthes, K. J. and Chaikoff, I. L. (1963). *Biochim. biophys. Acta* **70**, 357.

Aldridge, W. N. (1962). *Biochem. J.* **84**, 527.

Aldridge, W. N. and Johnston, M. K. (1959). *Biochem. J.* **73**, 270.

Allard, C., de Lamirande, G. and Cantero, A. (1957). *Cancer Res.* **17**, 862.

Anderson, N. G., Price, C. A., Fisher, W. D., Canning, R. E. and Burger, C. L. (1964). *Analyt. Biochem.* **7**, 1.

Ashworth, C. T., Luibel, F. J. and Stewart, S. G. (1963). *J. Cell Biol.* **17**, 1.

Avignon, J. and Steinberg, D. (1961). *J. biol. Chem.* **236**, 2898.

Axelrod, J. (1956). *Biochem. J.* **63**, 634.

Barbieri, G. P. and di Marco, A. (1963). *Expl Cell Res.* **30**, 193.

Baxter, C. E. Van Reen, R., Pearson, P. B. and Rosenberg, C. (1958). *Biochim. biophys. Acta* **27**, 584.

Beaufay, H. and de Duve, C. (1954). *Bull. Soc. Chim. biol.* **36**, 1551.

Binkley, F. (1961). *J. biol. Chem.* **236**, 1075.

Birbeck, M. S. C. and Reid, E. (1956). *J. biophys. biochem. Cytol.* **2**, 609.

Booth, J. and Boyland, E. (1957). *Biochem. J.* **66**, 73.

Börnig, H. (1964). Abstracts, *6th Int. Congr. Biochem.* 645.

Bouvet, C. and Moulé, Y. (1964). *Expl Cell Res.* **33**, 330.

Bradford, H. F., Swanson, P. D. and Gammack, D. B. (1964). *Biochem. J.* **92**, 247.
Brawerman, G. and Chargaff, E. (1955). *Biochim. biophys. Acta* **16**, 524.
Bremer, J. (1955). *Acta chem. scand.* **9**, 268.
Bremer, J. and Greenberg, D. M. (1961). *Biochim. biophys. Acta* **46**, 205 (see also 217).
Breuer, H., Petershof, I. and Knoppen, R. (1963). *Z. physiol. Chem.* **334**, 259.
Brosemer, R. W. and Rutter, W. J. (1961). *J. biol. Chem.* **236**, 1253–58.
Bucher, N. L. R., Overath, P. and Lynen, F. (1960). *Biochim. biophys. Acta* **40**, 491.
Burch, H. B. (1965). *Adv. Enzyme. Regulation* **3**, 185.
Busch, S., Weill, J. D. and Mandel, P. (1960). *C. r. Soc. Biol.* **54**, 798.
Campbell, P. N., Colper, C. and Hicks, M. (1964). *Biochem. J.* **92**, 247.
Carruthers, C. and Baumler, A. (1962). *Archs Biochem. Biophys.* **99**, 458.
Carruthers, C., Wernley, D. L., Baumler, A. and Lilga, K. (1960). *Archs Biochem. Biophys.* **87**, 266.
Chang, J. P. (1960). *In* "Cell Physiology of Neoplasia", M. D. Anderson Hospital and Tumor Institute. p. 471. Univ. of Texas Press, Austin.
Chauveau, J., Moulé, Y., Rouiller, C. and Schneebeli, J. (1962). *J. Cell Biol.* **12**, 17.
Claude, A. (1946). *J. exp. Med.* **84**, 51.
Clayton, R. B. (1965). *Biochem. J.* **96**, 19P.
Cohn, P. and Simson, P. (1963). *Biochem. J.* **88**, 206.
Coleman, R. and Finean, J. B. (1965). *Biochem. J.* **97**, 39P.
Coleman, R. and Hübscher, G. (1962). *Biochim. biophys. Acta* **56**, 479.
Colodzin, M. and Kennedy, E. P. (1964). *Fed. Proc.* **23**, 229.
Conney, A. H., Brown, R. R., Miller, J. A. and Miller, E. C. (1957). *Cancer Res.* **17**, 628.
Conway, E. J. (1960). *J. gen. Physiol.* **43**, Suppl. I, 17.
Cooksey, K. E. and Greenberg, D. M. (1961). *Biochem. biophys. Res. Commun.* **6**, 256.
Creasey, W. A. (1962). *Biochim. biophys. Acta* **64**, 559.
Dahlqvist, A. (1961). *Biochim. biophys. Acta* **50**, 55.
Dallner, G. (1963). *Acta path. microbiol. Scand.* Suppl. **166**, 1.
Dallner, G., Orrenius, S. and Bergstrand, A. (1963). *J. Cell Biol.* **16**, 426.
de Duve, C. (1964). *J. theoret. Biol.* **6**, 33.
de Duve, C., Berthet, J. and Beaufay, H. (1959). *Prog. Biophys. biophys. Chem.* **9**, 325.
de Duve, C., Pressman, B. C., Gianetto, R., Wattiaux, R. and Appelmans, F. (1955). *Biochem. J.* **60**, 604.
de Duve, C., Wattiaux, A. and Baudhuin, P. (1962). *Adv. Enzymol.* **24**, 291.
de Lamirande, G., Allard, C. and Cantero, A. (1958). *J. biophys. biochem. Cytol.* **3**, 373.
de Lamirande, G., Boileau, S. and Morais, R. (1966). *Can. J. Biochem.* **44**, 273.
de Pury, G. G. and Collins, F. D. (1962). *Nature, Lond.* **198**, 788.
Decken, A. von der and Campbell, P. N. (1961). *Biochem. J.* **80**, 8P.
Delhumeau de Ongay, G., Moulé, Y. and Frayssinet, C. (1965). *Expl Cell Res.* **38**, 187.
Dodgson, K. S., Rose, F. A., Spencer, B. and Thomas, J. (1957). *Biochem. J.* **66**, 363.
Dodgson, K. S., Spencer, B. and Thomas, J. (1955). *Biochem. J.* **59**, 29.
Dutton, G. J. and Storey, I. D. (1954). *Biochem. J.* **57**, 275.
Eichel, H. J. (1957). *J. biophys. biochem. Cytol.* **3**, 397.

El-Aaser, A. A. (1965). Ph.D. Thesis, University of London.
El-Aaser, A. A., Fitzsimons, J. T. R., Hinton, R. H., Reid, E., Klucis, E. and Alexander, P. (1966a). *Biochim. biophys. Acta* **127**. 553.
El-Aaser, A. A., Reid, E., Klucis, E., Alexander, P., Lett, J. T. and Smith, J. (1966b). *Natn. Canc. Inst. monogr.* **21**, 323.
Elliott, W. H. (1956). *Biochem. J.* **62**, 427.
Emanoil-Ravicovitch, R. and Herisson-Cavet, C. (1963). *Bull. Soc. Chim. biol.* **45**, 989.
Emmelot, P. and Bos., C. J. (1965). *Biochim. biophys. Acta* **99**, 580.
Emmelot, P. and Bos, C. J. (1966). *Biochim. biophys. Acta* **121**, 434.
Emmelot, P., Benedetti, E. L. and Rümke, Ph. (1964a). *In* "From Molecule to Cell", (P. Buffa, ed.), p. 253, Rome: Consiglio Nazionale delle Richerche.
Emmelot, P., Bos, C. J., Benedetti, E. and Rümke, Ph. (1964b). *Biochim. biophys. Acta* **90**, 126.
Emmelot, P., Mizrahi, I. J., Naccarato, R. and Benedetti, E. L. (1962). *Biochim. biophys. Acta* **12**, **177**.
Ericsson, J. L. E. (1966). *J. Histochem. Cytochem.* **14**, 366.
Ernster, L. and Jones, L. C. (1962). *J. Cell Biol.* **15**, 563.
Ernster, L. and Orrenius, S. (1965). *Fed. Proc.* **24**, 1190.
Ernster, L., Siekevitz, P. and Palade, G. E. (1962). *J. Cell Res.* **15**, 541.
Essner, E. and Novikoff, A. B. (1962). *J. Cell Biol.* **15**, 289.
Fawcett, D. W. (1955). *J. natn. Cancer Inst.* **15**, Suppl. 5, 1475.
Feuer, G., Golberg, L. and Gibson, K. I. (1965). *Biochem. J.* **97**, 29P.
Fouts, J. R. (1961). *Biochem. biophys. Res. Commun.* **6**, 373.
Fouts, J. R. and Rogers, L. A. (1965). *J. Pharmacol. exp. Therapy* **147**, 112.
Fouts, J. R. (1963). *Adv. Enzyme Regulation* **1**, 225.
Franz, I. D., Davidson, A. G., Dulit, E. and Mobberly, M. L. (1959). *J. biol. Chem.* **234**, 2290.
Gauguly, J. and Devel, H. J. (1953). *Nature, Lond.* **172**, 120.
Gilette, J. R. (1963). *Adv. Enzyme Regulation* **1**, 215.
Gilette, J. R., Brodie, B. B. and La Du, B. N. (1957). *J. Pharmac. exp. Therap.* **119**, 532.
Giuditta, A. and Strecker, H. J. (1959). *J. Neurochem.* **5**, 50.
Ghosh, N. C., Kar, N. C. and Chatterjee, I. (1963). *Nature, Lond.* **197**, 596.
Godson, G. N., Hunter, G. D. and Butler, J. A. V. (1961). *Biochem. J.* **81**, 59.
Goldfischer, S., Essner, E. and Novikoff, A. B. (1964). *J. Histochem. Cytochem.* **12**, 72.
Görlich, M. and Heise, E. (1962). *Z. Naturf.* **17b**, 465.
Görlich, M. and Heise, E. (1963). *Nature, Lond.* **197**, 698.
Gosselin, L., Podber-Wagner, E. and Waltregny, A. (1962). *Nature, Lond.* **193**, 252.
Goswami, P., Barr, G. C. and Munro, H. N. (1962). *Biochim. biophys. Acta* **55**, 408.
Goswami, P. and Munro, H. N. (1962). *Biochim. biophys. Acta* **55**, **410**.
Goutier, R. and Goutier-Pirotte, M. (1955). *Biochim. biophys. Acta* **16**, 366.
Green, D. E. (1952). *J. cell. comp. Physiol.* **39**, Suppl. 2, 75.
Green, D. E., Murer, E., Hultin, H. O., Richardson, S. H., Salmon, B., Brierly, G. P. and Baum, H. (1965). *Archs. Biochem. Biophys.* **112**, 635.
Green, D. E. and Perdue, J. F. (1966). *Proc. natln. Acad. Sci., U.S.A.*, **55**, 1295.
Hagen, P. (1955). *J. Physiol.* **129**, 50.
Haguenau, F. (1958). *Int. Rev. Cytol.* **7**, 425.
Hallinan, T. H. and Munro, H. N. (1964). *Biochim. biophys. Acta* **80**, 166.

Hallinan, T. H. and Munro, H. N. (1965). *Biochim. biophys. Acta* **108**, 285.
Hanson, H., Hermann, P. and Blech, W. (1959). *Z. physiol. Chem.* **315**, 201.
Harding, B. W. and Samuels, L. T. (1961). *Biochim. biophys. Acta* **54**, 42.
Hart, L. G. and Fouts, J. R. (1965). *Biochem. Pharmacol.* **14**, 263.
Hawtrey, A. O. and Schirren, V. (1962). *Biochim. biophys. Acta* **61**, 467.
Heise, E. and Görlich, M. (1963). *Nature, Lond.* **195**, 1311.
Heller, L. and Bargoni, N. (1950). *Arkiv. Kemi* **1**, 447.
Hendler, R. W. (1965). *Nature, Lond.* **207**, 1053, 1071.
Hers, H. G., Berthet, J., Berthet, L. and de Duve, C. (1951). *Bull. Soc. Chim. biol., Paris* **33**, 21.
Herzenberg, L. A. and Herzenberg, Leonore A. (1961). *Proc. natn. Acad. Sci., U.S.A.* **47**, 762.
Hess, E. L. and Lagg, S. E. (1963). *Biochemistry* **2**, 726.
Higashi, T. and Peters, T. (1963a). *J. biol. Chem.* **238**, 3945.
Higashi, T. and Peters, T. (1963b). *J. biol. Chem.* **238**, 3952.
Hilf, R., Breuer, C. and Borman, A. (1961). *Archs Biochem. Biophys.* **94**, 319.
Hird, H. J., McLean, E. J. T. and Munro, H. N. (1964). *Biochim. biophys. Acta* **87**, 219.
Hochstein, P. and Ernster, L. (1964). *In* "Cellular Injury", Ciba Found. Symp. p. 123, J. & A. Churchill, London.
Holtzer, R. L. and van Lancker, J. L. (1963). *Archs Biochem. Biophys.* **101**, 439.
Hosie, R. J. A. (1965). *Biochem. J.* **96**, 404.
Hübscher, G. (1962). *Biochim. biophys. Acta* **57**, 555.
Hübscher, G., West, G. R. and Brindley, D. N. (1965). *Biochem. J.* **97**, 629.
Hulsman, H. A. M. (1961). *Biochim. biophys. Acta* **54**, 1.
Hultin, T. (1957). *Expl Cell Res.* **12**, 290.
Hunter, A. R. and Korner, A. (1966). *Biochem. J.* **100**, 73P.
Jacobson, K. B. and Kaplan, N. O. (1957). *J. biophys. biochem. Cytol.* **3**, 31.
Järnefelt, J. (1964). Abstracts, *6th Int. Congr. Biochem.* 613.
Jones, A. L. and Fawcett, D. W. (1966). *J. Histochem. Cytochem.* **14**, 215.
Jungblut, P. W. (1963). *Biochem. Z.* **337**, 267 (see also 285).
Kamat, V. B. and Wallach, D. F. H. (1965). *Science, N.Y.* **148**, 1343.
Kamin, H., Masters, B. S. S., Gibson, Q. H. and Williams, C. H. (1965). *Fed. Proc.* **24**, 1164.
Kanfer, J., Burns, J. J. and Ashwell, G. (1959). *Biochim. biophys. Acta* **31**, 556.
Kar, N. C., Chatterjee, I. B., Ghosh, N. C. and Guha, B. C. (1962). *Biochem. J.* **84**, 16.
Kiessling, K. H. and Tilander, K. (1960). *Biochim. biophys. Acta* **43**, 335.
Kono, T. and Colowick, S. P. (1961). *Archs Biochem. Biophys.* **93**, 520.
Kornberg, A. and Pricer, W. E. (1954). *J. biol. Chem.* **204**, 345.
Krisch, K. (1963). *Biochem. Z.* **337**, 531.
Ku, K-Y., and Wang, C-T. (1963). *Acta Biol. exp. Sinica* **8**, 400. Cited (1964) *Chem. Abs.* **61**, 2123.
Kuff, E. L. and Dalton, A. J. (1957). *J. Ultrastr. Res.* **1**, 62.
Kuff, E. L. and Dalton, A. J. (1959). *In* "Subcellular Particles" (Hayashi, M. ed.), p. 114. Ronald Press, New York.
LaBella, F. S. and Brown, J. H. U. (1959). *J. biophys. biochem. Cytol.* **5**, 17.
Landon, E. J. and Norris, J. L. (1963). *Biochim. biophys. Acta* **71**, 266.
Lands, W. E. M. and Merkl, I. (1963). *J. biol. Chem.* **238**, 898.
Lea, M. A. and Walker, D. G. (1964). *Biochem. J.* **91**, 417.
Leadbeater, L. and Davies, D. R. (1964). *Biochem. Pharmacol*, **13**, 1607.

Leaf, A. (1960). *J. gen. Physiol.* **43**, Suppl. 1, 175.
Leloir, L. F. and Goldemberg, S. H. (1960). *J. biol. Chem.* **235**, 919.
Leonhäuser, S., Leybold, K., Krisch, K., Staudinger, Hj., Gale, P. H., Page, A. C. and Folkers, K. (1962). *Archs Biochem. Biophys.* **96**, 580.
Locke, M. (editor) (1964). "Cellular Membranes in Development", 382 pp. Academic Press, New York and London.
Loenig, U. E. (1961). *Biochem. J.* **81**, 254.
Luck, D. J. (1961). *J. biophys. biochem. Cytol.* **10**, 195.
Lundkvist, U. and Perlman, P. (1965). Abstracts, *2nd Mtg. Fed. Europ. Biochem. Socs.* p. 237.
Marsh, J. B. and James, A. T. (1962). *Biochim. biophys. Acta* **60**, 320.
Martin, E. M. and Morton, R. K. (1956). *Biochem. J.* **62**, 696.
Martin, E. M. and Morton, R. K. (1957). *Biochem. J.* **65**, 404.
Martonosi, A. and Feretos, R. (1964). *J. biol. Chem.* **239**, 648.
Mason, H. S., North, J. C. and Vaneste, M. (1965). *Fed. Proc.* **24**, 1172.
Maxwell, E. S. and Ashwell, G. (1953). *Archs Biochem. Biophys.* **43**, 389.
McCann, W. P. (1957). *J. biol. Chem.* **226**, 15.
McLeod, R. M., Farkas, W., Fridovitch, I. and Handler, P. (1961). *J. biol. Chem.* **236**, 1841.
Meister, A. (1952). *Science, N.Y.* **115**, 521.
Michell, R. H. and Hawthorne, J. N. (1965). *Biochem. biophys. Res. Commun.* **21**, 333.
Mizrahi, I. J. (1965). *Biochim. biophys. Acta* **108**, 419.
Molnar, J. and Lorand, L. (1962). *Archs Biochem. Biophys.* **98**, 356.
Morais, R. and de Lamirande, G. (1965). *Biochim. biophys. Acta* **95**, 40.
Morton, R. K. (1954a). *Biochem. J.* **57**, 231.
Morton, R. K. (1954b). *Biochem. J.* **57**, 595.
Moulé, Y. (1964). *In* "Cellular Membranes in Development" (M. Locke, ed.), p. 97. Academic Press, New York.
Moulé, Y., Bouvet, C. and Chauveau, J. (1963). *Biochim. biophys. Acta* **68**, 112.
Moulé, Y., Rouiller, C. and Chauveau, J. (1960). *J. biophys. biochem. Cytol.* **7**, 547.
Mullins, L. J. (1960). *J. gen. Physiol.* **43**, Suppl. 1, 105.
Muscatello, U., Anderson-Cedergren, E., Azzone, G. F. and Decken, A. von der (1961). *J. biophys. biochem. Cytol.* **10**, Suppl., 201.
Myers, L. T. and Worthen, H. G. (1961). *Fed. Proc.* **20**, 218.
Nebert, D. W. and Mason, W. S. (1964). *Biochim. biophys. Acta* **86**, 415.
Neu, H. C. and Heppel, C. A. (1964). *Biochem. biophys. Res. Commun.* **17**, 215.
Neville, D. N. (1960). *J. biophys. biochem. Cytol.* **8**, 413.
Nordlie, R. C. and Arion, W. H. (1964). *J. biol. Chem.* **239**, 1680.
Novikoff, A. B. (1960). *In* "Cell Physiology of Neoplasia", M. D. Anderson Hospital and Tumor Institute, p. 219. Univ. of Texas Press, Austin.
Novikoff, A. B. (1964). Abstracts, *6th Int. Congr. Biochem.* 609.
Novikoff, A. B. and Essner, E. (1962). *Fed. Proc.* **21**, 1130.
Novikoff, A. B. and Goldfischer, S. (1961). *Proc. natn. Acad. Sci., U.S.A.* **47**, 802.
Novikoff, A. B. and Heus, M. (1963). *J. biol. Chem.* **238**, 710.
Novikoff, A. B., Podber, E., Ryan, J. and Noe, E. (1953). *J. Histochem. Cytochem.* **1**, 27.
Omachi, A., Barnum, C. P. and Glick, D. (1948). *Proc. Soc. exp. Biol. Med.* **67**, 133.
Omura, T., Sato, R., Cooper, D. Y., Rosenthal, O. and Estabrook, R. W. (1965). *Fed. Proc.* **24**, 1181.
Orrenius, S., Ericsson, J. L. E. and Ernster, L. (1965). *J. Cell Biol.* **25**, 627.

Packer, L. and Rahman, M. M. (1962). *Texas Repts. Biol. Med.* **20**, 414.
Palade, G. E. and Siekevitz, P. (1956). *J. biophys. biochem. Cytol.* **2**, 171.
Palade, G. E. and Siekevitz, P. (1958a). *J. biophys. biochem. Cytol.* **4**, 309 (see also 203).
Palade, G. E. and Siekevitz, P. (1958b). *J. biophys. biochem. Cytol.* **4**, 557.
Persijn, J.-P., Daems, W. Th., de Man, J. C. H. and Meijer, A. E. F. H. (1961). *Histochemie* **2**, 372.
Petermann, M. L., Mizen, N. A. and Hamilton, M. G. (1956). *Cancer Res.* **16**, 496.
Peters, T. (1962). *J. biol. Chem.* **237**, 1181 (see also 1186).
Petrack, B., Greengard, P., Craston, A. and Kalinsky, H. J. (1963). *Biochem. biophys. Res. Commun.* **13**, 472.
Petrovic, S., Becarevic, A. and Petrovic, J. (1965). Abstracts, *2nd Mtg. Fedn. Europ. Biochem. Socs.* p. 331.
Phillips, A. H. and Langdon, R. G. (1962). *J. biol. Chem.* **237**, 2652.
Pitot, H. C. (1962). *Fed. Proc.* **21**, 1124.
Pitot, H. C. (1964). *Perspect. Med. Biol.* **8**, 50.
Pollak, J. K. and Shorey, C. D. (1964). *Biochem. J.* **93**, 36C.
Porteous, J. W. and Clark, B. (1965). *Biochem. J.* **96**, 159.
Quastel, J. H. (1964). *Can. J. Biochem.* **42**, 907.
Razzell, W. E. (1961). *J. biol. Chem.* **236**, 3028.
Redman, C. M., Siekevitz, S. and Palade, G. E. (1966). *J. biol. Chem.* **241**, 1150.
Reid, E. (1956). *J. Endocrinol.* **13**, 319.
Reid, E. (1961a). *Biochim. biophys. Acta* **49**, 218.
Reid, E. (1961b). *In* "Biochemists' Handbook" (Long, C., ed.), p. 814. E. F. N. Spon, London.
Reid, E. (1962). *Cancer Res.* **22**, 398.
Reid, E. (1965). "Biochemical Approaches to Cancer". Oxford, Commonwealth & Internat. Library.
Reid, E., El-Aaser, A. A., Turner, M. K. and Siebert, G. (1964). *Z. physiol. Chem.* **339**, 145.
Reid, E. and Stevens, B. M. (1958). *Nature, Lond.* **182**, 441.
Remmer, H. (1962). *In* "Enzymes and Drug Action" (J. L. Mongar and A. de Reuck, eds.), p. 398, Ciba Found. Symp., J. & A. Churchill, London.
Remmer, H. and Merker, H. J. (1963). *Science, N.Y.* **142**, 1657.
Robertson, J. D. (1959). *Biochem. Soc. Symp.* **16**, 3.
Roodyn, D. B. (1965). *Int. Rev. Cytol.* **18**, 99.
Rosenthal, O., Gottlieb, B., Gorry, J. D. and Vars, H. M. (1956). *J. biol. Chem.* **223**, 469.
Roth, J. S. (1960). *J. biophys. biochem. Cytol.* **8**, 665.
Rothschild, J. A. (1961). *Fed. Proc.* **20**, 145.
Rouiller, Ch. (1964). *In* "The Liver" (Ch. Rouiller, ed.), Vol. II. Academic Press, New York and London.
Ryter, A. and Jacob, F. (1966). *Ann. Inst. Pasteur* **110**, 53.
Sabatani, D. D., Bensch, V. G. and Barrnett, R. J. (1963). *J. Cell Biol.* **17**, 19.
Sabatani, D. D. Miller, F. and Barrnett, R. J. (1964). *J. Histochem. Cytochem.*, **12**, 57.
Sachs, H. (1958). *J. biol. Chem.* **233**, 650.
Schneider, W. C. (1963). *J. biol. Chem.* **238**, 3572.
Schneider, W. C. and Kuff, E. L. (1954). *Am. J. Anat.* **94**, 209.
Scholefield, P. G. (1964). *Can. J. Biochem.* **42**, 917.
Schotz, M. C., Rice, L. I. and Alfin-Slater, R. B. (1954). *J. biol. Chem.* **207**, 665.

Schwartz, A., Bachelard, H. S. and McIlwain, H. (1962). *Biochem. J.* **84**, 626.
Sellinger, O. Z. and Verster, F. de B. (1962). *J. biol. Chem.* **237**, 2836.
Senior, J. R. and Isselbacher, K. J. (1961). *Fed. Proc.* **20**, 245.
Share, L. and Hansrote, R. N. (1960). *J. biophys. biochem. Cytol.* **7**, 239.
Shibko, S. and Tappel, A. L. (1964). *Archs Biochem. Biophys.* **106**, 259.
Shull, K. H. (1959). *Nature, Lond.* **183**, 259.
Shulman, H. M. and Bonner, D. M. (1962). *Proc. natn. Acad. Sci. U.S.A.*, **48**, 53.
Siebert, G. and Humphrey, G. B. (1965). *Adv. Enzymol.* **27**, 239.
Siekevitz, P. (1962). *In* "The Molecular Control of Cellular Activity" (J. M. Allen, ed.), p. 143. New York: McGraw-Hill.
Siekevitz, P. (1955). *J. biophys. biochem. Cytol.* **1**, 477.
Siekevitz, P. (1965). *Fed. Proc.* **24**, 1153.
Siperstein, M. D. and Fagan, V. M. (1964). *Adv. Enzyme Regulation* **2**, 249.
Sjöstrand, F. S. (1963). *J. Ultrastr. Res.* **9**, 340.
Skipski, V. P., Barclay, M., Archibald, F. M., Terebus-Kekish, O., Reichman, E. S. and Good, J. J. (1965). *Life Sci.* **4**, 1673.
Smith, J. C., Foldes, V. and Foldes, F. F. (1960). *Fed. Proc.* **19**, 260.
Spiro, M. J. and Ball, E. G. (1961). *J. biol. Chem.* **236**, 231.
Stein, Y., Tietz, A. and Shapirc, B. (1957). *Biochim. biophys. Acta* **26**, 1256.
Stetten, M. R. (1964). *J. biol. Chem.* **239**, 3576.
Stoeckenius, W. (1964). Abstracts, *6th Int. Congr. Biochem.* 603.
Straus, W. (1962). *Expl Cell Res.* **27**, 80.
Strittmatter, P. (1965). *Fed. Proc.* **24**, 1156.
Strittmatter, P. and Velick, S. F. (1956). *J. biol. Chem.* **221**, 253.
Strominger, J. L., Maxwell, E. S., Axelrod, J. and Kalckar, H. M. (1957). *J. biol. Chem.* **224**, 79.
Struck, J. and Sizer, I. W. (1960). *Arch. Biochem.* **90**, 22.
Suzuki, K., Mano, Y. and Shimazono, N. (1960). *J. Biochem. Tokyo* **47**, 846.
Swanson, P. D., Bradford, H. F. and McIlwain, H. (1964). *Biochem. J.* **92**, 235.
Takanami, M. (1959). *J. Histochem. Cytochem.* **7**, 126.
Tanaka, R. and Strickland, K. P. (1965). *Archs Biochem. Biophys.* **111**, 583.
Tashiro, Y. (1957). *Acta Scholae Med. Univ. Kyoto* **34**, 238.
Tashiro, Y. (1958). *J. Biochem. Tokyo* **45**, 937.
Tashiro, Y. and Ogura, M. *Acta Scholae Med. Univ. Kyoto* **34**, 267.
Tedeschi, H., James, J. M. and Anthony, V. (1963). *J. Cell Biol.* **18**, 503.
Tepperman, J. and Tepperman, H. M. (1960). *Pharmacol. Revs.* **12**, 301.
Terroine, T. (1961). *Archs Sci. physiol.* **15**, 167.
Tice, L. W. and Barrnett, R. J. (1962). *J. Histochem. Cytochem.* **10**, 754.
Tomkins, G. M. (1959). *Ann. N.Y. Acad. Sci.* **82**, 836.
Toschi, G. (1959). *Expl Cell Res.* **16**, 232.
Trivus, R. H. and Spirtes, M. A. (1964). *Biochem. Pharmacol.* **13**, 1679.
Underhay, E., Holt, S. J., Beaufay, H. and de Duve, C. (1956). *J. biophys. biochem. Cytol.* **2**, 635.
Wachstein, M. (1963). *In* "The Liver" (Rouiller, Ch., ed.), Vol. I. p. 137. Academic Press, New York and London.
Wachstein, M. and Fernandez, C. (1964). *J. Histochem. Cytochem.* **12**, 40.
Wakid, N. W. (1960). *Biochem. J.* **76**, 88. (see also 95).
Walker, P. G. (1952). *Biochem. J.* **51**, 223.
Wallach, D. F. H. and Kamat, V. B. (1964). *Proc. natn. Acad. Sci. U.S.A.* **52**, 721.
Wallach, D. F. H. and Ullrey, D. (1964). *Biochim. biophys. Acta.* **88**, 620.
Wang, D. Y. (1962). *Biochem, J.*, **83**, 633.

Watson, M. L. (1960). *In* "Cell Physiology of Neoplasia", p. 129. M. D. Anderson Hospital & Tumor Institute. Univ, of Texas Press, Austin.

Weber, G., Singhal, R. L., Stamm, N. B., Fisher, E. A. and Mentendiek, M. A. (1964). *Adv. Enzyme Regulation* **2**, 1.

Weinstein, H., Roberts, E. and Kakefuda, T. (1963). *Biochem. Pharmacol.* **12**, 503.

Westrop, J. W. and Green, H. N. (1960). *Nature, Lond.* **186**, 350.

Whaley, W. G., Mollenhauer, H. H. and Leech, J. H. (1960), *Am. J. Bot.* **47**, **401**.

Whittaker, V. P. and Whittam, R. (1964). Abstracts, 6*th Int. Congr. Biochem.* 611.

Whittaker, V. P. (1963). *Biochem. Soc. Symp.* **23**, 109.

Whittaker, V. P., Michaelson, I. A. and Kirkland, R. J. A. (1964). *Biochem. J.* **90**, 293.

Whittam, R. (1964). Abstracts, 6*th Int. Congr. Biochem.* 611.

Wilgram, G. F. and Kennedy, E. P. (1963). *J. biol. Chem.* **238**, 2615.

Winkelman, J. and Lehninger, A. L. (1958). *J. biol. Chem.* **233**, 794.

Wolfe, L. S. (1964). *Can. J. Biochem.* **42**, 971.

Woodin, A. M. and Wieneke, A. A. (1966). *Biochem. J.* **99**, 493.

Wu, C. (1963). *Biochim. biophys. Acta* **77**, 482.

Yamada, K., Ishikawa, S. and Spimazono, N. (1959). *Biochim. biophys. Acta* **32**, 253.

Chapter 7

RIBOSOMAL ENZYMES†

DAVID ELSON

The Weizmann Institute of Science
Rehovoth, Israel

† This review was written during the tenure of research grants (GM 12588) from the U.S. Public Health Service, and (GB-1163) from the U.S. National Science Foundation.

I. Introduction

It has been hardly more than ten years since several broad lines of research, cytological, cytochemical, and biochemical, converged to give us our present concept of the ribosome and its place in the economy of the cell. The earlier cytological studies were culminated by the simultaneous and independent recognition by Brachet (1942, 1960) and Caspersson (1941) that the capacity of a cell to synthesize proteins was directly related to a substance with a high affinity for basic dyes, the basophilia, and that the basophilic material was RNA. This first pronouncement of the involvement of RNA in protein biosynthesis became a guiding principle of fundamental importance, setting the course of research for investigators in many disciplines.

At about this time Claude's discovery of the RNA-rich microsome (one of the most valuable artifacts of modern biology) opened a series of investigations by himself, Porter, Palade, Siekevitz and many others (see, e.g., Palade, 1958; Hogeboom and Schneider, 1955; Dounce, 1955) which led, through the separate or combined use of the then new techniques of cell fractionation and electron microscopy, to the development of a new cytology at a higher level of resolution. One consequence was the visualization of the endoplasmic reticulum, a cytoplasmic network of tubules and vesicles formed of lipid-containing membranes and found in a great variety of differentiated cells. It was later seen that portions of the reticulum were often covered with dense particles and that these particles were also found free in the cytoplasm both of such cells and of cells which contained no endoplasmic reticulum. (Fig. 1). These particles were shown to be made up of RNA and protein and were eventually given the name of "ribosome" (Roberts, 1958). Microsomes were recognized as heterodisperse fragments of the endoplasmic reticulum, formed when the cell was broken and consisting of two elements: pinched off segments of the reticular membrane, and ribosomes. When the membrane is dissolved (usually with sodium deoxycholate), pure ribosomes may be obtained. It is these particles which constitute the basophilic substance concerned with the synthesis of proteins, and they are located in both the nucleus and the cytoplasm.

In the meantime, similar particles had been isolated from many forms of living matter, and rapid progress was made in elucidating their physical, chemical and biological properties (McQuillen, 1962; Petermann, 1964),

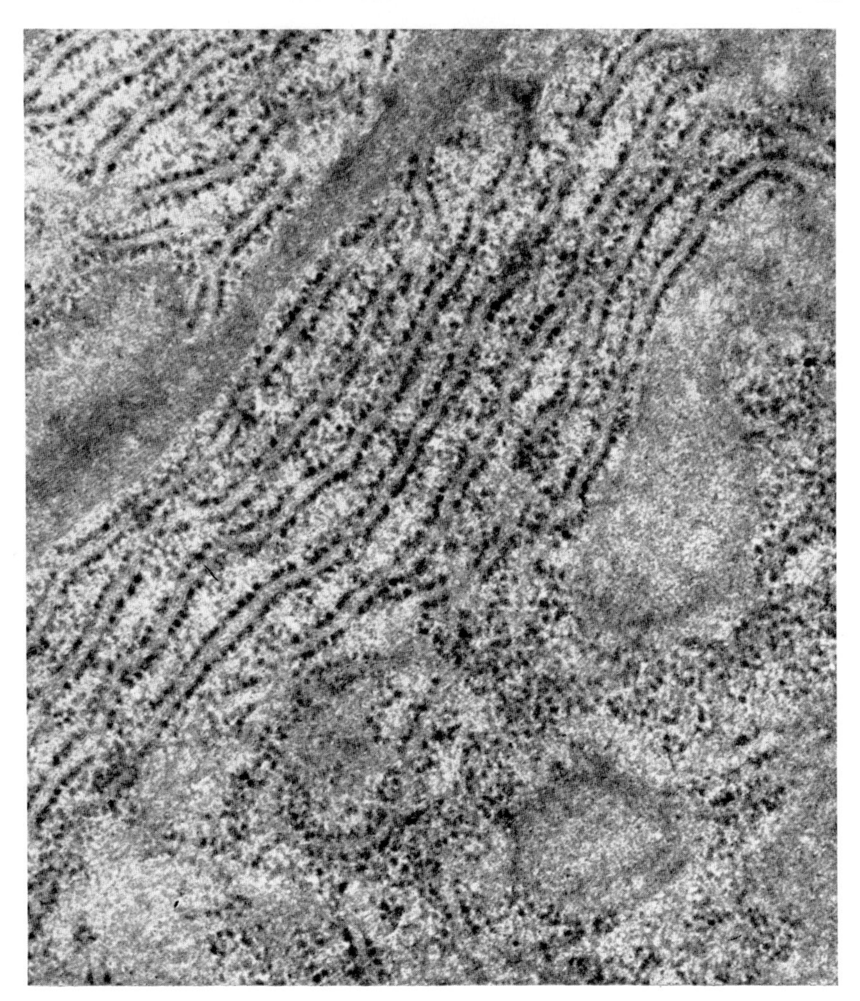

FIG. 1. Ribosomes and endoplasmic reticulum in a thin section of rat liver fixed with glutaraldehyde and osmium tetroxide and stained with uranyl acetate. The magnification is $\times$ 66,000. Electron micrograph by courtesy of Prof. D. Danon.

It is quite clear today that the ribosome occupies a central position in the biological production of proteins, and there is no substantial evidence that it has any other function.

It is my task to discuss in this chapter those enzymes which have been found to be associated with ribosomes, and a great many such enzymes are known. Ideally, this discussion should lead to a set of generalizations and principles by which the known data can be grouped, correlated and explained in terms of ribosomal function. Unfortunately, this cannot yet be done except on a very restricted scale. Most of the citations of ribosomal enzymes found in the literature are rather casual observations, often suggestive but hardly decisive, while many of the more extensive and critical studies have provided more questions than answers. The situation is, simply, that the field of ribosomal enzymes is still at a very early stage of development and some years must pass before a definitive treatise can be written on the subject. This review is therefore an interim report in which I shall try to discuss the present status of the field and point out certain inadequacies and unsolved problems.

In the following sections I shall discuss the structure and properties of the the ribosome, the nature of the process of protein biosynthesis, the place of the ribosome in this process, and finally, the various types of enzyme which have been found associated with ribosomes. At the end of the chapter I have listed all of the ribosomal enzymes of which I am aware (some have undoubtedly been overlooked). This list was not compiled in a critical frame of mind. It contains, among others, many citations based simply on the observation of enzymic activity in crude unwashed preparations of ribosomes. The single selective criterion applied was that the particle be reasonably well identified as a ribosome; references to microsomal enzymes, for example, have been omitted.

II. The Structure of the Ribosome

A. The Isolation of Ribosomes

As has been pointed out above, the ribosome is a ubiquitous particle found in virtually every cell, sometimes attached to a membrane and sometimes apparently free in the cytoplasm or nucleoplasm. Although techniques such as chemical purification (Parsons, 1953; Elson, 1959a) and ion exchange chromatography (Peterson and Kuff, 1961; McCarthy, 1961, cf. Roberts, 1964, p. 384) have been employed, the standard procedure for isolating ribosomes involves differential centrifuging in a preparative ultracentrifuge (e.g., Chao and Schachman, 1956; Petermann, 1964; Tissières, Watson, Schlessinger and Hollingworth, 1959). Cells are broken in an appropriate buffered medium and are centrifuged to remove debris, nuclei, mitochondria and any other structures which sediment more rapidly than microsomes and ribosomes. If microsomes are present, they are sedimented and their membranes are dissolved, or at least disrupted, usually with sodium deoxycholate, liberating the attached ribosomes. The ribosomes are then sedimented at high speed (e.g., for one or two hr at 100,000 *g*) and re-suspended, and aggregated material is spun down at low speed (10,000–30,000 *g* for 10 to 30 min). Purification is accomplished by repeating the cycle of high and low speed centrifuging as many times as is felt proper.

Sometimes further purification is achieved by precipitating the particles with Mg^{2+} salts, exposing them to relatively high ionic strengths, centrifuging them through high concentrations of sucrose, etc. (Petermann, 1964). The choice of the purification procedure will, of course, depend on the use to which the ribosomes are to be put, the nature of the impurities which are to be removed, the patience of the investigator, and so forth. Often a balance must be found between purification and over-handling of the ribosomes, which may destroy their usefulness, particularly in biological experiments. In general, there is no absolute standard of purity. Each preparation must be judged by an operational criterion: Is the preparation suitable for the experiment in which it will be employed?

B. PROPERTIES OF RIBOSOMES

When isolated in the usual way, the ribosome is a particle which is made up of two sub-particles, one about twice the size of the other. In the bacterium *Escherichia coli*, one of the most common sources of ribosomes for the biochemist, the sub-particles have sedimentation constants of 30 S and 50 S. As seen in the electron microscope, their dimensions are 95 × 170 Å for the asymmetric 30 S particle and 140 × 170 Å for the more nearly spherical 50 S particle (Hall and Slayter, 1959). Their masses are about 0·8–1·0 × 10^6 and 1·8–2·0 × 10^6, and each sub-particle consists of about 60–65% RNA by weight, the rest being protein.

When handled *in vitro* under controlled conditions, the particles can be made to undergo well defined association and dissociation reactions. Thus, conditions can be found where the 30 S and 50 S particles do not combine. If now the concentration of magnesium ions, for example, is raised, a 30 S and a 50 S particle will associate to form a single 70 S particle; and if the Mg^{2+} concentration is raised still further, two 70 S ribosomes will join to form a dimer of sedimentation constant 100 S (Figs. 2 and 3). In the test tube

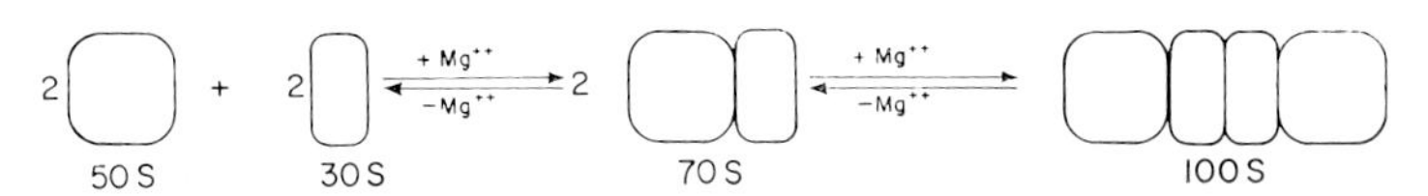

FIG. 2. The different states of association of *E. coli* ribosomes.

these association reactions are freely reversible, within certain limits, and the state of association of the ribosomes depends on their present environment and not on their past state of association. The state of association is governed not only by the concentration of magnesium ions but also by such factors as ionic strength, pH and temperature. The divalent cation dependence seems to be relatively specific; of several metals tested, only Ca^{2+} was found to be able to replace Mg^{2+}, and was far less effective (Chao and Schachman, 1956). These association reactions are not to be thought of as the random aggregation of macromolecular complexes. There appear to be specific sites on the surfaces of the particles at which interaction occurs; it is unlikely that the

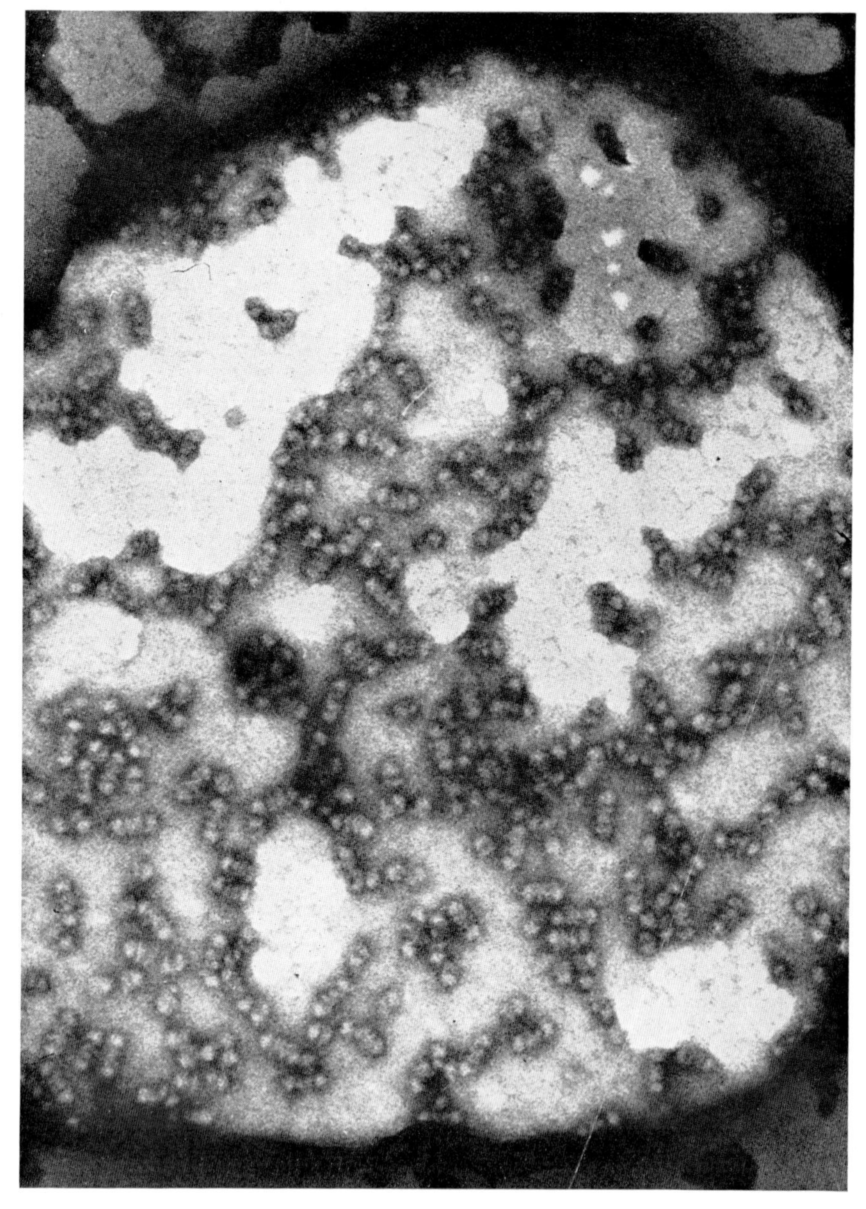

FIG. 3. An electron micrograph of *E. coli* ribosomes negatively stained with phosphotungstic acid and magnified $\times 133{,}000$. The ribosomes are almost entirely in the form of 70 S and 100 S particles. By courtesy of Prof. D. Danon and Mr. Y. Marikovsky. Reproduced with permission from Elson (1964 c).

100 S particles would present such a uniform appearance if this were not so (Fig. 3).

It is clear, of course, that the internal environment of the living cell is not identical to that of a test tube; and it is not yet clear to what extent the association-dissociation reactions observed *in vitro* mirror processes of physiological significance *in vivo*. The 70 S ribosome is the particle which participates directly in the synthesis of proteins. However, all of the degrees of association mentioned above have also been observed in untreated whole cell homogenates, and a correlation has been noted between the physiological state of cells and the state of association of their ribosomes (McCarthy, 1960). It is possible, therefore, that the association-dissocation reactions of ribosomes are of biological significance.

It should be remembered that the particles discussed here associate by direct interaction with each other. They should not be confused with another type of ribosomal aggregate, the polysome, which is distinctly different in nature. In the polysome, which will be discussed below, a number of 70 S particles are held together not because they interact with each other but because they are apparently attached to the same strand of RNA, presumably a strand of mRNA.

The specific interactions that the ribosomes undergo in these association reactions and in the precise and reversible binding of molecules during the process of protein synthesis (see section III) must be a function of the chemical fine structure of the ribosomal surface. As yet, we know nothing of this and, indeed, we do not know very much about the gross chemical structure of the ribosome. What is known is summarized briefly in the following two sections. (For general reviews see McQuillen, 1962; Roberts, Britten and McCarthy, 1963; Petermann, 1964).

C. RIBOSOMAL RNA

The 30 S ribosome contains a single RNA chain of molecular weight 0·5–0·6 × 10^6 (1500–1800 nucleotides). The 50 S particle contains about twice as much RNA, in some cases as a single continuous strand and in others as two strands of equal length, although it is not completely certain that these two strands are not produced by the breakage of a single strand during its isolation (Kurland, 1960; Petermann, 1964). The RNA moieties of the 30 S and 50 S ribosomes are distinct chemical entities; neither is derived from the other. They have been reported to differ slightly in gross nucleotide composition (Bolton, 1959a, cf. Roberts, 1964, p. 368) and definite differences have been discerned in their nucleotide sequences (Aronson, 1962), which appear to be prescribed by different genes (Yankofsky and Spiegelman, 1963).

D. RIBOSOMAL PROTEINS

Considerably less is known about the proteins of the ribosome. To a large degree this is due to their insolubility in aqueous media under conditions where their properties can be easily investigated. When the ribosomal RNA is degraded at neutral pH and low ionic strength, much of the ribosomal protein precipitates and cannot be wholly redissolved except in such inconvenient solvents as acid, alkali or concentrated urea. Although this is annoying to the chemist, it seems not unlikely that an understanding of this particular behaviour may in itself contribute to an understanding of the structure of the native ribosome.

Ribosomal proteins have been prepared in varying yield and purity (not always described) by extracting ribosomes with hot trichloroacetic acid (Crampton and Petermann, 1959) or cold 66% acetic acid (Waller and Harris, 1961; Spahr, 1962), dialysing them against urea solutions (Spahr, 1962) or concentrated neutral salt (Spitnik-Elson, 1962a) in order to allow an endogenous RNase to degrade the ribosomal RNA, or extracting them with LiCl (Curry and Hersh, 1962; Mathias and Williamson, 1964) or LiCl and urea (Leboy, Cox and Flaks, 1964; Spitnik-Elson, 1965).

The number of different amino end groups in such preparations is small, with methionine (and, in *E. coli*, also alanine) predominating (Waller and Harris, 1961; Waller, 1963, 1964; Mathias and Williamson, 1964; Petermann, 1964). The proteins appear to consist of relatively small molecules, and estimates of their molecular weight have ranged from 13,000 to 25,000, based on sedimentation data (Zillig, Krone and Albers, 1959; Curry and Hersh, 1962) and end group analyses (Waller and Harris, 1961; Mathias and Williamson, 1964).

Amino acid analyses show the ribosomal proteins of different organisms to be similar in composition (e.g. Ts'o *et al.*, 1958, summarized in Petermann, 1964). They contain large quantities of both basic and acidic amino acids. Since, however, a substantial proportion of the acidic amino acids are amidated, the proteins are predominantly basic. They appear to be somewhat less basic than the histones, which they resemble, but it is not yet clear to what extent these preparations contain proteins other than the true structural proteins of the ribosome. Consequently, a comparison of the basicities of the the histones and the ribosomal structural proteins cannot yet be made with certainty.

Although the ribosomal proteins are mainly basic, acidic molecules are also present. This was predicted on the basis of their solubility properties and later demonstrated by fractionation studies (Spitnik-Elson, 1962a, b, 1963, 1964). Acidic components have also been shown by gel electrophoresis and other techniques (Waller, 1964; Cox and Flaks, 1964; Leboy *et al.*, 1964; Duerre, 1964a). Comprising a small part of the total proteins (Spitnik-Elson, 1964), the acidic fraction can interact with the basic proteins to form insoluble electrostatic complexes which are difficult to dissociate.

It is, however, becoming increasingly apparent that the insolubility of the ribosomal proteins cannot be explained on the sole basis of electrostatic interactions (Petermann, 1964). Anomalies in the behaviour of the proteins during ion exchange chromatography and starch gel electrophoresis show that other forces are also operating; the acidic fraction, in particular, appears to exhibit a high degree of non-electrostatic aggregation (Spitnik-Elson, 1964, 1965; Duerre, 1964a).

The 30 S and 50 S ribosomes apparently contain different protein complements (Spahr, 1962; Waller, 1964; Cox and Flaks, 1964). There is general agreement that the ribosomal proteins are heterogeneous (see the preceding references, and Waller and Harris, 1961; Spitnik-Elson, 1962b, 1963, 1964; Duerre, 1964a), but the degree of heterogeneity is in doubt. Several investigators have demonstrated a very large number of protein bands in gel electrophoresis—so large, in fact, that it approaches the number of protein molecules in a ribosome (about 40 in a 70 S particle). Thus it is conceivable that no two protein molecules in one ribosome are the same. However, it must still be shown that this multiplicity of bands is not an artifact. A small number of aggregating proteins could, for example, give rise to a larger number of separable aggregates. Further, ribosomal protein preparations have been found, when examined with this in mind, to contain small amounts of RNA and non-RNA phosphorus (e.g., Spitnik-Elson, 1965). These might contribute different charges to different molecules of the same protein species and might also act as electrostatic cross-linking agents. Thus we really do not yet know how many different species of structural ribosomal protein there are.

In summary, the ribosomal proteins are small, and heterogeneous to an unknown degree. They are mainly basic but also contain acidic components. Once freed of the ribosomal RNA, they show a striking tendency to interact, both electrostatically and non-electrostatically. It seems not unreasonable to suspect that such interactions may play a decisive role in determining the structure of the intact ribosome.

E. THE STRUCTURE OF THE RIBOSOME

It was once felt that ribosomes might be similar to viruses in structure, but this is not the case. Both types of nucleoprotein have in common that they are made up of a long strand of nucleic acid and a number of relatively small protein molecules. In the virus, however, the nucleic acid forms an inner core which is insulated from the surrounding medium by an envelope of protein. In contrast, the ribosomal RNA is not covered in this way. Both the RNA and the protein of intact ribosomes are vulnerable to enzymic attack (Bolton, Hoyer and Ritter, 1958; Elson, 1959a, b; Petermann and Hamilton, 1961). Either both RNA and protein lie on the surface of the particle, or else the structure of the ribosome is open enough to allow enzymes to penetrate. In either case, the structure of the ribosome is clearly different from that of viruses.

We see, then, that the ribosome is a polynucleotide chain to which are attached a number of molecules of structural protein. That this attachment is independent of the three dimensional structure of the ribosome is suggested by recent experiments with ribosomes that had been exposed to a high concentration of salt. This treatment seems to loosen the structure of the particle, and when such ribosomes were suspended in distilled water, they showed the properties of a flexible linear polyelectrolyte. The ribosome had lost its structure and become an extended chain of RNA; however, the ribosomal proteins were still attached to it (Spirin, Kiselev, Shakulov and Bogdanov, 1963). The compact native particle would appear, then, to be formed by the coiling and folding of the chain, the exact structure being determined by specific protein-protein, RNA-RNA, and protein-RNA interactions. The marked tendency of the ribosomal proteins to aggregate in solution may be a reflection of such interactions, with the difference that in the intact ribosome the interactions are limited and controlled in such a way as to produce, together with the RNA, a specific structure.

It will be seen in the following section that the ribosome probably functions by forming a surface which presents a unique pattern of precisely aligned chemical groups. It can reasonably be assumed that the entire chemical makeup of the ribosome is adapted to this end. In this view the length and nucleotide sequence of the ribosomal RNA, the arrangement of the ribosomal proteins along the RNA strand, and the composition, sequence and folding of each ribosomal protein would be such as to ensure that these structural elements will interact specifically with each other to form a particle of unique structure and surface. This is simply to say that the forces which determine the structure of the ribosome are those which are believed to operate in other biological macromolecules such as proteins and DNA.

F. NON-BACTERIAL RIBOSOMES

This discussion has been based mainly on the ribosomes of *E. coli.* The ribosomes of most other bacteria seem to be very similar in almost all respects (Petermann, 1964). However, virtually all non-bacterial ribosomes differ from the bacterial particles in containing a somewhat larger strand of RNA and in having a higher protein content (about 50%, compared with about 40% for *E. coli*). Since the non-bacterial particles are larger, they have higher sedimentation constants: about 40 S, 60 S, 80 S and 120 S instead of the corresponding bacterial values of 30 S, 50 S, 70 S and 100 S. The particle weight of the 80 S ribosome is about $4–4{\cdot}5 \times 10^6$ as compared with about 3×10^6 for the equivalent bacterial 70 S particle. Except for this difference in size, non-bacterial ribosomes appear to be built in the same way as their bacterial counterparts. Under the same external influences they undergo the same reversible association-dissociation reactions.

Biologically, the two types of ribosome appear to be not only equivalent but even interchangeable, for there are many reports of mixed systems which are perfectly capable of effecting the incorporation of amino acids into polypeptides *in vitro*. Thus, bacterial ribosomes are capable of functioning in conjunction with non-bacterial soluble components and *vice versa*. In what follows I shall consider bacterial and non-bacterial ribosomes to be similar in structure and identical in function, and I shall not always distinguish between them.

III. The Function of the Ribosome

As noted above, the ribosome appears to have the sole function of presiding over the polymerization of amino acids into protein chains of precisely predetermined amino acid sequence. The involvement of the ribosome in this process was firmly established about a decade ago by the decisive experiments of Zamecnik and his colleagues (Littlefield, Keller, Gross and Zamecnik, 1955; Zamecnik, 1960), who followed the career of radioactive amino acids within the rat liver cell *in vivo* and showed by means of cell fractionation studies, that amino acids are first built into polypeptides while they are associated with ribosomes (Fig. 4). Many of the details of the process have since been worked out, particularly with the aid of cell-free systems which accomplish the incorporation of amino acids into polypeptides *in vitro*. This work is still in full progress today, but enough is already known to allow the construction of a coherent scheme with some measure of confidence that it is not too different from what actually takes place in nature (see, e.g., Crick, 1964).

In synthesizing protein molecules the cell employs an elaborate apparatus. The amino acids are first converted to an activated state; they are then fastened to carriers, brought to the site of polymerization (the ribosome), polymerized, and released into the cell medium. Each step is a chemical reaction catalysed by one or more specific enzymes. The early steps require different enzymes for the different amino acids, and since some twenty amino acids are involved, the number of enzymes participating is large.

In addition, there must be some provision to ensure that the various amino acids enter the protein chain in the required sequence. In considering this aspect, it is convenient to divide the steps into two groups, selective and non-selective. I shall use these terms in the following sense: a selective reaction is one which plays a direct role in determining the unique amino acid sequence of a protein; a non-selective reaction does not. In other words, both the selective and non-selective reactions are required for the polymerization of amino acids, but only the selective reactions determine that only one particular amino acid will occupy a given position in the protein chain

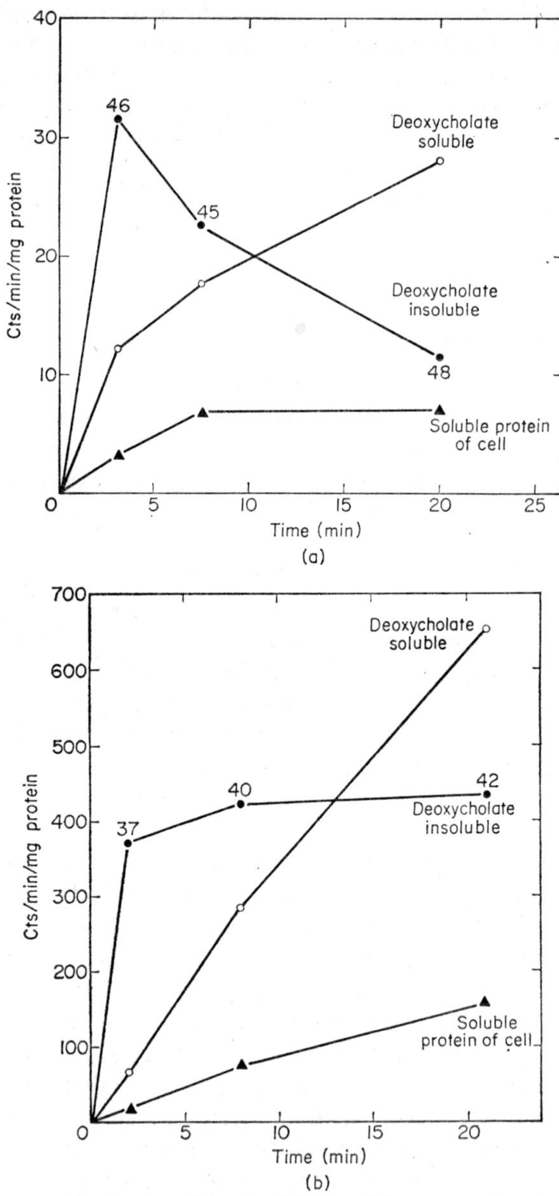

FIG. 4. The incorporation *in vivo* of leucine-^{14}C into different cytoplasmic fractions of the rat liver cell: the ribosomes ("deoxycholate insoluble"), the non-ribosomal components of the microsomes ("deoxycholate soluble"), and the soluble proteins. The RNA content (as weight per cent) of the ribosomal samples is shown. A small dose of radioactive amino acid was administered in *a* and a large dose in *b*. In both cases the ribosome-bound proteins are labelled most rapidly, after which the label passes into the other cell fractions. Reproduced with permission from Littlefield *et al.* (1955).

and nineteen others will not. In general, selectivity is a function of the nucleotide sequence of RNA, which is derived ultimately from the nucleotide sequence of the genetic material, DNA. Where an enzyme plays a selective role, it must recognize, whether directly or indirectly, a specific nucleotide sequence and relate this sequence to a specific amino acid.

Before continuing with this matter, let us return to review the biochemical reactions of protein synthesis in more detail. Figure 5 is an outline of the process. The first step is the activation of the amino acid through the enzymic formation of an anhydride bond between its carboxyl group and the phosphoryl group of the AMP moiety of ATP (Fig. 5, step 1). The same enzyme,

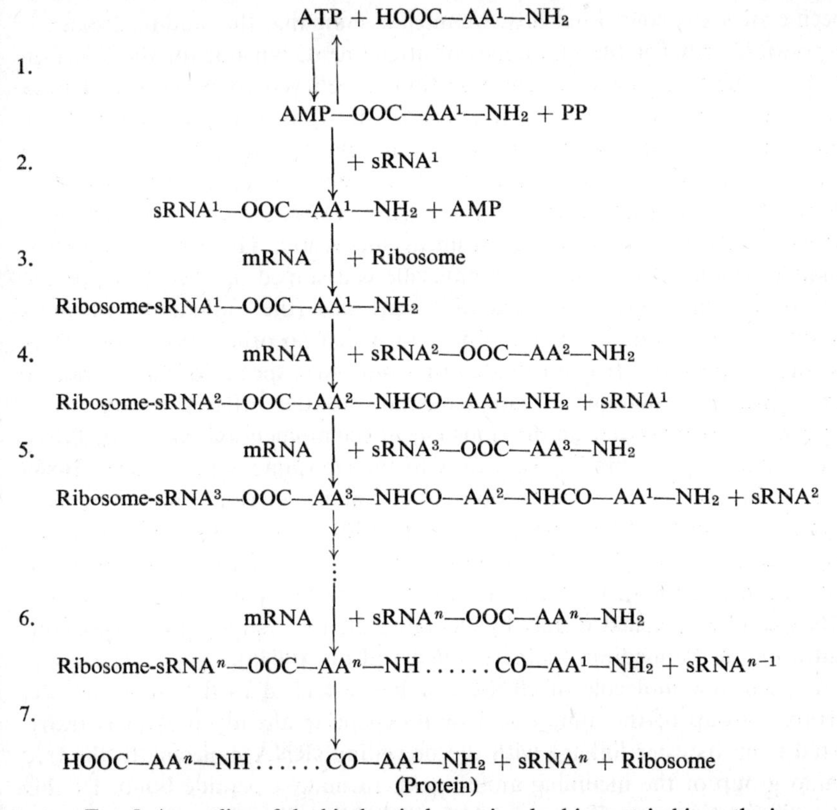

FIG. 5. An outline of the biochemical steps involved in protein biosynthesis.

apparently, then transfers the amino acid residue to a small molecule of RNA, the "soluble" or "transfer" RNA (sRNA) (Fig. 5, step 2). The attachment is again through the carboxyl group, which is esterified to the terminal ribose moiety of the sRNA. There are many different species of sRNA—probably fifty or more (since at least some of the 20 amino acids involved can be attached to more than one species of sRNA)—and each is reserved for one particular amino acid. The specificity of the responsible enzyme is such that it will act on only one species of amino acid and will attach it only to one of the sRNA molecules specific for that amino acid. This is one of the selective reactions: the pairing of specific amino acid and sRNA molecules.

Out of the pool of amino acid-charged sRNA molecules, one is now fixed to a site on the surface of the ribosome (Fig. 5, step 3). In the rabbit reticulocyte, at least, the binding appears to be catalysed by an enzyme (Arlinghaus, Shaeffer and Schweet, 1964). This too, is a selective reaction, for at any given time only one predetermined species of amino acid can be bound through its specific sRNA carrier. Logically, we must assume that the binding enzyme is responsible only for the mechanics of attachment, but not for the selection of the specific amino acid. The selection is believed to be governed by a molecule of mRNA which is attached to the ribosome. The mRNA is a complementary copy of the DNA of that gene which specifies the amino acid sequence of the protein in question. Its length is believed to consist of a series of codons (presumably adjacent trinucleotide sequences) which move across the face of the ribosome in linear succession. There are 64 different possible codons, and each sRNA molecule is assumed to have, in a specific part of the molecule, a trinucleotide sequence (the anticodon) which is complementary to one of the codons and which no other species of sRNA has in that position. It is postulated that only that species of sRNA can be bound whose anticodon is complementary to the mRNA codon which happens to be in register on the ribosome at the moment (Crick, 1958, 1964).

The nature of the binding of sRNA to the ribosome is not known. However, it involves more than codon-anticodon interaction since there is a binding site on the 50 S ribosome (Elson, 1961a, 1962a, b, 1964a; Gilbert, 1963b) and the mRNA appears to be attached to the 30 S ribosome (Okamoto and Takanami, 1963). Further, the binding is probably not covalent, for the sRNA can be detached merely by changing the ionic strength or Mg^{2+} concentration of the medium (Gilbert, 1963b; Elson, 1964a).

As each new molecule of sRNA becomes attached to the ribosome, the carboxyl group of the amino acid or polypeptide already present is transferred from its ester linkage with the preceding sRNA molecule to the free amino group of the incoming amino acid, forming a peptide bond. By this enzymic reaction (e.g., Traut and Monro, 1964; Arlinghaus *et al.*, 1964) the growing protein chain is handed from one molecule of sRNA to the next, being lengthened by one amino acid each time and always being bound to the ribosome by the sRNA molecule to which it is attached at the moment (Takanami, 1962; Gilbert, 1963b; Gierer, 1964). After a molecule of sRNA has given over the growing protein to its successor, it becomes detached from the ribosome. The process is shown in Fig. 5 for the first (step 3), second (step 4), third (step 5) and last (step 6) residues of a protein composed of *n* amino acids. Finally, the completed protein must be released from both the ribosome and the last molecule of sRNA (step 7), a reaction about which nothing is yet known except that it may require an enzyme (Morris, 1964).

I have given the above description in terms of a single ribosome interacting with a molecule of mRNA. It has, however, become apparent that proteins are usually synthesized by groups of ribosomes attached to the same molecule of mRNA and called, variously, polysomes, polyribosomes, ergosomes, clusters, etc. (Wettstein, Staehelin and Noll, 1963; Gierer, 1963; Gilbert, 1963a; Warner, Knopf and Rich, 1963; Marks, Burka and Schlessinger, 1962). It is believed that each ribosome becomes attached to the starting end of the mRNA and then moves along the successive codons. When it has gone far enough, a second ribosome is able to attach to the starting end, and so forth. All of the ribosomes in any one polysome would therefore turn out copies of the same protein, since they are being instructed by the same series of codons. The length of the mRNA would set an upper limit to the number of ribosomes in the polysome.

There is no apparent reason to believe that proteins must be made on polysomes. A single ribosome should function regardless of whether other ribosomes are accompanying it (Munro, Jackson and Korner, 1964). It may, however, be important in the economy of the cell to make the most efficient use of the mRNA. Electron micrographs of polysomes are shown in Fig. 6.

The scheme outlined here is the product both of much experimentation and much invention. It has gained general, though not universal, acceptance and it probably bears a substantial resemblance to what actually takes place in the living cell. In going through this scheme it will be seen that selectivity is not found in the ribosome. The high degree of selectivity required for the repeated manufacture of proteins of unique sequence resides in the activating enzymes which pair amino acids with molecules of sRNA in a unique way, and in the nucleotide sequences of the codons and anticodons of mRNA and sRNA.

In this scheme the ribosome cannot exercise selectivity, for it must process all species of amino acid and sRNA in the same way. Apparently, its function is to provide a surface on which the reactions take place. This surface must be capable of binding and releasing mRNA, sRNA, and a number of enzymes and cofactors in a precise steric arrangement and temporal sequence. The elucidation of its structure remains as one of the more intriguing unsolved problems in biological chemistry.

IV. Ribosomal Enzymes—General Considerations

A ribosomal enzyme would presumably be one which has a functional connection with the ribosome. In this review, however, I shall include any enzyme whose activity has been found in a preparation of ribosomes. This definition, broad to the point of absurdity, is nevertheless the most suitable for the present purpose at the present time. Since it says nothing about the purity of the preparation (except that it be a ribosomal and not, for example, a microsomal preparation), it will clearly include a number of enzymes which are simply impurities. On the other hand, our understanding of the various kinds of relationship that might exist between enzymes and ribosomes is still rudimentary. Thus, although a more restricted definition might be more

FIG. 6. (a) Polysomes *in situ*. A thin section of a rabbit reticulocyte fixed with glutaraldehyde and osmium tetroxide and stained with uranyl acetate. The magnification is ×66,000. In addition to many single ribosomes, clusters of two, three, four and five ribosomes can be seen; these are cross sections of polysomes. Electron micrograph by courtesy of Prof. D. Danon. Reproduced with permission from Elson (1964 c).

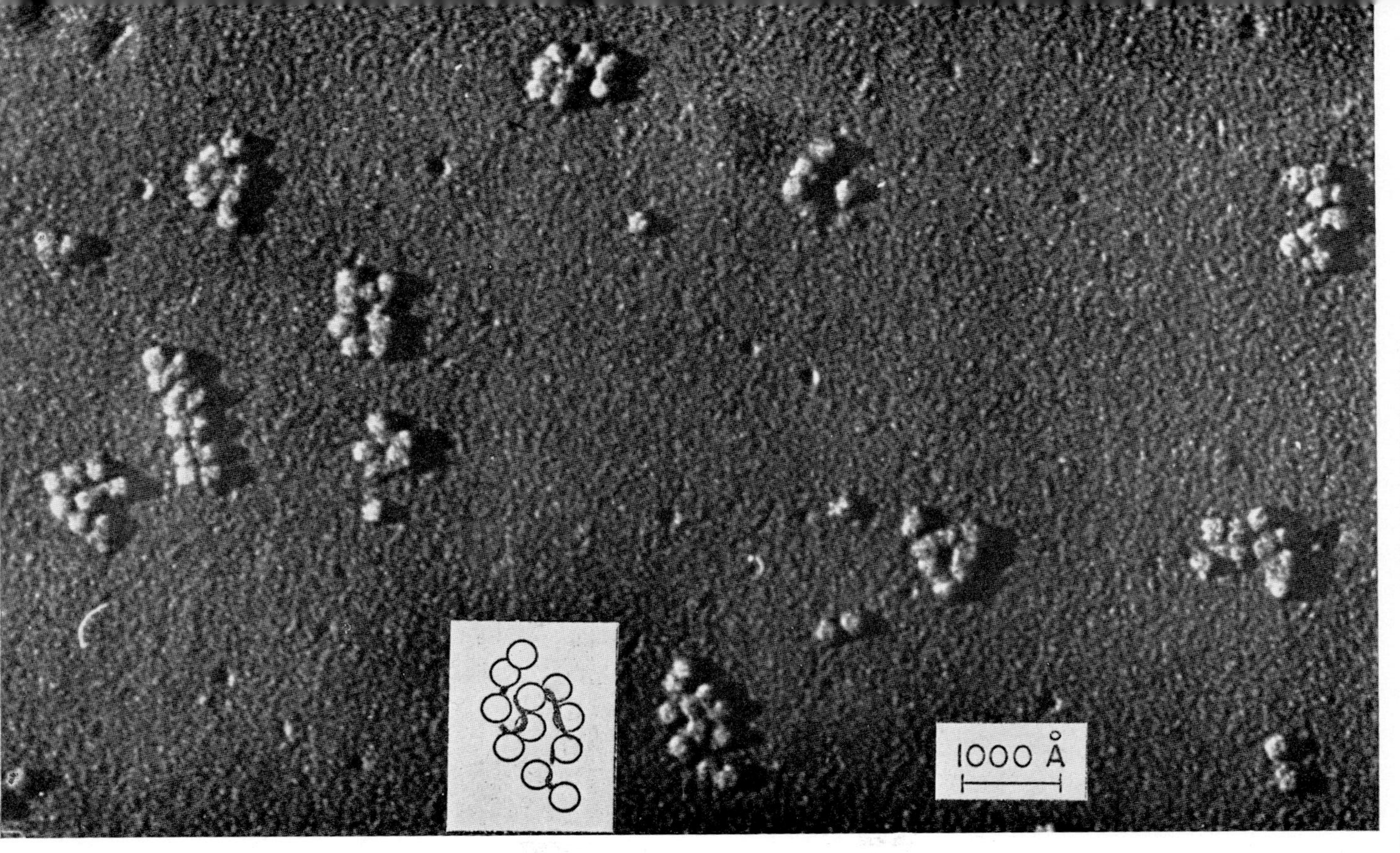

FIG. 6. (b) An electron micrograph of *E. coli* polysomes isolated by means of sucrose density gradient centrifuging, fixed with formaldehyde and shadowed with platinum. Reproduced with permission from Staehelin *et al.* (1963).

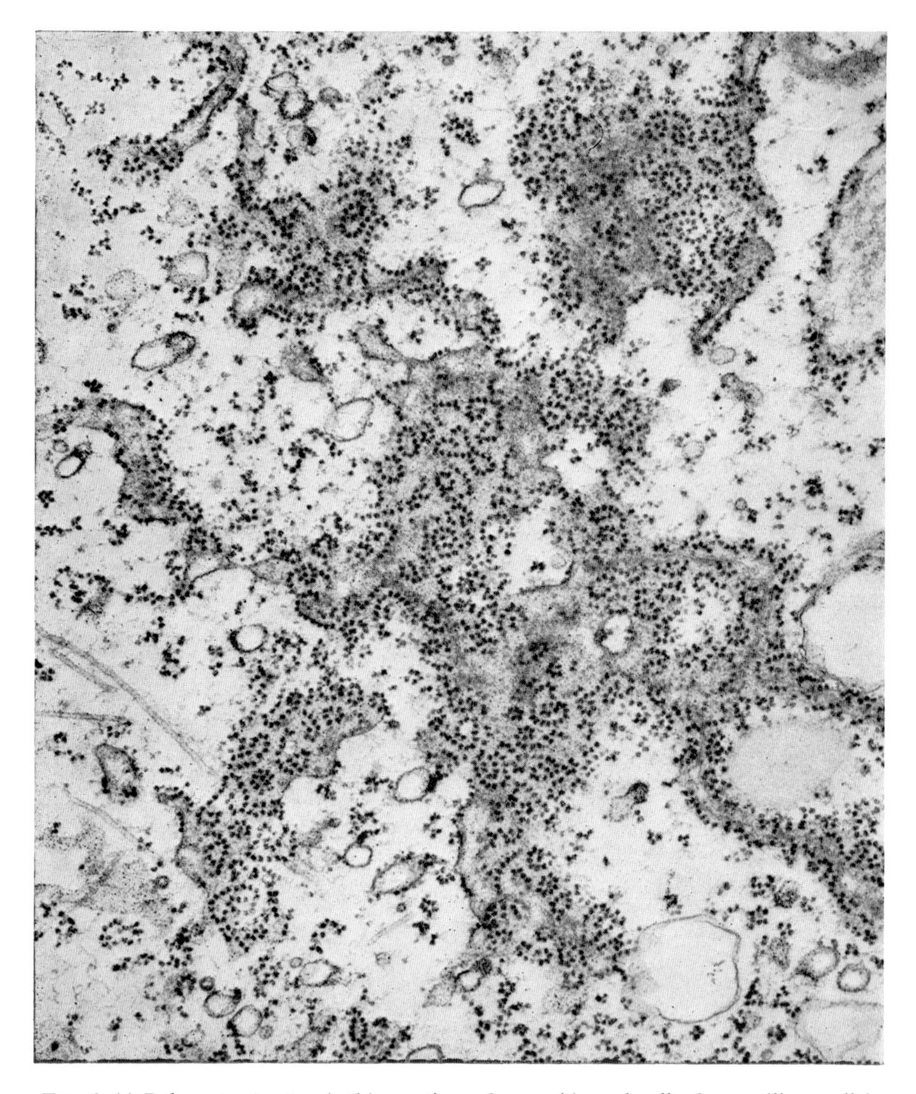

FIG. 6. (c) Polysomes *in situ*. A thin section of an epidermal cell of a seedling radish root fixed with glutaraldehyde and osmium tetroxide and stained with uranyl acetate and lead citrate. The magnification is $\times$ 37,000. Reproduced with permission from Bonnett and Newcomb (1965).

meaningful, it would also be unjustifiably arbitrary and would not cover a number of cases that, for one reason or another, should be covered.

The problem stems to a large extent from the structure of the ribosome and its place in the organization of the cell. For example, it is far less difficult, in principle if not always in practice, to define a mitochondrial enzyme, for this cell organelle is bounded by a membrane. Thus, an enzyme inside the membrane would be a mitochondrial enzyme, and one which is outside would not. Further, the location of an enzyme within the membrane is of itself a good reason for suspecting it to play a physiological role in the organelle. In practice it may sometimes be difficult to distinguish between an interior enzyme and one adsorbed on the surface of the membrane, and it may also prove necessary to stretch the definition in order to accommodate certain cases; but in any event, it is possible to employ a single broad definition as a starting point.

The ribosome, however, is not bounded by a membrane and its principal metabolic activities appear to be carried out on its surface. This surface is in contact with the cytoplasm or nucleoplasm, and the soluble components of the surrounding medium have free access to it. The ribosome has been shown to adsorb many substances very effectively, including proteins (e.g., Petermann, 1964). In particular, since its RNA moiety has a high density of negative changes, it has a high affinity for basic proteins; and it has repeatedly been demonstrated that when ribosomes and basic enzymes are mixed *in vitro*, ribosome-enzyme complexes are formed (e.g., Petermann, 1964; Petermann and Hamilton, 1961; Keller and Cohen, 1961; Keller, Cohen and Wade, 1963; Keck and Choules, 1962; Madison and Dickman, 1963a; Siekevitz, 1963; Neu and Heppel, 1964c). There is no question but that such artifacts may be formed when the cell is broken in order to isolate its components and such compartmentalization as may have existed is destroyed. Thus, its mere location on the ribosome is not proof that an enzyme is a true ribosomal enzyme. This is a serious and constantly present problem.

On the other hand, since the ribosomal function in the cell probably involves the attachment and detachment of certain enzymes, the adsorption so often seen *in vitro* might reflect a phenomenon which is biologically important *in vivo*. Particularly striking are several examples in which the adsorption exhibits a high specificity of location, namely, that the enzyme in question is bound to either the 30 S ribosome or the 50 S ribosome, but not to both (see ATPase (3.6.1.3.), RNase (2.7.7.–) and cysteinyl-glycine dipeptidase (3.4.3.5.) below).

In short, the problem is not only to decide whether adsorption may have taken place, but also to ascertain whether this adsorption points to a mere accident of chemical affinity or to an interaction of genuine physiological significance.

We may consider the same problem, but in the reverse sense. The physical absence of an enzyme from the ribosome is no proof that the enzyme has no functional connection with the ribosome. It is quite conceivable that an enzyme whose site of action is the ribosomal surface might still perform this action during only a fleeting contact with the ribosome. The binding between particle and enzyme might be so tenuous as not to survive the isolation procedure, and even in the intact cell only a small proportion of the total number of enzyme molecules might be situated on the ribosome at any given time. Functionally, however, such an enzyme could be considered to be a ribosomal enzyme.

It is clear, then, that this problem in its several aspects is a subtle one; and it is probable that individual solutions will perhaps have to be sought for quite a few enzymes.

In the sections that follow, a large number of ribosomal enzymes will be described, some in groups and others individually. There are two classes of enzyme which, according to our present knowledge of the function of the ribosome, we might expect to be associated with the ribosome. The first group includes those enzymes which participate in the ribosomal stages of protein biosynthesis and thus have a direct functional connection with the ribosome. The second group comprises the so-called nascent enzymes—enzyme molecules which, as proteins, are in the process of being synthesized on the ribosome and have not yet been released from the site of synthesis. The remaining enzymes do not appear to be nascent proteins and are of such a nature that there is no obvious reason to predict *a priori* that they would be attached to ribosomes. They have been divided arbitrarily into three additional groups. The first is ribonuclease I probably the first ribosomal enzyme to be detected and the one on which the most work has been done. The next group includes other enzymes which act in conjunction with a polynucleotide that serves as either substrate, template or primer for the enzyme. There happens to be considerable information on such enzymes, and it is convenient to consider them together. Finally, in the last group are summarily grouped all examples which do not fit into the preceding groups.

V. Ribosomal Enzymes—Enzymes of Protein Synthesis

The enzymes considered here participate in those stages of protein synthesis which take place in association with the ribosome. They catalyse such reactions as the attachment to, movement on, and detachment from the ribosomal surface of mRNA and sRNA; the transfer of nascent polypeptides from one molecule of sRNA to another and the concomitant formation of the peptide bond; perhaps the hydrolysis of GTP; and so on. There is as yet little evidence on the ribosomal location of such enzymes. Indeed, some of them have not yet been conclusively shown to exist, and none has been characterized more than perfunctorily.

This section is then little more than a statement to the effect that the general scheme of protein biosynthesis described earlier in this chapter implies that a number of enzymes must function on or near the ribosomal surface, since their substrates are located there. It is conceivable that in the interests of economy and efficiency such an enzyme might even be built into the permanent structure of the ribosome, but it is equally conceivable that it might not be attached to the ribosome with any degree of permanency. The matter must be examined experimentally in each case, with the attendant problems of non-specific adsorption, etc., that have already been mentioned.

A case in point is a study of the amino acid activating enzymes (6.1.1.–) of rat liver (Decken, 1961). These enzymes operate at an early stage of protein synthesis where the ribosome is not yet involved. Liver ribosomes, however, showed considerable amino acid activating activity, presumably owing to contamination. When, however, the ribosomes were washed with 0·6 M KCl–0·01 M $MgCl_2$, several other enzymic activities were completely removed, but a small amount of amino acid activating activity remained bound to the ribosomes and could not be taken off entirely even when the KCl concentration was raised as high as 3·2 M. Such situations are not uncommon, and the investigator may be pardoned if he abandons the matter at this inconclusive stage with the feeling that, for all that, the ribosome-bound activity probably represents contamination. If, however, the case should be one in which it is important to ascertain whether the activity is really ribosomal, then the problem becomes intricate and non-routine.

A somewhat similar situation has been reported for GTPase (3.6.1.–). It seems fairly certain in this case that such an enzyme is implicated in one or more reactions which directly involve the ribosome. GTP is invariably required for the incorporation of amino acids on the ribosome. In the reticulocyte the breakdown of GTP has been implicated in the binding of aminoacyl sRNA to the ribosome (Arlinghaus *et al.*, 1964) and also in the release of completed polypeptides from the ribosome (Morris, 1964). In this case there is considerable interest in the relation between the enzyme and the ribosome and the question is now being tenaciously pursued. Thus, a non-ribosomal protein factor of *E. coli*, active in stimulating ribosomal polypeptide synthesis (Allende, Monro and Lipmann, 1964), was shown to contain a ribosome-dependent GTPase activity (Conway and Lipmann, 1964). However, the factor could be further separated into two fractions: one stimulated polypeptide synthesis but lacked GTPase, and the other showed GTPase activity (still ribosome-dependent) but did not stimulate polypeptide synthesis (Nishizuka and Lipmann, 1965). The washed ribosomes employed in this experiment themselves exhibited a slight GTPase activity. Thus, the nature of the relationship among GTPase, ribosomes and protein synthesis is still obscure.

A number of investigators have shown that the incorporation of amino acids from sRNA into polypeptides, a process which takes place on the ribosome, is stimulated by two separable non-ribosomal protein factors. These factors, often called the transfer factors or transfer enzymes, produce maximal stimulation only when both are present together; apparently, each serves a different function. They have been observed and studied in *E. coli* (Allende *et al.*, 1964; Keller and Ferger, 1965), rat liver (Nathans and Lipmann, 1960; Fessenden and Moldave, 1962; Moldave and Gasior, 1965), rabbit reticulocytes (Arlinghaus, Favelukes and Schweet, 1963; Arlinghaus *et al.*, 1964; Lamfrom and Squires, 1962), and other tissues. A partial characterization of the reticulocyte factors has been achieved: one factor appears to catalyse the binding of sRNA charged with amino acids to the ribosome and the other may be the enzyme which forms the peptide bond (Arlinghaus *et al.*, 1964). As noted, these factors appear to be enzymes which act on substrates attached to the ribosomal surface, yet they are present in the soluble fraction of the cell. However, after reacting in the isolated reticulocyte ribosomal system *in vitro*, both factors became attached to the ribosomes to the extent that they sedimented with them in the centrifuge.

There is additional evidence as to the location of the enzyme which synthesizes the peptide bond. The antibiotic puromycin interferes with protein synthesis by causing the release of unfinished polypeptides from ribosomes. It is believed that it does this by substituting for a normal amino acid and being itself attached in peptide linkage to the growing end of the nascent polypeptide. Since, however, the puromycin is not attached to a molecule of sRNA, there is nothing to bind the polypeptide to the ribosome, and it comes off (Zamecnik, 1962). Thus, the release of polypeptides from ribosomes by puromycin appears to involve the enzymic formation of a peptide bond. The 50 S ribosomal sub-particle of *E. coli* can be isolated under conditions where it is devoid of mRNA and of the 30 S ribosome, but still has attached to it sRNA and nascent protein (Gilbert, 1963b; Elson, 1964a). Puromycin has been shown to cause extensive release of the nascent polypeptide from such 50 S particles (Traut and Monro, 1964). Since this would presumably require the formation of a peptide bond, it is possible that the enzyme which does this may be located on the 50 S ribosome.

Finally, an investigation of haemoglobin synthesis *in vitro* with a system derived from rabbit reticulocytes has yielded results suggesting that a specific enzyme may be responsible for the release of completed polypeptide chains from the ribosome, and that this enzyme may be located on the ribosome (Morris, 1964).

In sum, work on this group of enzymes is still at its beginning. Consequently, there is very little solid information on how, where and, above all, whether these enzymes are attached to the ribosome. For several of them, it is true, there are indications of a ribosomal attachment, either permanent or transient, but no systematic study of this matter has yet been brought to a decisive stage.

VI. Ribosomal Enzymes—Nascent Enzymes

The concept of nascent protein—unfinished protein molecules attached to their ribosomal site of synthesis—has been firmly established since the classic experiments of Zamecnik and his colleagues, which showed that amino acids are first assembled into polypeptides attached to ribosomes in the liver (Littlefield *et al.*, 1955; Zamecnik, 1960). Similar results were soon reported for the pancreas (Siekevitz and Palade, 1958b; 1960a), and were amplified by the work of McQuillen, Roberts and Britten (1959) on *E. coli.* These workers showed radioactive protein precursors to be rapidly incorporated into the ribosomes, which, if the exposure to precursor was brief enough, assumed the highest specific radioactivity of all cell fractions. When the incorporation of labelled precursor was stopped under conditions which allowed the synthesis of proteins to continue (by the addition of an excess of non-radioactive precursor) the labelled protein was seen to pass rapidly from the ribosomes to the soluble fraction (Fig. 7.) They concluded that "this nascent protein is a polypeptide strand which is formed on the ribosome and

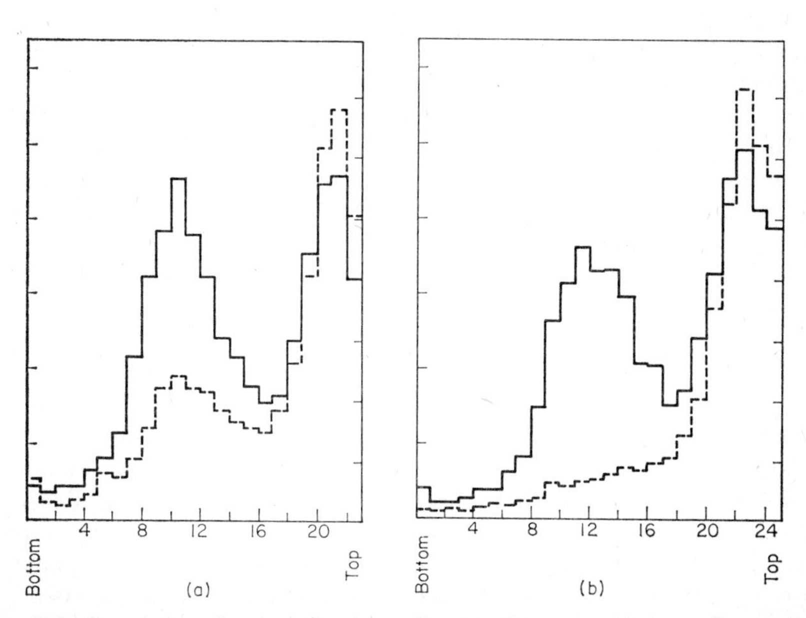

FIG. 7. Demonstration of nascent protein on ribosomes in sucrose density gradient centrifuge patterns of the cell juice of *E. coli.* The unbroken line shows the distribution of material absorbing at 260 mμ; the peak at the centre of the tube is the ribosomes. The broken line shows the distribution of radioactivity. (a). Cells incubated fifteen seconds with radioactive sulphate. (b). Cells incubated as in (a), followed by fifteen seconds with excess non-radioactive sulphate. Most of the radioactivity has left the ribosomes, indicating that the label represents protein molecules transiently bound to the ribosomes and removed from them by normal metabolic processes. Reproduced with permission from McQuillen *et al.* (1959).

is subsequently released as soluble protein" (see also Tissières, Schlessinger and Gros, 1960; Duerre, 1964b).

Later work showed that the nascent protein is attached to the ribosome by a molecule of sRNA (see section 3). This attachment is not very firm, and the sRNA and protein are released from the ribosome when the Mg^{2+} concentration is reduced or the ionic strength is raised (Gilbert, 1963b; Elson, 1964a). They appear to become re-attached when the change in the medium is reversed (Schlessinger and Gros, 1963; Tal and Elson, 1963a), but it is not known whether they become re-attached to the same site and in the same way. Obviously the lability of the attachment of the nascent protein to the ribosome can make it difficult to distinguish between a nascent and an adsorbed enzyme.

Since every protein is presumably synthesized on a ribosome, the nascent proteins should, by definition, include precursors of every enzyme then being synthesized in the cell. It is not evident that a nascent enzyme will show enzymic activity at this stage in its career. It is conceivable, however, that some might acquire the ability to act as catalysts before being released from the ribosome. This would probably depend on the structural properties of each individual enzyme: one might be able to fold into an active tertiary configuration before being released, while another might not.

If the rate of synthesis of an enzyme is known, the amount which might reasonably be expected to occur in nascent form can be calculated. It has been estimated, for example, that of any enzyme in a steady state culture of *E. coli* growing exponentially with a generation time of 50 minutes, about 0·6% would be nascent and bound to the ribosomes (McQuillen *et al.*, 1959; Cowie, Spiegelman, Roberts and Duerksen, 1961). Since this estimate includes both active and inactive forms of the enzyme, the amount of active nascent enzyme, if any, would be far less.

The matter is further complicated by the fact that many enzymes are polymeric, being made up of more than one polypeptide chain, and the subunits may not necessarily be synthesized on the same ribosome. For such an enzyme to be both nascent and active, polymerization would have had to occur while one enzyme subunit was still attached to a ribosome. This has been observed to occur *in vitro* in the case of *E. coli* β-galactosidase (3.2.1.23) by Zipser and Perrin, (1963). Whether it occurs *in vivo* is not known. In a study of the synthesis *in vivo* of the L-histidine ammonia-lyase (4.3.1.3) of *B. subtilis*, Hartwell and Magasanik (1964) have concluded that the polymerization step which is necessary for the enzyme to become active does not take place on the ribosome. If this should be generally true, then the finding of an active polymeric enzyme attached to ribosomes would be an artifact, albeit, perhaps, a useful and instructive artifact.

It is apparent that every enzymic activity found in a preparation of ribosomes might be due, wholly or partly, to nascent enzymes. In this section, however, only those ribosomal enzymes are discussed which may reasonably be suspected of being nascent or which have been investigated with this possibility in mind.

A. PANCREATIC ENZYMES AND ZYMOGENS

One of the more widely studied mammalian secretory organs, the pancreas, is adapted for the large scale synthesis of a relatively small number of enzymes and zymogen precursors of enzymes. These proteins have, in the main, been well characterized and considerable work has been done on their intracellular distribution. We might expect to find in such a tissue a rather high nascent ribosomal level of those few enzymes and zymogens which are synthesized in large quantities.

The earliest work on this subject showed that guinea pig pancreas ribosomes are rich in two types of protein synthesized for secretion: RNase and protease zymogens ("trypsin-activable protease") (Siekevitz and Palade, 1958a). Later work in the same laboratory concentrated on α-chymotrypsinogen and added amylase (3.2.1.-) to the list of such enzymes (Siekevitz and Palade, 1960a, b). A study of the time course of labelling of the cellular proteins showed that at short times after the injection of a radioactive amino acid the α-chymotrypsinogen of those ribosomes which had formerly been attached to the membranes of the endoplasmic reticulum was more highly labelled than in any other cell fraction. This is a clear indication that ribosomal α-chymotrypsinogen is a precursor of the rest of the zymogen in the cell.

It is difficult, however, to recognize true nascent enzymes in this system. This becomes evident when the quantities of ribosomal enzymes are considered. The guinea pig pancreas ribosomes employed carried some three to 7% of their total protein as RNase and from three to ten per cent as chymotrypsinogen and trypsinogen, and these proteins seemed quite firmly attached to the isolated ribosomes (Siekevitz and Palade, 1960b, 1962). Clearly, only a minute fraction could actually be nascent protein, and its properties would be obscured. The observed preferential labelling of the ribosomal α-chymotrypsinogen can perhaps be explained by postulating that the most recently synthesized molecules remained in the closest proximity to the ribosomes, in the endoplasmic reticulum, and were preferentially adsorbed by them when the reticulum and later the microsomes were destroyed.

The problem of adsorption has been directly pointed out in a number of investigations of the ribosomes of beef pancreas. The cell juice and zymogen granules of this tissue contain large quantities of the basic proteins RNase, trypsinogen and chymotrypsinogen A, and the acidic proteins DNase (3.1.4.-), chymotrypsinogen B and procarboxypeptidases A and B. Although the acidic proteins outnumber the basic ones by about two to one in the cell and therefore, presumably, in the nascent protein, isolated ribosomes contained only the basic ones, which amounted to nearly 9% of the ribosomal protein (Keller *et al.*, 1963). Further, when labelled chymotrypsinogen A was added to ribosomes, it was adsorbed and was not displaced by subsequently added unlabelled chymotrypsinogen A (Keller and Cohen, 1961).

Thus the promised advantages of the pancreas as a source of nascent ribosomal enzymes have been far outweighed by the great quantities of non-ribosomal enzymes in the tissue, at least some of which are adsorbed in large amounts by the ribosome. It will be necessary to find effective ways of distinguishing between adsorbed and nascent enzyme or of selectively removing adsorbed enzyme from the ribosome before a conclusive study of the pancreatic nascent enzymes can be carried out. There are, in fact, a number of accounts of the removal of enzymes and zymogens from pancreatic ribosomes by various means: e.g., by washing the ribosomes with salt (Keller, *et al.*, 1963; Madison and Dickman, 1963a; Siekevitz, 1963), precipitating them with Mg^{2+} ions (Gazzinelli and Dickman, 1962; Siekevitz, 1963), or treating them with spermine (Siekevitz and Palade, 1962) or antiserum against the adsorbed enzyme (Madison and Dickman, 1963b; Gazzinelli and Dickman, 1962). The enzymic activities of the ribosomes have been substantially reduced by such treatments; however, a low level of activity usually remains. Whether this remaining ribosomal activity represents an enrichment in nascent enzyme has not yet been examined in detail.

B. OTHER STUDIES OF NASCENT ENZYMES

The literature contains a number of examples of ribosomal enzymes which have been studied specifically to probe the possibility of their being nascent enzymes. Some of these investigations have been perfunctory, others, more extensive and searching. Among the criteria employed to distinguish nascent enzymes have been: that the enzyme is not easily removed from the ribosome, that specific antiserum against the enzyme precipitates not only the enzyme but also a small fraction of the ribosomes, that the ribosome-bound activity is only a small fraction of the total cellular activity, and that the ribosomal enzyme shows the characteristics expected of a precursor of the other enzyme molecules of the same species. Since this is relatively new ground, it is not surprising that some of these criteria have been found inadequate and others have proved unexpectedly difficult to apply. The work summarized below represents, essentially, a beginning.

1. Viral Neuraminidase (3.2.1.18)

When influenza virus is grown in allantoic membranes or monkey kidney cells, the enzyme neuraminidase (sialidase), an enzyme which forms part of the protein complement of the virus, is found associated with host cell ribosomes. Possibly nascent protein, the enzyme is released from the ribosomes as a molecule of molecular weight 2×10^5, and is later incorporated into larger complexes and built into the viral envelope (Noll, 1962).

2. Triosephosphate Dehydrogenase (1.2.1.-)

Well washed yeast ribosomes have been shown to contain a low but significant level of triosephosphate dehydrogenase activity (Warren and Goldthwait, 1962), although crystalline enzyme added to ribosomes was easily washed away. Four to eight % of the washed ribosomes were precipitated by rabbit antiserum against the crystalline enzyme, and the precipitate was somewhat enriched in enzyme activity. The implication is, of course, that the ribosomes were precipitated by the antiserum because they were attached to nascent enzyme molecules.

3. Tryptophan Synthetase (4.2.1.20)

The tryptophan synthetase of *E. coli* is made up of two separate and dissimilar protein chains, designated A and B. Employing immunological techniques to distinguish between the two proteins, Horibata and Kern (1964) have presented evidence which indicates that some ribosomes carry only the A protein and others only the B protein. Other ribosomes, however, contained both proteins. This is a matter of some interest, since the A and B proteins are governed by different but adjacent structural genes and there is a possibility that both proteins are synthesized by the same ribosome from a single polycistronic mRNA (i.e., a single molecule of RNA in which is transcribed the information contained in more than one structural gene). The authors suggest that this is what occurs and that those ribosomes which carried only one of the proteins originally contained both but lost one. However, it is entirely possible that the ribosome-A-B complexes result from an interaction between ribosome-A and free B or ribosome-B and free A, as has been observed in the case of β-galactosidase (Zipser and Perrin, 1963). The data presented to eliminate this possibility are not convincing, and it is not certain that those A and B chains found on the same ribosome were in fact both synthesized there.

4. Pencillinase (3.5.2.6)

Duerksen and O'Connor (1963) have studied the inducible penicillinase of ribosomes isolated from lysozyme spheroplasts of *B. cereus*. The ribosomes were not washed, unfortunately, but the enzymic activity sedimented with the particles in a sucrose density gradient. The authors have calculated that there is one ribosome-bound enzyme molecule per uninduced cell, ten per fully induced cell, and fifteen per cell in a constitutively positive strain.

5. Lactate Dehydrogenase (1.1.1.27)

Some of the difficulties of distinguishing a true nascent enzyme are pointed out in a thoughtful paper by Keck and Choules (1962), who have investigated the lactate dehydrogenase isoenzymes of guinea pig liver and kidney in an attempt to identify nascent precursors still attached to ribosomes. Ribosomes from these organs did contain isoenzymes, but predominantly the basic ones, suggesting the adsorption of free enzyme on the acidic ribosome. The point was not conclusive, since non-basic isoenzymes were also present. However, the level of ribosomal activity was such that if it were all due to nascent protein, at least one of every thirty kidney ribosomes would have had to be engaged in synthesizing this enzyme, and this does not seem reasonable. Consequently, the authors concluded that they were dealing with adsorbed rather than nascent enzymes, even where non-basic isoenzymes were involved.

6. *β-Glucosidase* (3.2.1.21)

The work of Halvorson and his colleagues (Hauge, MacQuillan, Cline and Halvorson, 1961; Kihara, Halvorson and Bock, 1961; Kihara, Hu and Halvorson, 1961) on the β-glucosidase of a diploid strain of yeast constitutes one of the earliest and yet one of the fullest studies of a nascent enzyme now available. These workers showed that when ribosomes of a constitutively positive strain were washed by repeated sedimentation in sucrose density gradients, their specific β-glucosidase activity fell to a level not further reduced by continued washing. Control ribosomes from a constitutively negative strain could be washed free of added soluble enzyme. The identity of the ribosomal enzyme was examined by degrading the ribosomes and so releasing the enzyme (this treatment also results in a several fold rise in activity). The released enzyme was found not to differ from non-ribosomal β-glucosidase in a number of physico-chemical properties and its affinity for a number of β-glucosides.

p-Fluorophenylalanine was next employed in an attempt to determine whether the ribosomal enzyme could be a precursor of the soluble enzyme. This amino acid analogue is incorporated into proteins in place of phenylalanine and the cell produces an enzymically inactive version of β-glucosidase. When the analogue was administered to a culture of cells, the quantity of non-ribosomal enzyme ceased to rise, and remained constant; but the amount of ribosomal activity dropped by a factor of ten. Upon the addition of excess phenylalanine to the culture, enzyme synthesis was resumed and the ribosome-bound activity rose to its former level. These results are consistent with the proposition that the ribosomal enzyme is the nascent precursor of the soluble enzyme.

7. *β-Galactosidase*

One of the most extensively studied enzymes known, the inducible β-galactosidase of *E. coli*, has also been widely employed in investigations of nascent enzymes. Early work by Kameyama and Novelli (1960, 1962) demonstrated the presence of a particle-bound β-galactosidase in *E. coli* which differed from the soluble enzyme in several respects and appeared likely to be a precursor of the soluble enzyme. The quantity of particle bound enzyme was, however, much too high to allow it to be identified with nascent protein, and the particles examined were not purified ribosomes but something much more complex.

Perhaps the first detailed study with purified ribosomes was that of Cowie *et al.* (1961). When they washed the ribosomes of a constitutively positive strain by repeated sedimentation, the ribosomal activity was reduced to a low but constant and significant level, and this activity sedimented with the ribosomes in sucrose density gradients. In the control, a mixture of ribosomes from a strain genetically incapable of producing the enzyme and free enzyme from a different strain, essentially all of the adsorbed activity was washed off (Fig. 8). Inducible but non-induced cultures were estimated to contain one ribosome-bound active enzyme molecule per cell. Constitutively positive

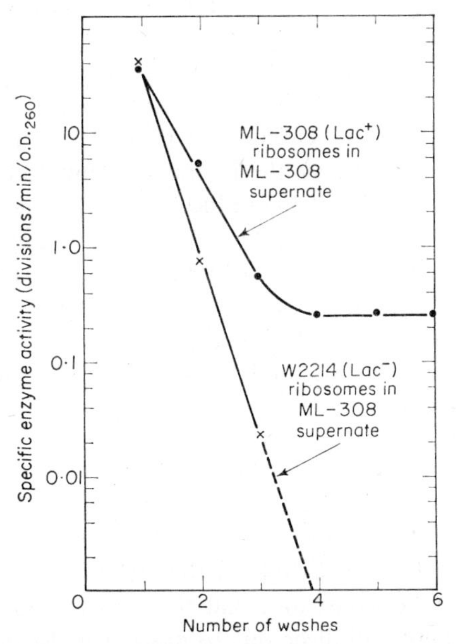

FIG. 8. The effect of successive washings by centrifuging on the specific β-galactosidase activity of ribosomes from genetically positive (●) and negative (×) strains of *E. coli*. Both sets of ribosomes were initially suspended in extracts containing large amounts of active enzyme. Reproduced with permission from Cowie *et al.* (1961).

cells carried about ten times more and induced cells, about twenty five. Rabbit anti-β-galactosidase serum reacted with the ribosomes but did not precipitate them. When, however, this treatment was followed with chicken anti-rabbit serum, about 1% of the ribosomes precipitated and carried with them over 90% of the ribosomal β-galactosidase activity.

Certain of these results have been confirmed in other laboratories (Proctor,

1961; Fogel and Elson, 1964; Kiho and Rich, 1964), and in addition, the ribosomal β-galactosidase has been shown to be located mainly in the polysome fraction, the ultimate site of protein synthesis (Kiho and Rich, 1964). The indications were, then, that the ribosomal β-galactosidase activity might well represent nascent molecules in a stage of synthesis sufficiently advanced to allow enzymic activity.

Almost from the beginning, however, it became apparent that the matter was more complex than the results summarized above might suggest. In following up their earlier investigation, Duerksen and Cowie (1961; cf. Roberts, 1964, p. 407) studied the kinetics of the appearance of ribosomal β-galactosidase during induction. If all the ribosomal enzyme were a nascent precursor of soluble enzyme, then the amount of ribosomal activity should follow the rate of enzyme synthesis. It should rise rapidly at the beginning of induction, reach a constant maximum level when the rate of enzyme synthesis becomes constant (about two and a half minutes after addition of the inducer), maintain this level during induction, and decline when the inducer is removed. The experimental observations were, however, quite different. The quantity of ribosomal enzyme continued to rise throughout the entire period of enzyme synthesis and did not decline when synthesis was stopped by the removal of the inducer. Thus the quantity of ribosomal enzyme ran parallel not with the rate of enzyme synthesis, but with the total amount of enzyme in the cell. Furthermore, the quantity of ribosomal enzyme was different when the cells were broken by different techniques. These findings led to the conclusion that "the newly formed enzyme can only be a minor fraction of the total quantity found to sediment at the same rate as the ribosomes."

These results have been amplified by Zipser (1963), who was able to detect two kinds of ribosomal β-galactosidase: one fraction which remains bound to the ribosomes and increases the longer the cells have been synthesizing the enzyme, and another fraction which leaves the ribosomes after the removal of the inducer and could be the precursor of the soluble enzyme.

There are other indications, too, that the β-galactosidase activity of well washed ribosomes may not all be due to nascent molecules. For one, it has been reported that when such purified ribosomes are sedimented in a caesium chloride gradient of density 1·3–1·6 (over 2 M), the amount of ribosome-bound enzyme is reduced forty times. (Lederberg, Rotman and Lederberg, 1964). This does not prove that only 2 or 3% of the ribosomal enzyme is nascent, since high ionic strength will detach nascent protein from ribosomes (Elson, 1964a), but it does suggest that not all of the ribosomal enzyme is bound to the ribosome in the same way. Also when purified ribosomes were separated into 30 S and 50 S particles by sucrose density gradient centrifuging, both particles showed activity (Fogel and Elson, 1964) although nascent protein remains bound only to the 50 S ribosome (Gilbert, 1963b).

Finally, another factor must be fitted into the picture before it will be possible to understand the nature of ribosomal β-galactosidase. Like many other enzymes β-galactosidase is polymeric and is active only in the polymeric form. Although β-galactosidase subunits can polymerize *in vitro* while one is still attached to a ribosome, it is not known whether this occurs *in vivo* or whether the process takes place at a completely post-ribosomal stage of protein synthesis. If the latter, then any association of β-galactosidase activity with ribosomes is an artifact. This is still an open question (see the opening discussion of this section).

It seems likely that the true nascent β-galactosidase has so far eluded our grasp. The considerable amount of work, sometimes both ingenious and profound, which has been done has only cleared the approaches to the problem.

8. *Latency of Nascent Enzymes*

Several ribosomal enzymes are activated by one treatment or another, treatments which either release them from the ribosome or apparently cause some change in their attachment to the particle. Before such treatment the enzyme is presumably bound to the ribosome in such a way as to reduce or prevent its enzymic action. Of the enzymes discussed in this section, two have been reported to be latent in this sense.

As mentioned above, the β-glucosidase of yeast becomes more active when it is liberated through the destruction of the ribosome (Kihara, Hu and Halvorson, 1961). There are conflicting reports concerning the ribosomal β-galactosidase, to the effect that the destruction of the ribosome does (Proctor, 1961) or does not (Cowie *et al.*, 1961; Fogel and Elson, 1964) lead to an increase in enzymic activity. A different treatment, however, does cause the activity to increase several times. This is the treatment of the ribosomes with rabbit anti-β-galactosidase serum, which reacts with the ribosomes but neither precipitates them nor causes their destruction (Cowie *et al.*, 1961). This effect has now been further studied by Lederberg *et al.* (1964). Using a technique sensitive enough to detect single molecules of β-galactosidase (Rotman, 1961), they succeeded in showing that the increase in activity is caused not by an increase in the activity of already active ribosomes, but by the appearance of enzyme activity on ribosomes which were previously inactive. It is to be hoped that the nature of this interesting phenomenon will soon be elucidated.

C. SUMMARY

These examples of work aimed, sometimes more and sometimes less persistently, at characterizing the nascent enzymes show that we are still far from understanding the relationship between the ribosome and the growing polypeptide chain, which at some point in its growth acquires the ability to catalyse a specific biochemical reaction. We have, in fact, not quite achieved the first necessary objective, the ability to recognize such a nascent enzyme and to distinguish it from other forms of the same enzyme.

The criterion of ribosomal location is a necessary one but is not satisfactory alone, since it is easy to recognize a high level of adsorbed enzyme but can be quite difficult to discern a small amount. Also, the enzyme must be identified and measured by its activity. This probably makes it impossible to deal with certain enzymes in their nascent stages and introduces numerous uncertainties. Where polymeric enzymes are involved, the question of the polymerization step and its place in the normal sequence of events is a complicating factor whose severity cannot yet be accurately assessed.

It would seem desirable to stress two criteria in this sort of investigation. The first is that a true nascent protein must have the properties of a precursor to other forms of the enzyme (even here there are pitfalls: see the section on pancreatic enzymes). The second is a criterion that has only recently become available and has not yet been applied, namely, that a nascent protein molecule should be attached to a molecule of sRNA (see section III).

VII. Ribosomal Enzymes—Ribonuclease (2.7.7.-)

The ribonucleases are a ubiquitous family of enzymes which differ in several respects from other enzymes which degrade RNA. The RNases are endonucleases which produce mono- and oligonucleotide fragments whose terminal phosphate residue is attached to the 3′ position of the nucleoside ribose. Most other enzymes which degrade RNA are exonucleases, and most of them form 5′-mononucleotides. What appears to distinguish the RNases absolutely from the others is their mechanism of action, which proceeds in two steps. The first step, a transfer reaction, involves the splitting of an internucleotide phosphoester bond between phosphate and the 5′ position of the adjacent nucleoside, and the simultaneous formation of a phosphoester bond at the 2′ position of the nucleoside to which the phosphoryl group is already attached. Thus, the first degradation product is a fragment whose terminal phosphate group is esterified to both the 2′ and 3′ positions of the same ribose moiety. The enzyme subsequently hydrolyses this cyclic diester to form the 3′-phosphomonoester. This mechanism is specific and it would seem proper to reserve the name "ribonuclease" for those enzymes which form the cyclic diester intermediate. The various RNases share this reaction mechanism; they differ in that some will split only those bonds which are adjacent to certain nucleosides but not to others.

In practice, of course, the reaction mechanism is rarely defined, and it is common to identify RNases by another property which distinguishes them from all other nucleases known at present, namely, that they do not require Mg^{2+} or any other divalent cation. Thus, a nuclease which degrades RNA in the presence of an excess of a complexing agent such as EDTA is generally considered to be an RNase. (For reviews see, e.g., Anfinsen and White, 1961; Roth, 1963).

RNase is probably the first enzyme shown to be specifically associated with ribosomes. In 1950, Pirie (1950, 1957) reported ribonucleoproteins prepared from tobaco leaf to contain the enzyme in a tightly bound form. This unambiguous observation did not become generally known, and some seven or eight years later, after a number of intensive investigations of the ribosome had been undertaken, the phenomenon was inevitably rediscovered in several laboratories (Elson, 1958, 1959b; Bolton, 1959b, cf. Roberts, 1964, p. 362; Siekevitz and Palade, 1958a; Stenesh, 1958; Tashiro, 1958; Ts'o, Bonner and Vinograd, 1958; Zillig *et al.*, 1959; Roth, 1960a, b).

These findings stemmed generally from the common observation that ribosomes, which are quite stable at low temperature and ionic strength, become unstable when either the temperature or the ionic strength is raised or the concentration of Mg^{2+} ions is reduced (e.g., Elson, 1959a). At such times the ribosomal RNA is degraded and an enzyme, soon identified as an RNase, was found to be responsible (Elson, 1958, 1959a; Bolton, 1959b, cf. Roberts, 1964, p. 362; Zillig *et al.*, 1959; Tashiro, 1958). The fact that the ribosomes carry with them an enzyme capable of catalysing their own destruction was startling enough to provoke several investigators into rather detailed examinations of the matter.

By now, an association between RNase and ribosomes has been reported for a great many different organisms (see Table I at the end of this chapter). These observations are so numerous and so widespread thoughout nature that at first it seemed that the phenomenon might be universal (Elson, 1961b). There are, however, exceptions, some better and some less well established, and these appear to be significant, as will be discussed below.

A. PANCREATIC RIBOSOMAL RNASE

It is clear that the association between ribosome and enzyme is of at least two different types. One extreme is typified by the ribosomes of the pancreas (see section VI.A.) RNase is one of several enzymes synthesized and secreted in large amounts by this organ. The pancreas contains a high concentration of the enzyme, and so do its ribosomes. Pancreatic ribosomes have, in fact, been reported to contain from 1–16% of their protein as RNase (Madison and Dickman, 1963a; Keller, Cohen and Wade, 1963, 1964; Siekevitz and Palade, 1960b; Siekevitz, 1963). This is probably due virtually entirely to non-specific adsorption. A basic protein, RNase would inevitably be bound by the acidic RNA of the ribosomal surface. It has been directly demonstrated that extraneous RNase is adsorbed by pancreatic and other ribosomes (e.g., Madison and Dickman, 1963a; Siekevitz, 1963), and that most and often virtually all of the RNase of pancreatic ribosomes can be rather easily removed from the particles by salt (Madison and Dickman, 1963a; Siekevitz, 1963), magnesium ions (Keller, *et al.*, 1964; Gazzinelli and Dickman, 1962; Siekevitz, 1963), spermine (Siekevitz and Palade, 1962), bentonite (Keller *et al.*, 1964), anti-RNase serum (Gazzinelli and Dickman, 1962; Madison and Dickman, 1963b), and other agents.

In short, the presence of large amounts of RNase in ribosomes isolated from the pancreas is probably of no direct physiological significance, reflecting only the normal affinity between cationic and anionic macromolecules. The same may well be true of at least a portion of the RNase activity of ribosomes from other organs.

B. *E. coli* RIBOSOMAL RNASE

In certain other types of cell, however, a different relationship between ribosome and RNase can be discerned. This type is exemplified by *E. coli*, on which considerable work has been done. These cells contain comparatively little RNase; however, most of it is attached to ribosomes isolated in the usual way, and it is attached so firmly that it can hardly be removed without destroying the ribosome (Elson, 1959b, 1961b; Bolton, 1959b, cf. Roberts, 1964, p. 362; Spahr and Hollingworth, 1961). This striking relationship, further documented below, has led to speculation that the function of the enzyme is connected with the function of the ribosome (e.g., Elson, 1961b; Zillig, *et al.*, 1959; Tashiro, 1958; Roth, 1960b; Spahr and Hollingworth, 1961, Tal and Elson, 1963b), namely that it has to do with protein biosynthesis. Recent developments, however, have cast doubt on both the ribosomal location of the enzyme and its proposed biological function (Neu and Heppel, 1964a,b,c; Gesteland, 1965). I shall treat this work in roughly chronological order, outlining first the findings which appeared to establish RNase as a true ribosomal enzyme and then those which have placed these conclusions and the attendant speculations in question.

1. *Identity*

In the early work with the RNase of *E. coli*, evidence was soon accumulated which indicated that the enzyme was a true RNase. Several other possibilities were eliminated by direct test; and the enzyme was shown to resemble pancreatic RNase in heat stability, in having no requirement for divalent cations, and in producing 3′-nucleotides (Elson, 1958, 1959b; Bolton, 1959b, cf. Roberts, 1964, p. 362; Zillig *et al.*, 1959). Later work with a purified enzyme confirmed these observations, and showed also that the enzyme degrades RNA, via cyclic phosphodiester intermediates, completely to mononucleotides, thus differing from bovine pancreatic RNase and resembling certain plant RNases in specificity (Spahr and Hollingworth, 1961).

2. Latency

A phenomenon early noted was the latency of this ribosomal enzyme (Elson, 1958, 1959b; Bolton, 1959b, cf. Roberts, 1964, p. 362). This concept has been widely misunderstood and should, perhaps, be clarified here. It was initially observed that the ribosomes were stable in the cold at low ionic strength. That is, the ribosomal RNase did not attack the ribosomal RNA under these conditions, nor did it attack added free RNA. Active enzyme was liberated by a variety of treatments: e.g. urea, increased ionic strength, tryptic digestion, EDTA (which removes the Mg^{2+} ions which stabilize the nucleoprotein structure), and increased temperature (to 37°). Following such treatments, which disrupt the structure of the ribosome, the enzyme became able to attack

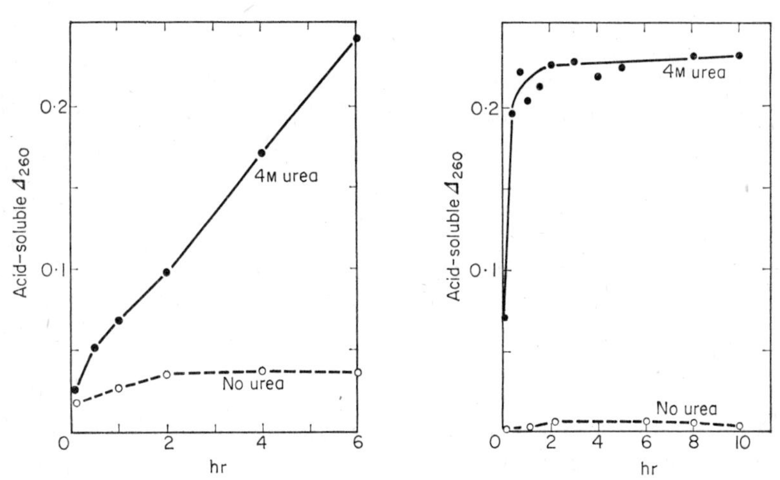

FIG. 9. The latency of ribonuclease in a highly purified ribonucleoprotein from *E. coli*, and the activation of the enzyme by 4M urea. "Acid-soluble Δ_{260}" is a measure of the degradation of the ribosomal RNA by the enzyme. Conditions: *a.* 0°; 0·075 M *tris* buffer, pH 8·0. *b.* 37°; 0·005 M $MgSO_4$. Reproduced with permission from Elson (1959 b).

both the ribosomal RNA and added free RNA under conditions of temperature, pH, ionic strength and Mg^{2+} concentration where intact ribosomes showed no RNase activity (Elson, 1958, 1959b; Bolton, 1959b cf. Roberts, 1964, p. 362; Spahr and Hollingworth, 1961) (Fig. 9). Latency was thus defined in the following way: that there exist conditions under which the RNase can attack RNA but does not when the enzyme is associated with ribosomes (Elson, 1959b). This is not true of all conditions. For example, under the conditions usually employed for the assay of RNase (e.g., neutral pH, ionic strength of about 0·1–0·2, 30° or 37°, Mg^{2+} concentration of 0·01 M or less) ribosomes will generally break down rapidly, their RNase will be liberated, and its latency will not be seen. Latency can be demonstrated only under conditions where the ribosomes are stable, and these are not the conditions of maximum enzyme activity.

The conditions prevailing inside a living cell of *E. coli* are not those under which the ribosomal RNase would remain inactive outside the cell, and the demonstrable latency *in vitro* does not, of course, imply that the enzyme is inactive *in vivo*. It has been suggested, however, that there is an implication that the action of this enzyme *in vivo* may be controlled and limited by virtue of its association with the ribosome (Elson, 1961b).

3. Ribosomal Location

The early work on the RNase of *E. coli* strongly suggested that virtually all of the enzyme was associated with the ribosomes (the cell debris was not examined in these investigations). This was shown in a variety of ways. When ribonucleoproteins were purified by a relatively drastic chemical procedure, all of the RNase activity of the crude cell extract remained in the nucleoprotein (Elson, 1959b). When the ribosomes were sedimented in the centrifuge, the supernatant fluid contained little or no RNase (Bolton, 1959b, cf. Roberts, 1964, p. 362; Spahr and Hollingworth, 1961); and when a crude extract was resolved in a sucrose density gradient, RNase activity was found almost exclusively in the fractions containing ribosomes (Tal and Elson, 1963b; Elson, 1964b). Later reports indicate that the soluble fraction may contain as much as 11% of the RNase activity of the crude extract (Shortman and Lehman, 1964; Anraku and Mizuno, 1965) and, further, that the cell debris (presumbably fragments of cell wall and cell membrane) contains an appreciable quantity of an RNase which may be different from that of the other fractions (Anraku and Mizuno, 1965). In any event, all of these investigations showed that the bulk, if not virtually all, of the RNase was associated with the ribosomes.

The observed concentration of the RNase of *E. coli* in the ribosomes prompted a close examination of the localization of the enzyme, and it was then found to be localized exclusively in the 30 S ribosome (Elson and Tal, 1959). This has been confirmed in a number of ways. Purified 50 S ribosomes contained only a low level of RNase activity and this could be accounted for by a small amount of contaminating 30 S particles (Elson and Tal, 1959; Tal and Elson, 1963b). Further, the RNase activity of mixtures of 30 S and 50 S ribosomes varied directly with the proportion of 30 S ribosomes in the mixture (Elson and Tal, 1959; Tal and Elson, 1963b; Spahr and Hollingworth, 1961) and extrapolated to zero activity for pure 50 S particles (Fig. 10). Again, when 30 S and 50 S particles were resolved by centrifuging in a sucrose density gradient, only the 30 S ribosomes

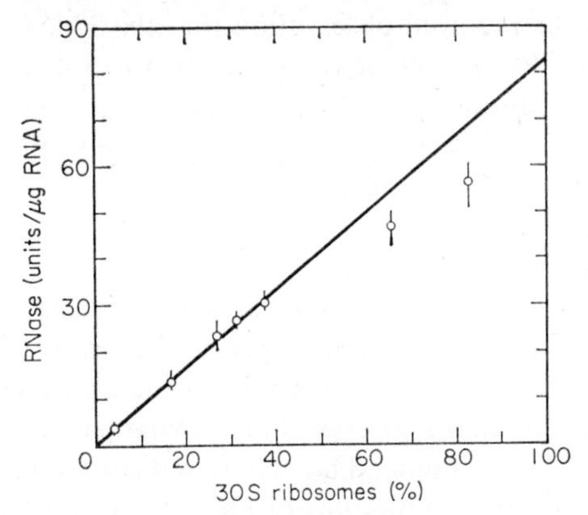

FIG. 10. The intraribosomal location of ribonuclease in *E. coli*; a plot of the ribonuclease activities of 30 S and 50 S ribosomes mixed in different proportions. The plot extrapolates to zero activity for pure 50 S particles. Reproduced with permission from Tal and Elson (1963 b).

showed RNase activity (Tal and Elson, 1963b; Elson, 1964b) (Fig. 11). Finally, when the proteins derived from separate preparations of 30 S and 50 S ribosomes were subjected to electrophoresis in starch gel, only the 30 S particles showed a band with RNase activity (Waller, 1964). Opposed

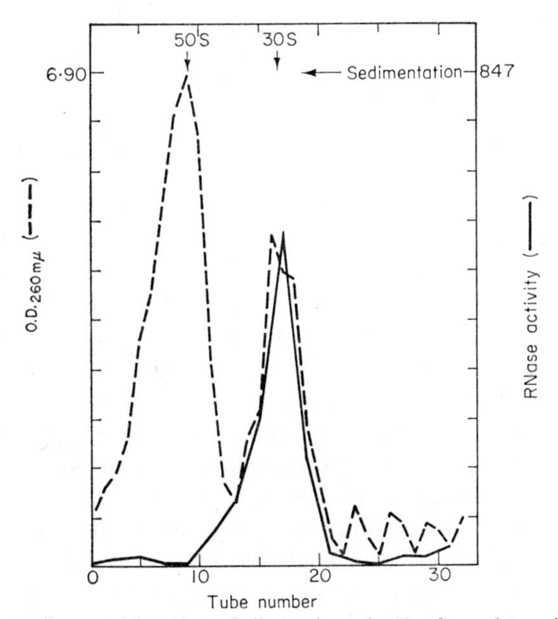

FIG. 11. The intraribosomal location of ribonuclease in *E coli*, as determined by sucrose density gradient centrifuging. The broken line shows the location of the ribosomes and the unbroken line shows that of the enzyme, which is only in the 30 S particle. Reproduced with permission from Tal and Elson (1963 b).

to these results is the reported observation that the 30 S and 50 S ribosomes show equal levels of RNase activity (Bolton, 1959b, cf. Roberts 1964, p. 362). The reason for this discrepancy is not apparent; however, the bulk of the available evidence indicates strongly that the affinity of *E. coli* ribosomes for RNase is restricted to the 30 S particle.

4. Amount in the Ribosomes

Since the RNase of *E. coli* has not been fully characterized, it has not been possible to determine unequivocally the quantity of enzyme in the ribosomes. What has been done is to compare the ribosomal RNase activity with that of known amounts of crystalline bovine pancreatic RNase. Three such estimates have been made. The results indicate that the ribosomes of *E. coli* contain about 0·08 molecules of enzyme per 70 S ribosome (Elson, 1959b), between 0·1 to 1·0 (Bolton, 1959b, cf. Roberts, 1964, p. 362) or 0·1 (Spahr and Hollingworth, 1961). The first two estimates were based on the assumption that the *E. coli* enzyme is identical with pancreatic RNase in both molecular weight and specific activity. The third estimate, however, was made with a purified enzyme whose specific activity had been approximately determined, although here too the molecular weight could not be measured.

The concensus is, then, that no more than one ribosome in about ten would carry a molecule of RNase. In the case of yeast, an estimate of one molecule per four ribosomes has been published (Ohtaka and Uchida, 1963).

5. Firmness of Binding

The enzyme is bound to the ribosome very firmly, so that, until recently, its removal had not been accomplished without the destruction of the ribosome. For example, in early chemical purifications in my laboratory, ribonucleoproteins of *E. coli* were exposed to 1·25% sodium deoxycholate, precipitated as the barium salt in 50% ethanol, redissolved with sodium sulphate, and repeatedly precipitated with ammonium sulphate in the saturation range of 25–60% (1·2–2·6 M, ionic strength 3·6–7·8). This harsh and varied procedure was designed to remove extraneous proteins from the ribonucleoprotein; yet the entire complement of RNase remained attached (Elson, 1959a, b). *E. coli* ribosomes have also been exposed to high concentrations of caesium chloride (approximately 2 to 5 M) without losing RNase activity (Lederberg, *et al.*, 1964).

In view of these observations, a milder salt treatment might not be expected to remove the ribosomal RNase of *E. coli.* There are, however, reports that prolonged washing with 0·5 M ammonium chloride de-activates the ribosomal RNase (Jardetzky and Julian, 1964) or removes it (Spirin *et al.*, 1963), although data on the effectiveness of the procedure were not published. In our hands the procedure described by Spirin *et al.* (1963)

has given variable results; some activity has been removed, but in all cases so far the ribosomes have retained a significant level of RNase (P. Spitnik-Elson, personal communication; D. Elson, unpublished results). In contrast, when large quantities of bovine pancreatic RNase were adsorbed on *E. coli* ribosomes, a brief exposure to 0·5 M ammonium chloride reduced the ribosomal activity to its low basal level. Apparently, the pancreatic enzyme was completely removed, while the endogenous *E. coli* RNase remained attached (D. Elson, unpublished results).

However, what appears to be an effective technique for removing RNase from *E. coli* ribosomes has now been described by Salas, Smith, Stanley, Wahba and Ochoa (1965). After two prolonged washings with 0·5 M ammonium chloride, the ribosomes, still containing a high level of RNase activity, were chromatographed on DEAE-cellulose. This treatment reduced the RNase activity by about 99%. The ribosomes were active in promoting the incorporation of amino acids into peptides, and were successfully employed in an experiment in which the virtual absence of RNase activity was critical.

Taken together, these observations show that RNase can be removed from the ribosomes of *E. coli* without causing their inactivation; the enzyme is, however, bound very firmly to the ribosomes.

6. Location—Observations with Spheroplasts of E. coli

The foregoing observations indicated that the bulk of the RNase of *E. coli* is bound firmly and specifically to the surface of the 30 S ribosome in such a way as to affect the action of the enzyme. This is certainly true after the breakage of cells and the isolation of ribosomes in the usual way. It has been assumed that the same is true of the intact cell, but this assumption has now been put in doubt by the findings of Neu and Heppel (1964a, b, c, d; 1965).

It had previously been shown by Malamy and Horecker (1961) that *E. coli* cells release their entire content of alkaline phosphatase (3.1.3.1) into the external medium when converted to spheroplasts with lysozyme and EDTA in a sucrose-*tris* medium. A number of other enzymes remain within the spheroplast. This observation was interpreted in terms of a compartmentalization of enzymes. It was postulated that the alkaline phosphatase is situated on the periphery of the cell, either associated with the cell wall or lying between it and the cell membrane, while the unreleased enzymes are in the interior of the cell.

When Neu and Heppel (1964a, b, c; 1965) performed similar experiments, they found that several additional enzymes leaked out of the spheroplasts, among them RNase. Spheroplasts made from cells in stationary phase released about 45–60% of their RNase; those from exponentially growing cells, up to 90%. The release of enzyme was rapid, reaching completion in a few minutes at room temperature. Ribosomes were not released from the spheroplasts under these conditions and, compared with ribosomes isolated from cells broken in a different way, the spheroplast ribosomes were lower in RNase activity by an amount roughly equal to that released from the spheroplasts. It was not apparent at first whether the RNase might not, for all that, have been attached to the ribosomes within the spheroplast and have been released because of structural damage caused by the withdrawal of Mg^{2+} by the EDTA in the spheroplast medium (Neu and Heppel, 1964a). However, later control experiments showed that the spheroplast ribosomes have a normal ratio of RNA to protein and, by this criterion, are not grossly degraded (Neu and Heppel, 1964c). Furthermore, ribosomes isolated from broken whole cells did not release RNase when exposed to the reagents employed to form spheroplasts (EDTA, *tris*, sucrose and lysozyme), either singly or in combination (Neu and Heppel, 1964c; D. Elson, unpublished results).

There is no indication, then, that the release of RNase from spheroplasts is caused by a direct effect on the ribosomes; at least, such an effect has not been observed *in vitro*. Accordingly, Neu and Heppel (1964b, c) have proposed that the RNase of *E. coli* may not, or may not all, be located on the ribosomes but may be segregated near the surface of the cell, while the ribosomes are situated in the interior. As an alternative they suggest that the enzyme may be concentrated on those ribosomes or polysomes which are attached to the cell membrane, and that this fraction of the ribosomal RNase is, somehow, easily detached.

If the enzyme is not bound to the ribosome *in vivo*, it would appear somewhat difficult to account for the several strikingly specific features of its ribosomal attachment *in vitro*. That they can nevertheless be accounted for is shown by the following observation (Neu and Heppel, 1964c). Purified *E. coli* RNase, added to *E. coli* ribosomes, became bound to the particles. The added enzyme was not removed during washing in sucrose; it was enzymically inactive unless the ribosomes were destroyed; and when 30 S and 50 S ribosomes were separated in a sucrose density gradient, the newly attached enzyme was found only in the 30 S particle (Fig. 12). Thus, the earlier noted characteristics of the binding between RNase and ribosome were reproduced merely by mixing the two components together. While this does not prove that the ribosomal location of the *E. coli* RNase is an artifact produced when the cell is broken, it clearly indicates that such could be the case.

The release of enzyme from spheroplasts has several still puzzling features. For example, the six enzymes (including RNase) presently known to be released from EDTA-lysozyme spheroplasts (Neu and Heppel, 1964d, 1965) are not released from spheroplasts prepared by other methods (Neu and Heppel, 1964c). The other five enzymes are, however, released from cells which have been exposed to a shock treatment which does not convert them to spheroplasts, but in this case no RNase is released (Neu and Heppel,

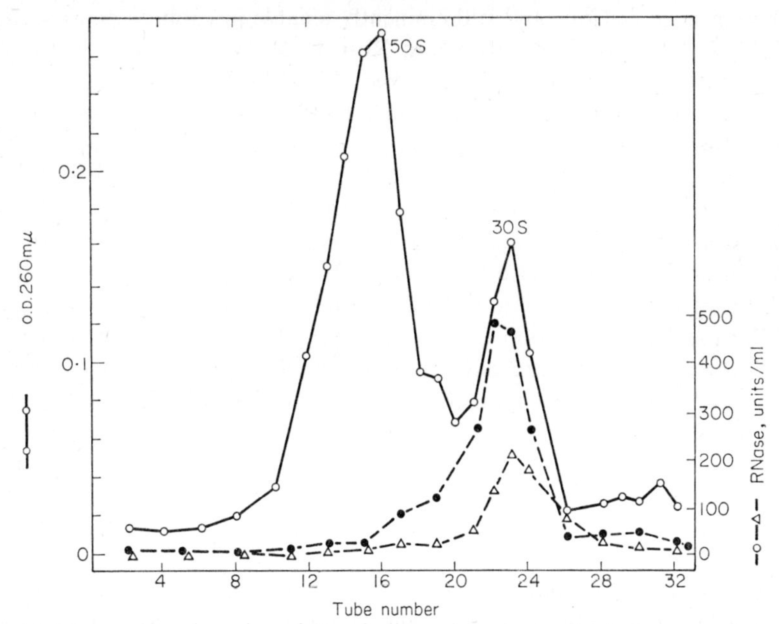

FIG. 12. The specific adsorption of *E. coli* ribonuclease by the 30 S ribosome of *E. coli.* Sucrose density gradient centrifuge patterns show the distribution of ribosomes (○), of the ribonuclease activity of untreated ribosomes (△), and of the enzymic activity of ribosomes which had been mixed with soluble ribonuclease (●). Reproduced with permission from Neu and Heppel (1964 c).

1964d, 1965). Furthermore, although the balance studies indicate that it is the "ribosomal" enzyme which is released from EDTA-lysozyme spheroplasts, this has been questioned by Anraku and Mizuno (1965), who suggest that it may be a different RNase, which they have discovered in the cell envelope of *E. coli* and which can be adsorbed by the ribosomes although it is not usually found in them. It has also been pointed out that the RNase released from spheroplasts is not chromatographically identical with RNase isolated from ribosomes (Neu and Heppel, 1964b; Anraku and Mizuno, 1965; Anderson and Carter, 1965), although this may mean that different treatments have wrought different changes in the molecules.

From all of this, it is obvious that we still have much to learn about the intracellular location of RNase in *E. coli*, and the matter may not be fully clarified until new techniques have been applied to the problem. It seems clear that the enzyme has a high and specific affinity for the ribosome, but how much, if any, of the enzyme is attached to the ribosomes in the intact cell, is at present an open question.

C. BIOLOGICAL FUNCTION OF RNASE

There has been much speculation on the function of RNase. A scavenger function is an obvious and still eminently tenable possibility, and a number of others have also been proposed. Roth (1954) was probably the first to suggest that the enzyme might have the task of breaking down the template RNA in order to free the proteins which had been made on the template (see also Elson, 1959b), and I shall return later to a modernized version of this proposal. It has also been suggested that RNase acts in uncovering RNA templates (Leslie, 1961) or in modulating the interaction between ribosome and sRNA (Bosch, Huizinga and Bloemendal, 1962). A correlation between growth and RNase has also been noted (Roth, 1963).

The apparent localization of the *E. coli* RNase in the ribosomes pointed to a ribosomal site of action for this enzyme. As noted above, the enzyme seemed to be bound exclusively to the 30 S ribosome with exceptional firmness and in such a way as to influence its activity. The properties of the bound RNase did not seem to be those of a nascent protein or of a non-specifically adsorbed protein. There were indications in the literature that a similar situation might hold for ribosomes and RNase in other organisms Bosch, *et al.*, 1962; Datta, Bhattacharyya and Ghosh, 1964; Roth, 1960b; Burdon, 1963; Stavy, Feldman and Elson, 1964; Danner and Morgan, 1963; Ohtaka and Uchida, 1963). Taken as a whole, this body of data suggested a connection between the physiological role of RNase and that of the ribosome, that is, that the enzyme might have to do with the biosynthesis of proteins. The later observations of Neu and Heppel, mentioned above, which cast doubt on the ribosomal location of RNase do not eliminate this possibility, for it remains true that the enzyme exhibits a high and specific affinity for the ribosome. This may imply a functional relationship, whether or not the enzyme is permanently located on the ribosome.

With the advent of messenger RNA and the elaboration of a detailed mechanism of protein synthesis, the number of acceptable ways in which RNase might participate in protein synthesis became rather strictly limited. Furthermore, it soon became evident that whatever its function might be, the enzyme could not be essential for the polymerization of amino acids. This arose from the observation that rabbit reticulocyte ribosomes, if washed, have very little RNase and quite possibly have none at all (Williamson and Mathias, 1963; Stavy *et al.*, 1964; Mathias, Williamson, Huxley and Page, 1964); nevertheless, they synthesize protein. This, together with another feature of the mammalian reticulocyte, suggested a role for the ribosomal RNase. In bacteria the mRNA is rapidly broken down and renewed (Levinthal, Keynan and Higa, 1962). This does not appear to take place in the mammalian reticulocyte (Marks, Willson, Kruh and Gros, 1962) and, indeed, it could hardly do so, since the cell lacks a nucleus and mRNA cannot be synthesized in the absence of DNA (Hurwitz and August, 1963). This suggested a correlation between RNase and the metabolic turnover of mRNA.

It had been indicated that RNase is not the enzyme responsible for the gross degradation of mRNA, since the end products of this degradation appear to be nucleoside-5′-phosphates (Sekiguchi and Cohen, 1963). We suggested, therefore, that the ribosomal RNase might inactivate mRNA, perhaps by cleaving no more than a single internucleotide bond, and so render the molecule unable to attach to ribosomes (Tal and Elson, 1963b, Elson, 1964b). The mRNA would then be exposed to the degradative action of other nucleases. The location of RNase on the 30 S ribosome of *E. coli* was in keeping with this suggestion, since the mRNA appears to be attached to the same particle (Okamoto and Takanami, 1963). It has also been pointed out that a molecule of mRNA survives, on the average, long enough to promote the synthesis of about ten to twenty protein molecules (in *B. subtilis*) and about one in ten ribosomes carries RNase (in *E. coli*). Thus, the lifetime of a molecule of mRNA might be governed by the random probability of its becoming attached to a ribosome which carries RNase (Artman and Engelberg, 1964).

Based on circumstantial evidence—affinity between RNase and the ribosome—this proposal requires that cells which possess a metabolically labile mRNA should also possess RNase. It can be argued, of course, that the function in question might be performed by RNase in certain organisms and by a different nuclease in others, but such an argument would seem less likely to be valid in the case of closely related organisms. With the discovery of bacterial strains which appear to lack RNase, a direct test of the matter became possible.

There are several reports of bacteria which lack RNase. They include *B. subtilis* and *St. albus* (Neu and Heppel, 1964d) and *Ps. fluorescens* (Wade and Robinson, 1963). No data are available for the first two, but the case of *Ps. fluorescens* was investigated rather thoroughly, in that activity was sought over a wide pH range in the presence of EDTA, and was not found. On the other hand, ribosomal RNase has been reported in *Ps. fluorescens* (McCarthy, 1959, cf. Roberts, 1964, p. 370), and a cursory examination of *B. subtilis* ribosomes in my laboratory indicated the presence of an RNase active in excess EDTA (B. Benish and D. Elson, unpublished results). These would appear to be instances where RNase is present in one strain and absent in another of the same bacterial species. More recently three RNase-less strains of *E. coli* have been reported. Two, isolated in the laboratory of J. D. Watson, are related mutants of the same strain of *E. coli* (Gesteland, 1965; see also Haruna and Spiegelman, 1965). Both lack RNase and one also lacks polynucleotide phosphorylase. The third is an unrelated strain of *E. coli* shown, in the laboratory of H. E. Wade, to lack RNase (Cammack and Wade, 1965; Wade and Robinson, 1963).

What remained was to examine the metabolic lability of mRNA in these RNaseless bacteria. This has been done with the three strains of *E. coli* mentioned above (T. Kivity-Vogel and D. Elson, 1966). The method of

Kepes (1963) was employed to measure the half life of the mRNA of induced β-galactosidase. The cultures examined contained no more than a small fraction of the RNase activity of a control culture of *E. coli* B. Nevertheless their mRNA became inactivated at essentially the same rate as that of the control. Thus, the lack of RNase did not cause a derangement of mRNA catabolism, and the enzyme would not appear to be involved in the inactivation of mRNA in *E. coli.* (one of the strains also appeared to lack polynucleotide phosphorylase, (2.7.7.8.).

We have seen, then, that a number of biological functions have been proposed for RNase and have, with the accumulation of new data, been ruled out. Certain possibilities still remain. One—that RNase is concerned with the regulation of the concentration of ribosomes in the cell—is now being investigated in my laboratory. But so far no ribosomal function has been demonstrated, and it remains possible that the striking affinity of RNase for the ribosome is merely a chemical accident of no physiological significance.

VIII. Ribosomal Enzymes—Other Enzymes which Act on Polynucleotides

The enzymes of this rather arbitrarily defined group have in common that they act in conjunction with a polynucleotide, which serves as either substrate, template or primer. Unlike RNase, these enzymes require magnesium or other divalent cations. These two features may not be unrelated to the presence of the enzymes in ribosomal preparations, a point that will be elaborated on later in this section.

A. Deoxyribonuclease (3. 1. 4. –)

DNase activity was first observed in a ribonucleoprotein prepared from *E. coli* by a drastic chemical technique designed not to preserve the fine structure of the ribosome but rather to remove as much extraneous protein as possible (Elson, 1959a, b). The enzyme was latent, in that conditions could be found under which it was inactive until something was done to disrupt the structure of the nucleoprotein (Fig. 13).

Ribosomes prepared conventionally by differential ultracentrifuging also contained DNase activity, but in varying amount. This was eventually found to depend on the degree of dissociation of the ribosomes, which is in turn controlled by the concentration of magnesium ions in the medium (Tal and Elson, 1961, 1963a). Ribosomes isolated in the form of 70 S and 100 S particles contained about 8 or 9% of the total DNase activity of the crude cell extract (from which cell debris had previously been removed), and their activity was not reduced by repeated washing in the same medium. When such purified ribosomes were dissociated to 30 S and 50 S particles

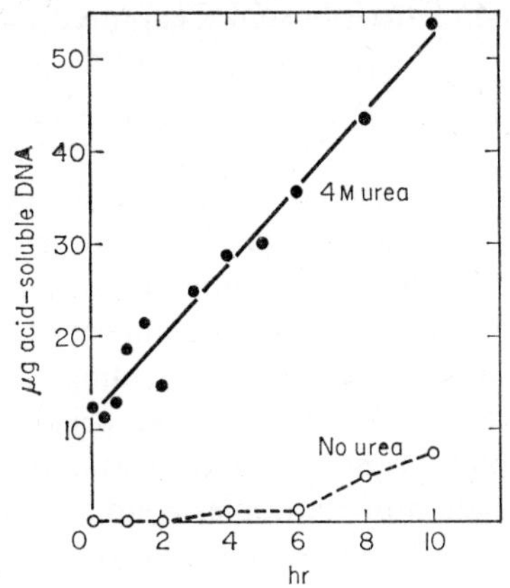

FIG. 13. The latency of deoxyribonuclease in a highly purified ribonucleoprotein from *E. coli*, and the activation of the enzyme by 4 M urea. The ribonucleoprotein was incubated with DNA at 37° in the presence of 0·005 M $MgSO_4$. In the absence of urea the enzyme became active only after six hours of incubation. In 4 M urea there was maximum activity from the beginning. Reproduced with permission from Elson (1959 b).

by lowering the Mg^{2+} concentration, about 90% of their DNase activity was released into the medium. When these dissociated ribosomes were re-converted to 70 S and 100 S particles in the presence of the previously released DNase, the enzyme became re-attached to the ribosomes, apparently as firmly as before (Fig. 14); and when the process was carried out with an

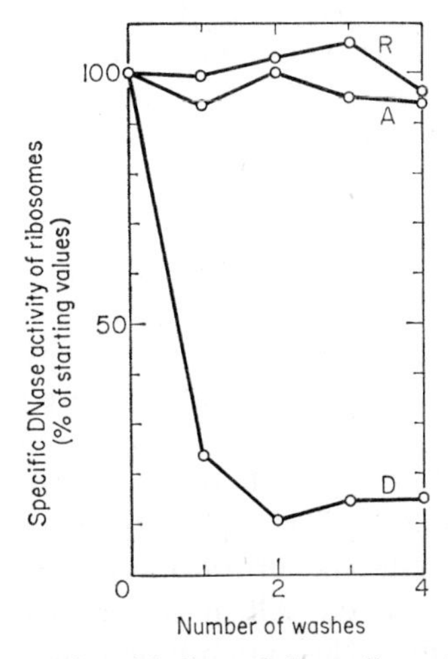

FIG. 14. The magnesium-dependent binding of deoxyribonuclease to *E. coli* ribosomes. The ribosomes were treated as follows: A. Maintained as 70 S and 100 S particles in 0·01 M magnesium acetate—0·001 M *tris* buffer, pH 7·4; D. Dissociated to 30 S and 50 S particles by dialysis against 0·001 M *tris*; R. First dissociated as in D, then re-associated in the medium of A. After treatment the ribosomes were washed repeatedly by centrifuging. Reproduced with permission from Tal and Elson (1963 a).

excess of released enzyme, the ribosomes captured more than their original content of DNase.

At intermediate concentrations of Mg^{2+}, which caused only partial dissociation of the ribosomes, the release of DNase was also partial and was roughly equivalent to the degree of ribosomal dissociation. Further, when the concentration of Mg^{2+} was shifted either upward or downward, DNase activity moved into or out of the ribosomes accordingly. In short, like ribosomal dissociation and association, the attachment of DNase to the ribosome is freely reversible in both directions, depending only on the final Mg^{2+} concentration. It is still not clear whether the primary factor governing the attachment is Mg^{2+} concentration or ribosomal association. It is, however, evident that the dissociation of ribosomes is not the only controlling factor, for enzyme release is much slower than ribosomal dissociation, and when complete dissociation is achieved in different media, the amounts of DNase released are also different (Tal and Elson, 1963a).

Except for the quantity of DNase involved this phenomenon of reversible release is reminiscent of the behaviour of nascent protein. As mentioned above, nascent protein is attached to the ribosomes by sRNA, and both the ribosomal sRNA and the nascent protein have been shown to become detached from the ribosomes at low concentrations of Mg^{2+} and to return to them at high Mg^{2+} concentrations (Tal and Elson, 1963a; Schlessinger and Gros, 1963; Elson, 1964a). However, nearly 10% of the total DNase participates in this reversible release, and this is too much to be nascent protein, by several orders of magnitude. It seems more likely, then, that if the phenomenon is of biological significance (and it is not certain that it is), its significance may not be general but may be restricted to a small group of ribosomal enzymes, and perhaps only to DNase. It might, for example, have to do with ribosomal participation in the regulation of DNA metabolism—perhaps with its co-ordination with protein synthesis—but this is pure conjecture, and more prosaic suggestions can also be advanced (see below).

There are four known DNases in *E. coli*, one endonuclease and three exonucleases, each distinguishable from the other by virtue of their properties and specificity (Lehman, 1963). It would be instructive to know whether all or only certain of these enzymes undergo the reversible release. Unwashed ribosomes from *E. coli* have been found to carry all four enzymes, each in an amount ranging from 5–14% of the total activity of that enzyme in the crude cell extract (Shortman and Lehman, 1964). Since the endonuclease is by far the predominant DNase in the cell, it is almost certainly involved in the observed reversible binding. This supposition is strengthened by the work of Weissbach and Korn (1964) who extracted ribosomes from *E. coli* in the absence of magnesium (presumably as 30 S and 50 S particles) and found them to contain only a low level of DNase activity. When the extraction was carried out with 0·01 M Mg^{2+}, the ribosomes (70 S and 100 S in this medium) carried much more DNase; this DNase was latent and appeared to be the endonuclease. Whether the level of ribosomal exonucleases also changes with the Mg^{2+} concentration is not known.

Upon the infection of *E. coli* with phage T4r, the quantity of ribosome-bound DNase rises markedly (Weissbach and Korn, 1964). The new ribosomal enzyme, whose synthesis is induced by the infecting phage, resembles *E. coli* exonuclease II. This does not, of course, shed light on the situation in the normal uninfected cell.

A comment may be inserted here concerning both the latency of the ribosomal DNase and the possible nature of its attachment to the ribosome. RNA was found long ago to inhibit DNase (Kozloff, 1953), and specifically the endonuclease (Lehman, 1963). Consequently, this enzyme is known to bind to RNA, and this might well account both for its binding to the RNA-rich ribosome and for its latency when so bound. Those conditions which have been employed to activate the ribosomal DNase are also conditions which activate the ribosomal RNase and lead to the degradation of the ribosomal RNA (Elson, 1959b; Weissbach and Korn, 1964). Magnesium is necessary for the action of DNase. Its function is unknown, but it might conceivably be required for the binding of the enzyme to its substrate, DNA. If this is so, then Mg^{2+} might function in a similar way to bind the enzyme to the RNA of the ribosome, and this might account for the reversible binding of DNase to the ribosome and for the fact that the extent of binding is a direct function of the magnesium ion concentration. This suggests that the binding of the DNase to the ribosome may be non-specific and that the association and dissociation of the ribosomes may not play a direct role in the binding. Whether the phenomenon is of biological value is, of course, an open question.

It should be noted that the above suggestion has been framed essentially for the endonuclease. As for the exonucleases, there is no evidence as to whether or not their binding to the ribosome is reversible or magnesium-dependent. Further, since they are not inhibited by RNA, it is not certain that they bind to this substance.

B. DNA POLYMERASE (2. 7. 7. 7)

Unwashed 70 S and 100 S ribosomes of *E. coli* (extracted with 0·01 M Mg^{2+}) carry up to 40% of the DNA polymerase (DNA nucleotidyltransferase) activity of the crude cell extract (Weissbach and Korn, 1964). When the extract contains no magnesium ions, the ribosomal activity drops to less than 2%. Cells infected with T4r phage produce a new DNA polymerase. The phage-induced enzyme has a higher affinity for ribosomes, which now carry 30% of the total activity in the absence of magnesium ions and over 50% in 0·01 M Mg^{2+}. The normal enzyme thus shows a magnesium-dependent affinity for ribosomes reminiscent of that of DNase. The greater affinity between ribosomes and the phage-induced enzyme appears less sensitive to the concentration of magnesium ions.

C. POLYRIBONUCLEOTIDE POLYMERASES (2.7 7. –)

A magnesium-dependent enzyme which catalyses the synthesis of polyribo-adenylic acid from ATP in the presence of RNA has been observed to be concentrated in *E. coli* ribosomes. Some seventy to ninety per cent of the activity is ribosomal (August, Ortiz and Hurwitz, 1962; Ortiz, 1963).

An enzyme which incorporates nucleotides from ATP or UTP into polyribonucleotides has been found on the ribosomes of the nuclei of Landschutz ascites tumour cells (Burdon, 1963).

D. AMINOACYL-sRNA SYNTHETASES (6. 1. 1. –)

The enzymes which activate amino acids and attach them to sRNA are soluble enzymes. As mentioned above, low levels of activity have been found in ribonucleoproteins prepared from rat liver (von der Decken, 1961). Believed to be a contamination, the activity nevertheless resisted strenuous efforts to remove it.

E. RNA METHYLASES

When the methionine-requiring mutant K 12 W-6 of *E. coli* is starved of methionine, it ceases to synthesize protein. Normally, the synthesis of RNA would also be blocked, but this strain carries a second mutation which causes an impaired control of RNA synthesis, and the production of RNA continues (Srinivasan and Borek, 1964). Under these conditions the cells accumulate RNA in the form of particles which resemble and may be identical with the particles produced in the presence of chloramphenicol (Nomura and Watson, 1959). These particles appear to be defective ribosomes which carry the full complement of ribosomal RNA, but less than the full complement of ribosomal proteins.

Ribosomal RNA contains a very small amount of methylated purines and pyrimidines; these bases are methylated by specific enzymes which act on the fully polymerized RNA (Srinivasan and Borek, 1964; Gordon and Boman, 1964). Since, however, the source of methyl groups is a molecule derived from methionine (Srinivasan and Borek, 1964), the RNA of the particles lacks the methylated bases. When the purified particles are incubated with the methyl donor, they are able to accomplish the methylation of their own RNA in the absence of any soluble factors. This indicates that the required methylating enzymes are present on the defective ribosomes (Gordon and Boman, 1964).

F. POLYNUCLEOTIDE PHOSPHORYLASE (2.7.7.8.)

Ribosomes (not always washed) of *E. coli* and *Ps. aeruginosa* have been found to contain as much as 40% of the total polynucleotide phosphorylase activity of crude cell extracts (Wade, 1961; Wade and Lovett, 1961; Kimhi and Littauer, 1962; Strasdine, Hogg and Campbell, 1962; Grunberg-Manago, 1963). There are also indications of activity in ribosomes isolated from *B. subtilis* and *St. aureus* (B. Benish and D. Elson, unpublished results). On the other hand, the free ribosomes of lysozyme protoplasts of *Strep. faecalis* are devoid of activity and the enzyme appears to be concentrated on the cell membrane (Abrams and McNamara, 1962).

The considerable variation in the amounts of polynucleotide phosphorylase found in different ribosome preparations indicates that the attachment is rather labile. Thus, Y. Kimhi and U. Z. Littauer (1962, and personal communication) have washed 70 S and 100 S *E. coli* ribosomes to constant activity and found them to contain from 5–15% of the total cellular activity. When such particles were centrifuged in a sucrose gradient of the same magnesium concentration, much of the residual enzyme was removed. When 70 S and 100 S ribosomes which had undergone either or both of the above treatments were transferred to a medium of lower magnesium ion concentration, most of the remaining ribosomal activity was released. Thus, the polynucleotide phosphorylase content of the ribosomes is greatly influenced by the way in which the particles are treated. Further, the binding of enzyme to ribosome appears to be magnesium-dependent, as previously noted for DNase and DNA polymerase.

G. PHOSPHODIESTERASES (3. 1. 4. 1)

Phosphodiesterase activity has been found in ribosomes of goat brain (Datta and Ghosh, 1963b) and *E. coli* (Wade and Lovett, 1961), but is probably absent from those of rabbit reticulocytes (Stavy *et al.*, 1964). The recently discovered potassium-activated phosphodiesterase has been isolated from the soluble fraction of *E. coli*, although ribosomal activity was not ruled out (Spahr and Schlessinger, 1963).

H. THE BINDING TO RIBOSOMES OF ENZYMES WHICH ACT ON POLYNUCLEOTIDES

The enzymes listed in this section all interact with a polymeric nucleic acid substrate, template, or primer. It is not surprising, therefore, that they are often found attached to the RNA-rich ribosomes. This may simply follow from their necessary affinity for nucleic acid, and may or may not be significant biologically. Much remains to be learned of the firmness and extent of binding. In some cases this has been studied with some thoroughness, but in others, in which this matter has not been the main burden of the work, there is little more than an estimate of enzyme activity in unwashed preparations of ribosomes.

Magnesium ions are required for the action of all of these enzymes. The role of this ion is not known. If it should promote the binding of enzyme to polynucleotide, then we might have a basis for understanding the effect observed with three of these enzymes—DNase, DNA polymerase and polynucleotide phosphorylase—namely, that the degree of their binding to the ribosome depends on the magnesium ion concentration of the medium. If this is true, we might expect the binding of other enzymes of this group to be magnesium-dependent, but this has not yet been tested.

IX. Ribosomal Enzymes—Other Examples

A. Phosphatases (3. 1. 3. –)

Phosphatases, both acid and alkaline, have been found in ribosomes isolated from swine and rat kidney (Binkley, 1961), goat brain (Datta and Ghosh, 1963a), and *E. coli* (Spahr and Hollingworth, 1961; Warren and Goldthwait, 1961). Except for the kidney particles, the ribosomes had been purified by repeated sedimentation at low ionic strength. This treatment brought the alkaline phosphatase activity of *E. coli* ribosomes to a constant level under conditions where purified enzyme was not adsorbed (Warren and Goldthwait, 1961); this ribosomal activity was presumed to represent nascent enzyme molecules. The acid phosphatase activity of *E. coli* ribosomes was not removed by centrifuging through 50% sucrose. It was present in both 30 S and 50 S ribosomes, with the 30 S particle showing a higher level of activity (Spahr and Hollingworth, 1961).

B. Peptidases

1. Leucine Aminopeptidase (3.4.1.1.)

The first of several ribosomal peptidases found, the leucine aminopeptidase of *E. coli*, has also been the most extensively studied (Bolton and McCarthy, 1959, cf. Roberts, 1964, p. 365; Matheson, 1963). Unwashed ribosomes carry from 10–80% of the total cellular activity, and the quantity appears to depend on the physiological state of the cells. Thus, cells grown in glucose-rich media have a higher proportion of ribosomal enzyme than cells grown in media poorer in glucose (Bolton and McCarthy, 1959, cf. Roberts, 1964, p. 365). During the transition of a culture from exponential growth to stationary phase the enzymic activity of the free ribosomes falls and that of the membrane fraction rises; it was suggested that this may come about through a transfer of enzyme to membrane-bound ribosomes or a binding of free ribosomes to the membrane (Matheson, 1963).

This ribosomal enzyme appears to be distributed equally between well washed 30 S and 50 S ribosomes. Its specificity may not be identical with that of the soluble enzyme. It has therefore been suggested that the ribosomal peptidase may play a role in protein synthesis (Bolton and McCarthy, 1959, cf. Roberts, 1964, p. 365).

Leucine aminopeptidase has also been found to be attached to the ribosomes of *Ps. fluorescens*, *B. megatherium* and *A. aerogenes*, but not to those of yeast (Bolton and McCarthy, 1959, cf. Roberts, 1964, p. 365).

2. *Cysteinyl-glycine Dipeptidase*

An enzyme which hydrolyses L-cysteinylglycine was first observed by Binkley (1961) to sediment with the ribosomes of swine and rat kidney. This enzyme was studied further in *E. coli* by McCorquodale (1963), who showed that ribosomes purified by repeated centrifuging through 10% sucrose at low ionic strength reached a constant activity at which they still carried some 30% of the total cellular activity. When the ribosomes were dissociated, the enzyme appeared exclusively in the 50 S ribosome. When, however, the magneisum concentration was raised and the ribosomes underwent re-association to form a mixture of 70 S, 85 S, 100 S and larger aggregates, an unanticipated result was obtained. The 70 S ribosomes were apparently devoid of the peptidase; the enzyme was found only in 85 S and larger particles (Fig. 15). This was interpreted as indicating that those 50 S particles which carry the enzyme are incorporated into 70 S ribosomes which have a higher tendency to form larger aggregates. The nature of the 85 S particle is obscure (Roberts *et al.*, 1963), but the 100 S particle is a dimer formed by the joining of two 70 S ribosomes through their 30 S sub-particles (Hall and Slayter, 1959; see also Figs. 1 and 2). It might be, then, that those 50 S ribosomes which carry the enzyme belong to a class of particles which selectively pair with those 30 S ribosomes possessing a high tendency to interact with each other. There is a hint here of a multiple relationship which might be interesting to clarify.

3. *Iminodipeptidase* (3.4.3.6)

An enzyme which splits L-prolylglycine but not glycylglycine or poly-L-proline has been found in chemically purified ribonucleoproteins as well as in conventionally prepared ribosomes from *E. coli* (Elson, 1961b, 1962b, and unpublished results).

C. NAD PYROPHOSPHORYLASE (2. 7. 7. 1)

The NAD pyrophosphorylase of rat liver and Krebs-2 tumour is a nuclear enzyme. It has been shown to be associated with the nuclear ribosomes, while cytoplasmic ribosomes do not carry the enzyme (Traub,

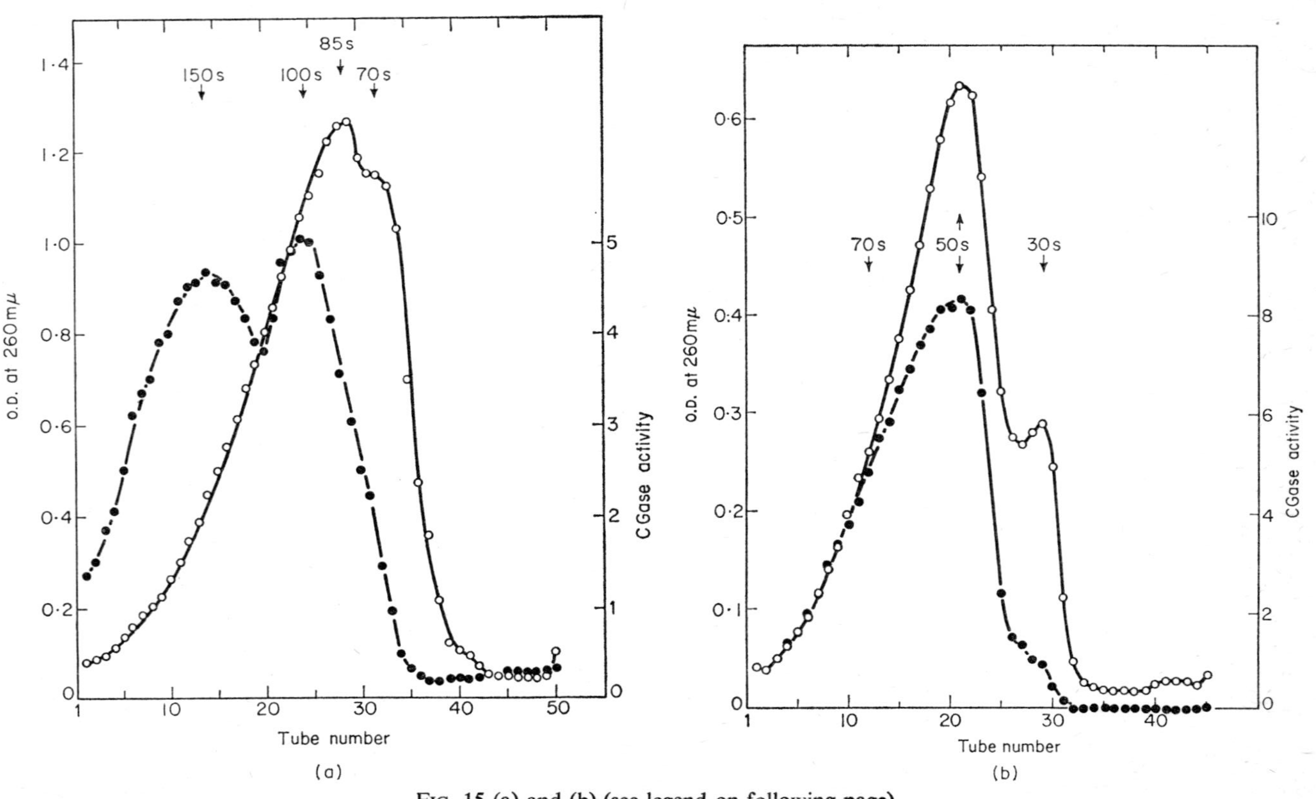

FIG. 15 (a) and (b) (see legend on following page)

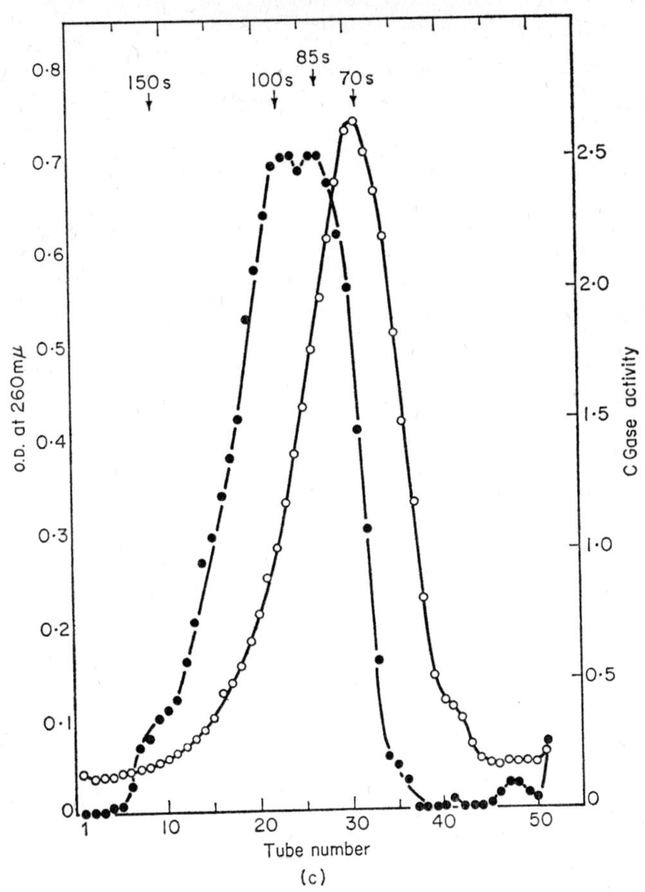

FIG. 15. (a), (b) and (c). The location of cysteinyl-glycine dipeptidase in *E.* coli ribosomes. The sucrose density gradient centrifuge patterns show the distribution of ribosomes (○) and enzymic activity (●). (a). Associated ribosomes in 0·01 m Mg^{2+}. (b). Dissociated ribosomes in 0·1 m Mg^{2+}. (c). Ribosomes which had been dissociated and then re-associated in the solvents of b. and a. Reproduced with permission from McCorquodale (1963).

Kaufmann and Ginzburg-Tietz, 1964). More recently, the localization has been narrowed down further and the enzyme has been shown to reside in the ribosomes of the nucleolus (Kaufmann, Traub and Ginzburg-Tietz, 1964). Because of its location, the enzyme has been used as a marker by which nuclear ribosomes can be distinguished from cytoplasmic particles. Thus, when Krebs-2 tumour cells are infected with encephalomyocarditis virus, about half of the total NAD pyrophosphorylase activity moves from the nucleus to the cytoplasm, suggesting that the viral infection causes a release of nuclear ribosomes to the cytoplasm (Ginzburg-Trietz, Kaufmann and Traub, 1964).

D. ADENOSINE TRIPHOSPHATASE (3.6.1.3)

While studying amino acid incorporation *in vitro* in an isolated system derived from *E. coli*, Schlessinger (1964) found that the ribosomes contained a high level of ATPase activity. When the ribosomes were dissociated and separated by sucrose density gradient centrifuging, the enzyme was located exclusively in the 30 S particles (Fig. 16).

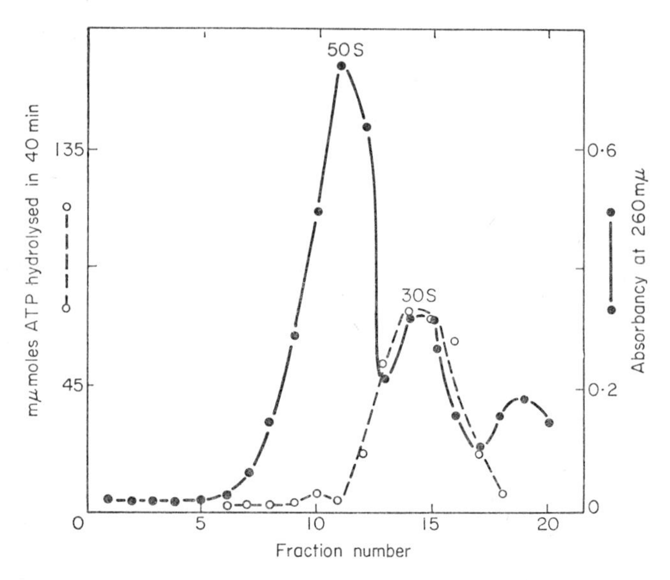

FIG. 16. The intraribosomal location of ATPase in *E. coli*, as shown by sucrose density gradient centrifuging. Reproduced with permission from Schlessinger (1964).

E. OTHER ENZYMES FOUND IN RIBOSOME PREPARATIONS

There are a number of references to ribosomal enzymic activities which have not been mentioned. Some of them are: proteolytic activity in *E. coli* (Chaloupka, 1961); amylase (3.2.1.–) in well washed goat brain ribosomes (Datta and Ghosh, 1963a); glutamate dehydrogenase (1.4.1.–) and β-glucuronidase (3.2.1.31) in calf thymus nuclei (Wang and Wang, 1962); traces of aldolase activity (4.1.2.7) in *E. coli* (Elson, 1961b; 1962b); peroxidase (1.11.1.7) in pea seedlings and rabbit liver (Matsushita and Ibuki, 1960b) and in wheat embryo (Lanzani and Galante, 1964); and ribonucleoside triphosphatases (3.6.1.–) in pea seedlings (Raacke, 1963) and *E. coli* (Raacke and Fiala, 1964). In the last mentioned case the enzymes appeared to be bound only to polysomes. They were not found in 70 S ribosomes produced by breaking down the polysomes.

X. Summary

In this chapter I have tried to outline what is known and what may reasonably be inferred about the function and structure of the ribosome, and to describe the current, somewhat chaotic, status of ribosomal enzymology. The enzymes considered here fall into two broad classes: those which, from our present understanding of the role of the ribosome in protein synthesis, we might expect to be associated with ribosomes; and other enzymes which have been found associated with ribosomes although there seems no reason *a priori* to expect them to be there.

The first class includes both the enzymes which catalyse those reactions of protein synthesis which take place on the ribosome and the nascent enzymes, the unreleased products of ribosomal protein synthesis. Many investigators have touched on the question of nascent enzymes, yet in no case has a nascent enzyme been unequivocally identified and its properties fully elucidated. What has been done so far has served, essentially, to define the problem, to clarify the nature of certain difficulties, and to indicate experimenta approaches which may possibly be effective.

As for the enzymes which mediate the ribosomal stages of protein synthesis, they are largely a matter of conjecture. Insofar as we know the reactions of protein biosynthesis, we assume that certain enzymes are involved. Since our knowledge is incomplete, we do not yet know exactly how many such enzymes there are. There is a still meagre body of indirect experimental evidence which testifies to the existence of some of these enzymes, but none of them has yet been isolated or characterized. Furthermore, there is no firm evidence as to whether these enzymes are actually bound to the ribosome. Our guiding principle here states merely that there is a functional association between the ribosome and these enzymes, but it says nothing about the physical or chemical nature of the association. Thus, even if we were to have a complete list of these enzymes, we would still be ignorant of exactly how they work. This would require a detailed understanding of the ribosomal structure' particularly its surface, and the complex interactions among ribosome, enzyme, substrate, etc., and these are matters which will remain unclear for some time to come.

The second broad class of ribosomal enzymes includes those which have been reported to be associated with ribosomes but which do not seem to fit into the first class; that is, they are enzymes for whose ribosomal association we have experimental evidence but no *a priori* guiding principle. Quite a few such enzymes are listed. It is certain that the list would be shorter if all the ribosome preparations involved had been washed thoroughly. It would be, however, unjustifiably glib to dismiss this entire class of ribosomal enzymes as adsorption artifacts. Some of them—such as, for example, ATPase, RNase and peptidases—may eventually find their way into the list of enzymes which participate in ribosomal protein synthesis. Others may conceivably teach us about other presently unsuspected aspects of ribosome function, for it is not impossible that the ribosome, by virtue of its abundance, makeup and ability to interact with many kinds of molecule, may play a role in the regulation and co-ordination of metabolic processes other than protein synthesis. If this is true, this role may emerge in the course of time as a generalization from a mass of individual observations of which each, in itself, may not seem to be of general significance.

TABLE I. *A List of References to Enzymic Activity in Preparations of Ribosomes*

Enzyme	Enzyme Commission Number	Source of Ribosomes	References
Adenosine triphosphatase	3.6.1.3	*E. coli*	Raacke and Fiala (1964), Schlessinger (1964)
		Pea	Raacke (1963)
Alcohol dehydrogenase	1.1.1.1	Yeast	Warren and Goldthwait (1962)
Aldolase	4.1.2.7	*E. coli*	Elson (1961b, 1962b)
Aminoacyl-sRNA synthetase	6.1.1.–	Rat liver	von der Decken (1961)
Amylase	3.2.1.–	Goat brain	Datta and Ghosh (1963a)
		Guinea pig pancreas	Siekevitz and Palade (1960b, 1962)
		Not specified	Siekevitz (1965)
Cytidine triphosphatase	3.6.1.–	*E. coli*	Raacke and Fiala (1964)
		Pea	Raacke (1963)
Deoxyribonuclease	3.1.4.–	*E. coli*	Elson (1959b, 1961b, 1962b, 1964b), Lehman (1963), Shortman and Lehman (1964), Tal and Elson (1961, 1963a), Weissbach and Korn (1964)
DNA polymerase	2.7.7.7	*E. coli*	Weissbach and Korn (1964)
β-Galactosidase	3.2.1.23	*E. coli*	Chaloupka (1961), Duerksen and Cowie (1961), cf. Roberts (1964), p. 407 Elson (1964b), Fogel and Elson (1964), Kiho and Rich (1964), Lederberg,, Rotman and Lederberg (1964), Madison and Dickman (1963b), Zipser (1963), Zipser and Perrin (1963),
		Not specified	Proctor (1961)

TABLE I—*continued.*

Enzyme	Enzyme Commission Number	Source of Ribosomes	References
β-Glucosidase	3.2.1.21	Yeast	Hauge, MacQuillan, Cline and Halvorson (1961), Kihara, Halvorson and Bock (1961), Kihara, Hu and Halvorson (1961)
β-Glucuronidase	3.2.1.31	Calf thymus nuclei	Wang and Wang (1962)
Glutamate dehydrogenase	1.4.1.–	Calf thymus nuclei	Wang and Wang (1962)
Guanosine triphosphatase	3.6.1.–	*E. coli*	Conway and Lipmann (1964), Nishizuka and Lipmann (1965), Raacke and Fiala (1964)
		Pea	Raacke (1963)
Lactate dehydrogenase	1.1.1.27	Guinea pig kidney	Keck and Choules (1962)
		Guinea pig liver	Keck and Choules (1962)
NAD pyrophosphorylase	2.7.7.1	Krebs-2 tumour nucleoli	Ginzburg-Tietz, Kaufmann and Traub (1964), Kaufmann, Traub and Ginzburg-Tietz (1964), Traub, Kaufmann and Ginzburg-Tietz (1964),
		Rat liver nucleoli	Ginzburg-Tietz, Kaufman and Traub (1964), Kaufman, Traub and Ginzburg-Tietz (1964), Traub, Kaufmann and Ginzburg-Tietz (1964),
Neuraminidase	3.2.1.18	Influenza virus	Noll (1962)
Penicillinase	3.5.2.6	*B. cereus*	Duerksen and O'Connor (1963)
Peptidases			
Cysteinyl-glycine dipeptidase	3.4.3.5	*E. coli*	McCorquodale (1963)
		Rat kidney	Binkley (1961)
		Swine kidney	Binkley (1961)
Iminodipeptidase	3.4.3.6	*E. coli*	Elson (1961b, 1962b)

TABLE I—*continued.*

Enzyme	Enzyme Commission Number	Source of Ribosomes	References
Leucine aminopeptidase	3.4.1.1	*A. aerogenes*	McCarthy (1959), and cf. also Roberts, 1964 p. 370
		B. megatherium	McCarthy (1959) and cf. also Roberts, 1964 p. 370
		E. coli	Bolton and McCarthy (1959); also cf. Roberts (1964, p. 365); Matheson (1963)
		Ps. fluorescens	McCarthy (1959) and cf. also Roberts, 1964 p. 370
Peroxidase	1.11.1.7	Pea	Matsushita and Ibuki (1960b)
		Rabbit liver	Matsushita and Ibuki (1960b)
		Wheat embryo	Lanzani and Galante (1964)
Phosphatases	3.1.3.–	*E. coli*	Spahr and Hollingworth (1961), Warren and Goldthwait (1961)
		Goat brain	Datta and Ghosh (1963a)
		Rat kidney	Binkley (1961)
		Swine kidney	Binkley (1961)
Phosphodiesterases	3.1.4.1	*E. coli*	Wade (1961), Wade and Lovett (1961)
		Goat brain	Datta and Ghosh (1963b)
Polynucleotide phosphorylase	2.7.7.8	*B. subtilis*	B. Benish and D. Elson, unpublished
		E. coli	Grunberg-Manago (1963), Kimhi and Littauer (1962), Wade (1961), Wade and Lovett (1961)
		Ps. aeruginosa	Grunberg-Manago (1963), Strasdine, Hogg and Campbell (1962)
		St. aureus	B. Benish and D. Elson, unpublished
Polyribonucleotide-synthesizing enzymes		*E. coli*	August, Ortiz and Hurwitz (1962), Ortiz (1963)
		Landschutz ascites tumour nuclei	Burdon (1963)
Protease		*E. coli*	Chaloupka (1961)

TABLE I—*continued.*

Enzyme	Enzyme Commission Number	Source of Ribosomes	References
Protease zymogens			
Chymotrypsinogen		Beef pancreas	Dickman, Holtzer and Gazzinelli (1962); Keller and Cohen (1961), Keller, Cohen and Wade (1963), Madison and Dickman (1963a)
		Guinea pig pancreas	Siekevitz and Palade (1960a, 1962)
"Trypsin-activable protease"		Guinea pig pancreas	Siekevitz and Palade (1958a, (1960b)
Trypsinogen		Beef pancreas	Keller and Cohen (1961), Keller, Cohen and Wade (1963)
Protein synthesis, enzymes of		*E. coli*	Traut and Monro (1964)
		Rabbit reticulocyte	Arlinghaus, Shaeffer and Schweet (1964), Morris (1964)
Ribonuclease	2.7.7.–	*A. aerogenes*	McCarthy (1959), cf. Roberts, 1964, p. 370
		Apple leaves	Kessler and Engelberg (1962)
		Beef pancreas	Dickman, Holtzer and Gazzinelli (1962), Gazzinelli and Dickman (1962), Keller and Cohen (1961), Keller, Cohen and Wade (1963, 1964), Madison and Dickman (1963a, 1963b), Siekevitz (1963)
		B. megatherium	McCarthy (1959), cf. Roberts, 1964, p.370
		B. subtilis	B. Benish and D. Elson, unpublished
		Calf liver	Siekevitz (1963)
		Calf thymus nuclei	Wang and Wang (1962)
		E. coli	Anderson and Carter (1965), Anraku and Mizuno (1965), Artman and Engelberg (1964, 1965), Beer, Highton and McCarthy (1960), Bolton (1959b) (also, cf. Roberts, 1964, p. 362),

TABLE I—*continued.*

Enzyme	Enzyme Commission Number	Source of Ribosomes	References
Ribonuclease (continued)			Elson (1958, 1959b, 1961b, 1962b, 1964b), Elson and Tal (1959), Gesteland (1965), Jardetzky and Julian (1964), Lederberg, Rotman and Lederberg (1964), Lehman (1963), McQuillen (1961), Neu and Heppel (1964a, b, c, d), Shortman and Lehman (1964), Spahr and Hollingworth (1961), Spirin, Kiselev, Shakulov and Bogdanov (1963), Spitnik-Elson (1962a), Tal and Elson (1963b), Thomas and Herbst (1963), Wade (1961), Waller (1964), Zillig, Krone and Albers (1959)
		Goat brain	Datta, Bhattacharyya and Ghosh (1964), Datta and Ghosh (1963a, c; 1964)
		Guinea pig liver	Martin, England, Turkington and Leslie (1963), Siekevitz (1963)
		Guinea pig pancreas	Siekevitz (1963), Siekevitz and Palade (1958a, 1960b, 1962)
		Landschutz ascites tumour nuclei	Burdon (1963)
		Pea	Lett and Takahashi (1962), Matsushita and Ibuki (1960a), Ts'o, Bonner and Vinograd (1958b)
		Ps. fluorescens	McCarthy (1959) and cf. Roberts, 1964, p. 370, Wade and Robinson (1963)
		Rabbit liver	Stavy, Feldman and Elson (1964)
		Rabbit lymph node	D. Elson, unpublished

TABLE I—*continued.*

Enzyme	Enzyme Commission Number	Source of Ribosomes	References
Ribonuclease (continued)		Rabbit reticulocyte	Elson (1964b), Farkas, Singer and Marks (1964), Mathias, Williamson, Huxley and Page (1964), Stavy, Feldman and Elson (1964), Williamson and Mathias (1963)
		Rabbit spleen	D. Elson, unpublished
		Rat Jensen sarcoma	Petermann and Pavlovec (1963a)
		Rat liver	Bosch, Huizinga and Bloemendal (1962), Madison (1962), Madison and Dickman (1963a), Petermann and Pavlovec (1963b), Roth (1960a, b), Stavy, Feldman and Elson (1964), Tashiro (1958)
		Rat spleen	Wust and Novelli (1964)
		Rat uterus	Greenman and Kenney (1963)
		Tobacco leaf	Pirie (1950, 1957), Whitfeld and Williams (1963)
		Yeast	Danner and Morgan (1963), Kihara, Halvorson and Bock (1961), McCarthy (1959) and cf. Roberts, 1964, p. 370, Ohtaka and Uchida (1963), Stenesh (1958)
		Zoospore nuclear cap	Lovett (1963)
RNA methylases		*E. coli*	Gordon and Boman (1964)
Triosephosphate dehydrogenase	1.2.1.–	Yeast	Warren and Goldthwait (1961, 1962)
Tryptophan synthetase	4.2.1.20	*E. coli*	Horibata and Kern (1964)
Uridine triphosphatase	3.6.1.–	*E. coli*	Raacke and Fiala (1964)
		Pea	Raacke (1963)

REFERENCES

Abrams, A., and McNamara, P. (1962). *J. biol. Chem.* **237**, 170.

Allende, J. E., Monro, R., and Lipmann, F. (1964). *Proc. natn. Acad. Sci., U.S.A.* **51**, 1211.

Anderson, J. H., and Carter, C. E. (1965). *Biochemistry* **4**, 1102.

Anfinsen, C. B. and White, Jr., F. H. (1961). *In* "The Enzymes" (P. D. Boyer, H. Lardy, and K. Myrbäck, eds.) 2nd edition, vol. 5, p. 95, Academic Press, New York and London.

Anraku, Y. and Mizuno, D. (1965). *Biochem. biophys. Res. Comm.* **18**, 462.

Arlinghaus, R., Favelukes, G. and Schweet, R. (1963). *Biochem. biophys. Res. Comm.* **11**, 92.

Arlinghaus, R., Shaeffer, J. and Schweet, R. (1964). *Proc. natn. Acad. Sci., U.S.A.* **51**, 1291.

Aronson, A. I. (1962). *J. mol. Biol.* **5**, 453.

Artman, M. and Engelberg, H. (1964). *Biochim. biophys. Acta* **80**, 517.

Artman, M. and Engelberg, H. (1965). *Biochim. biophys. Acta* **95**, 687.

August, J. T., Ortiz, P. J. and Hurwitz, J. (1962). *J. biol. Chem.* **237**, 3786.

Beer, M., Highton, P. J. and McCarthy, B. J. (1960). *J. mol. Biol.* **2**, 447.

Binkley, F. (1961). *J. biol. Chem.* **236**, 1075.

Boezi, J. A., Bolton, E. T., Britten, R. J., Cowie, D. B., McCarthy, B. J., Midgeley, J. E. and Roberts, R. B. (1961). "Carnegie Institution of Washington Year Book," 60.

Bolton, E. T. (1959a). *In* "Carnegie Institution of Washington Year Book," 58, p. 274.

Bolton, E. T. (1959b). *In* "Carnegie Institution of Washington Year Book" 58 p. 276.

Bolton, E. T. and McCarthy, B. J. (1959). *In* "Carnegie Institution of Washington Year Book, 58, p. 278.

Bolton, E. T., Hoyer, B. H. and Ritter, D. B. (1958). *In* "Microsomal Particles and Protein Synthesis" (R. B. Roberts, ed.) p. 18. Pergamon Press, London.

Bolton, E. T., Britten, R. J., Cowie, D. B., McCarthy, B. J., McQuillen, K. and Roberts, R. B. (1959). "Carnegie Institution of Washington Year Book," 58, p. 259.

Bonnett, H. J., Jr. and Newcomb, E. H. (1965). *J. Cell Biol.* **27**, 243.

Bosch, L., Huizinga, F. and Bloemendal, H. (1962). *Biochim. biophys. Acta* **61**, 220.

Brachet, J. (1942). *Arch. biol. (Liège)* **53**, 207.

Brachet, J. (1960). "The Biological Role of Ribonucleic Acids", Elsevier Publishing Co., Amsterdam.

Burdon, R. H. (1963). *Biochem. biophys. Res. Commun.* **11**, 472.

Cammack, K. A. and Wade, H. E. (1965). *Biochem. J.* **96**, 671.

Caspersson, T. (1941). *Naturwissenschaften* **29**, 33.

Chaloupka, J. (1961). *Nature, Lond.* **189**, 512.

Chao, F.-C. and Schachman, H. K. (1956). *Archs Biochem. Biophys.* **61**, 220.

Conway, T. W. and Lipmann, F. (1964). *Proc. natn. Acad. Sci. U.S.A.* **52**, 1462.

Cowie, D. B., Spiegelman, S., Roberts, R. B. and Duerksen, J. D. (1961). *Proc. natn. Acad. Sci. U.S.A.* **47**, 114.

Cox, E. C. and Flaks, J. G. (1964). *Fed. Proc.* **23**, 220.

Crampton, C. F. and Petermann, M. L. (1959). *J. biol. Chem.* **234**, 2642.

Crick, F. H. C. (1958). *Symp. Soc. exp. Biol.* **12**, 138.

Crick, F. H. C. (1964). Sixth International Congress of Biochemistry, Proceedings of the Plenary Sessions, p. 109.

Curry, J. B. and Hersh, R. T. (1962). *Biochem. biophys. Res. Commun.* **6**, 415.
Danner, J. and Morgan, R. S. (1963). *Biochim. biophys. Acta* **76**, 652.
Datta, R. K. and Ghosh, J. J. (1963a). *J. Neurochem.* **9**, 463.
Datta, R. K. and Ghosh, J. J. (1963b). *J. Neurochem.* **10**, 285.
Datta, R. K. and Ghosh, J. J. (1963c). *Biochem. Pharmacol.* **12**, 1355.
Datta, R. K. and Ghosh, J. J. (1964). *J. Neurochem.* **11**, 595.
Datta, R. K., Bhattacharyya, D. and Ghosh, J. J. (1964). *J. Neurochem.* **11**, 87.
Decken, A. von der (1961). *Expl Cell Res.* **23**, 517.
Dickman, S. R., Holtzer, R. L. and Gazzinelli, G. (1962). *Biochemistry* **1**, 574.
Dounce, A. L. (1955). *In* "The Nucleic Acids" (E. Chargaff and J. N. Davidson, eds.) Vol. II, p. 93, Academic Press, New York and London.
Duerksen, J. D. and Cowie, D. B. (1961). *In* "Carnegie Institution of Washington Year Book" 60, p. 322.
Duerksen, J. D. and O'Connor, M. L. (1963). *Biochem. biophys. Res. Commun.* **10**, 34.
Duerre, J. A. (1964a). *Biochim. biophys. Acta* **86**, 490.
Duerre, J. A. (1964b). *J. Bact.* **88**, 130.
Elson, D. (1958). *Biochim. biophys. Acta* **27**, 216.
Elson, D. (1959a). *Biochim. biophys. Acta* **36**, 362.
Elson, D. (1959b). *Biochim. biophys. Acta* **36**, 372.
Elson, D. (1961a). *Biochim. biophys. Acta* **53**, 232.
Elson, D. (1961b). *In* "Protein Biosynthesis" (R. J. C. Harris, ed.) p. 291, Academic Press, New York and London.
Elson, D. (1962a). *Biochim. biophys. Acta* **61**, 460.
Elson, D. (1962b). *In* "The Molecular Basis of Neoplasia", p. 535, University of Texas Press, Austin, Texas, U.S.A.
Elson, D. (1964a). *Biochim. biophys. Acta* **80**, 379.
Elson, D. (1964b). *In* "New Perspectives in Biology" (M. Sela, ed.) p. 92, Elsevier Publishing Co., Amsterdam.
Elson, D. (1964c). *Discovery* **25**, 24.
Elson, D. and Tal, M. (1959). *Biochim. biophys. Acta* **36**, 281.
Farkas, W. R., Singer, M. S. and Marks, P. A. (1964). *Fed. Proc.* **23**, 220.
Fessenden, J. M. and Moldave, K. (1962). *Biochemistry* **1**, 485.
Fogel, Z. and Elson, D. (1964). *Biochim. biophys. Acta* **80**, 601.
Gazzinelli, G. and Dickman, S. R. (1962). *Biochim. biophys. Acta* **61**, 980.
Gesteland, R. F. (1965). *Fed. Proc.* **24**, 293.
Gierer, A. (1963). *J. mol. Biol.* **6**, 148.
Gierer, A. (1964). *In* "New Perspectives in Biology" (M. Sela, ed.) p. 106, Elsevier Publishing Co., Amsterdam.
Gilbert, W. (1963a). *J. mol. Biol.* **6**, 374.
Gilbert, W. (1963b). *J. mol. Biol.* **6**, 389.
Ginzburg-Tietz, Y., Kaufmann, E. and Traub, A. (1964). *Expl Cell. Res.* **34**, 384.
Gordon, J. and Boman, H. G. (1964). *J. mol. Biol.* **9**, 638.
Greenman, D. L. and Kenney, F. T. (1963). *Fed. Proc.* **22**, 581.
Grunberg-Manago, M. (1963). *In* "Progress in Nucleic Acid Research" (J. N. Davidson and W. E. Cohn, eds.) Vol. 1, p. 93, Academic Press, New York.
Hall, C. E. and Slayter, H. S. (1959). *J. mol. Biol.* **1**, 329.
Hartwell, L. H. and Magasanik, B. (1964). *J. mol. Biol.* **10**, 105.
Haruna, I. and Spiegelman, S. (1965). *Proc. natn. Acad. Sci. U.S.A.* **54**, 579.
Hauge, J. G., MacQuillan, A. M., Cline, A. L. and Halvorson, H. O. (1961). *Biochem. biophys. Res. Commun.* **5**, 267.

Hogeboom, G. H. and Schneider, W. C. (1955). *In* "The Nucleic Acids" (E. Chargaff and J. N. Davidson, eds.) Vol. II, p. 199, Academic Press, New York and London.

Horibata, K. and Kern, M. (1964). *Proc. natn. Acad. Sci. U.S.A.* **51**, 218.

Hurwitz, J. and August, J. T. (1963). *In* "Progress in Nucleic Acid Research" (J. N. Davidson and W. E. Cohn, eds.) Vol. 1, p. 59, Academic Press, New York and London.

Jardetzky, O. and Julian, G. R. (1964). *Nature, Lond.* **201**, 397.

Kameyama, T. and Novelli, G. D. (1960). *Biochem. biophys. Res. Commun.* **2**, 393.

Kameyama, T. and Novelli, G. D. (1962). *Archs Biochem. Biophys.* **97**, 529.

Kaufmann, E., Traub, A. and Ginzburg-Tietz, Y. (1964). *Israel J. Chem.* **2**, 252.

Keck, K. and Choules, E. A. (1962). *Archs Biochem. Biophys.* **99**, 205.

Keller, E. B. and Ferger, M. F. (1965). *Fed. Proc.* **24**, 283.

Keller, P. J. and Cohen, E. (1961). *J. biol. Chem.* **236**, 1407.

Keller, P. J., Cohen, E. and Wade, R. D. (1963). *Biochemistry* **2**, 315.

Keller, P. J., Cohen, E. and Wade, R. D. (1964). *J. biol. Chem.* **239**, 3292.

Kepes, A. (1963). *Biochim. biophys. Acta* **76**, 293.

Kessler, B. and Engelberg, N. (1962). *Biochim. biophys. Acta* **55**, 70

Kihara, H. K., Halvorson, H. and Bock, R. (1961). *Biochim. biophys. Acta* **49**, 221.

Kihara, H. K., Hu, A. S. L. and Halvorson, H. O. (1961). *Proc. natn. Acad. Sci., U.S.A.* **47**, 489.

Kiho, Y. and Rich, A. (1964). *Proc. natn. Acad. Sci., U.S.A.* **51**, 111.

Kimhi, Y. and Littauer, U. Z. (1962). *Bulletin of the Research Council of Israel* **11A**, 77.

Kivity-Vogel, T. and Elson, D. (1966). **Biohim. biophys. Acta**, (In press.)

Kozloff, L. (1953). *Cold Spr. Harb. Symp. Quant. Biol.* **18**, 209.

Kurland, C. G. (1960). *J. mol. Biol.* **2**, 83.

Lamfrom, H. and Squires, R. F. (1962). *Biochim. biophys. Acta* **61**, 421.

Lanzani, G. A. and Galante, E. (1964). *Archs Biochem. Biophys.* **106**, 20.

Leboy, P. S., Cox, E. C. and Flaks, J. G. (1964). *Proc. natn. Acad. Sci. U.S.A.* **52**, 1367.

Lederberg, S., Rotman, B. and Lederberg, V. (1964). *J. biol. Chem.* **239**, 54.

Lehman, I. R. (1963). *In* "Progress in Nucleic Acid Research" (J. N. Davidson and W. E. Cohn, eds.) Vol. 2, p. 83, Academic Press, New York and London.

Leslie, I. (1961). *Nature, Lond.* **189**, 260.

Lett, J. T. and Takahashi, W. N. (1962). *Archs Biochem. Biophys*, **96**, 569.

Levinthal, C., Keynan, A. and Higa, A. (1962). *Proc. natn. Acad. Sci. U.S.A.* **48**, 1631.

Littlefield, J. W., Keller, E. B., Gross, J. and Zamecnik, P. C. (1955). *J. biol. Chem.* **217**, 111.

Lovett, J. S. (1963). *J. Bact.* **85**, 1235.

Madison, J. T. (1962). Ph.D. Dissertation, University of Utah.

Madison, J. T. and Dickman, S. R. (1963a). *Biochemistry* **2**, 321.

Madison, J. T. and Dickman, S. R. (1963b). *Biochemistry* **2**, 326.

Malamy, M. and Horecker, B. L. (1961). *Biochem. biophys. Res. Commun.* **5**, 104.

Marks, P. A., Burka, E. R. and Schlessinger, D. (1962). *Proc. natn. Acad. Sci. U.S.A.* **48**, 2163.

Marks, P. A., Willson, C., Kruh, J. and Gros, F. (1962). *Biochem. biophys. Res. Commun.* **8**, 9.

Martin, S. J., England, H., Turkington, V. and Leslie, I. (1963). *Biochem. J.* **89**, 327.

Matheson, A. T. (1963). *Can J. Biochem. Physiol.* **41**, 9.

Mathias, A. P. and Williamson, R. (1964). *J. mol. Biol.* **9**, 498.

Mathias, A. P., Williamson, R., Huxley, H. E. and Page, S. (1964). *J. mol. Biol.* **9**, 154.
Matsushita, S. and Ibuki, F. (1960a). *Biochim. biophys. Acta* **40**, 358.
Matsushita, S. and Ibuki, F. (1960b). *Biochim. biophys. Acta* **40**, 540.
McCarthy, B. J. (1959). *In* "Carnegie Institution of Washington Year Book," 58, p. 281.
McCarthy, B. J. (1960). *Biochim. biophys. Acta* **39**, 563.
McCarthy, B. J. (1961). *In* "Carnegie Institution of Washington Year Book", 60, p. 328.
McCorquodale, D. J. (1963). *J. biol. Chem.* **238**, 3914.
McQuillen, K. (1961). *In* "Protein Biosynthesis" (R. J. C. Harris, ed.) p. 263. Academic Press, New York and London.
McQuillen, K. (1962). *In* "Progress in Biophysics and Biophysical Chemistry" (J. A. V. Butler, H. E. Huxley and R. E. Zirkle, eds.) Vol. 12, p. 67, Pergamon Press, New York.
McQuillen, K., Roberts, R. B. and Britten, R. J. (1959). *Proc. natn. Acad. Sci. U.S.A.* **45**, 1437.
Moldave, K. and Gasior, E. (1965). *Biochim. biophys. Acta* **95**, 679.
Morris, A. J. (1964). *Biochem. J.* **91**, 611.
Munro, A. J., Jackson, R. J. and Korner, A. (1964). *Biochem. J.* **92**, 289.
Nathans, D. and Lipmann, F. (1960). *Biochim. biophys. Acta* **43**, 126.
Neu, H. C. and Heppel, L. A. (1964a). *Biochem. biophys. Res. Commun.* **14**, 109.
Neu, H. C. and Heppel, L. A. (1964b). *J. biol. Chem.* **239**, 3893.
Neu, H. C. and Heppel, L. A. (1964c). *Proc. natn. Acad. Sci., U.S.A.* **51**, 1267.
Neu, H. C. and Heppel, L. A. (1964d). *Biochem. biophys. Res. Commun.* **17**, 215.
Neu, H. C. and Heppel, L. A. (1965). *Fed. Proc.* **24**, 349.
Nishizuka, Y. and Lipmann, F. (1965). *Fed. Proc.* **24**, 283.
Noll, H. (1962). *Cold Spr. Harb. Symp. Quant. Biol.* **27**, 256.
Nomura, M. and Watson, J. D. (1959). *J. mol. Biol.* **1**, 204.
Ohtaka, Y. and Uchida, K. (1963). *Biochim. biophys. Acta* **76**, 94.
Okamoto, T. and Takanami, M. (1963). *Biochim. biophys. Acta* **68**, 325.
Ortiz, P. J. (1963). *Dissertation Abstracts* **24**, 1823.
Palade, G. E. (1958). *In* "Microsomal Particles and Protein Synthesis" (R. B. Roberts, ed.) p. 36, Pergamon Press, New York.
Parsons, Jr., C. H. (1953). *Archs Biochem. Biophys.* **47**, 76.
Petermann, M. L. (1964). "The Physical and Chemical Properties of Ribosomes" Elsevier Publishing Co., Amsterdam.
Petermann, M. L. and Hamilton, M. G. (1961). *In* "Protein Biosynthesis" (R. J. C. Harris, ed.) p. 233, Academic Press, New York and London.
Petermann, M. L. and Pavlovec, A. (1963a). *J. biol. Chem.* **238**, 318.
Petermann, M. L. and Pavlovec, A. (1963b). *J. biol. Chem.* **238**, 3717.
Peterson, E. A. and Kuff, E. L. (1961). *Fed. Proc.* **20**, 390.
Pirie, N. W. (1950). *Biochem. J.* **47**, 614.
Pirie, N. W. (1957). *Biokhimiya* (English translation) **22**, 133.
Proctor, M. H. (1961). *Biochem. J.* **80**, 27P.
Raacke, I. D. (1963). *Fed. Proc.* **22**, 348.
Raacke, I. D. and Fiala, J. (1964). *Proc. natn. Acad. Sci. U.S.A.* **51**, 323.
Roberts, R. B. (ed.) (1958). "Microsomal Particles and Protein Synthesis" Pergamon Press, New York.
Roberts, R. B. (ed.) (1964). "Studies of Macromolecular Biosynthesis" Carnegie Institution of Washington Publication 624, Washington.

Roberts, R. B., Britten, R. J. and McCarthy, B. J. (1963). *In* "Molecular Genetics" (J. H. Taylor, ed.) Part 1, p. 292, Academic Press, New York and London.
Roth, J. S. (1954). *Nature, Lond.* **174**, 129.
Roth, J. S. (1960a). *J. biophys. biochem. Cytol.* **7**, 443.
Roth, J. S. (1960b). *J. biophys. biochem. Cytol.* **8**, 665.
Roth, J. S. (1963). *Cancer Res.* **23**, 657.
Rotman, B. (1961). *Proc. natn. Acad. Sci. U.S.A.* **47**, 1981.
Salas, M., Smith, M. A., Stanley, Jr., W. M., Wahba, A. J. and Ochoa, S. (1965). *J. biol. Chem.* **240**, 3988.
Schlessinger, D. (1964). *Biochim. biophys. Acta* **80**, 473.
Schlessinger, D. and Gros, F. (1963). *J. mol. Biol.* **7**, 350.
Sekiguchi, M. and Cohen, S. S. (1963). *J. biol. Chem.* **238**, 349.
Shortman, K. and Lehman, I. R. (1964). *J. biol. Chem.* **239**, 2964.
Siekevitz, P. (1963). *Ann. N.Y. Acad. Sci.* **103**, 773.
Siekevitz, P. (1965). *Fed. Proc.* **24**, 293.
Siekevitz, P. and Palade, G. E. (1958a). *J. biophys. biochem. Cytol.* **4**, 309.
Siekevitz, P. and Palade, G. E. (1958b). *J. biophys. biochem. Cytol.* **4**, 577.
Siekevitz, P. and Palade, G. E. (1960a). *J. biophys. biochem. Cytol.* **7**, 619.
Siekevitz, P. and Palade, G. E. (1960b). *J. biophys. biochem. Cytol.* **7**, 631.
Siekevitz, P. and Palade, G. E. (1962). *J. Cell Biol.* **13**, 217.
Spahr, P. F. (1962). *J. mol. Biol.* **4**, 395.
Spahr, P. F. and Hollingworth, B. R. (1961). *J. biol. Chem.* **236**, 823.
Spahr, P. F. and Schlessinger, D. (1963). *J. biol. Chem.* **238**, PC 2251.
Spirin, A. S., Kiselev, N. A., Shakulov, R. S. and Bogdanov, A. A. (1963). *Biokhimiya* (English translation) **28**, 765.
Spitnik-Elson, P. (1962a). *Biochim. biophys. Acta* **55**, 741.
Spitnik-Elson, P. (1962b). *Biochim. biophys. Acta* **61**, 624.
Spitnik-Elson, P. (1963). *Biochim. biophys. Acta* **74**, 105.
Spitnik-Elson, P. (1964). *Biochim. biophys. Acta* **80**, 594.
Spitnik-Elson, P. (1965). *Biochem. biophys. Res. Commun.* **18**, 557.
Srinivasan, P. R. and Borek, E. (1964). *Science, N.Y.* **145**, 548.
Staehelin, T., Brinton, C. C., Wettstein, F. O. and Noll, H. (1963). *Nature, Lond.* **199**, 865.
Stavy, L., Feldman, M. and Elson, D. (1964). *Biochim. biophys. Acta* **91**, 606.
Stenesh, J. J. (1958). Ph.D. Dissertation, University of California, Berkeley.
Strasdine, G. A., Hogg, L. A. and Campbell, J. J. R. (1962). *Biochim. biophys. Acta* **55**, 231.
Takanami, M. (1962). *Biochim. biophys. Acta* **61**, 432.
Tal, M. and Elson, D. (1961). *Biochim. biophys. Acta* **53**, 227.
Tal, M. and Elson, D. (1963a). *Biochim. biophys. Acta* **72**, 439.
Tal, M. and Elson, D. (1963b). *Biochim. biophys. Acta* **76**, 40.
Tashiro, Y. (1958). *J. Biochem.* (*Tokyo*) **45**, 937.
Thomas, G. H. and Herbst, E. J. (1963). *Fed. Proc.* **22**, 349.
Tissières, A., Watson, J. D., Schlessinger, D. and Hollingworth, B. R. (1959). *J. mol. Biol.* **1**, 221.
Tissières, A., Schlessinger, D. and Gros, F. (1960). *Proc. natn. Acad. Sci. U.S.A.* **46**, 1450.
Traub, A., Kaufmann, E. and Ginzburg-Tietz, Y. (1964). *Expl Cell Res.* **34**, 371.
Traut, R. R. and Monro, R. E. (1964). *J. mol. Biol.* **10**, 63.
Ts'o, P. O. P., Bonner, J. and Dintzis, H. M. (1958a). *Archs Biochem. Biophys.* **76**, 225.

Ts'o, P. O. P., Bonner, J. and Vinograd, J. (1958b). *Biochim. biophys. Acta* **30**, 570.
Wade, H. E. (1961). *Biochem. J.* **78**, 457.
Wade, H. E. and Lovett, S. (1961). *Biochem. J.* **81**, 319.
Wade, H. E. and Robinson, H. K. (1963). *Nature, Lond.* **200**, 661.
Waller, J.-P. (1963). *J. mol. Biol.* **7**, 483.
Waller, J.-P. (1964). *J. mol. Biol.* **10**, 319.
Waller, J.-P. and Harris, J. I. (1961). *Proc. natn. Acad. Sci., U.S.A.* **47**, 18.
Wang, T.-Y. and Wang, K. M. (1962). *Biochim. biophys. Acta* **55**, 392.
Warner, J. R., Knopf, P. M. and Rich, A. (1963). *Proc. natn. Acad. Sci. U.S.A.* **49**, 122.
Warren, W. and Goldthwait, D. (1961). *Fed. Proc.* **20**, 144.
Warren, W. A. and Goldthwait, D. A. (1962). *Proc. natn. Acad. Sci. U.S.A.* **48**, 698.
Weissbach, A. and Korn, D. (1964). *Biochim. biophys. Acta* **87**, 621.
Wettstein, F. O., Staehelin, T. and Noll, H. (1963). *Nature, Lond.* **197**, 430.
Whitfeld, P. R. and Williams, S. (1963). *Virology* **21**, 156.
Williamson, R. and Mathias, A. P. (1963). *Biochem. J.* **89**, 13P.
Wust, C. J. and Novelli, G. D. (1964). *Archs Biochem. Biophys.* **104**, 185.
Yankofsky, S. A. and Spiegelman, S. (1963). *Proc. natn. Acad. Sci. U.S.A.* **49**, 538.
Zamecnik, P. C. (1960). *Harvey Lectures* **54**, 256.
Zamecnik, P. C. (1962). *Biochem. J.* **85**, 257.
Zillig, W., Krone, W. and Albers, M. (1959). *Z. physiol. Chem.* **317**, 131.
Zipser, D. (1963). *J. mol. Biol.* **7**, 739.
Zipser, D. and Perrin, D. (1963). *Cold Spr. Harb. Symp. Quant Biol.* **28**, 533.

Chapter 8

THE SOLUBLE PHASE OF THE CELL[1]

NORMAN G. ANDERSON AND JOHN G. GREEN

Molecular Anatomy Section, Biology Division[2], Oak Ridge National Laboratory, and Technical Division[2], Oak Ridge Gaseous Diffusion Plant, Oak Ridge, Tennessee U.S.A.

I. INTRODUCTION

The soluble or diffusing phase (the cell sap) is the main stream for the metabolic commerce of the cell and is the milieu in which cellular formed elements persist and function. Its composition reflects so much of cell biochemistry that a complete review of its role is no longer feasible. The

[1] Work supported by the U.S. Atomic Energy Commission and by the Joint National Institutes of Health-Atomic Energy Commission Molecular Anatomy Program supported by the National Cancer Institute, the National Institute for Allergy and Infectious Diseases, and the U.S. Atomic Energy Commission.

[2] Operated for the U.S. Atomic Energy Commission by the Nuclear Division of Union Carbide Corporation.

present chapter will therefore be concerned with a discussion of only a few salient aspects of the chemistry, physics, and physiology of this cell fraction.

The soluble phase (SP) has been studied from several viewpoints which are difficult to interrelate. These may be grouped under physical and chemical analysis, immunoanalysis, enzymology, and physiology. Thus, the SP proteins may be grouped into ultracentrifugal or electrophoretic classes, may be studied as specific antigens, assayed for enzymic activity, or related to structural changes in cells. Unfortunately, rather few cytoplasmic proteins have been studied from each of these viewpoints.

In examining a cell fraction which is related to such a wide spectrum of interests we shall adopt the following negative viewpoints and then examine them in the light of the available experimental results.

1. The SP is an intractable, irresolvable mixture including hundreds and perhaps thousands of substances.

2. Any given physical or chemical property is randomly distributed among these constituents. When examined by any physical or chemical separation method, essentially Gaussian results will be obtained.

3. (As a corollary to 1 and 2) Any substance isolated from the SP will, on more refined analysis, be found to be a mixture.

Results which run contrary to these postulates must therefore indicate that the constituents of the SP may ultimately be reduced to some degree of order.

II. Definition of the Soluble Phase

The soluble phase is defined operationally as the supernatant remaining after the sedimentation of all particulate material. The definition applies *in vivo* to centrifugally stratified cells (Zalokar, 1960) and *in vitro* to centrifuged cell homogenates. The central question is: What relation does the isolated soluble phase bear to the interparticulate fluid of the intact cell? Any consideration of this question must deal first with the problem of whether particulate and non-particulate (soluble) cell constituents can be sharply defined (i.e., are two mutually exclusive sets of entities) and whether any interchange between these two sets occurs before, during, or after cell disruption. If, in a cell brei, there exists a sedimentation continuum (i.e., particles of all sedimentation coefficients are present) then the soluble phase can only be defined arbitrarily. However, if a discontinuity is observed in a plot of sedimentation coefficients *vs* mass in a brei, then a rational basis for both the definition and the separation of the soluble phase exists. For the most part the constituents of the soluble phase may be considered to be single "molecular" entities. However, numerous soluble proteins are known to be composed of covalently linked chains held together, in many instances, by secondary valencies.

A few large complexes have been described in cells including, for example, the 19S thyroglobulin (Salvatore, Vecchio, Salvatore, Cahnmann, and Robbins, 1965) the 36 S α-ketoglutarate dehydrogenase (Mukherjee, Matthews, Horney, and Reed, 1965) and 58 S pyruvate dehydrogenation complex (Koike, Reed, and Carroll, 1963) from *E. coli*. There is no reason to believe that other complex micro-organelles may not be found. Indeed, many of the "soluble" enzyme systems may exist in the form of labile aggregates in the cell. Part of the responsiveness of cells to alterations in the external environment may be due to the reversible formation of such aggregates and may give rise to changes in the consistency or viscosity of the cytoplasm (Gross, Philpott, and Nass, 1960), a view supported by the experimental demonstration of particle formation in soluble phase preparations (Anderson, 1959).

A. CLASSIFICATION OF COMPONENTS

The weights in daltons of a number of cell constituents are given in Fig. 1 These are taken from convenient sources. The plot suggests that the soluble

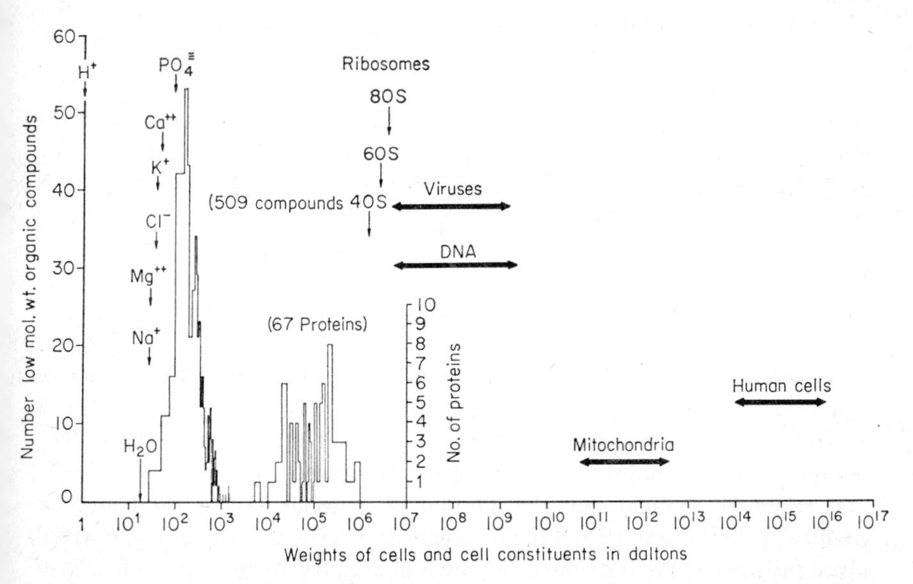

FIG. 1. Plot of molecular weights of classes of particles found in cells and of cells themselves. Note that few substances in the region of 5×10^3 Daltons have been described in cells. Microsomes are omitted since analytical data on individual particles are not available.

phase may consist of at least three general subfractions which are:

1. Micromolecular subfraction, including water, inorganic ions, and dissolved gases.

2. Mesomolecular subfraction, including all metabolic intermediates, lipids, sugars, nucleotides, nucleotide derivatives, and other low-information-content molecules.

3. Free macromolecules. Chiefly proteins and RNA (including sRNA).

The smallest well-defined particle which is generally described in electron micrographs is the 70–80 S ribosome. The ribosomal subunits (30–40 and 50–60 S) do not appear to exist in appreciable quantities in most cells. In vaccinia infected HeLa cells about 10% of the ribosomal mass appears as free subunits (Joklik and Becker, 1965). Since most studies with the soluble phase are done with the Spinco No. 40 rotor (or its equivalent) operating at 40,000 rev/min for one hour, it is of interest to consider what would be sedimented under these conditions. Neglecting the accelerated removal of particles by collision with the wall of an angle head rotor tube, the smallest particle completely sedimented according to the simple equation:

$$T_s = \frac{1}{S} \frac{\ln R_{\max} - \ln R_{\min}}{\omega^2}$$

where T_s = time in seconds
S = sedimentation coefficient
ω = angular velocity in radians per second
R = radius

would be 120 S. Smaller particles are sedimented to a lesser degree, and the pellet is enriched with all sedimenting molecules, however small, to some extent. Under these conditions in an angle-head rotor, the majority of 80 S ribosomes would be sedimented. However, if the distribution of an enzyme known to be associated with ribosomes is being studied, the separations should be made in a zonal ultracentrifuge (Anderson, 1966) where the ribosomes and ribosome subunits can be quantitatively removed from the soluble phase.

As shown in Fig. 2, however, with rat liver a surprisingly flat plateau is reached after about 20 minutes at 105,000*g*. This suggests that if a continuum of sedimentation coefficients exists, it must reach a very deep minimum at some point to give the plateau observed.

It might be thought that the amount of material having sedimentation coefficients in the 100–1000 S region could be determined using the analytical ultracentrifuge. This instrument is not well adapted to the study of heterodisperse preparations, however. For example, if 5–10 % of the cell mass was distributed uniformly over the range 20–1000 S, it would be difficult to detect, i.e., only a small shift in base line would be observed. The zonal ultracentrifuge (Anderson, 1966) is better suited to this problem since both preparative and analytical studies may be done in the same rotor on one experiment.

It is emphasized here that the relation of the isolated fraction to the diffusible

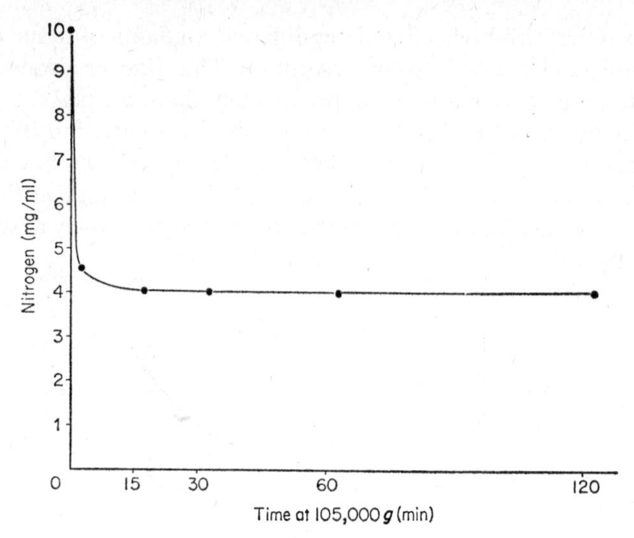

FIG. 2. Sedimentation of rat liver brei nitrogen as a function of time in a Spinco No. 40 angle head rotor (Anderson, 1956a).

sap of the intact cell remains to be elucidated. However, as will be discussed in a subsequent section, rather simple techniques are available for studying interactions between the formed elements of the cell and the cell sap. It is concluded that experimentally the soluble phase is a fairly well defined entity which may be easily and reproducibly isolated.

We may ask whether the SP is resolvable experimentally into the subgroups suggested by Fig. 1. The results obtained with rat liver SP on Sephadex columns suggest that it is. Very little ultraviolet absorbing material is observed in the range between the macromolecular and mesomolecular peaks. Salts and micromolecules or ions are also a fairly discrete group. The soluble phase is therefore resolvable into rather well defined subgroups on the basis of molecular size.

Unfortunately, preoccupation with studies on the distribution of substances and activities between centrifugally isolated fractions has obscured the fact that the first question to be settled is simply this: is the substance of interest bound to particles or not? Since complex isolation procedures obviously may give rise to artifacts, it would appear wise to determine the distribution between the sediment and the supernatant in the shortest possible period of time. Subcellular particles are more likely to be in a physiologically native state in the original homogenate than in any subsequent isolation medium. We therefore require very simple and extremely rapid methods for homogenizing tissues, sedimenting particulate material, and (when required) freezing the supernatant. This is especially true when labile metabolic intermediates are being studied. In addition, simple particle-supernatant distribution studies done with breis containing different amounts of tissue can give clues to adsorption or (more likely) desorption. Thus if, over a wide range of homogenate tissue concentrations approximately the same percentage of an enzyme is found in the SP, then the curve may be extrapolated to 100% tissue (i.e., to the intracellular state) with some confidence as shown in curve A in the illustrative diagram in Fig. 3. However, if the percentage in the supernatant increases on dilution (curve B) then the activity is loosely adsorbed on surfaces in the cell.

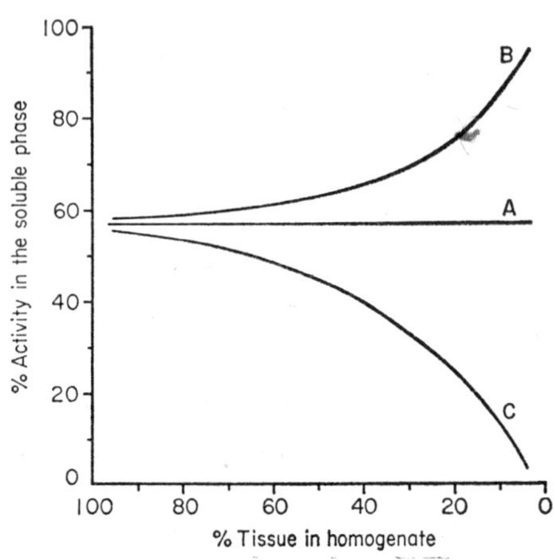

FIG. 3. Diagramatic representation of changes in enzymic activity found in the soluble phase as a function of tissue concentration in the homogenate. Interpretation of curves described in text.

If the percentage decreases on dilution (curve C) altered surfaces of subcellular components produced as artifacts of homogenization are removing previously soluble materials from solution. An example of a study of this type is shown in Fig. 4 where acid soluble nucleotides of the SP (expressed in terms of μmoles of adenylic acid per gram of liver) are plotted against the percentage of liver in a brei. It is evident that appreciable changes in the nucleotide composition of the SP occur when a homogenate is diluted. Unfortunately, studies of this type have rarely been done, and the chief contribution of this chapter is to indicate in some detail the present unsatisfactory state of research in this area.

An additional problem arises from the rapid turnover rate and lability of many of the compounds found in the soluble phase and their effect on the stability of some of the proteins found in it. In studies on brain metabolites it has been found necessary to freeze the tissue very rapidly, and to make acid extracts of frozen powdered tissue as thawing occurs (Lowry, Passoneau, Hasselberger, and Schulz, 1964). Marked changes were observed, especially in the ATP level, if circulation was interrupted for a few seconds. It is obvious that very large changes in the metabolite composition of the soluble phase may occur during the isolation procedures presently in use.

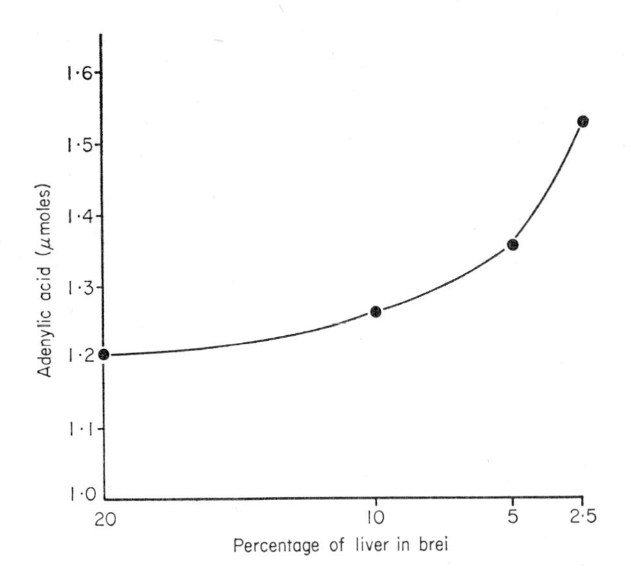

FIG. 4. Ultraviolet absorbing material from supernatant after centrifugation at 105,000*g* for one hour of liver breis of different concentration. Results expressed as equivalent amount of adenylic acid per gram of liver.

B. INTERPRETATION OF RESULTS

For many comparative purposes the method used to tabulate analytical results is of little consequence. However, if we desire to develop concepts concerning the composition of the cell as a functional unit then the results might be presented in terms of the number of molecules found in a cell. Unfortunately rat liver, which is the most widely studied tissue, contains many binucleate cells. For this reason, it is suggested that results be presented in terms of a 1 N cell mass equivalent. For a diploid cell this number would be doubled; for a tetraploid cell it would be quadrupled to give the quantity present per cell.

Table I lists the number of molecules of several substances per 1 N liver cell mass. For B_{12} only 25,000 molecules are present. The concentration which a substance present in a tissue would have when only one molecule is present per cell (the limiting molarity) is given by the equation

$$M_{\text{min}} = \frac{10^{15}}{6{\cdot}023 \times 10^{23} \times V_c} = \frac{1{\cdot}66 \times 10^{-9}}{V_c}$$

TABLE I

Number of molecules of various compounds in 1 N liver cell mass equivalent[1]

Compound	Number
Inorganic phosphate	$4{\cdot}8 \times 10^9$
Glycerophosphate	$1{\cdot}2 \times 10^9$
AMP	$4{\cdot}5 \times 10^8$
ADP	$2{\cdot}5 \times 10^9$
ATP	$1{\cdot}1 \times 10^9$
NAD^+	$4{\cdot}2 \times 10^8$
NADH	$2{\cdot}3 \times 10^8$
$NADP^+$	$6{\cdot}0 \times 10^6$
NADPH	$2{\cdot}1 \times 10^8$
Glucose-1-phosphate	$4{\cdot}5 \times 10^7$
Glucose-1:6-diphosphate	$9{\cdot}8 \times 10^6$
Uridine diphosphoglucose	$1{\cdot}1 \times 10^8$
Vitamin A	2×10^8
Vitamin C	9×10^8
Vitamin E	$3{\cdot}8 \times 10^7$
Thiamine	$1{\cdot}1 \times 10^7$
Riboflavin	$6{\cdot}0 \times 10^7$
Nicotinic acid	$1{\cdot}2 \times 10^9$
Biotin	$3{\cdot}0 \times 10^6$
Choline	$7{\cdot}9 \times 10^6$
Inositol	$2{\cdot}7 \times 10^6$
B_{12}	$2{\cdot}5 \times 10^4$

[1] Data from Biochemists Handbook, C. Long, ed., D. Van Nostrand Co., Inc., Princeton, New Jersey (1961). Calculations based on the assumption that 1 N DNA = 3 picograms, and that 100 mg of fresh tissue contains 208 μg DNA.

where $M_{\min}$ is the limiting molarity and V_c is the cell volume in cubic microns. For an 800 μ^3 cell the limiting molarity would be 2×10^{-12} when one molecule is present per cell. For an amino acid with a molecular weight of 120, $2{\cdot}4 \times 10^{-10}$ grams would be present in a kilogram of cells. This quantity is well below the limits of present detection methods. The interesting question arises, is a single molecule of any substance except DNA or viral RNA functionally important in a cell? If so, then present lists of the total number of compounds present in cells may be very far from complete.

A second point regarding presentation of data concerns the possibility of constructing a balance sheet of cell constituents. Does the sum of the known acid soluble nitrogen-containing compounds equal and account for the total acid-soluble nitrogen found? What percentage of soluble cytoplasmic proteins are accounted for by known enzymes? Until such balance sheets are constructed it is difficult to determine what work remains to be done, and whether, as is the case in blood plasma, a large fraction of proteins are present which have functions other than enzymic ones.

III. The Composition of the Soluble Phase

The soluble phase contains between 23–67% of the total nitrogen, depending on the tissue and the procedure used (Reid, 1961). The percentage in rat liver is 43% of which 87·8% is non-dialysable (Anderson, 1956a). While the concentration of a large number of substances in liver and other tissues is known (Long, 1961), with the exception of enzyme activities the percentage of each found in the soluble phase is known in only a few instances. Since many reviews listing cell fractionation results have appeared (Roodyn, 1965; de Duve, Wattiaux, and Baudhuin, 1962; Claude, 1948; de Duve and Berthet, 1954; Hogeboom, Kuff, and Schneider, 1957; and Roodyn, 1959) the distribution of enzymic activities between cell particles and the soluble phase will not be tabulated in detail here.

A. Micromolecules

The inorganic ion content of cells reflects the balance between rate of gain and rate of loss of individual ions (Ussing, 1960; Rothstein, 1961). Analytical data on inorganic substances has been reviewed (Long, 1961; Ussing, 1960), and the distribution of ions between plasma and tissue cells examined. Monovalent ions are thought to be largely free in the cell, while divalent cations such as magnesium and calcium appear to be chelated by nucleic acids and acid polysaccharides. Since the intracellular ion composition reflects that of the external medium, the effect of the contents of the extracellular spaces in tissues used in cell fractionation must be considered. If perfusion is used to remove blood, then the ionic composition, or the lack of ions when sucrose is used for perfusion, may cause changes in intracellular ionic composition. In addition, the degree of binding of inorganic ions to subcellular particles is concentration dependent, i.e., there appears to be an equilibrium between bound and unbound calcium, for example. Dilution during homogenization may be expected to disturb such equilibria. In addition, selective accumulation of ions in subcellular particles may occur; for example, K^+, Na^+, Ca^{2+}, and Mg^{2+} levels are maintained or increased in isolated mitochondria under conditions where oxidative phosphorylation is taking place. Since this is dependent on the physical state of the mitochondria and the supply of ATP (Lehninger, 1964) alterations during or after homogenization may produce either leakage or additional uptake. Studies on the inorganic ion composition of the soluble phase should therefore be plotted both as a function of tissue dilution and of time after cell breakage, and should take into consideration the amount of extracellular fluid included in the SP.

Since subcellular particles exist in the SP originally, we would like to devise an isolation medium which approximates to the SP in its composition and effects on subcellular components. Solutions containing the same salts found in ashed cells are notoriously poor isolation media, whereas the most successful ones are ion free. In fact, the SP of the intact cell may contain only traces of diffusible inorganic anions, chiefly the bicarbonate ion, while the remainder of the inorganic anionic sites appear to be fixed. Both chloride and free orthophosphate are present in extremely low concentrations in liver cells, the diffusible anions being chiefly low molecular weight organic acids. The problem of approximating to the SP with a solution in which the subcellular particles remain freely suspended and in which these particles experience the same ionic environment found in the cytoplasm remains to be solved.

The hydrogen ion concentration of the cell (and therefore of the SP) has been extensively reviewed (Wiercinski, 1955; Caldwell, 1956; Bittar, 1964). The pH of soluble phase in the cell in most mammalian tissues appears to be very close to or slightly less than 7·0. Since at pH 7·0, one cubic micron contains approximately 60 hydrogen ions, local variations in pH in cells highly compartmentalized by cytomembranes is probable. The soluble phase as usually isolated in sucrose from rat liver has a pH of 6·45 (Anderson, Makinodan, and Norris, 1961). The pH of the SP of mammalian tissues needs to be re-examined in very rapidly prepared samples. (e.g. after 20–30 sec.)

B. MESOMOLECULES

Cellular mesomolecules are generally grouped by the analytical techniques used for their analysis. Most group analyses depend on sensitive separation techniques coupled to sensing systems which measure the concentration of a chemical group, element, or may measure total mass. Thus, automated analytical systems exist for amino acids and ninhydrin-positive compounds (Spackman, Stein, and Moore, 1958; Piez and Morris, 1960; Hamilton, 1963), nucleotides and ultraviolet absorbing derivatives (Anderson, Green, Barber, and Ladd, 1963; Green, Nunley, and Anderson, 1966) and sugars (Green and Anderson, 1965; Green, 1966). The detectors used in gas chromatography may be sensitive to mass, or to specific groups or elements (Juvet and Nogare, 1964; Giuffrida, 1964). Pre-fractionation or selective volatilization often allow known classes of substances to be studied. In the case of glycolytic intermediates specific fluorometric techniques may be preferable because of their extreme sensitivity (see Lowry and Passoneau, 1964), although chromatographic methods are available for their analysis (Schmitz and Walpurger, 1959).

1. Nucleotides and Related Compounds

A large number of nucleotides and nucleotide derivates have been reported in acid extracts of mammalian tissues (Tunn and Dumazert, 1962; Gregoire, 1958; Henderson and LePage, 1958; Potter, 1958; Strominger, 1960; Brady and Tower, 1960; Mandel, 1964). The distribution of nucleotides among cell fractions, however, has been less frequently studied. The problem again is the rapid breakdown of the triphosphates during the centrifugal separations. The rate at which this may occur is illustrated by the studies of Lowry *et al.* (1964) who followed the change in ATP and glycolytic intermediates with time in ischemic brain. It is evident from their studies that marked technical advances will be required before the SP can be isolated without appreciable change in the nucleotide triphosphate content. The nucleotides found in tissues are not catalogued here; however, a summary of their functions is given in Fig. 5. Automated high-resolution chromatographic systems for nucleotides, nucleosides, and purine and pyrimidine bases (Anderson, *et al.*

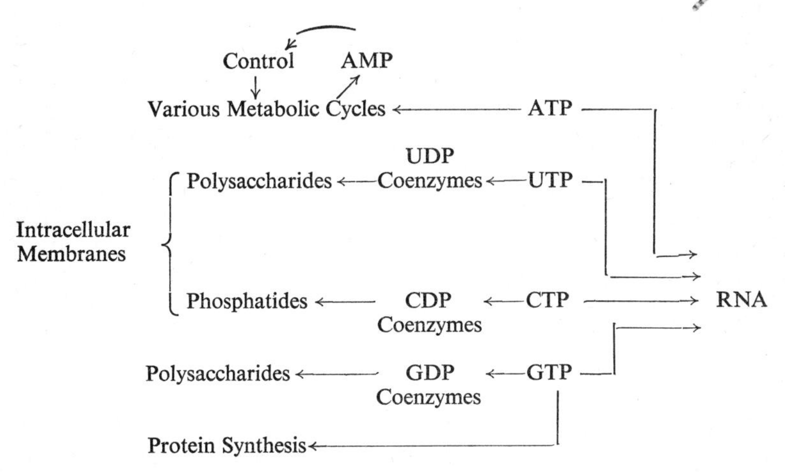

FIG. 5. Summary of the roles of the nucleotide di- and triphosphates in different metabolic cycles (From P. Mandel, 1964)

1963; Green, *et al.* 1966) will allow a larger number of these substances to be determined in tissue fractions than has been the case heretofore. The key role of adenylates as intracellular regulatory effectors has recently been reviewed by Atkinson (1965). The chromatography of extractable nucleotides has recently been reviewed by Saukkonen (1965).

2. Metabolic Intermediates

No complete study of metabolic intermediates in rapidly prepared SP has appeared. It is evident that the levels of metabolic intermediates of the glycolytic pathway found in the soluble phase reflect not only the condition of the cells at the moment of rupture, but the metabolic changes which occur during preparation of the fraction. It is probable, however, that tissue analyses for glycolytic intermediates such as the fluorometric procedures used by Lowry and Passoneau (1964) reflect the concentration of these substances in the SP. The glucose level of the SP of the intact cell has a profound effect on the glycolytic pathway. In Ehrlich ascites tumor cell homogenates the glycolytic activity is controlled by the glucose supply at concentrations below 1·2 mM, although anaerobic glycolysis is already maximal at 0·2 mM glucose (Wu, 1965a). Under aerobic conditions inhibition of phosphofructokinase (2.7.1.11) occur after a short interval. Orthophosphate stimulates both aerobic and anaerobic glycolysis and phosphofructokinase. The effects of the intracellular inorganic phosphate level appear to be mediated through hexokinase, (2.7.1.1) phosphofructokinase and glyceraldehyde-3-phosphate dehydrogenase (1.2.1.9) in a co-ordinated manner. The effect of P_i on phosphofructokinase may be mainly responsible for the Pasteur effect (Wu, 1965b). Automated chromatographic systems for the glycolytic and Krebs' cycle intermediates which extend and make quantitative presently available techniques (Passera, Pedrotti, and Ferrari, 1964; Stern, Gramet, Trabal, Mennis, and Zinsser, 1965; Schmitz and Walpurger, 1959) are urgently needed.

3. Sugars

While the glucose level of tissues has been repeatedly studied (Long, 1961) sensitive quantitative methods for all the sugars which might be found in mammalian tissues have not previously been available. With the development of automated carbohydrate analysers (Green and Anderson, 1965; Green, 1966), this problem can be examined in detail using media other than sucrose for the isolation of the soluble phase.

4. Amino Acids (and Related Compounds)

Amino acids are probably the most studied mesomolecules of the cell. This volume of experimental effort has resulted largely from the availability of simple analytical techniques such as paper and column chromatography and the use of simple extraction procedures which yield reproducible results. A complete review and discussion of the vast body of literature available is not within the mission of this article, and when one applies rigidly the definition of the soluble phase much of the literature is not applicable. Enough information is available, however, to permit discussion of the problems involved in the study of amino acids in the soluble phase (SP). For additional information the reader is referred to extensive reviews on the various aspects of amino acid research including analytical methodology (Light and Smith, 1963), occurrence in tissues (Christensen, 1964; Roberts and Simonsen, 1960; Mesister, 1965) and the pool concept (Holden, 1962).

(*a*) *Current problems*—Although numerous technical problems have been met and solved in the field of amino acid biochemistry, only the problems of terminology, sample preparation, and data interpretation will be considered here, since these problems relate directly to the consideration of amino acids in the soluble phase of the cell.

Generally, amino acids from natural sources are placed in two classes with respect to the manner of isolation. "Free" amino acids occur in extracellular fluids such as sera, urine, or culture media, or can be removed from cells and tissues by mild extraction procedures with such materials as hot or cold water, alcohol solutions or dilute acids. Amino acids which are not extracted easily are classified as "bound" amino acids. We propose to revise the terminology used as follows:

Free amino acids; This classification would be restricted to amino acids derived from clearly extracellular sources. Derived terms include free peptides, free amino sugars, and free amino acids (FAA); however, the abbreviated form is the same one designating a commonly used fixative.

Extractable amino acids; This classification would include amino acids obtained by extraction of cells and tissues and includes much sample material now called free amino acids. Derived terms include modifiers depicting the extractant, e.g., acid-extractable; application of the term to other amino compounds, e.g., extractable peptides; and the abbreviated form, extractable amino acids (EAA). The term "easily extractable" amino acids or an equivalent form has appeared with increasing frequency in recent literature to define accurately the fraction within the designation of free amino acids.

Soluble phase amino acids; This arbitrary classification would be used to designate amino acids obtained by sampling techniques directed specifically toward obtaining materials identical to their state *in vivo*. Derived terms include such items as soluble phase peptides, as well as the abbreviated form soluble phase amino acids (SPAA).

Bound amino acids; This category includes amino acids associated with other materials such as protein in a manner preventing their appearance in other fractions. Derivatives include such terms as protein-bound, or lipid-bound amino acids or peptides, and it may be desirable to designate whether the compounds are intracellular or extracellular.

This or similar classifications present some inadequacy. For example, it is conceivable that amino acids associated with subcellular particles may be "bound" during isolation of the particle yet be extractable in subsequent procedures. Nevertheless, the system permits more exact description of experimental work and facilitates evaluation of the vast literature on the basis of title alone. Comparable terminology may be used for other low molecular weight substances such as nucleotides, sugars, and metabolic intermediates.

Experimentation has not established the true nature of intracellular extractable amino acids. Recent studies, for example, have shown that (a) very rapid post-mortem, pre-extracting changes occur in tissues which may be modified by techniques involving freeze quenching, and (b) the proportions of free and bound amino acids in homogenates of brain and other tissues vary with pH, ionic composition and temperature of the extraction medium (De Ropp and Snedeker, 1960; Stephen and Steinbauer, 1963; Mitchell and Simmons, 1962; Kolousek, Jiracek and Cizinsky, 1964; Elliott, Khan, Bilodeau and Lovell, 1965). Evidence has also been presented indicating that in crustacean neural tissue glutamic and aspartic acids appear to exist in a physiologically inactive form and that in the case of aspartic acid the physiologically active form is very unstable (Florey, 1964). The problems involved have been ably discussed by Holden (1962).

Some attempts have been made to apply methods involving cryological techniques to the study of cell constituents to preserve their true state during sample preparation. They include freezing, sectioning, and extraction of the tissue (Tee, Wang and Watkins, 1964), freezing, pulverizing, and centrifugal extraction (Kolousek, *et al.*, 1964), centrifugation and freezing of homogenates in hematocrit tubes (Wittgenstein and Rowe, 1965), and freeze quenching, sectioning, and extraction at subfreezing temperatures (Bucher, Krejei, Russman, Schnitger and Wesemann, 1964). The results of applying such techniques to the study of the amino acids in cells have not yet appeared in the literature. Another approach to the problem which has not yet been applied involves the sampling of material from living cells by microsurgical techniques (Chambers and Chambers, 1961). In one such study, human tumor cells were stratified by centrifugation but only particulate materials were withdrawn by micropuncture for further study (Mateyko and Kopac, 1964). In similar studies *Neurospora* hyphae (Zalokar, 1960) or *Paracentrotus* eggs (Pasteels, Castiaux and Vandermeerssche, 1958) were centrifuged and subjected to cytochemical testing, but here, too, no effort was made to assay mesomolecular cellular components. In the case of amino acids, the correlative problem of minute samples has been reduced by the development of techniques for quantitative analysis of concentrations as small as 10^{-11} molar (Hamilton, 1962).

Sampling techniques are only part of the problem of determining amino acids in the soluble phase of the cell. The thoroughness of extraction techniques has been rarely checked. In potato tissue, amino acids are removed selectively, repeated extraction is required for quantitative removal, and the degree of difficulty and order of selectivity of extraction varies with different tissues (Talley, Carter and Porter, 1958). Although major findings have been obtained through the use of paper chromatography, the procedure lacks precision. In one example, paper chromatography showed the existence of 5 or 6 peptides in homogenates of larval and rabbit brain tissue but column chromatography yielded over 600 peptides and amino derivatives in the same tissues (Mitchell and Simmons, 1962). In another example, findings from paper chromatography agreed poorly with data obtained from column chromatography and microbiological assay, although the results with the latter two techniques were in close agreement (Dunn, Sakamoto, Sutaria and Murphy, 1963).

In spite of the promise of new and improved techniques, the state of intracellular amino acids has not been adequately described. The question is, what, of the vast amount of data found in the literature, is applicable to this problem? It is logical to conclude that although much data may be subject to reinterpretation in the light of subsequent findings and discoveries yet to come, certain concepts and facts have been derived which will condition future study and interpretation of the soluble phase. Therefore, it is appropriate to consider some of these concepts including analytical data, the pool concept and the discovery of complex forms.

(*b*) *Important findings*—The vast literature giving extractable amino acid contents of tissues will not be reviewed here. Reviews listed previously contain many such tabulations and additional sources are widely distributed (see e.g., Biochemist's Handbook). However, certain general observations merit citation. It has been reported that extractable amino acids exist in a concentration of 15–30 mM in tissues and that 60–120 times that quantity is bound as protein (Christensen, 1964). It has been found that the number of easily extracted amino acids in tissues considerably exceeds the number found in proteins (Greenstein and Winitz, 1961; Steward and Pollard, 1962). In fact, it has been noted that new amino acids are being reported at the rate of ten a year; whereas the spectrum of amino acids found in proteins has not changed in 25 years (Meister, 1965). Some stable complexes of amino acids have been isolated by extraction procedures. In *Escherichia coli* covalently bonded RNA-amino acid complexes constitute about 5% of the extractable amino acid pool (Lacks and Gros, 1959), and acid soluble cytidine nucleotide linked amino acids have been reported from rabbit liver extracts (Agren, 1960). Amino acid-phospholipid complexes have been isolated from ascites tumour cells (Bailey and Woodson, 1963), and lipid-bound amino acids have been recovered from extracts of broad bean leaves (Brady, 1964), and from hen oviduct (Hendler, 1961; Hendler, 1962). A very high turnover rate has been observed in the latter fraction (Hendler, 1962).

Perhaps the most important principle evolved in the interpretation of extractable amino acids is the concept of intracellular amino acid pools. Although the subject has been extensively reviewed (Holden (ed.), 1962) it is appropriate to consider several aspects of this concept here. The amino acid pool is operationally defined to include the quantity of low molecular weight material which is extracted from the cell under conditions which exclude the contribution of material degraded from macromolecular constituents (Britten and McClure, 1962). The principal features of the amino acid pool in *E. coli*, where it has been most extensively studied, are given by Britten and McClure (1962) as follows:

1. Pool passage is an obligatory step for incorporation of extracellular amino acids into proteins.
2. Pool amino acids are incorporated at random into protein regardless of the length of time they have been in the pool.
3. An energy source is required for formation of a pool at normal rates but not for maintenance of the pool.
4. Specific pool formation mechanisms exist for each amino acid or structurally similar group.
5. For each amino acid, there exists a maximum pool size for saturated external concentrations.
6. Exchange between pool and external amino acids occurs at high rates when steady flow is occurring through the pool and when pool formation is being suppressed.

The dynamic nature of the pool is attested by the observation that in *E. coli* the pool is easily lost under conditions such as water washing and osmotic shock which do not cause cell death and that the pool in yeast actually increases under conditions such as exposure to ultraviolet radiation which induce leakage of nucleotides and amino acids from the cells (Siegel and Swenson, 1964). It has been suggested that maintenance of the pool including amino acid transport, binding phenomena and uptake is closely associated with alkaline phosphatase activity (3.1.3.1.) in rat brain (de Waart, 1964). The limitations of the pool concept are evident and some of them are given by Christensen (1964) as follows:

1. Cellular amino acids do not represent a functionally homogenous pool due to compartmentalization, e.g., mitochondrial pools.
2. The pool is not always diluted by amino acid precursors during incorporation into bound forms.
3. The pool is defined on the basis of kinetic relationships rather than on the basis of physical barriers.
4. Undetected chemical differences such as labile complexes may exist between forms of an amino acid in the pool.

Although many questions remain unanswered about the function, nature, and constituency of the amino acid pool, it is now the major vehicle for interpreting the findings about easily extractable amino acids and future results will undoubtedly be interpreted in light of the pool concept.

This section would not be complete without some consideration of the role of amino acids in the soluble phase. A comprehensive discussion of the biochemistry of amino acids has appeared (Meister, 1965) but some functions merit citation here. The role of amino acids in protein synthesis is well known. In addition, they may serve as substrates for the synthesis of peptides (e.g., glutathione) and related substances and, as respiratory substrates (via transamination), resulting in yields of ATP (Krebs, 1964). Considerable effort has been expended to associate extractable amino acid contents of tissue with osmoregulatory activity, especially in aquatic animals (Shaw, 1958; Jeuniaux, Duchateau-Besson and Florkin, 1961; Allen, 1961; Lange, 1963 and in bacteria (Kuczynski-Halmann, Avi-Dor and Mager, 1958). It has been suggested that lysine may replace lost intracellular potassium in maintaining cation-anion balance in potassium defficient tissue in *Fundulus* (Hanlon, 1960) and that glutamic and asparic acids may function as transmitter substances in crustacean nerve tissue (Florey, 1964). In plants, extractable amino acid and amide content has been correlated with cold hardening in winter wheat (Zech and Pauli, 1962). The great spectrum of amino compounds found in cells and tissues suggests that many more roles for them remain to be elucidated.

Determination of the state of amino acids in the soluble phase of the cell *in vivo* presents a most pressing challenge. New approaches and new and refined techniques even now promise to solve this problem.

C. MACROMOLECULES

In this section macromolecules of the soluble phase will be considered from a physical, chromatographic, and immunochemical viewpoint.

1. Proteins

(*a*) *Ultracentrifug alanalysis*—Ultracentrifug alanalysis of mammalian tissue extracts reveals evidence for the existence of several classes of proteins. In rat liver three major and at least two minor electrophoretic peaks are distinguishable (Sorof and Cohen, 1951a, b; Anderson and Canning, 1959). Schlieren patterns for rat liver, kidney, brain, and testis are shown in Fig. 6. The most important finding is that half or more of all the soluble proteins of brain, kidney, and liver have sedimentation coefficients between 3·4 and 4·2 S. In testis the amount of 4 S protein is slightly less, accounting for an average of 39·1 % of the total. The most characteristic differences between the various organs were in the amount and in the sedimentation coefficients of the more rapidly sedimenting material. The macroglobulin-like constituents of brain and testis deserve special attention since their concentration is much higher than in rat plasma. In rat liver, protein-bound amino azo dye derivatives of several hepato-carcinogens are found in the 3·6 S fraction (Sorof, Golder, and Ott, 1954).

The lack of preparative centrifugal systems having a resolution comparable to that of the analytical systems has made the task of determiningthe position of SP enzymes among the ultracentrifuge peaks difficult, although some progress has been made using swinging bucket rotors (Martin and Ames, 1961), most of which, however, use smaller samples than the analytical ultracentrifuge. With the development of higher speed zonal centrifuges

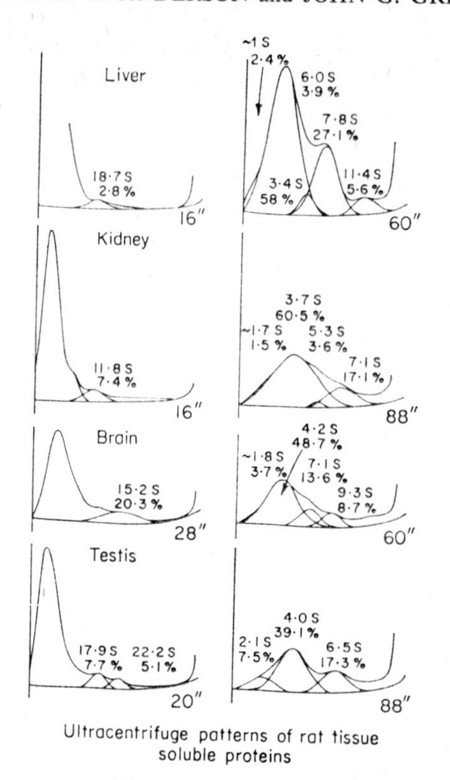

FIG. 6. Ultracentrifugal patterns of soluble phase proteins from rat tissues (Anderson and Canning, 1959).

the problem of identifying the members of the various sedimentation groups should be much easier.

The findings of ultracentrifugal classes among tissue proteins indicates that size is not a randomly distributed property. This suggests that (a) tissue proteins fall into a number of general classes, (b) the peaks observed are caused by the presence of a disproportionately large amount of a few specific proteins that obscure a broad distribution of the majority of protein types, or (c) the observed patterns are a combination of the two.

(*b*) *Electrophoretic analysis*—Electrophoretic studies on soluble phase proteins done before 1960 have been reviewed by Anderson and Swanson (1961). If a large number of proteins are present in low concentration, and if electrophoretic mobilities are randomly distributed, then the electrophoretic pattern should also be a Gaussian curve. Examination of rat liver, brain, testis, and kidney soluble proteins in the Tiselius apparatus at pH 7·5 reveals the presence of distinct peaks, some of which are common to several tissues, and also characteristic differences between the preparations (Fig. 7). The presence of distinct peaks therefore suggests either (a) extensive complexing occurs in

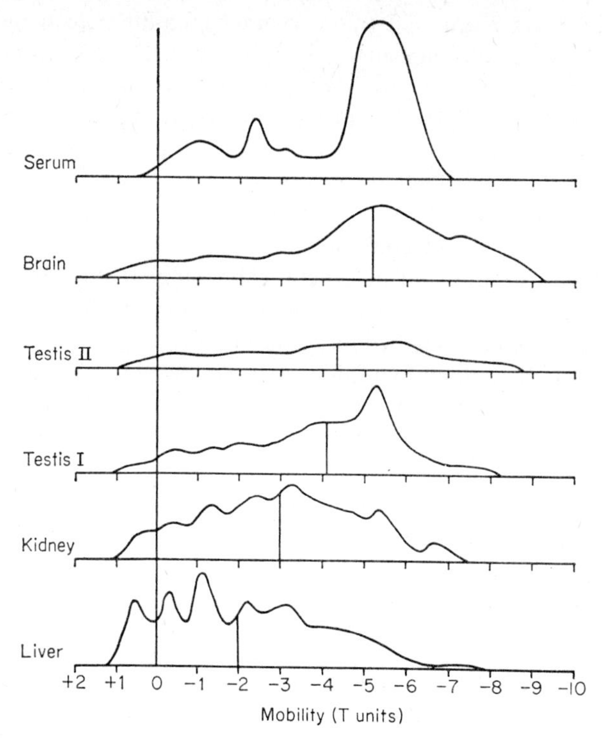

FIG. 7. Electrophoretic analysis of soluble phase proteins of rat liver, kidney, brain, testis (I with, and II without tubular fluid), pH 7·5, 0·1 ionic strength. Rat serum proteins included for comparison (Anderson and Canning, 1961).

solution giving rise to aggregates with distinctive average mobilities; (b) a few proteins are present in fairly high concentration and mask the random distribution of mobilities in the remainder of the proteins present, or (c) families of proteins with similar mobilities are present. Ultracentrifugal analyses tend to rule out (a). The choice between (b) and (c) cannot as yet be made.

Inter-organ differences are indicated both by the shape of the Schlieren pattern and by the differences in median mobilities (liver, –2·0; kidney, –3·0; testis, –4·1; and brain, –5·2). A peak of low positive mobility is present in liver and kidney and almost totally absent in testis and brain. The location o catalase (1.11.1.6), rhodanese (2.8.1.1), acid phosphatase (3.1.3.2), nucleoside phosphorylase, esterase, adenosine deaminase (3.5.4.4), glutathione reductase (1.6.4.2), and β-glucuronidase (3.2.1.31) in the electrophoretic pattern have been determined (Anderson, Fisher, and Anderson, 1961).

The hypothesis that nuclear volume is related to the level of slightly basic cytoplasmic proteins present has been proposed (Anderson and Swanson, 1961; Swanson, 1960a). If this is correct, cells with large nuclei should have proteins which are more acid (more rapidly migrating electrophoretically) than cells of the same organism with small nuclei. The comparative experiments in rat tissue, and electrophoretic analysis of oocyte SP and its comparison with embryonic extracts supports this view. In addition, evidence has been presented that the nuclear globulins are the more basic proteins of the soluble phase (Barton, 1960). The view has been expressed in some detail (Anderson, 1956b) that DNA is functional in an extended state, and non-functional in a condensed condition and has given rise to recent studies linking histone to repression of genetic expression. If nuclear volume of a somatic cell reflects the sum of the volumes of the genes, then factors controlling nuclear volume may be important in differentiation and gene control (Swanson, 1960b). Radioautographic evidence has been presented by Goldstein (1963) and Prescott and Bender (1963) for the existence of a protein which is predominantly localized in the nucleus during interphase, is uniformly distributed throughout the cell during division and returns to the nucleus after telophase. In addition, nuclear transplantation studies have shown that a nucleus transplanted into an enucleate host assumes some of the characteristics of the host (Lorch and Danielli, 1950) further supporting the view that the cytoplasm may produce physical changes in the nucleus, presumably through the mediation of slightly basic soluble cytoplasmic proteins. It is of interest in this regard that the hormonally-controlled "male" protein is among the slightly basic proteins of the soluble phase (Bond, 1962).

Electrophoresis in starch (Smithies, 1962) or acrylamide gels (Raymond and Weintraub, 1959; Ornstein, 1964) depends on the electrophoretic mobility of the proteins being studied and their interaction with the stationary medium (Tombs, 1965). The result is a very high resolution separation which has been widely used to demonstrate isoenzymes and to evaluate purity of isolated protein preparations. As the resolution of this method is improved and preparative counterparts developed, the problem of identifying each of the bands observed may be approached.

(*c*) *Chromatographic analysis*—In addition, chromatographic studies of soluble proteins on ion exchange media have been widely done as part of enzyme purification procedures. However, the initial extracts are often made in ways which do not minimize damage to subcellular components. In addition, interest in isolation leads to the development of isolation methods giving maximum purification of the enzyme in question and not to techniques for achieving the maximum resolution of the entire mixture. The latter problem has been studied in detail by Bond (1965) and an elution program for high resolution separation of rat liver SP on DEAE columns developed. This work led to the discovery of a protein in male rat liver which is absent or present in only very small amount in female liver (Bond, 1960, 1962; Barzilai and Pincus, 1965). It is induced in both sexes by androgens and repressed by estrogens. Its function is unknown.

The detailed analysis of the distribution of enzymes of the SP by high resolution DEAE chromatography and by gel filtration remains to be done. With the introduction of low cross-linked beads, filtration through columns of particulate gels (Flodin, 1962) has been extended to the separation of large molecules.

(*d*) *Immunological analysis*—Immunochemical and immuno-electrophoretic analyses of soluble phase antigens of normal and tumour tissues have been reported by many workers, (e.g. Perlmann and Hultin, 1958; Perlmann and D'Amelio, 1958; Perlmann, Hultin, D'Amelio, and Morgan, 1959; D'Amelio and Perlmann, 1960; Gazzaniga, Di Macco and Sonnino, 1963; Rossowski, Weinrander, and Hierowski, 1963; Baldwin, 1964). The results support the view that in liver a relatively small number of proteins are present in rather high concentration against a background of many which account for the biochemical versatility of the organ. A minimum of 5 immunologically distinct non-plasma proteins can always be recognized in rat liver soluble phase (D'Amelio and Perlmann, 1960). The "h" protein described by Sorof (Sorof, Young, and Ott, 1958), is one of these, and is also found in liver microsomal fraction. A small number of common antigens occur in the deoxycholate extract of microsomal fraction and the cell sap. Several soluble phase antigens are deleted from liver SP during aminoazo dye carcinogenesis, while one antigen was shown to increase fourfold in the tumour (Baldwin, 1964). An increase in the number of soluble antigens in liver from the eighteenth day of intrauterine life to three months after birth has been demonstrated (Rossowski, *et al.*, 1963). Using immuno-electrophoresis 7 to 10 lines are observed with rat liver soluble phase. In regenerating liver three of the lines were frequently absent, one was always absent, three showed no change, and two increased in density (Gazzaniga, *et al.*, 1963). The correlaion of immunologically identified proteins with fractions isolated by centriugation, electrophoresis, and chromatography and with enzymic activities known to occur in the cell is urgently needed.

2. *Lipids and Lipoproteins*

Studies on cell lipids have generally started with extracts of either whole tissues or isolated cell fractions excluding the fatty layer seen when whole breis are ultracentrifuged, or have utilized the surface lipids obtained after high speed centrifugation of whole homogenates. Evidence for a spectrum of particles having flotation rates sufficient to cause them to move centripetally has been repeatedly obtained in the zonal ultracentrifuge. The centrifugal techniques for the isolation of plasma-lipoproteins (Lindgren, Elliott, and Gofman, 1951; De Lalla, Elliott and Gofman, 1954; Gustafson, Alanpovic, and Furman, 1965; Bobbitt and Levy, 1965) are also applicable to the soluble phase but thus far do not appear to have been used.

3. Ribonucleic Acids

(*a*) *Separation of RNAs*—While amino acyl transfer RNAs (sRNA or tRNA) are largely located in the soluble phase, not all SP RNA appears to be active sRNA. Isolation of the soluble phase or at least removal of the major fraction of particulate material is generally the first step in sRNA isolation. The percent of total cell RNA found in the soluble phase varies widely depending on the tissue used, its physiological state, and the isolation methods employed. Values ranging from as little as 5% to as much as 71% of the total tissue RNA have been recorded (Reid, 1961). With the development of methods using the zonal centrifuge (Hastings, Parish, Kirby, and Klucis, 1965), and gel electrophoresis (Richards, Coll, and Gratzer, 1965) for separating RNA according to size, the question of the nature of soluble phase RNA may now be re-examined.

The number of species of sRNA present in a tissue is larger than the number of amino acids found in proteins (Morton and Rogers, 1965) making the problem of resolving all the species present more difficult than initially thought. The complete nucleotide sequence of alanine sRNA has been published (Holley, *et al.*, 1965). Operationally, ribosomal subunits, which may account for 10% of the total cytoplasmic ribosomal material in HeLa cells (Joklik and Becker, 1965), would be found in the soluble phase as ordinarily prepared, and their characteristic RNAs would therefore be found in total RNA extracts of this fraction.

IV. Enzymes of the Soluble Phase

Excellent reviews and tabulations of the enzymes found in centrifugally isolated fractions have appeared (Roodyn, 1965; de Duve, Wattiaux and Baudhuin, 1962). A listing of all enzymes which have been reported to occur in the soluble phase will therefore not be presented here.

A. ENZYME GROUPS OF THE SOLUBLE PHASE

1. Protein Synthesis

The amino acid activating enzymes are almost exclusively located in the SP (Gasior and Moldave, 1965) and in addition, are precipitable iso-electrically at pH 5 (see later section on pH stability of the soluble phase).

2. Nucleic Acid Synthesis

All of the enzymes necessary to incorporate thymidine into DNA are present in the soluble phase of Walker 256 carcinoma, Flexner-Jobling carcinoma, rat thymus, spleen, small intestine, kidney, normal liver, brain, lung, testis, heart muscle, pancreas, skeletal muscle, regenerating liver (Bollum and Potter, 1958), and in adenovirus infected cells (Green, Pina, and Chagoya, 1964). Thymidine kinase (2.7.1.21), thymidylate kinase (2.7.4.9), thymidine diphosphate kinase, thymidylate synthetase and deoxycytidylate deaminase have all been found to increase during cell proliferation (Myers, Hemphill and Townsend, 1961), and decrease during stationary phase (Kit, Dubbs and Frearson, 1965). In addition, part of the soluble DNA-dependent RNA polymerase (2.7.7.6) of rat prostate and liver nuclei may be recovered from isolated nuclei in a soluble form. The enzyme appears to be normally bound to its substrate, however (Doly, Ramuz, Mandel, and Chambon, 1965). Both replicative deoxynucleotidyl transferase (deoxyribonucleic acid polymerase) (2.7.7.7) and terminal deoxynucleotidyl transferase (end addition enzyme) are recoverable in a soluble fraction from calf thymus (Yoneda and Bollum, 1965).

3. *Fatty Acid Synthesis*

The enzymes responsible for fatty acid synthesis appear to be largely confined to the soluble phase (Langdon, 1957).

4. *Glycolysis*

In all animal cells thus far examined the majority of the glycolytic activity is present in the soluble phase. Recoveries range from 75–100% of the total homogenate activity in rat liver (Bucher and McGarrahan, 1956; Kennedy and Lehninger, 1949), rat brain (Abood, Gerard, Banks, and Tschirgi, 1952; Balazs and Lagnado, 1959), rat anterior pituitary (McShan, Rozich, and Meyer, 1953), and Flexner-Jobling sarcoma (Le Page and Schneider, 1948).

5. *Pentose-Phosphate Oxidative Cycle*

Studies on the enzymes of this cycle suggest that the major enzymes are found in the soluble phase of rabbit liver and kidney prepared with saline diluents in place of sucrose (Newburgh and Cheldelin, 1956).

6. *Glycogen Metabolism*

The two most important enzymes in the synthesis and breakdown of glycogen (UDP-glucose-α glucan glucosyltransferase, 2.4.1.11; and α glucan phosphorylase, 2.4.1.1) are bound to the polymer itself in fed animals, and released into the soluble phase during starvation (Leloir and Goldemberg, 1960; Tata, 1964). Phosphoglucomutase (2.7.5.1) is characteristically an enzyme of the soluble phase (Hers, Berthet, Berthet, and de Duve, 1951), while glucose-6-phosphatase (3.1.3.9) is bound to cytomembranes (de Duve, Pressman, Gianetto, Wattiaux, and Appelmans, 1955) (see Chapter 6).

B. PROBLEMS IN INTERPRETATION OF ENZYME DATA

From a physiological or functional viewpoint the objective of enzyme studies on cell fractions is to find out where a given reaction occurs in the cell. Provided the natural substrates and cofactors and an approximation of intracellular pH and ionic environment obtain, it is of little interest whether more than one enzyme is being measured. The question asked is where the cell carries out a particular process. However, if an enzyme assay is used to define one species of protein molecule, then it is of interest to show that only one is being measured. In addition, where similar enzymic activities are observed in several cell fractions, the question of leakage, adsorption, or cross contamination is easily solved if the enzyme molecules in different fractions differ in some physical property. The discovery of isoenzymes (Hunter and Markert, 1957; Shaw, 1965) adds both complexity and new experimental possibilities to cell fractionation studies. Quantitative methods for measuring the activities of many of the possible isoenzymes are not yet available, however.

The question of whether enzymes tend to be characteristic of one cell fraction may be approached by plotting the number of enzymes against increments in the percentage found in the soluble phase (Fig. 8). There

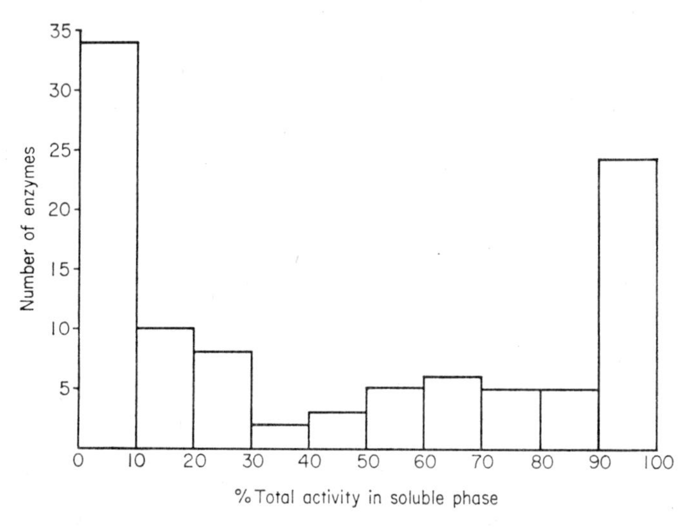

FIG. 8. Plot of % total activity of 102 enzymes found in the soluble phase of animal tissues (chiefly rat liver). 0–10% group includes instances where no activity was observed, 90–100% group includes studies where more activity was found in the isolated soluble phase than was observed in the total homogenate. (Data from Roodyn, 1965). Biphasic distribution will probably be accentuated when specific isoenzymes are measured rather than total enzymic activity.

is a marked tendency for an activity to be present (90% or more) or absent (less than 10%) from the soluble phase.

1. Presence of Low Activity Levels

Enzymes which are largely present in one cell fraction are usually stressed. A very small amount of an enzyme in the soluble phase however may be adequate to deal with the substrate levels present. The basic concept of the lysosome—that a series of otherwise deleterious enzymes are and *must* be compartmented within special bodies—hinges on this point. For this and other reasons low percentage recovery of enzyme activity in the soluble phase is often disregarded and attributed to isolation artifacts.

2. *Localization of Activity*

Non-isoparticulate distributions may be attributed to one of several causes which include: (a) transport (the enzyme is being moved from site of synthesis to site of action), (b) equilibrium (the structure containing an enzyme is in equilibrium with its constituents in solution), or (c) isolation artifacts.

Rather few enzymes have been reported to be *totally* absent from this fraction. These include such enzymes as aryl-4-hydroxylase (1.14.1.1) (Brodie, *et al.*, 1955), nucleoside diphosphatase (3.6.1.6) (Novikoff and Heus, 1963) using ADP and CDP as substrates, thyroxine degrading activity of rat liver (Wynn, Gibbs, and Royster, 1962), 4-monomethyl aminoazobenzene demethylase activity (Hultin, 1957), and the synthesis of ascorbic acid from glucuronolactone (Chatterjee, Chatterjee, Ghosh, Ghosh, and Guha, 1960). It would be of interest to re-check these very carefully for evidence of inhibitors, or very low level activity to settle the point of whether any enzymic activity is ever totally excluded from a cell fraction.

The question of whether the same enzymic forms or isoenzymes are being studied in the soluble and a particulate phase is directly relevant to the problem of whether an activity observed in the SP is particle derived Rat liver β-galactosidase (3.2.1.23) from lysosomes and from the soluble phase appear to be identical (Furth and Robinson, 1965). However, differences in soluble and mitochondrial aspartate aminotransferase (2.6.1.1) (Sheid, Morris, and Roth, 1965), phosphoenolpyruvate carboxykinase (2.7.1.40) (Holten and Nordlie, (1965), and alanine amino transferase (2.6.1.2) (Swick, Barnstein, and Stange, 1965) have been described suggesting but not proving that the SP enzymes were not bound to mitochondria in the cell (see Chapter 3).

C. CONTROL OF ENZYMIC ACTIVITY IN THE CELL

A few enzymes appear to be especially responsive to changes in the concentration of one or more metabolites related in some way to their own activity. Such enzymes appear to be composed of subunits (Gerhart and Schachman, 1965), undergo conformational changes when exposed to "effectors" (Changeux, 1964), and have a sigmoid-shaped substrate saturation curve under certain conditions (Atkinson, 1965). The "effectors" produce a conformational change or allosteric transition (Monod, Changeux, and Jacob, 1963) by binding at a site distinct from the enzymic site. The result may be either activation or inhibition. Generally, the first enzyme of a sequence is inhibited by the end product of that pathway. Where several isoenzymes are found, one may be a regulatory enzyme while another may not. Thus, in rabbit muscle lactate dehydrogenase (1.1.1.27) isoenzyme 5 is an allosteric protein and a regulatory enzyme while lactate dehydrogenase isoenzyme 1 from rabbit heart is neither (Fritz, 1965). As higher resolution separations are attained, surprisingly large numbers of electrophoretically separable forms of a single enzyme may be found (Fritz and Jacobson, 1965) adding to the possibilities of selective regulation. In addition to feedback inhibition by single metabolites, a "concerted" feedback inhibition dependent on simultaneous presence of threonine and lysine has been described (Datta and Gest, 1964) for a bacterial aspartate kinase (2.7.2.4) while a contrasting form of multiple response has been observed in *E. coli* glutamine synthetase (6.3.1.2) where eight metabolites were found to have small additive effects, termed "cumulative" feedback inhibition (Woolfolk and Stadtman, 1964). An interplay between these forms of control, enzyme repression, and enzyme induction to achieve a fairly steady *concentration* of a series of metabolites while varying *rate* at which they flow through a metabolic pathway has been discussed by Atkinson (1965). The control of the level of specific isoenzymes by hormones is outside the scope of this review.

Enzymes have long been known to be exquisitely sensitive to their chemical environment, and variation in pH, ionic strength, and cofactor concentration are routinely examined in the development of enzyme assays. Since it now appears that many alterations in enzymic activity are mediated through sites on the protein molecule other than the enzymic site, an enzyme molecule may, in fact, be considered as a sensor of the intracellular environment translating one or many variations in milieu into a change in catalytic activity. The advantages of such direct and compact control mechanisms are at once apparent. It is not unlikely that all enzymes are in a wider sense regulatory enzymes, although a few have this function developed to a much higher degree.

V. Physiology of the Soluble Phase

A. The Effect of Heat

Proteins of rat liver soluble phase are surprisingly unstable (Anderson, Makinodan, and Norris, 1961). At 40°C approximately 10% of the total nitrogen is precipitated in two hours, while at 100°C 86% of the total nitrogen is flocculated. Precipitation rate at 40°C is rapidly diminished as the pH of the SP is raised from 6·4 to 8·2. However, the flocculation observed, especially around 40°C, is not attributable to a decrease in pH during incubation. When turbidity formation is followed an initial lag in the formation of a precipitate is observed suggesting that a protective substance initially present may be destroyed. Addition of ATP inhibited precipitate formation, but the effect disappears during incubation, possibly because of hydrolysis of ATP. Addition of polyanions (RNA, heparin, polymanuronic acid sulphate) markedly decreases the rate of precipitate formation, while RNase, $CaCl_2$, and $MgCl_2$ promote precipitation.

The observation that a large fraction of the soluble proteins of rat liver precipitate *in vitro* during incubation at temperatures not lethal for an intact animal (Anderson, Makinodan, and Norris, 1961) raises several interesting questions. Is the precipitation due to enzymic digestion, thermal denaturation, or the removal of protective groups or substances? What is the relation, if any, of this effect to physical changes in the ground substance of cells during cell division, after fertilization, in response to chemical, physical and electrical stimuli, during amoeboid movement, and in incipient cytolysis? Work in this area has suffered from a lack of valid isolated model systems, and the effects observed in the SP deserve attention since they contain at least some of the reactants. The effect of incubating rat liver soluble phase prepared in 0·25 M sucrose at various temperatures is shown in Fig. 9. (Note that heating at times and temperatures empirically determined has long been used in the

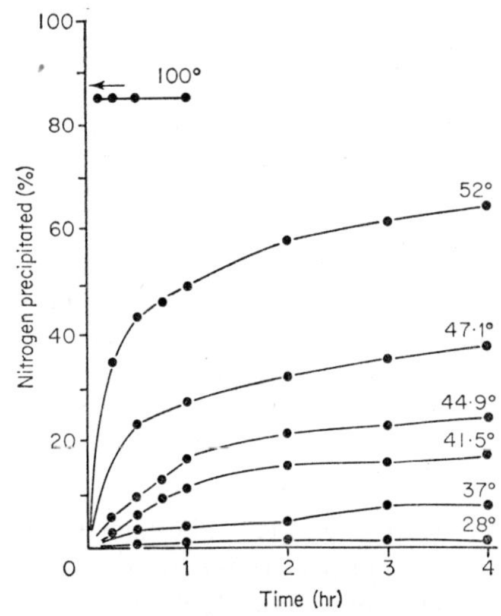

FIG. 9. Percentage of total nitrogen precipitated from soluble rat liver protein solutions as a function of time and temperature. Arrow along ordinate indicates percentage of total nitrogen *not* lost during dialysis (Anderson, *et al.*, 1961). 100% = 3·76 of N/ml Av.

isolation of enzymes from soluble tissue extracts). Sulfhydryl compounds and inhibitors of proteolytic enzymes and colchicine were without protective effect.

B. THE EFFECT OF pH

In phosphate buffers the SP of rat liver is stable for short periods at pH values between 1·26 and 3·5 and between 6·3 and 11. Between pH 3·5 and 6·3 marked precipitation occurs with a maximum of 24% of rat liver SP nitrogen precipitated near pH 5 (Fig. 10). At acid pH values in the presence of HCl voluminous precipitates are obtained which are not seen with phosphoric

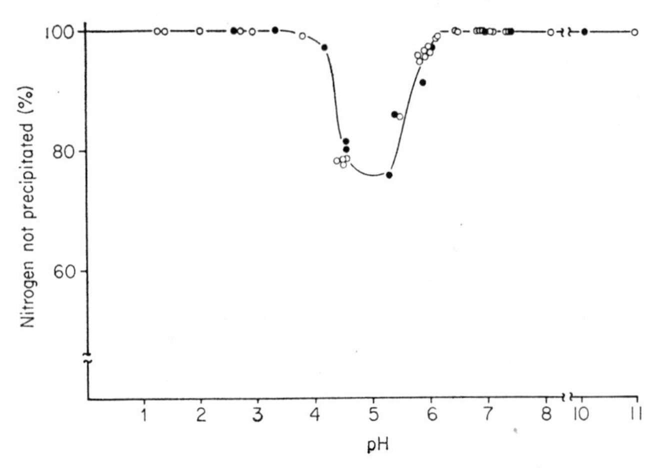

FIG. 10. Percentage of total nitrogen of rat liver soluble phase nitrogen precipitated as a function of pH.

acid. The precipitation of part of the soluble phase at pH 5·0 has long been used in the purification of enzymes from tissues. Among those which precipitate are enzymes which incorporate amino acids into sRNA (Hoagland, Keller, and Zamecnik, 1956) and dihydrofolate reductase (Bertino, Perkins, and Johns, 1965). The percentage of soluble proteins iso-electrically precipitated at pH 5·0 in various tissues has not been examined in detail; however, voluminous precipitates have been observed in supernatants from Ehrlich ascites tumour (E. C. Horn, personal communication).

The effect of pH on the SP is of interest because parallel effects are seen in cells where a slight decrease in pH produces a marked increase in viscosity. After injury, cell pH often drops and the cytoplasm is observed to coagulate (Heilbrunn, 1956). The precipitation observed *in vitro* parallels the intracellular change in that precipitation is initially easily reversed, but becomes irreversible with time. It is evident that if cell fractionation of a homogenate is attempted in the neighborhood of pH 5·0 that gross cross contamination with precipitated protein will occur. The nuclear isolation methods developed by Dounce (1955) involved the use of solutions on either side of the precipitation region shown in Fig. 10. The proteins of the mitotic apparatus are thought to be iso-electric at a pH below neutrality on the basis of studies on proteins solubilized from the isolated mitotic apparatus (Went and Mazia, 1959) and because of a sharpening of the mitotic apparatus observed in the light microscope in dividing cells exposed to slightly acid solutions. If the proteins of the mitotic apparatus persist in a soluble form in the interphase cell (Went and Mazia, 1959), they may well be found in the SP fraction isoelectrically precipitated at pH 5·0.

C. THE ROLE IN MEMBRANE AND PARTICLE FORMATION

The proposal that some cell structures may exist in equilibrium with their constituent macromolecules present in the SP (Anderson, 1956a, b, 1959) is based on the concept of coacervation (Bungenberg de Jong, 1949). This mechanism would allow changes in structure to occur in response to alterations in the composition of the SP during differentiation, growth, starvation, cell division, or in response to external agents or stimuli. If specific structures are to form in place of simple coacervates, then molecular surface specificities must exist on structure forming elements which, in response to discrete micro-environments in the cell, result in the formation of characteristic structures in their proper places. The alternative to this view is that the molecular elements of subcellular structures are formed *in situ.* All presently available evidence suggests that this is not the case but that protein synthesis for example is confined to specific structures devoted to that purpose. The movement of structure-associated proteins to their structure-position must therefore occur by diffusion in the SP. Their concentration will depend on the binding equilibrium constant. If this is shifted far in the direction of a stable structure, the amount of structure-forming protein in the SP will be very small. Destruction of formed elements may then require special intracellular digestion mechanisms. Conceivably structure-forming proteins may, like fibrinogen, require activation to form insoluble configurations.

Experimentally, the formation of structures not unlike those seen in homogenates has been observed in the soluble phase of rat liver (Anderson, 1959) and in clear bacterial extracts after the addition of magnesium (Bolton, *et al.*, 1957). Somewhat similar structures have also been reported from mixtures of haemoglobin and DNA (Nelson, 1958a and b). For these observations to have any relevance to events *in vivo* it must be shown that the same molecular species are involved. The observation that the plasma membrane of the cell can rapidly reform after cell rupture (Heilbrunn, 1956) and that antisera against isolated cell particles cross react with the SP suggest that this is the case. From the point of view taken here the only interesting and unexpected finding in a cell fractionation study would be the total absence of a protein from the soluble phase.

VI. CONCLUSIONS

With the use of high resolution separation techniques including zonal ultracentrifugation, automated chromatographic systems, and gel electrophoresis and immunoelectrophoresis, the molecular anatomy of the soluble phase may now be worked out in a manner analogous to blood plasma. A purely descriptive approach is essential initially since the functions of many of the constituents observed will not be known. On the basis of work already done it is concluded that the soluble phase is a tractable mixture which should now be resolved.

If the results are to reflect the intracellular composition it is imperative that extremely rapid methods for the isolation of the soluble phase be worked out, if possible compressing homogenization, centrifugation, and freezing into a single process requiring only a few seconds. Centrifuge systems on which to base such studies have been described (Kolousek, Jiracek, and Cizinsky, 1964).

Progress on the elucidation of homeostatic mechanisms which control the intracellular environment will depend on the development of rapid analytical methods which allow the quantitation of mesomolecular families of compounds. The way in which enzymes respond to this total intracellular environment may then be examined.

References

Abood, L. G., Gerard, R. W., Banks, J. and Tschirgi, R. D. (1952). *Am. J. Physiol.* **168,** 728.

Agren, Gunnar (1960). *Acta chem. scand.* **14**, 2241.

Allen, Kenneth (1961). *Am. Zool.* **1**, 253.

Anderson, N. G. (1956a). *Expl Cell Res.* **11**, 186.

Anderson, N. G. (1956b). *Quart. Rev. Biol.* **31**, 169.

Anderson, N. G. (1959). *Expl. Cell Res.* **16**, 42.

Anderson, N. G. (1966). *J. natn. Cancer Inst. Monograph* No. 21., 9.

Anderson, N. G. and Canning, R. E. (1959). *Expl. Cell Res.* **17**, 465.

Anderson, N. G., Fisher, W. D. and Anderson, M. L. (1961). *Expl. Cell Res.* **23,** 318.

Anderson, N. G., Green, J. G., Barber, M. L. and Ladd, Sr. F. C. (1963). *Analyt. Biochem.* **6,** 153.

Anderson, N. G., Makinodan, T. and Norris, C. B. (1961). *Expl Cell Res.* **22**, 526.

Anderson, N. G. and Swanson, H. D. (1961). *Expl Cell Res.* **23**, 58.

Atkinson, D. E. (1965). *Science, N.Y.* **150**, 851.

Bailey, J. M. and Woodson, M. (1963). *Fed. Proc.* **22**, 417.

Balazs, R. and Lagnado, J. R. (1959). *J. Neurochem.* **5**, 1.

Baldwin, R. W. (1964). *Brit. J. Cancer.* **18**, 285.

Barton, A. D. (1960). *In* "The Cell Nucleus, Proceedings of the Faraday Society", p. 142. Butterworths, London.

Barzilai, D. and Pincus, G. (1956). *Proc. Soc. exp. Biol. Med.* **118**, 57.

Bertino, J. R., Perkins, J. P. and Johns, D. G. (1965). *Biochemistry* **4**, 839.

Bittar, E. E. (1964). "Cell pH". Butterworths, Washington.

Bobbitt, J. L. and Levy, R. S. (1965). *Biochemistry* **4**, 1282.

Bollum, F. J. and Potter, V. R. (1958). *J. biol. Chem.* **233**, 478.

Bolton, E. T., Britten, R. J., Cowie, D. B., Leahy, J. J., McClure, F. T. and Roberts, R. B. (1957). "Carnegie Institution of Washington Year Book 1956", pp. 140–142.

Bond, H. E. (1960). *Biochem. biophys. Res. Commun.* **3**, 53.
Bond, H. E. (1962). *Nature, Lond.* **196**, 242.
Bond, H. E. (1965). "Chromatographic methods for studying soluble tissue proteins". USNRDL-TR-848.
Brady, C. J. (1964). *Biochem. J.* **91**, 105.
Brady, R. O. and Tower, D. B., eds. (1960). "The Neurochemistry of Nucleotides and Amino Acids". Symposium of the Section on Neurochemistry, American Academy of Neurology. Wiley, New York, 292 pp.
Britten, R. J. and McClure, F. T. (1962). *Bact. Rev.* **26**, 292.
Brodie, B. B., Axelrod, J., Cooper, J. R., Gaudette, L., La Du, B. N., Mitoma, C. and Udenfriend, S. (1955). *Science, N.Y.* **121**, 603.
Bucher, N. L. R. and McGarrahan, K. (1956). *J. biol. Chem.* **222**, 1.
Bücher, Th., Krejei, K., Rüssman, W., Schnitger, H. and Wesemann, W. (1964). *In* "Rapid Mixing and Sampling Techniques in Biochemsitry" (Britton Chance, Rudolf Eisenhardt, Quentin H. Gibson and K. Karl Lonberg-holm, eds.) pp. 255–264. Academic Press, New York and London.
Bungenberg De Jong, H. G. (1949). *In* "Colloid Science", Vol. II, "Reversible Systems" (H. R. Kruyt, ed.) pp. 433–482. Elsevier Publishing Co., New York.
Caldwell, P. C. (1956). *Int. Rev. Cytol.* **5**, 229.
Chambers, R. and Chambers, E. L. (1961). "Explorations into the nature of the living cell". Harvard University Press, Cambridge, Mass. 352 pp.
Changeux, J. P. (1964). *Brookhaven Symp. Biol.* **17**, 232.
Chatterjee, I. B., Chatterjee, G. C., Ghosh, N. C., Ghosh, J. J. and Guha, B. C. (1960). *Biochem. J.* **76**, 279.
Christensen, Halvor N. (1964). *In* "Mammalian Protein Metabolism", Vol. 1 (H. N. Munro and J. B. Allison, eds.) pp. 105–124. Academic Press, Inc., New York and London.
Claude, A. (1948). *Harvey Lectures* **43**, 121.
Datta, P. and Gest, H. (1964). *Proc. natn. Acad. Sci. U.S.A.* **52**, 1004.
D'Amelio, V. and Perlmann, P. (1960). *Expl Cell Res.* **19**, 383.
de Duve, C. and Berthet, J. (1954). *Int. Rev. Cytol.* **3**, 225.
de Duve, C., Pressman, B. C., Gianetto, R., Wattiaux, R. and Appelmans, F. (1955). *Biochem. J.* **60**, 604.
de Duve, C., Wattiaux, R. and Baudhuin, P. (1962). *Adv. Enzymol.* **24**, 291.
De Lalla, O. F., Elliott, H. A. and Gofman, J. W. (1954). *Am. J. Physiol.* **179**, 333.
De Ropp, R. S. and Snedeker, E. H. (1960). *Analyt. Biochem.* **1**, 424.
de Waart, C. (1964). *6th Int. Congr. Biochem.* **32**, 648.
Doly, J., Ramuz, M., Mandel, P. and Chambon, P. (1965). *Life Sci.* **4**, 1961.
Dounce, A. L. (1955). *In* "The Nucleic Acids" Vol. II (E. Chargaff and J. N. Davidson, eds.) pp. 93–153. Academic Press, Inc., New York and London.
Dunn, M. S., Sakamoto, K., Sutaria, P. and Murphy, E. A. (1962). *Proc. Soc. exp. Biol. Med.* **110**, 475.
Elliott, K. A. C., Khan, R. T., Bilodeau, F. B. and Lovell, R. A. (1965). *Can. J. Biochem.* **43**, 407.
Florey, E. (1964). *In* "Major Problems in Neuro Endocrinology", pp. 17–41. S. Karger, Basel, Switzerland and New York.
Flodin, Per (1962). "Dextran gels and their applications in gel filtration." *Pharmacia*, Uppsala, Sweden, pp. 1–85.
Fritz, P. J. (1965). *Science, N.Y.* **150**, 364.
Fritz, P. J. and Jacobson, K. B. (1965). *Biochemistry* **4**, 282.

Furth, A. J. and Robinson D. (1965). *Biochem. J.* **97**, 59.
Gasior, E. and Moldave, K. (1965). *J. biol. Chem.* **240**, 3346.
Gazzaniga, P. P., DiMacco, G. and Sonnino, F. R. (1963). *Experientia* **19**, 419.
Gerhart, J. C. and Schachman, H. K. (1965). *Biochemistry* **4**, 1054.
Giuffrida, L. (1964). *J. Assoc. Off. Agr. Chemists* **47**, 293.
Goldstein, L. (1963). *J. cell comp. Physiol.* **62**, (Suppl. No. 1): 190.
Green, J. G. (1966). *In* "The Development of Zonal Centrifuges" (N. G. Anderson, ed.). J. Nat. Cancer Inst. Monograph No. 21., 447.
Green, J. G. and Anderson, N. G. (1965). *Fed. Proc.* **24**, 606.
Green, J. G., Nunley, C. E. and Anderson, N. G. (1966). *In* "The Development of Zonal Centrifuges", (N. G. Anderson, ed.). J. Nat. Cancer Inst. Monograph No. 21., 431.
Green, M., Piña, M. and Chagoya, Victoria (1964). *J. biol. Chem.* **239**, 1188.
Greenstein, J. P. and Winitz, M. (1961). *In* "Chemistry of the Amino Acids", Vol. V. p. 25. J. Wiley and Sons, New York.
Gregoire, J. (1958). *Bull. Soc. Chim. biol.* **40**, 1245.
Gross, P. R., Philpott, D. E. and Nass, S. (1960). *J. biophys. biochem. Cytol.* **7**, 135.
Gustafson, A., Alanpovic, P. and Furman, R. H. (1965). *Biochemistry* **4**, 596.
Hamilton, P. B. (1962). *Ann. N.Y. Acad. Sci.* **102**, 55.
Hamilton, P. B. (1963). *Analyt. Chem.* **35**, 2055.
Hanlon, David P. (1960). *Biol. Bull.* **118**, 79.
Hastings, J. R. B., Parish, J. H., Kirby, K. S. and Klucis, E. S. (1965). *Nature, Lond.* **208**, 645.
Heilbrunn, L. V. (1956). "The Dynamics of Living Protoplasm". Academic Press, New York and London.
Henderson, J. F. and LePage, G. A. (1958). *Chem. Rev.* **58**, 645.
Hendler, R. W. (1961). *Biochim. biophys. Acta* **49**, 297.
Hendler, R. W. (1962). *In* "Amino Acid Pools" (J. T. Holden, ed.), pp. 750–758. Elsevier Publishing Company, New York.
Hers, H. G., Berthet, J., Berthet, L. and de Duve, C. (1951). *Bull. Soc. Chim. biol.* **33**, 21.
Hoagland, M. B., Keller, E. B. and Zamecnik, P. C. (1956). *J. biol. Chem.* **218**, 345.
Hogeboom, G. H., Kuff, E. L. and Schneider, W. C. (1957). *Int. Rev. Cytol.* **6**, 425.
Holden, J. T. (1962). *In* "Amino Acid Pools" (J. T. Holden, ed.), pp. 73–108. Elsevier Publishing Company, New York.
Holley, R. W., Apgar, J., Everett, G. A., Madison, J. T., Marquisee, M., Merrill, S. H., Penswick, J. R. and Zamir, A. (1965). *Science, N.Y.* **147**, 1462.
Holten, D. D. and Nordlie, R. C. (1965). *Biochemistry* **4**, 723.
Hultin, T. (1957). *Expl. Cell Res.* **13**, 47.
Hunter, R. L. and Markert, C. L. (1957). *Science, N.Y.* **125**, 1294.
Jeuniaux, Ch., Duchateau-Bosson, Gh. and Florkin, Marcel (1961). *J. Biochem.* (*Tokyo*) **49**, 527.
Joklik, W. K. and Becker, Y. (1965). *J. mol. Biol.* **13**, 496.
Juvet, R. S. and Nogare, S. D. (1964). *Analyt. Chem.* **36**, 36R.
Kennedy, E. P. and Lehninger, A. L. (1949). *J. biol. Chem.* **179**, 957.
Kit, S., Dubbs, D. R. and Frearson, P. M. (1965). *J. biol. Chem.* **240**, 2565.
Koike, M., Reed, L. J. and Carroll, W. R. (1963). *J. biol. Chem.* **238**, 30.
Kolousek, J., Jiracek, V. and Cizinsky, B. (1964). *J. Neurochem.* **11**, 541.

Krebs, H. A. (1964). *In* "Mammalian Protein Metabolism", Vol. I. (H. N. Munro and J. B. Allison, eds.), pp. 125–176. Academic Press, Inc., New York and London.

Kuczynski-Halmann, M., Avi-Dor, Y. and Mager, J. (1958). *J. gen. Microbiol.* **18**, 364.

Lacks, S. and Gros, F. (1959). *J. mol. Biol.* **1**, 301.

Langdon, R. G. (1957). *J. biol. Chem.* **226**, 615.

Lange, R. (1963). *Comp. Biochem. Physiol.* **10**, 173.

Lehninger, A. L. (1964). "The Mitochondrion. Molecular basis of structure and function". W. A. Benjamin, Inc., New York.

Leloir, L. F. and Goldemberg, S. H. (1960). *J. biol. Chem.* **235**, 919.

LePage, G. A. and Schneider, W. C. (1948). *J. biol. Chem.* **176**, 1021.

Light, Albert and Smith, E. L. (1963). *In* "The Proteins". Vol. I, 2nd Ed. (Hans Neurath, ed.), pp. 1–44. Academic Press, New York and London.

Lindgren, F. T., Elliott, H. A. and Gofman, J. W. (1951). *J. Phys. Coll. Chem.* **55**, 80.

Long, C. N. H., ed. (1961). "Biochemist's Handbook". pp. 640–838. D. Van Nostrand, Inc., New York.

Lorch, I. J. and Danielli, J. F. (1950). *Nature, Lond.* **166**, 329.

Lowry, O. H. and Passonneau, J. V. (1964). *J. biol. Chem.* **239**, 31.

Lowry, O. H., Passonneau, J. V., Hasselberger, F. X. and Schulz, D. W. (1964). *J. biol. Chem.* **239**, 18.

Mandel, P. (1964). *In* "Progress in Nucleic Acid Research", Vol. 3, pp. 299–334. Academic Press, Inc., New York and London.

Martin, R. G. and Ames, B. N. (1961). *J. biol. Chem.* **236**, 1372.

Mateyko, G. M. and Kopac, M. J. (1964). *Prog. exp. Tumor Res.* **4**, 27.

McShan, W. H., Rozich, R. and Meyer, R. K. (1953). *Endocrinology* **52**, 215.

Meister, A. (1965). "Biochemsitry of the Amino Acids", Vols. 1 & 2. Academic Press, Inc., New York and London.

Mitchell, H. K. and Simmons, J. R. (1962). *In* "Amino Acid Pools" (J. T. Holden, ed.). pp. 136–146. Elsevier Publishing Company, New York.

Monod, J., Changeux, J. P. and Jacob, F. (1963). *J. mol. Biol.* **6**, 306.

Morton, M. J. and Rogers, W. I. (1965). *Analyt. Biochem.* **13**, 108.

Mukherjee, B. B., Matthews, J., Horney, D. L. and Reed, L. J. (1965). *J. biol. Chem.* **240**, PC2268.

Myers, D. K., Hemphill, C. A. and Townsend, C. M. (1961). *Can. J. Biochem. Physiol.* **39**, 1043.

Nelson, E. L. (1958a). *J. exp. Med.* **107**, 755.

Nelson, E. L. (1958b). *J. exp. Med.* **107**, 769.

Newburgh, R. W. and Cheldelin, V. H. (1956). *J. biol. Chem.* **218**, 89.

Novikoff, A. B. and Heus, M. (1963). *J. biol. Chem.* **238**, 710.

Ornstein, L. (1964). *Ann. N.Y. Acad. Sci.* **121**, 321.

Passera, C., Pedrotti, A. and Ferrari, G. (1964). *J. Chromatog.* **14**, 289.

Pasteels, J. J., Castiaux, P. and Vandermeerssche, G. (1958). *J. biophys. biochem. Cytol.* **4**, 575.

Perlmann, P. and D'Amelio, V. (1958). *Nature, Lond.* **181**, 491.

Perlmann, P. and Hultin, T. (1958). *Nature, Lond.* **182**, 1530.

Perlmann, P., Hultin, T., D'Amelio, V. and Morgan, W. S. (1959). *Expl Cell Res.* Suppl. 7, 279.

Piez, K. A. and Morris, L. (1960). *Analyt. Biochem.* **1**, 187.

Potter, V. R. (1958). *Fed. Proc.* **17**, 691.

Prescott, D. M. and Bender, M. A. (1963). *J. cell comp. Physiol.* Suppl. No. 1, **62**, 175.

Raymond, S. and Weintraub, L. (1959). *Science, N.Y.* **130**, 711.

Reid, E. (1961). *In* "Biochemist's Handbook" (C. N. H. Long, ed.), pp. 816–838. D. Van Nostrand Company, Inc., New York.

Richards, E. G., Coll, J. A. and Gratzer, W. B. (1965). *Analyt. Biochem.* **12**, 452.

Roberts, E. and Simonsen, Daisy G. (1960). *In* "Amino Acids, Proteins and Cancer Biochemistry", Jesse P. Greenstein Memorial Symposium, Sept. 16, 1959 (John T. Edsall, ed.). Academic Press, Inc., New York and London.

Roberts, E. and Simonsen, Daisy G. (1962). *In* "Amino Acid Pools", (J. T. Holden, ed.), pp. 284–349. Academic Press, New York and London.

Roodyn, D. B. (1959). *Int. Rev. Cytol.* **8**, 279.

Roodyn, D. B. (1965). *Int. Rev. Cytol.* **18**, 99.

Rossowski, W., Weinrander, H., and Hierowski, M. (1963). *Bull. Acad. Polon. Sci., Ser. Sci. Biol.* **11**, 579.

Rothstein, A. (1961). *In* "Regulation of the Inorganic Ion Content of Cells", pp. 53–64. Ciba Foundation Study Group No. 5. Little, Brown and Company, Boston.

Salvatore, G., Vecchio, G., Salvatore, M., Cahnmann, H. J., and Robbins, J. (1965). *J. biol. Chem.* **240**, 2935.

Saukkonen, J. J. (1965). *Chromatog. Rev.* **6**, 53.

Schmitz, H. and Walpurger, G. (1959). *Angew. Chem.* **71**, 549.

Shaw, C. R. (1965). *Science, N.Y.* **149**, 936.

Shaw, J. (1958). *J. exp. Biol.* **35**, 920.

Sheid, B., Morris, H. P. and Roth, J. S. (1965). *J. biol. Chem.* **240**, 3016.

Siegel, S. J. and Swenson, P. A. (1964). *J. cell. comp. Physiol.* **63**, 253.

Smithies, O. (1962). *Archs Biochem. Biophys.* Suppl. 1, 125.

Sorof, S. and Cohen, P. P. (1951a). *J. biol. Chem.* **190**, 311.

Sorof, S. and Cohen, P. P. (1951b). *Cancer Res.* **11**, 376.

Sorof, S., Golder, R. H. and Ott, Marilyn (1954). *Cancer Res.* **14**, 190.

Sorof, S., Young, E. M. and Ott, M. G. (1958). *Cancer Res.* **18**, 33.

Spackman, D. H., Stein, W. H. and Moore, S. (1958). *Analyt. Chem.* **30**, 1190.

Stephen, W. P., Steinbauer, A. L. (1963). *Proc. Entomol. Soc. Wash.* **65**, 99.

Stern, F., Grumet, G., Trabal, F., Mennis, A. and Zinsser, H. H. (1965). *J. Chromatog.* **19**, 130.

Steward, F. C. and Pollard, J. K. (1962). *In* "Amino Acid Pools" (J. T. Holden, ed.), pp. 25–42. Elsevier Publishing Co., New York.

Strominger, J. L. (1960). *Physiol. Rev.* **40**, 55.

Swanson, H. D. (1960a). Ph.D. Dissertation, University of Tennessee, U.S.A.

Swanson, H. D. (1960b). *Genetics* **45**, 1014.

Swick, R. W., Barnstein, P. L. and Stange, J. L. (1965). *J. biol. Chem.* **240**, 3334.

Talley, E. A., Carter, F. L. and Porter, W. L. (1958). *J. Agric. Fd. Chem.* **6**, 608.

Tata, J. R. (1964). *Biochem. J.* **90**, 284.

Tee, D. E. H., Wang, M. and Watkins, J. (1964). *Nature, Lond.* **204**, 682.

Tombs, M. P. (1965). *Analyt. Biochem.* **13**, 121.

Tunn, V. and Dumazert, C. (1962). *Bull. Soc. Pharm. Marseille* **11**, 11.

Ussing, H. (1960). "The Alkali Metal Ions in Biology". 280 pp. Springer, Berlin.

Went, H. A. and Mazia, D. (1959). *Expl Cell Res.* Suppl. 7, 200.

Wiercinski, F. J. (1955). "The pH of Animal Cells." *Protoplasmatologia.* Springer-Verlag, Vienna.

Wittgenstein, Eva and Rowe, K. W., Jr. (1965). *Clin. Chem.* **11**, 155.

Woolfolk, C. A. and Stadtman, E. R. (1964). *Biochem. biophys. Res. Commun.* **17**. 313.
Wu, R. (1965a). *J. biol. Chem.* **240**, 2373.
Wu, R. (1965b). *J. biol. Chem.* **240**, 2827.
Wynn, J., Gibbs, R. and Royster, B. (1962). *J. biol. Chem.* **237**, 1892.
Yoneda, M. and Bollum, F. J. (1965). *J. biol. Chem.* **240**, 3385.
Zalokar, M. (1960). *Expl Cell Res.* **19**, 559.
Zalokar, M. (1960). *Expl Cell Res.* **19**, 114.
Zech, A. C. and Pauli, A. W. (1962). *Crop Sci.* **2**, 421.

APPENDIX

SOME SUGGESTIONS FOR TERMINOLOGY IN ENZYME CYTOLOGY

In all rapidly advancing subjects there is a danger that confusions in terminology may arise, and impede communication between workers in the field. The list that follows is not intended as an arbitrary "ruling" on such questions, since all language is to some extent subjective and individualistic. Also some of the terms at the moment escape precise definition and other terms, although inadequate or even incorrect (e.g. homogenate) have become firmly entrenched in our parlance. What I have tried to do is to present what seem to be reasonable definitions, derived partly from my own predudices, and also by abstraction from the various chapters in "Enzyme Cytology". Many of the terms are more fully explained by the various authors, and the reader is referred to the appropriate chapter for a full discussion.

I am grateful to the various authors for their suggestions and comments, and I fully accept the responsibility for any violent reactions to this attempt to discipline our language. If this list does nothing more than provoke controversy, at least something will have been achieved.

Throughout the list I have tried to distinguish between "cytological" and "operational" terms (see Chapter 1). The superscript(c) after a term indicates that the component is being described in cytological terms, and the superscript(o) indicates an operational description (for example "the material sedimenting after 1 hr at 100,000 *g*). The operational component may have a cytological counterpart (e.g. "mitochondrial fraction" and "mitochondria") and the two terms may then be used interchangeably with little confusion. However, it may have little or no cytological meaning (e.g. "fluffy layer" or "pH 5 fraction"). Conversely the cytological term may have no operational counterpart (e.g. "Golgi apparatus"). A complete separation of operational and cytological terms is of course impossible because of the great overlap between the two fields of differential centrifuging and cytology, but in some cases the difference is clear and should be stressed. Standardization in the use of the two terminologies at the present time may also prevent future confusion.

D.B.R.

TERMINOLOGY LIST

Agranular cytomembranes(c) *See* "smooth surface cytomembranes"

Aleuroplast(c) Leucoplast in which protein granules predominate as a storage product.

Amyloplast(c) Leucoplast in which starch predominates as a storage product.

Area of focal degeneration(c) *See* "autophagic vacuole"

Autolysosome(c) *See* "autophagic vacuole".

Autophagic vacuole(c) Enlarged lysosome containing mitochondria or other cell components. Essentially the same as "cytolysome", "area of focal degeneration", "cytosegresome", "composite bodies" or "autolysosomes".

Cell sap(c) Interparticulate fluid of the cell. Equivalent to hyaloplasm(c) and soluble phase(c)(o) and not to be used synonymously with soluble fraction(o).

Chloroplast[c] Cell particle characterized by a limiting double membrane and an internal, chlorophyll-containing lamellar structure embedded in a protein-rich stroma. The chloroplast is capable of light-dependent photophosphorylation and of photosynthetic CO_2 fixation.
Chloroplast fraction[o] Preparation of cell particles consisting predominantly of chloroplasts.
Chromatin[c] Material that includes spiralized (condensed chromatin) and despiralized (extended chromatin) parts of chromosomes.
Chromatin threads [o] Chromosomal material obtained by differential centrifuging of disrupted nuclei.
Chromatophores[o] Particles isolated from photosynthetic bacteria and containing photosynthetic pigments.
Chromoplast[c] Coloured plastid.
Chromosome[c] This term should only be used to refer to the discrete chromosome. In other cases the term "chromatin" should be used.
Cisterna of cytomembranes[c] Space between paired cytomembranes.
Composite body[c] *See* "autophagic vacuole"
Cristae of mitochondria[c] Characteristic internal double membrane structure of mitochondria, generally thought to be continuous with the inner membrane.
Cytolysome[c] *See* "autophagic vacuole".
Cytomembrane[c][o] Membrane, or membrane system, in the cytoplasm. The term is probably more meaningful and less ambiguous than endoplasmic reticulum. May consist of paired structures separated by a space (cisterna). Divided into rough surface (α, *granular*) cytomembranes, and smooth surface (β, γ agranular) cytomembranes. β *cytomembranes* are membranes which appear to represent invaginations of the plasma membrane. γ *cytomembranes* are a cluster (or "complex") of membranes making up the Golgi cytomembranes.
Cytoplasmic fraction[o] It is a frequent practice to remove unbroken cells, cell debris and nuclei from tissue homogenates before fractionation. Since this is not strictly a total homogenate, the fraction may best be called the "cytoplasmic fraction".
Cytosegresome[c] *See* "autophagic vacuole".
Cytosome[c] *See* lysosome.
Differential centrifuging This is a general term for the fractionation of tissue homogenates by centrifugal methods.
Dispersate[o] The term "homogenate" is preferred.
Dispersion[o] Again, the term "homogenate" is preferred, although "tissue dispersion" is probably a more correct definition of the material than "tissue homogenate". *See* below, under "homogenate".
DNP-fibril [c] (deoxyribonucleoprotein fibril) A fibrillar structure 100–250Å in diameter consisting of deoxyribonucleoprotein. It may also be called an "elementary chromosomal fibril".
Elaioplast [c] Leucoplast in which oil predominates as the storage product.
Electron transport particle [o] : *ETP* A particle derived from mitochondria, and capable of carrying out electron transport from suitable substrates to oxygen. In general the use of abbreviations to describe cell particles is to be avoided, in order to prevent difficulty in reading papers with too many such abbreviations. However, ETP has such general use that it should be retained.
Endoplasmic reticulum [c] A system of cytoplasmic membranes, generally considered to consist of two components: 1. smooth surface endoplasmic reticulum, which are arrays of smooth membranes and 2. rough surface endoplasmic reticulum, arrays

of membrane studded with ribosomes. Endoplasmic reticulum is not synonymous with microsomes, or microsomal fraction (o)—*see* "microsomes" below.
Enzyme cytology The study of the intracellular localization of enzymes.
E.R. or e.r. This abbreviation is sometimes used for endoplasmic reticulum, or ergastoplasm, but is best avoided or at least restricted to the labelling of electron micrographs.
Ergastoplasm (c) System of cytoplasmic membranes. Cytomembranes or endoplasmic reticulum preferred. Not synonymous with microsomes or microsomal fraction (o)—(*see* "microsomes" below.) Generally does not include smooth surface cytomembranes.
Ergosomes (c) (o) *See* "polysomes".
Fluffy layer (o) Material lying above the densely packed mitochondrial pellet after decanting the supernatant fluid. It is undoubtedly a heterogenous fraction and should not be described as a separate sub-cellular component. If possible, the term is best avoided, since it is very difficult to prepare the fraction in a reproducible way.
Fraction Material separated from a homogenate by differential centrifuging. Since some of the fractions have a clear content of well defined cell components (e.g. the mitochondrial fraction) there has been a very great tendency to describe the fraction in terms of the most abundant sub-cellular component present and even to use the terms synonymously (e.g. nuclei for nuclear fraction, ribosomes for microsomal fraction, lysosomes for lysosome-rich mitochondrial fractions etc.) Since there is usually cross-contamination between components during isolation of the fractions, and since they are frequently damaged more or less severely by the isolation procedure, a clear distinction must be made between the isolated sub-cellular fraction and the original cell component. Although somewhat cumbersome, it is better to use the composite term "nuclear fraction", "mitochondrial fraction", "soluble fraction" etc. where any danger of such confusion exists.
g Taken to mean the g_{av} at the centre of the tube. A precise description of centrifugal forces employed would include the r_{max} and r_{min} values, the type of rotor, and the forces developed during acceleration and deceleration.
g *min* The product of the integrated centrifugal force in **g** and the time it was applied, in minutes—*see* "***Kg*** min" below.
Golgi apparatus (c) A cytological term referring to a specialized system of cytoplasmic membranes. Not to be used synonymously with smooth membrane fraction (o) unless the fraction is derived exclusively from the Golgi apparatus. Golgi cytomembranes are a cluster (or "complex") of so-called "γ-cytomembranes".
Gradient differential centrifuging Centrifugal fractionation based on differences in sedimentation rates of cell components in a density gradient medium. The gradient is generally employed to stabilize the various layers and the particles do not reach density equilibrium with the medium—*see* "isopycnic gradient centrifuging" below.
Gradients Some standardization of the presentation of gradients would be desirable. For example, it might be suggested that the figures are always drawn with the lightest fractions to the right and the heaviest to the left i.e. the density-gradient increases from right to left. Also, the position of fractions along the gradient are best marked in some meaningful manner e.g. cm from meniscus or, better cm from the axis of rotation, rather than simply in fraction numbers.
Granular cytomembranes See "rough surface cytomembranes".
Heavy microsomal fraction See "microsomal fractions".
Homogenate (o) Preparation of disrupted cells suspended in an appropriate medium. The term is strictly incorrect, since no homogenates are homogeneous. However,

the very wide use of the term must justify its retention. It need not strictly be applied to preparations of disrupted cells, since it is convenient, for example, to talk of a homogenate of disrupted mitochondria. The homogenate of cells or whole tissue is best called a "cell-homogenate" or "tissue homogenate".

Hyaloplasm (c) The inter-particulate fluid of the cell, equivalent to "cell sap" and "soluble phase". Not to be used synonymously with soluble fraction (o).

Inner compartment of mitochondria (c) Space bounded by cristal and inner membrane. Equivalent to "matrix".

Intracellular localization, This term should only be applied to the distribution of a given enzyme or chemical substance in sub-cellular fractions if these fractions have been examined cytologically. Preferably, its use should be restricted to methods in which the actual localization can be examined (e.g. microspectrophotometry) and not used at all for experiments based on differential centrifuging. "Subcellular distribution" or "Distribution in subcellular fractions" would be better terms to describe work based on fractions derived from homogenates.

Isoenzymes, Preferred to "isozymes" since it retains the original term "enzymes".

Isopycnic gradient centrifuging, Centrifugal fractionation in which cell components reach density equilibrium in the density gradient. This should be distinguished from gradient differential centrifuging (*see* above).

Isotonic sucrose, The true tonicity of sub-cellular particles is generally not known, and it is best to avoid this term in cell fractionation studies, stating instead the exact concentration of solute.

Isozymes, *See* isoenzymes.

Kg Centrifugal force in $g \times 10^{-3}$.

Kg** min*, The product of the integrated centrifugal force in ***Kg and the time it was applied, in minutes. (***Kg*** hr for hours). These terms are useful in providing a simple way of describing the procedure used in the isolation of cell components, and may be used to distinguish sub-cellular fractions. If used in conjunction with suitable standard symbols for the common sub-cellular fractions, considerable information can be conveyed. For example, two mitochondrial fractions isolated after 10 and 20 min at 5000 ***g*** could conveniently be designated 50 M and 100 M fractions respectively, indicating that they were mitochondrial fractions isolated after 50 and 100 ***Kg*** min.

Leucoplast (c) Colourless plastid.

Light microsomal fraction, *See* "microsomal fraction".

Lysate, Homogenate obtained by lysis of cells or cell particles; except where there is danger of ambiguity, "homogenate" is preferred.

Lysophagosome (c) *See* "phagolysosome".

Lysosome (c) A particle rich in lytic enzymes, and showing latency of these enzymes. A young lysosome may be called a "protolysosome". The term "cytosome" is probably equivalent to lysosome.

Lysosomal fraction (o) Preparation of cell particles consisting predominantly of lysosomes. It should be clearly distinguished from mitochondrial preparations rich in lysosomes.

Matrix of mitochondrion, *See* "inner compartment of mitochondrion".

Microbodies (c) Catalase-rich cell particles, distinct from lysosomes and small mitochondria. They have a characteristic dense core and belong to the general class of particles known as "peroxisomes".

Microbody fraction (o) Preparation of cell particles consisting predominantly of microbodies.

Microsomal fraction (o) Sub-cellular fraction obtained by centrifuging the supernatant fraction obtained after sedimenting the mitochondria. It usually consists of a mixture of ribosomes and pinched off elements of the reticular membrane, but is an entirely operational component and should never be used synonymously with endoplasmic reticulum, ergastoplasm or cytomembranes.

Microsomal subfractions (o) Phospholipid-rich sub-fractions may be called "heavy" or "light" microsomal fractions. The final microsomal fraction, poor in phospholipid, may be called the "post-microsomal fraction".

Microsomes, There is no such structure as a microsome and the term should be avoided. If used, it should be synonymous with microsomal fraction, and not endoplasmic reticulum, ergastoplasm, or cytomembranes.

Mitochondria (c) Cell particles characterized by double outer and inner membranes (cristae) and capable of oxidizing suitable substrates to CO_2 and water. Since most mitochondrial fractions consist predominantly of mitochondria, it is usually reasonably safe to use the two terms synonymously. If there is some doubt as to whether the respiratory activity observed in the isolated cell particles is due to mitochondria, the term "respiratory particles" is preferred (*see* below).

Mitochondrial fraction (o) Preparation of cell particles consisting predominantly of mitochondria.

Mitochondrial sap (c) Intramitochondrial fluid with dissolved substances.

Mitochondrial soluble fraction (o) Easily extractable soluble components of the mitochondrion.

Mitochondrial supernatant (o) This term is sometimes used to describe the fraction remaining after sedimenting the mitochondria. However, it implies that the supernatant contains mitochondria and some alternative is preferred.

Nuclear envelope (c) Preferred to nuclear membrane in order to emphasize the complexity of the structure and the fact that it is not a simple lipoprotein membrane.

Nuclear fraction (o) Isolated sub-cellular fraction consisting predominantly of nuclei. Because of difficulties experienced in the isolation of pure nuclei until fairly recently, many so-called nuclear fractions were in fact grossly contaminated with cell debris and unbroken cells, and are best called "whole cell-nuclear fractions" (*see* below).

Nuclear sap (c) Intranuclear fluid in which nuclear structures are bathed.

Nuclear soluble fraction (o) Soluble components of the nucleus not bound to the fibrillar structures of chromosomes and nucleoli. This roughly corresponds to material extracted from nuclei by dilute salt media.

Nucleolar fraction (o) Fraction obtained by differential centrifuging of disrupted nuclei. Enriched in nucleoli, but generally containing significant non-nucleolar contamination.

Nucleolar-chromosomal complex (o) The main structural material of the nucleus, containing chromosomal and nucleolar nucleoproteins. It corresponds to material (apart from membrane) remaining after removal of the nuclear soluble fraction.

Nucleoli (c) Nuclear organelles containing ribonucleoproteins, usually organized in fibrillar structures. Active in RNA synthesis.

Nucleolonema (c) (pl. nucleolonemata) Fibrillar structures present in nucleoli, 500–1000 Å in diameter, and possessing high electron density. They consist of fibrils and granules (*see* nucleonema).

Nucleonema (c) Term suggested by Professor Georgiev and Dr. Chentsov for fibrillar structures containing 100 Å fibrils (probably protein in nature) with attached ribosome-like granules. These elements are found in nucleoli (where the bundles of nucleonemata form nucleolonemata), and in chromosomes.

Nucleoplasm (c) The extra nucleolar zone of the interphase nucleus. It includes despiralized chromosomes and nuclear sap and is not identical to "nuclear sap".
Nucleolus associated chromatin (c) (o) Chromatin closely associated with the nucleolus, and usually isolated with the nucleolar fraction. It is not necessarily concerned with nucleolar function and should not be confused with the "nucleolar organizer".
Outer compartment of mitochondria (c) (o) Space between inner and outer mitochondrial membranes.
Pelleted, Sedimented is preferred.
Peroxisome (c) (o) Particle rich in enzymes of peroxide metabolism, in particular catalase, urate oxidase and D amino acid oxidase. Microbodies are peroxisomes and have a characteristic internal core. This may or may not be present in all peroxisomes.
pH 5 fraction (o) Sediment obtained after adjusting soluble fractions to pH 5.
Phagolysosome (c) Fused lysosome and phagosome. Also called lysophagosome or telolysosome.
Phagosome (c) Phagocytic or pinocytic vacuole.
Polyribosome, See polysome.
Polysome (c) (o) Aggregate of ribosomes bound by a strand of RNA, presumably mRNA. A special example of a ribosomal aggregate (*see* below).
Post-microsomal fraction (o) *See* microsomal fraction.
Protolysosome, See lysosome.
Relative concentration, The amount of an enzyme or chemical substance per unit weight of the sub-cellular fraction divided by the amount per unit weight in the homogenate. The term may also be used to describe the properties of fractions derived from sub-cellular components e.g. sub-mitochondrial fractions may be defined using the mitochondrial homogenate as standard. Relative concentration $\times 100 = \%$ specific activity (*see* below).
Respiratory particle (o) An operational term to describe any sub-cellular particle isolated from cell-homogenates that can carry out the oxidation of suitable substrates. Respiratory particles may be intact or damaged mitochondria, sub-mitochondrial particles (e.g. ETP) or non-mitochondrial material derived, for example, from bacteria.
Ribosomal aggregate (o) Group of ribosomes, not necessarily bound by a strand of RNA.
Ribosomal fraction (o) Preparation consisting predominantly of isolated ribosomes. Since most ribosomal fractions have a relatively high standard of purity, the term is usually synonymous with ribosomes.
Ribosome (o, c) Discrete particle consisting of RNA and protein. It should not be used to describe any RNA-rich sediment.
RNP particle (c) (o) (Ribonucleoprotein particle) General term for all ribonucleoprotein particles. It should be used when the precise nature of the RNA rich particle is not known.
Rough surface cytomembranes (c) System of cytomembranes lined with attached ribosomes. It should not be used to describe the microsomal fraction.
Rough surface endoplasmic reticulum, See "rough surface cytomembranes".
Sarcoplasmic reticulum (c) Endoplasmic reticulum system in muscle.
Smooth-membrane fraction (o) Preparation consisting predominantly of cell membranes free of ribosomes.
Smooth surface cytomembranes (c) Agranular cytomembranes forming irregular ramifications near the outer area of the cytoplasm. β cytomembranes represent

invaginations of the plasma membrane and γ cytomembranes make up the Golgi cytomembranes.
Soluble fraction (o) The non-sedimentable material in a tissue homogenate may also be referred to as the "soluble phase". A usual definition of non-sedimentable is remaining in the supernatant after at least 1 hr at 100 *Kg*. Not to be used synonymously with cell sap or hyaloplasm (*see* above).
Soluble phase (c) (o) (S.P.) Cytologically, the soluble phase is the interparticulate fluid of the cell. Operationally it is defined as the supernatant remaining after the sedimentation of all particulate material. If it is intended to refer to the cytological meaning, the term "cell sap" is suggested. Conversely, "soluble fraction" is the unambigious operational term.
Specific activity, The amount of an enzyme or chemical substance per unit weight of the sub-cellular fraction.
% *Specific activity* The specific activity of the fraction divided by that of the homogenate $\times$ 100. (*see* relative concentration above).
Supernate "Supernatant fluid" is preferred. "Supernatant" should strictly only be used as an adjective, but its use as a noun is so common that it is probably acceptable.
Tissue breis, Alternative term for homogenate.
Total homogenate (o) Homogenate of the entire cell or tissue, to be distinguished from "cytoplasmic fraction" (*see* above) which is a homogenate from which the whole cells and nuclei have been removed.
Whole cell-nuclear fraction (o) Fraction obtained by centrifuging the homogenate at low speed for short time. It should be clearly distinguished from "nuclear fraction" (*see* above) which is reserved for fractions consisting predominantly of nuclei.

AUTHOR INDEX

Numbers in italics refer to the page at the end of papers where complete references can be found

Z*

C

E

G

H

I

J

M

N

O

S

Y

Z

SUBJECT INDEX

(The numbers in brackets after certain enzymes refer to their Enzyme Commission numbers—see "List of Enzyme Numbers", at the end of the Index. These numbers are also given in the text, on the first occasion that a given enzyme is mentioned in each chapter).

A

B

C

D

G

H

I

K

L

O

P

Q

R

S

T

U

Y

Z

LIST OF ENZYME NUMBERS

(This is a list of enzymes cited in the book, arranged according to their number in "Enzyme Nomenclature", Recommendations (1964), of the International Union of Biochemistry on the Nomenclature and Classification of Enzymes, Elsevier Publishing Company, 1965. Details of the enzymes may be found by referring to the main Subject Index).

Number	Enzyme
1.1.1.1.	Alcohol dehydrogenase
1.1.1.8.	Glycerol-3-phosphate dehydrogenase
1.1.1.20	Glucuronolactone reductase
1.1.1.26	Glyoxylate reductase
1.1.1.27	Lactate dehydrogenase
1.1.1.30	3-Hydroxybutyrate dehydrogenase
1.1.1.34	Hydroxymethylglutaryl-CoA reductase
1.1.1.35	3-Hydroxyacyl-CoA dehydrogenase
1.1.1.37	Malate dehydrogenase
1.1.1.41	Isocitrate dehydrogenase
1.1.1.42	Isocitrate dehydrogenase (NADP)
1.1.1.43	Phosphogluconate dehydrogenase
1.1.1.44	Phosphogluconate dehydrogenase (decarboxylating)
1.1.1.49	Glucose-6-phosphate dehydrogenase
1.1.2.3	Lactate dehydrogenase (cytochrome b_2)
1.1.3.1.	Glycollate oxidase
1.1.99.1	Choline dehydrogenase
1.1.99.5	Glycerolphosphate dehydrogenase
1.2.1.8	Betaine aldehyde dehydrogenase
1.2.1.12	Triosephosphate dehydrogenase
1.2.1.13	Triosephosphate dehydrogenase (NADP)
1.2.3.2	Xanthine oxidase
1.2.4.1	Pyruvate dehydrogenase
1.2.4.2	Oxoglutarate dehydrogenase
1.3.99.1	Succinate dehydrogenase
1.3.99.3	Acyl-CoA dehydrogenase
1.4.1.2	Glutamate dehydrogenase
1.4.1.3	Glutamate dehydrogenase (NAD(P))
1.4.1.4	Glutamate dehydrogenase (NADP)
1.4.3.2	L-Aminoacid oxidase
1.4.3.3	D-Aminoacid oxidase
1.4.3.4	Monoamine oxidase
1.5.1.2	Pyrroline-5-carboxylate reductase
1.5.1.4	Dihydrofolate dehydrogenase
1.6.1.1	NAD(P) transhydrogenase
1.6.2.1	NADH cytochrome *c* reductase
1.6.2.2	NADH cytochrome b_5 reductase
1.6.2.3	NADPH cytochrome *c* reductase
1.6.4.1	Cystine reductase
1.6.4.2	Glutathione reductase
1.6.4.3	Lipoyl dehydrogenase
1.6.6.2	Nitrate reductase (NAD(P))
1.6.6.4	Nitrite reductase
1.6.6.7	Azobenzene reductase
1.6.99.1	NAD(P)H dehydrogenase
1.6.99.3	NADH-cytochrome *c* reductase
1.7.3.3	Urate oxidase
1.8.3.1	Sulphite oxidase
1.9.3.1	Cytochrome oxidase
1.10.3.1	*o*-Diphenol oxidase
1.11.1.2	NADPH peroxidase
1.11.1.6	Catalase
1.11.1.7	Peroxidase
1.14.1.1	Aryl 4-hydroxylase
2.2.1.1	Transketolase
2.3.1.12	Lipoate acetyltransferase
2.3.1.15	Glycerolphosphate acyltransferase
2.3.1.16	Acetyl-CoA acyltransferase
2.3.1.20	Diglyceride acyltransferase
2.4.1.1	α-Glucan Phosphorylase
2.4.1.11	UDPglucose-glycogen glucosyltransferase
2.4.1.13	UDPglucose-fructose glucosyltransferase
2.4.1.14	UDPglucose-fructose-phosphate glucosyltransferase
2.4.1.18	Q-enzyme, branching factor

2.4.2.1	Purine nucleoside phosphorylase
2.4.2.12	Nicotinamide phosphoribosyl-transferase
2.6.1.1	Aspartate aminotransferase
2.6.1.2	Alanine aminotransferase
2.6.1.5	Tyrosine aminotransferase
2.6.1.12	Alanine-ketoacid aminotransferase
2.7.1.1	Hexokinase
2.7.1.11	Phosphofructokinase
2.7.1.40	Pyruvate kinase
2.7.2.3	Phosphoglycerate kinase
2.7.3.2	Creatine kinase
2.7.4.3	Adenylate kinase
2.7.4.6	Nucleosidediphosphate kinase
2.7.5.1	Phosphoglucomutase
2.7.7.1	NAD pyrophosphorylase
2.7.7.6	RNA polymerase
2.7.7.7	DNA polymerase
2.7.7.8	Polynucleotide phosphorylase
2.7.7.9	UDPG pyrophosphorylase
2.7.7.16	Ribonuclease
2.7.7.17	Ribonuclease
2.7.7.19	Polyadenylate nucleotidyl-transferase
2.7.8.2	Cholinephosphotransferase
2.8.1.1	Thiosulphate sulphurtransferase
2.8.3.5	3-Ketoacid CoA-transferase
3.1.1.1	Carboxylesterase
3.1.1.2	Arylesterase
3.1.1.3	Lipase
3.1.1.5	Lysophospholipase
3.1.1.7	Acetylcholinesterase
3.1.1.9	Benzoylcholinesterase
3.1.1.12	Vitamin A esterase
3.1.1.13	Cholesterol esterase
3.1.1.14	Chlorophyllase
3.1.1.19	Uronolactonase
3.1.2.3	Succinyl-CoA deacylase
3.1.3.1	Alkaline phosphatase
3.1.3.2	Acid phosphatase
3.1.3.4	Phosphatidate phosphatase
3.1.3.5	5′-Nucleotidase
3.1.3.9	Glucose-6-phosphatase
3.1.3.11	Fructose 1,6 diphosphatase
3.1.3.16	Phosphoprotein phosphatase
3.1.4.1	Phosphodiesterase
3.1.4.3	Phospholipase C
3.1.4.4	Phospholipase D
3.1.4.5	Deoxyribonuclease
3.1.4.6	Deoxyribonuclease II
3.1.6.1	Arylsulphatase
3.2.1.1	α-Amylase
3.2.1.2	β-Amylase
3.2.1.18	Neuraminidase
3.2.1.20	α-Glucosidase
3.2.1.21	β-Glucosidase
3.2.1.23	β-Galactosidase
3.2.1.24	α-Mannosidase
3.2.1.26	β-Fructofuranosidase
3.2.1.31	β-Glucuronidase
3.2.2.5	NAD nucleosidase
3.4.1.1.	Leucine aminopeptidase
3.4.1.2	Aminopeptidase
3.4.1.3	Aminopeptidase
3.4.3.5	Cysteinyl-glycine dipeptidase
3.4.3.6	Iminodipeptidase
3.4.4.5	Chymotrypsin
3.4.4.9	Cathepsin
3.4.4.19	Collagenase
3.5.2.6	Pencillinase
3.5.3.1	Arginase
3.5.4.4	Adenosine deaminase
3.6.1.1	Inorganic pyrophosphatase
3.6.1.3	ATPase
3.6.1.6	Nucleosidediphosphatase
3.6.1.9	NAD pyrophosphatase
4.1.1.31	Phosphopyruvate carboxylase
4.1.1.32	Phosphopyruvate carboxylase
4.1.1.39	Ribulosediphosphate carboxylase
4.1.2.7	Aldolase
4.1.3.1	Isocitrate lyase
4.1.3.2	Malate synthase
4.1.3.4	Hydroxymethylglutaryl-CoA lyase
4.1.3.5	Hydroxymethylglutaryl-CoA synthase
4.1.3.7	Citrate synthase
4.2.1.2	Fumarase
4.2.1.3	Aconitate hydratase (aconitase)
4.2.1.11	Enolase
4.2.1.20	Tryptophan synthase
4.2.99.1	Hyaluronidase
4.3.1.3	Histidine ammonia-lyase
5.1.3.1	Ribulosephosphate 3-epimer ase
5.3.1.1	Triosephosphate isomerase
5.3.1.6	Ribosephosphate isomerase

5.3.1.9 Glucosephosphate isomerase
5.4.2.1 Phosphoglycerate mutase
6.1.1.1-
6.1.1.11 Aminoacyl-sRNA synthetases
6.2.1.1 AcetylCoA synthetase
6.2.1.3 AcylCoA synthetase
6.2.1.7 Cholyl-CoA synthetase
6.3.1.2 Glutamine synthetase